MINNESOTA STUDIES IN THE PHILOSOPHY OF SCIENCE

Minnesota Studies in the
PHILOSOPHY OF SCIENCE

VOLUME III
Scientific Explanation, Space, and Time

EDITED BY

HERBERT FEIGL AND GROVER MAXWELL

FOR THE MINNESOTA CENTER FOR THE PHILOSOPHY OF SCIENCE

UNIVERSITY OF MINNESOTA PRESS, MINNEAPOLIS

PRINTED IN THE UNITED STATES OF AMERICA

Second printing 1966
Third printing 1971

Library of Congress Catalog Card Number: 57-12861

PUBLISHED IN GREAT BRITAIN, INDIA, AND PAKISTAN BY THE
OXFORD UNIVERSITY PRESS, LONDON, BOMBAY, AND KARACHI
AND IN CANADA BY THOMAS ALLEN, LTD., TORONTO

Preface

The contents of this third volume of *Minnesota Studies in the Philosophy of Science* include much that is relevant for a general logic and methodology of the empirical sciences. There is considerable emphasis, however, on the philosophy of the *physical* sciences. The Minnesota Center for Philosophy of Science, after devoting several years to the foundations of psychology, has shifted its attention to the philosophy of physics. This does not mean that we have abandoned our interest in psychology, or that work in that area has been terminated. Very likely, we shall return to it full force at a later date. In the meantime a large number of articles[*] published elsewhere represent our continuing endeavors in the philosophy of psychology and other areas. Even in the present volume there are contributions, such as those by C. G. Hempel, M. Scriven, M. Brodbeck, W. W. Rozeboom, and W. Sellars, which, in varying degrees, are relevant also for the philosophy of psychology and the philosophy of history.

As in the preceding volumes of our series, almost all contributions have either resulted from, or been modified by, intensive discussions held at Center conferences of varying duration at various times. Also, as before,

[*] H. Feigl, "Philosophical Embarrassments of Psychology," *American Psychologist*, 14:115–128 (1959); "Other Minds and the Egocentric Predicament," *Journal of Philosophy*, 55:978–987 (1958); "Mind-Body—Not a Pseudoproblem," in S. Hook, ed., *Dimensions of Mind* (New York: New York University Press, 1959), pp. 24–36; "Critique of Intuition," *Philosophy East and West*, 8:1–16 (1958); "On the Vindication of Induction," *Philosophy of Science*, 28:211–216 (1961); G. Maxwell and H. Feigl, "Why Ordinary Language Needs Reforming," *Journal of Philosophy*, 55: 488–498 (1961); H. Feigl, "Matter Still Largely Material," *Philosophy of Science* (forthcoming, 1962); "Reduction of Psychology to Neurophysiology," *Daedalus* (forthcoming, 1962); G. Maxwell, "An Analytic Vindication of Induction," *Philosophical Studies*, 12:43–45 (1961); "Theories, Frameworks, and Ontology," *Philosophy of Science* (forthcoming); as well as a number of essays in H. Feigl and G. Maxwell, eds., *Current Issues in the Philosophy of Science* (New York: Holt, Rinehart, and Winston, 1961). See also the essays by P. K. Feyerabend in the forthcoming *University of Pittsburgh Studies in the Philosophy of Science*.

Preface

some basic divergencies in philosophical outlook could not be removed even by prolonged interaction. We invite the readers to form their own independent judgment.

We wish to express our deep appreciation to the many scholars who have collaborated with the Center either by way of longer visits, or shorter conferences, or by way of correspondence. Among them are Professors R. Carnap, Ernest Nagel, N. R. Hanson, R. C. Buck, Henry Mehlberg, P. E. Meehl, and Kurt Baier.

It is with pleasure and sincere gratitude that we acknowledge the financial support received from the Louis W. and Maud Hill Family Foundation in St. Paul, the National Science Foundation, and the University of Minnesota.

Thanks are also due to Dr. Bruce Aune of Oberlin College, who assisted us in the editing of manuscripts, and to Mr. John A. Visvader and Mr. Michael Radner for preparing the indexes of the present volume.

Herbert Feigl, *Director*
Grover Maxwell, *Research Associate*
MINNESOTA CENTER FOR PHILOSOPHY OF SCIENCE

November 1961

Synopsis

The brief summaries that follow will point up some of the main issues discussed in the papers in this volume.

1. *The Ontological Status of Theoretical Entities:* GROVER MAXWELL. The thesis of this paper, bluntly put, is that electrons, photons, and even electromagnetic fields are just as real, and exist in the same full-blooded sense, as chairs, tables, or sense impressions. Influential views to the contrary often assert (1) that theoretical terms such as 'electron' are useful in science only as calculating devices, having, like carpenter's tools, a legitimate use but no significant reference, and (2) that it is ontologically illuminating to remove such terms from the language of science by explicit definition, by the introduction of Ramsey sentences, or by the use of Craig's theorem. Assertions of this sort are criticized in detail, and an attempt is made to clarify the notions of reality, existence, and observability, which, misunderstood, supply a major stimulus to instrumentalist interpretations of scientific theories.

2. *Explanation, Reduction, and Empiricism:* P. K. FEYERABEND. This article contains an exposition and a criticism of two principles which contemporary empiricism shares with some very influential traditional philosophies, such as Platonism and Cartesianism. The first principle, the *principle of derivability*, asserts that explanation is by derivation, that when an explanation is given, the explanandum is derivable from its explanans without any change in its original formulation. The second principle, the *principle of meaning invariance*, asserts that the process of explanation leaves, or should leave, meanings unchanged. It is shown (1) that these principles are inconsistent both with actual scientific practice and with a reasonable, nondogmatic empiricism, (2) that contemporary empiricism which has adopted these principles thereby decreases the empirical content of scientific theories, making them less empirical and more

Synopsis

dogmatic, (3) that the difficulties which emerge when one attempts to solve such major philosophical problems as the mind-body problem, or the problem of the existence of the external world, are due to the fact that the two principles are made the *conditio sine qua non* of their solution, and (4) that a formal account of explanation is impossible. Finally, an attempt is made to present the outlines of a disinfected empiricism, one no longer bound by these two principles.

3. *Deductive-Nomological vs. Statistical Explanation:* CARL G. HEMPEL. The essay sets forth a comparative analysis of two basic types of explanation found in empirical science: explanation by deductive subsumption under laws of strictly universal form, and statistical, or probabilistic, explanation by means of laws of statistical form. Part I presents the concept of deductive-nomological explanation and examines some questions that have been raised about it in the recent literature. Part II is a first attempt to determine the distinctive logical features of probabilistic explanation. Foremost among these is its inductive character: the explanandum does not follow logically from the explanans, but receives more or less strong inductive support from it. As a consequence, probabilistic explanation is seen to differ in several important respects from its deductive counterpart. The search for criteria of acceptability for a probabilistic explanation finally leads to some problems concerning the notion of rational belief; these are tentatively dealt with in the light of current probabilistic theories of decision-making.

4. *Explanations, Predictions, and Laws:* MICHAEL SCRIVEN. This paper attempts to provide the most comprehensive treatment of the logic of explanation in the literature. An attempt is made to comment on every plausible analytical comment on the subject made in recent years; but the main aim of the paper is not to criticize, it is to synthesize a new account. The criticism of highly plausible views is a way of clearing the underbrush so that a fresh crop may grow, and a number of seedlings are planted in this essay. The argument is extended to cover "statistical" explanations and the paper by Hempel in the present volume in which some interesting propositions on this topic are put forward. An important new theme in the present essay is the attempt to give an analysis of scientific rigor which avoids the requirement of strict deduction. A point of special interest is the production of a putative proof that an analogue of the "requirement of total evidence" is necessary for all explanations, whether deductions from universal laws or merely probabilistic inferences.

Synopsis

5. *Explanation, Prediction, and "Imperfect Knowledge"*: MAY BROD-BECK. This paper examines certain recent criticisms of the deductive model of scientific explanation and prediction and the associated rejection of the hypothetico-deductive account of scientific theories, of valid deduction, and of formal logic. These criticisms and rejections are shown to arise, first, from a failure to distinguish between the elliptical, context-bound use of language for communication and the context-free use of language for description of the world. More fundamentally, the view that there are nondeductive explanatory connections among individual facts, as well as the rejection of the account of scientific theories as deductive connections among contingent empirical laws, rests on a philosophically untenable notion of "conceptual analysis." This notion is shown to result in a theory of meaning that causes language to lose all contact with the world that it is supposed to be about. Positively, the paper discusses the philosophical reasons why scientific explanation must be deductive and, using the notion of "imperfect knowledge," tries to show how the deductive model can account for the common-sense fact that in everyday life, in history, and in the social sciences generally we frequently do explain facts after they occur that we were not able to predict beforehand.

6. *The Factual Content of Theoretical Concepts*: WILLIAM W. ROZE-BOOM. This article attempts (1) to set forth in formal precision the belief-commitments which are entailed by acceptance of a scientific theory; (2) to clarify the currently problematic cognitive status of observationally irreducible theories; (3) to disclose a serious, heretofore unrecognized, semantic obstacle to uniting an empiricist epistemology with a realistic interpretation of theoretical terms; and (4) to suggest the framework of a solution to this difficulty. By formalizing certain plausible assumptions about the manner in which theories are actually used, together with carefully weakened forms of conventional semantic postulates, it becomes possible to deduce the factual commitments of an accepted theory without prejudging whether the theory is itself cognitively meaningful. A subsequent attempt to construe such theories as genuine, semantically proper assertions, however, is seen to require liberalization of the concept of "designation," resulting in a generalized semantics of which classical semantic theory is merely a limiting case.

7. *The Analytic and the Synthetic*: HILARY PUTNAM. If "analytic" statements are those which could not turn out to be false unless some change in the meaning of a term (affecting even the extension) first occurred, and

"synthetic" statements are those which could be subjected to experimental test (and this, it is contended, is how philosophers, in large part, are actually using these terms today, regardless of how they may formally define them), then many statements (e.g., some of the fundamental laws of physics) do not fall happily into either category. Rather than try to "stretch" the terms to cover the recalcitrant cases, this paper tries to analyze this situation in terms of a distinction between *single-criterion* concepts (e.g., "bachelor"), and *multiple-criterion* concepts (e.g., "man," "acid," "energy"), with the latter being subdivided into ordinary *cluster concepts* ("man," "swan") and *law cluster* concepts ("acid," "energy").

8. *The Necessary and the Contingent:* GROVER MAXWELL. The traditional segregation of sentences which are necessary (analytic) from those which are contingent (synthetic) is defended. Analyticity in the broad sense arises from linguistic rules expressed by sets of *meaning postulates* which *implicitly define* some of the terms which they contain. Necessity in both natural and constructed languages is discussed, including considerations of *context dependence*. Selection of particular meaning postulates is justified by actual linguistic usage and by other nonlogical factors such as simplicity, general usefulness, and personal preferences. The analytic-synthetic distinction among theoretical postulates is given special attention.

9. *Geometry, Chronometry, and Empiricism:* ADOLF GRÜNBAUM. In what precise sense do the metric geometry of physical space and the chronometry ingredient in the laws of physics have a *factual* warrant? The epistemological status of geochronometry is clarified via scrutiny of the concepts of rigid body and isochronous clock in the light of Riemann's conception that the metric is *not* intrinsic to the continua of space and time. The resulting statement of the conventionality of congruence, which is espoused by both Carnap and Reichenbach, is coupled with a critique of some parts of their philosophy of geometry.

To elaborate the import of the thesis of the conventionality of congruence, the following rival objections to it are refuted: (i) the system of physical laws countenanced by it fails to provide *causal* understanding of dynamical phenomena, (ii) it is false as shown by an analysis of the logic of measurement (Russell-Whitehead objection) and by the deliverances of sense perception (Whitehead's objection), and (iii) it is a mere truism of the elementary theory of signs and of the theory of models of formal calculi, but its exponents misconstrue this triviality in a highly misleading way (objection of Eddington and Putnam).

The alternative metrizability of space and of time vouchsafed by the conventionality of congruence is shown to issue in a *linguistic* interdependence of geochronometry and physics which must be carefully distinguished from an *inductive* (epistemological) interdependence between them in the sense of P. Duhem. The latter's claim that no explanans constituting a part of a wider scientific theory can ever be conclusively falsified is criticized as not only a *non sequitur* but also false. And Einstein's espousal of this Duhemian conception in regard to the epistemological status of physical geometry is then rejected. For it is demonstrated that the initial epistemological inseparability of geometry and physics need not preclude the accessibility of the geometry itself to experimental determination.

10. *Time and the World Order:* WILFRID SELLARS. This paper is a dialectical exploration of temporal discourse in an attempt to throw new light on a number of classical philosophical puzzles about time. Part I is concerned with such topics as (a) the 'timelessness' of truths or facts about temporal episodes; (b) the clash between substance ontologies and 'event' ontologies; (c) the connection between tensed verbs and temporal predicates; (d) the meaning of 'exists' and the existence of past and future events; (e) the status of becoming. Part II is concerned with issues pertaining to the connection (if any) between truth-or-falsity and determinism. It includes some tentative remarks on the philosophical significance of quantum mechanics and on the interpretation of three-valued 'logics.'

Contents

Contents

Contents

MINNESOTA STUDIES IN THE PHILOSOPHY OF SCIENCE

The Ontological Status of Theoretical Entities

That anyone today should seriously contend that the entities referred to by scientific theories are only convenient fictions, or that talk about such entities is translatable without remainder into talk about sense contents or everyday physical objects, or that such talk should be regarded as belonging to a mere calculating device and, thus, without cognitive content—such contentions strike me as so incongruous with the scientific and rational attitude and practice that I feel this paper should turn out to be a demolition of straw men. But the instrumentalist views of outstanding physicists such as Bohr and Heisenberg are too well known to be cited, and in a recent book of great competence, Professor Ernest Nagel concludes that "the opposition between [the realist and the instrumentalist] views [of theories] is a conflict over preferred modes of speech" and "the question as to which of them is the 'correct position' has only terminological interest."[1] The phoenix, it seems, will not be laid to rest.

The literature on the subject is, of course, voluminous, and a comprehensive treatment of the problem is far beyond the scope of one essay. I shall limit myself to a small number of constructive arguments (for a radically realistic interpretation of theories) and to a critical examination of some of the more crucial assumptions (sometimes tacit, sometimes explicit) that seem to have generated most of the problems in this area.[2]

[1] E. Nagel, The Structure of Science (New York: Harcourt, Brace, and World, 1961), Ch. 6.

[2] For the genesis and part of the content of some of the ideas expressed herein, I am indebted to a number of sources; some of the more influential are H. Feigl, "Existential Hypotheses," Philosophy of Science, 17:35–62 (1950); P. K. Feyerabend, "An Attempt at a Realistic Interpretation of Experience," Proceedings of the Aristotelian Society, 58:144–170 (1958); N. R. Hanson, Patterns of Discovery (Cambridge: Cambridge University Press, 1958); E. Nagel, loc. cit.; Karl Popper, The Logic of Scientific Discovery (London: Hutchinson, 1959); M. Scriven, "Definitions, Explanations, and Theories," in Minnesota Studies in the Philosophy of Science,

Grover Maxwell

The Problem

Although this essay is not comprehensive, it aspires to be fairly self-contained. Let me, therefore, give a pseudohistorical introduction to the problem with a piece of science fiction (or fictional science).

In the days before the advent of microscopes, there lived a Pasteur-like scientist whom, following the usual custom, I shall call Jones. Reflecting on the fact that certain diseases seemed to be transmitted from one person to another by means of bodily contact or by contact with articles handled previously by an afflicted person, Jones began to speculate about the mechanism of the transmission. As a "heuristic crutch," he recalled that there is an obvious *observable* mechanism for transmission of certain afflictions (such as body lice), and he postulated that all, or most, infectious diseases were spread in a similar manner but that in most cases the corresponding "bugs" were too small to be seen and, possibly, that some of them lived inside the bodies of their hosts. Jones proceeded to develop his theory and to examine its testable consequences. Some of these seemed to be of great importance for preventing the spread of disease.

After years of struggle with incredulous recalcitrance, Jones managed to get some of his preventative measures adopted. Contact with or proximity to diseased persons was avoided when possible, and articles which they handled were "disinfected" (a word coined by Jones) either by means of high temperatures or by treating them with certain toxic preparations which Jones termed "disinfectants." The results were spectacular: within ten years the death rate had declined 40 per cent. Jones and his theory received their well-deserved recognition.

However, the "crobes" (the theoretical term coined by Jones to refer to the disease-producing organisms) aroused considerable anxiety among many of the philosophers and philosophically inclined scientists of the day. The expression of this anxiety usually began something like this: "In order to account for the facts, Jones must assume that his crobes are too small to be seen. Thus the very postulates of his theory preclude

Vol. II, H. Feigl, M. Scriven, and G. Maxwell, eds. (Minneapolis: University of Minnesota Press, 1958); Wilfrid Sellars, "Empiricism and the Philosophy of Mind," in *Minnesota Studies in the Philosophy of Science,* Vol. I, H. Feigl and M. Scriven, eds. (Minneapolis: University of Minnesota Press, 1956), and "The Language of Theories," in *Current Issues in the Philosophy of Science,* H. Feigl and G. Maxwell, eds. (New York: Holt, Rinehart, and Winston, 1961).

their being observed; they are *unobservable in principle.*" (Recall that no one had envisaged such a thing as a microscope.) This common prefatory remark was then followed by a number of different "analyses" and "interpretations" of Jones' theory. According to one of these, the tiny organisms were merely convenient fictions—*façons de parler*—extremely useful as heuristic devices for facilitating (in the "context of discovery") the thinking of scientists but not to be taken seriously in the sphere of cognitive knowledge (in the "context of justification"). A closely related view was that Jones' theory was merely an instrument, useful for organizing observation statements and (thus) for producing desired results, and that, therefore, it made no more sense to ask what was the nature of the entities to which it referred than it did to ask what was the nature of the entities to which a hammer or any other tool referred.[3] "Yes," a philosopher might have said, "Jones' theoretical expressions are just meaningless sounds or marks on paper which, when correlated with observation sentences by appropriate syntactical rules, enable us to predict successfully and otherwise organize data in a convenient fashion." These philosophers called themselves "instrumentalists."

According to another view (which, however, soon became unfashionable), although expressions containing Jones' theoretical terms were genuine sentences, they were translatable without remainder into a set (perhaps infinite) of observation sentences. For example, 'There are crobes of disease X on this article' was said to translate into something like this: 'If a person handles this article without taking certain precautions, he will (probably) contract disease X; and if this article is first raised to a high temperature, then if a person handles it at any time afterward, before it comes into contact with another person with disease X, he will (probably) not contract disease X; and . . .'

Now virtually all who held any of the views so far noted granted, even insisted, that theories played a useful and legitimate role in the scientific enterprise. Their concern was the elimination of "pseudo problems" which might arise, say, when one began wondering about the "reality of supraempirical entities," etc. However, there was also a school of thought, founded by a psychologist named Pelter, which differed in an

[3] I have borrowed the hammer analogy from E. Nagel, "Science and [Feigl's] Semantic Realism," *Philosophy of Science*, 17:174–181 (1950), but it should be pointed out that Professor Nagel makes it clear that he does not necessarily subscribe to the view which he is explaining.

interesting manner from such positions as these. Its members held that while Jones' crobes might very well exist and enjoy "full-blown reality," they should not be the concern of medical research at all. They insisted that if Jones had employed the correct methodology, he would have discovered, even sooner and with much less effort, all of the observation laws relating to disease contraction, transmission, etc. without introducing superfluous links (the crobes) into the causal chain.

Now, lest any reader find himself waxing impatient, let me hasten to emphasize that this crude parody is not intended to convince anyone, or even to cast serious doubt upon sophisticated varieties of any of the reductionistic positions caricatured (some of them not too severely, I would contend) above. I am well aware that there are theoretical entities and theoretical entities, some of whose conceptual and theoretical statuses differ in important respects from Jones' crobes. (I shall discuss some of these later.) Allow me, then, to bring the Jonesean prelude to our examination of observability to a hasty conclusion.

Now Jones had the good fortune to live to see the invention of the compound microscope. His crobes were "observed" in great detail, and it became possible to identify the specific kind of *microbe* (for so they began to be called) which was responsible for each different disease. Some philosophers freely admitted error and were converted to realist positions concerning theories. Others resorted to subjective idealism or to a thoroughgoing phenomenalism, of which there were two principal varieties. According to one, the one "legitimate" observation language had for its descriptive terms only those which referred to sense data. The other maintained the stronger thesis that *all* "factual" statements were *translatable* without remainder into the sense-datum language. In either case, any two non-sense data (e.g., a theoretical entity and what would ordinarily be called an "observable physical object") had virtually the same status. Others contrived means of modifying their views much less drastically. One group maintained that Jones' crobes actually never had been unobservable in principle, for, they said, the theory did not imply the impossibility of finding a means (e.g., the microscope) of observing them. A more radical contention was that the crobes were not observed at all; it was argued that what was seen by means of the microscope was just a shadow or an image rather than a corporeal organism.

The Observational-Theoretical Dichotomy

Let us turn from these fictional philosophical positions and consider some of the actual ones to which they roughly correspond. Taking the last one first, it is interesting to note the following passage from Bergmann: "But it is only fair to point out that if this . . . methodological and terminological analysis [for the thesis that there are no atoms] . . . is strictly adhered to, even stars and microscopic objects are not physical things in a literal sense, but merely by courtesy of language and pictorial imagination. This might seem awkward. But when I look through a microscope, all I see is a patch of color which creeps through the field like a shadow over a wall. And a shadow, though real, is certainly not a physical thing."[4]

I should like to point out that it is also the case that if this analysis is strictly adhered to, we cannot observe physical things through opera glasses, or even through ordinary spectacles, and one begins to wonder about the status of what we see through an ordinary windowpane. And what about distortions due to temperature gradients—however small and, thus, always present—in the ambient air? It really does "seem awkward" to say that when people who wear glasses describe what they see they are talking about shadows, while those who employ unaided vision talk about physical things—or that when we look through a windowpane, we can only infer that it is raining, while if we raise the window, we may "observe directly" that it is. The point I am making is that there is, in principle, a continuous series beginning with looking through a vacuum and containing these as members: looking through a windowpane, looking through glasses, looking through binoculars, looking through a low-power microscope, looking through a high-power microscope, etc., in the order given. The important consequence is that, so far, we are left without criteria which would enable us to draw a non-arbitrary line between "observation" and "theory." Certainly, we will often find it convenient to draw such a to-some-extent-arbitrary line; but its position will vary widely from context to context. (For example, if we are determining the resolving characteristics of a certain microscope, we would certainly draw the line beyond ordinary spectacles, probably

[4] G. Bergmann, "Outline of an Empiricist Philosophy of Physics," *American Journal of Physics*, 11:248–258; 335–342 (1943), reprinted in *Readings in the Philosophy of Science*, H. Feigl and M. Brodbeck, eds. (New York: Appleton-Century-Crofts, 1953), pp. 262–287.

beyond simple magnifying glasses, and possibly beyond another micro-scope with a lower power of resolution.) But what ontological ice does a mere methodologically convenient observational-theoretical dichotomy cut? Does an entity attain physical thinghood and/or "real existence" in one context only to lose it in another? Or, we may ask, recalling the con-tinuity from observable to unobservable, is what is seen through specta-cles a "little bit less real" or does it "exist to a slightly less extent" than what is observed by unaided vision?[5]

However, it might be argued that things seen through spectacles and binoculars look like ordinary physical objects, while those seen through microscopes and telescopes look like shadows and patches of light. I can only reply that this does not seem to me to be the case, particularly when looking at the moon, or even Saturn, through a telescope or when looking at a small, though "directly observable," physical object through a low-power microscope. Thus, again, a continuity appears.

"But," it might be objected, "theory tells us that what we see by means of a microscope is a real image, which is certainly distinct from the object on the stage." Now first of all, it should be remarked that it seems odd that one who is espousing an austere empiricism which re-quires a sharp observational-language/theoretical-language distinction (and one in which the former language has a privileged status) should need a theory in order to tell him what is observable. But, letting this pass, what is to prevent us from saying that we still observe the object on the stage, even though a "real image" may be involved? Otherwise, we shall be strongly tempted by phenomenalistic demons, and at this point we are considering a physical-object observation language rather than a sense-datum one. (Compare the traditional puzzles: Do I see one physical object or two when I punch my eyeball? Does one object split into two? Or do I see one object and one image? Etc.)

Another argument for the continuous transition from the observable to the unobservable (theoretical) may be adduced from theoretical con-

[5] I am not attributing to Professor Bergmann the absurd views suggested by these questions. He seems to take a sense-datum language as his observation language (the base of what he called "the empirical hierarchy"), and, in some ways, such a position is more difficult to refute than one which purports to take an "observable-physical-object" view. However, I believe that demolishing the straw men with which I am now dealing amounts to desirable preliminary "therapy." Some nonrealist interpreta-tions of theories which embody the presupposition that the observable-theoretical distinction is sharp and ontologically crucial seem to me to entail positions which correspond to such straw men rather closely.

siderations themselves. For example, contemporary valency theory tells us that there is a virtually continuous transition from very small molecules (such as those of hydrogen) through "medium-sized" ones (such as those of the fatty acids, polypeptides, proteins, and viruses) to extremely large ones (such as crystals of the salts, diamonds, and lumps of polymeric plastic). The molecules in the last-mentioned group are macro, "directly observable" physical objects but are, nevertheless, genuine, single molecules; on the other hand, those in the first mentioned group have the same perplexing properties as subatomic particles (de Broglie waves, Heisenberg indeterminacy, etc.). Are we to say that a large protein molecule (e.g., a virus) which can be "seen" only with an electron microscope is a little less real or exists to somewhat less an extent than does a molecule of a polymer which can be seen with an optical microscope? And does a hydrogen molecule partake of only an infinitesimal portion of existence or reality? Although there certainly *is* a continuous transition from observability to unobservability, any talk of such a continuity from full-blown existence to nonexistence is, clearly, nonsense.

Let us now consider the next to last modified position which was adopted by our fictional philosophers. According to them, it is only those entities which are *in principle* impossible to observe that present special problems. What kind of impossibility is meant here? Without going into a detailed discussion of the various types of impossibility, about which there is abundant literature with which the reader is no doubt familiar, I shall assume what usually seems to be granted by most philosophers who talk of entities which are unobservable in principle— i.e., that the theory(s) itself (coupled with a physiological theory of perception, I would add) entails that such entities are unobservable.

We should immediately note that if this analysis of the notion of unobservability (and, hence, of observability) is accepted, then its use as a means of delimiting the observation language seems to be precluded for those philosophers who regard theoretical expressions as elements of a calculating device—as meaningless strings of symbols. For suppose they wished to determine whether or not 'electron' was a theoretical term. First, they must see whether the theory entails the sentence 'Electrons are unobservable.' So far, so good, for their calculating devices are said to be able to select genuine sentences, provided they contain no theoretical terms. But what about the selected "sentence" itself? Suppose that 'electron' is an observation term. It follows that the expression is a genuine sentence and asserts that electrons are unobservable. But this entails that 'electron' is *not* an observation term. Thus if 'electron' is an observation term, then it is *not* an observation term. Therefore it is not an observation term. But then it follows that 'Electrons are unobservable' is not a genuine sentence and does not assert that electrons are unobservable, since it is a meaningless string of marks and does not assert anything whatever. Of course, it could be stipulated that when a theory "selects" a meaningless expression of the form 'Xs are unobservable,' then 'X' is to be taken as a theoretical term. But this seems rather arbitrary.

But, assuming that well-formed theoretical expressions are genuine sentences, what shall we say about unobservability in principle? I shall begin by putting my head on the block and argue that the present-day status of, say, electrons is in many ways similar to that of Jones' crobes before microscopes were invented. I am well aware of the numerous theoretical arguments for the impossibility of observing electrons. But suppose new entities are discovered which interact with electrons in such a mild manner that if an electron is, say, in an eigenstate of position, then, in certain circumstances, the interaction does not disturb it. Suppose also that a drug is discovered which vastly alters the human perceptual apparatus—perhaps even activates latent capacities so that a new sense modality emerges. Finally, suppose that in our altered state we are able to perceive (not necessarily visually) by means of these new entities in a manner roughly analogous to that by which we now see by means of photons. To make this a little more plausible, suppose that the energy eigenstates of the electrons in some of the compounds present in the relevant perceptual organ are such that even the weak interaction with the new entities alters them and also that the cross sections, relative to the new entities, of the electrons and other particles of the gases of the air are so small that the chance of any interaction here is negligible. Then we might be able to "observe directly" the position and possibly the approximate diameter and other properties of some electrons. It would follow, of course, that quantum theory would have to be altered in some respects, since the new entities do not conform to all its principles. But however improbable this may be, it does not, I maintain, involve any logical or conceptual absurdity. Furthermore, the

10

modification necessary for the inclusion of the new entities would not necessarily change the meaning of the term 'electron.'[6]

Consider a somewhat less fantastic example, and one which does not involve any change in physical theory. Suppose a human mutant is born who is able to "observe" ultraviolet radiation, or even X rays, in the same way we "observe" visible light.

Now I think that it is extremely improbable that we will ever observe electrons directly (i.e., that it will ever be reasonable to assert that we have so observed them). But this is neither here nor there; it is not the purpose of this essay to predict the future development of scientific theories, and, hence, it is not its business to decide what actually is observable or what will become observable (in the more or less intuitive sense of 'observable' with which we are now working). After all, we are operating, here, under the assumption that it is theory, and thus science itself, which tells us what is or is not, in this sense, observable (the 'in principle' seems to have become superfluous). And this is the heart of the matter; for it follows that, at least for this sense of 'observable,' there are no a priori or philosophical criteria for separating the observable from the unobservable. By trying to show that we can talk about the *possibility* of observing electrons without committing logical or conceptual blunders, I have been trying to support the thesis that any (nonlogical) term is a *possible* candidate for an observation term.

There is another line which may be taken in regard to delimitation of the observation language. According to it, the proper term with which to work is not 'observable' but, rather 'observed.' There immediately comes to mind the tradition beginning with Locke and Hume (No idea without a preceding impression!), running through Logical Atomism and the Principle of Acquaintance, and ending (perhaps) in contemporary positivism. Since the numerous facets of this tradition have been extensively examined and criticized in the literature, I shall limit myself here to a few summary remarks.

Again, let us consider at this point only observation languages which contain ordinary physical-object terms (along with observation predicates, etc., of course). Now, according to this view, all descriptive terms of the observation language must refer to that which has been observed.

[6] For arguments that it is possible to alter a theory without altering the meanings of its terms, see my "Meaning Postulates in Scientific Theories," in *Current Issues in the Philosophy of Science*, Feigl and Maxwell, eds.

11

How is this to be interpreted? Not too narrowly, presumably, otherwise each language user would have a different observation language. The name of my Aunt Mamie, of California, whom I have never seen, would not be in my observation language, nor would 'snow' be an observation term for many Floridians. One could, of course, set off the observation language by means of this awkward restriction, but then, obviously, not being the referent of an observation term would have no bearing on the ontological status of Aunt Mamie or that of snow.

Perhaps it is intended that the referents of observation terms must be members of a *kind* some of whose members have been observed or instances of a *property* some of whose instances have been observed. But there are familiar difficulties here. For example, given any entity, we can always find a kind whose only member is the entity in question; and surely expressions such as 'men over 14 feet tall' should be counted as observational even though no instances of the "property" of being a man over 14 feet tall have been observed. It would seem that this approach must soon fall back upon some notion of simples or determinables vs. determinates. But is it thereby saved? If it is held that only those terms which refer to observed simples or observed determinates are observation terms, we need only remind ourselves of such instances as Hume's notorious missing shade of blue. And if it is contended that in order to be an observation term an expression must at least refer to an observed determinable, then we can always find such a determinable which is broad enough in scope to embrace any entity whatever. But even if these difficulties can be circumvented, we see (as we knew all along) that this approach leads inevitably into phenomenalism, which is a view with which we have not been concerning ourselves.

Now it is not the purpose of this essay to give a detailed critique of phenomenalism. For the most part, I simply assume that it is untenable, at least in any of its translatability varieties.[7] However, if there are any unreconstructed phenomenalists among the readers, my purpose, insofar as they are concerned, will have been largely achieved if they will grant what I suppose most of them would stoutly maintain anyway, i.e., that theoretical entities are no worse off than so-called observable physical objects.

[7] The reader is no doubt familiar with the abundant literature concerned with this issue. See, for example, Sellars' "Empiricism and the Philosophy of Mind," which also contains references to other pertinent works.

Nevertheless, a few considerations concerning phenomenalism and related matters may cast some light upon the observational-theoretical dichotomy and, perhaps, upon the nature of the "observation language." As a preface, allow me some overdue remarks on the latter. Although I have contended that the line between the observable and the unobservable is diffuse, that it shifts from one scientific problem to another, and that it is constantly being pushed toward the "unobservable" end of the spectrum as we develop better means of observation—better instruments —it would, nevertheless, be fatuous to minimize the importance of the observation base, for it is absolutely necessary as a confirmation base for statements which do refer to entities which are unobservable at a given time. But we should take as its basis and its unit not the "observational term" but, rather, the quickly decidable sentence. (I am indebted to Feyerabend, *loc. cit.*, for this terminology.) A quickly decidable sentence (in the technical sense employed here) may be defined as a singular, nonanalytic sentence such that a reliable, reasonably sophisticated language user can very quickly decide[8] whether to assert it or deny it when he is reporting on an occurrent situation. 'Observation term' may now be defined as a 'descriptive (nonlogical) term which may occur in a quickly decidable sentence,' and 'observation sentence' as a 'sentence whose only descriptive terms are observation terms.'

Returning to phenomenalism, let me emphasize that I am not among those philosophers who hold that there are no such things as sense contents (even sense data), nor do I believe that they play no important role in our perception of "reality." But the fact remains that the referents of most (not all) of the statements of the linguistic framework used in everyday life and in science are *not* sense contents but, rather, physical objects and other publicly observable entities. Except for pains, odors, "inner states," etc., *we do not usually observe sense contents*; and although there is good reason to believe that they play an indispensable role in observation, *we are usually not aware of them when we visually or tactilely) observe physical objects.* For example, when I observe a distorted, obliquely reflected image in a mirror, I may seem to be seeing a baby elephant standing on its head; later I discover it is an image of Uncle Charles taking a nap with his mouth open and his hand in a peculiar position. Or, passing my neighbor's home at a high rate of

[8] We may say "noninferentially" decide, provided this is interpreted liberally enough to avoid starting the entire controversy about observability all over again.

13

speed, I observe that he is washing a car. If asked to report these observations I could quickly and easily report a baby elephant and a washing of a car; I probably would not, without subsequent observations, be able to report what colors, shapes, etc. (i.e., what sense data) were involved.

Two questions naturally arise at this point. How is it that we can (sometimes) quickly decide the truth or falsity of a pertinent observation sentence? and, What role do sense contents play in the appropriate tokening of such sentences? The heart of the matter is that these are primarily scientific-theoretical questions rather than "purely logical," "purely conceptual," or "purely epistemological." If theoretical physics, psychology, neurophysiology, etc., were sufficiently advanced, we could give satisfactory answers to these questions, using, in all likelihood, the physical-thing language as our observation language and treating sensations, sense contents, sense data, and "inner states" as theoretical (yes, theoretical!) entities.[9]

It is interesting and important to note that, even before we give completely satisfactory answers to the two questions considered above, we can, with due effort and reflection, train ourselves to "observe directly" what were once theoretical entities—the sense contents (color sensations, etc.)—involved in our perception of physical things. As has been pointed out before, we can also come to observe other kinds of entities which were once theoretical. Those which most readily come to mind involve the use of instruments as aids to observation. Indeed, using our painfully acquired theoretical knowledge of the world, we come to see that we "directly observe" many kinds of so-called theoretical things. After listening to a dull speech while sitting on a hard bench, we begin to become poignantly aware of the presence of a considerably strong gravitational field, and as Professor Feyerabend is fond of pointing out, if we were carrying a heavy suitcase in a changing gravitational field, we could observe the changes of the $G_{\mu\nu}$ of the metric tensor.

I conclude that our drawing of the observational-theoretical line at any given point is an accident and a function of our physiological make-

[9] Cf. Sellars, "Empiricism and the Philosophy of Mind." As Professor Sellars points out, this is the crux of the "other-minds" problem. Sensations and inner states (relative to an intersubjective observation language, I would add) are theoretical entities (and they "really exist") and not merely actual and/or possible behavior. Surely it is the unwillingness to countenance theoretical entities—the hope that every sentence is translatable not only into some observation language but into the physical-thing language—which is responsible for the "logical behaviorism" of the neo-Wittgensteinians.

14

p, our current state of knowledge, and the instruments we happen to have available and, therefore, that it has no ontological significance whatever.

What If We COULD Eliminate Theoretical Terms?

Among the candidates for methods of eliminating theoretical terms, three have received the lion's share of current attention: explicit definability, the Ramsey sentence,[10] and implications of Craig's theorem.[11] Today there is almost (not quite) universal agreement that not all theoretical terms can be eliminated by explicitly defining them in terms of observation terms. It seems to have been overlooked that even if this could be accomplished it would not necessarily avoid reference to unobservable (theoretical) entities. One example should make this evident. Within the elementary kinetic theory of gases we could define 'molecules' as 'particles of matter (or stuff), not large enough to be seen even with a microscope, which are in rapid motion, frequently colliding with each other, and are the constituents of all gases.' All the (nonlogical) terms in the definiens are observation terms, and still the definition itself, as well as kinetic theory (and other theoretical considerations), implies that molecules of gases are unobservable (at least for the present).

It seems to me that a large number—certainly not all, however; for example, 'photon,' 'electromagnetic field,' 'ψ-function'—of theoretical terms could be explicitly defined wholly in terms of observation terms, but this would in no way avoid a reference to unobservable entities. This important fact seems to have been quite generally overlooked. It is an important oversight because philosophers today are devoting so much attention to the meaning of theoretical terms (a crucially important problem, to be sure), while the ontological stomach-aches (ultimately unjustifiable, of course) concerning theories seem to have arisen from the fact that the *entities* rather than the *terms* were nonobservational. Implicit, of course, is the mistaken assumption that terms referring to unobservable entities cannot be among those which occur in the observation language (and also, perhaps, the assumption that the referent of a defined term always consists of a mere "bundle" of the entities which are referents of the terms of the definiens).

[10] Frank P. Ramsey, *The Foundations of Mathematics* (New York: Humanities, 1931).
[11] William Craig, "Replacement of Auxiliary Expressions," *Philosophical Review*, 65:38–55 (1956).

15

Surprisingly enough, both the Ramsey sentence and Craig's theorem provide us with genuine (in principle) methods for eliminating theoretical terms provided we are interested only in the deductive "observational" consequences of an axiomatized theory. That neither can provide a viable method for avoiding reference to theoretical entities has been pointed out clearly by both Hempel and Nagel.[12] I shall discuss these two devices only briefly.[13]

The first step in forming the Ramsey sentence of a theory is to take the conjunction of the axioms of the theory and conjoin it with the so-called correspondence rules (sentences containing both theoretical and observational terms—the "links" between the "purely theoretical" and the observational). This conjunction can be represented as follows:

$$--- P --- Q --- \ldots$$

where the dashes represent the sentential matrixes (the axioms and C-rules) containing the theoretical terms (which are, of course, almost always predicates or class terms) 'P,' 'Q,' '. . .'; the theoretical terms are then "eliminated" by replacing them with existentially quantified variables. The resulting "Ramsey sentence" is represented, then, by

$$(\exists f)(\exists g) \ldots (--- f --- g --- \ldots).$$

Or, consider an informal illustration. Let us represent schematically an oversimplified axiomatization of kinetic theory by

> All gases are composed entirely of molecules. The molecules are in rapid motion and are in frequent collision, etc., etc.

And for simplicity's sake, suppose that 'molecules' is the only theoretical term. The Ramsey sentence would be something like the following:

> There is a kind of entity such that all gases are composed entirely of these entities. They are in rapid motion and are in frequent collision, etc., etc.

Now it is a simple matter to demonstrate that any sentence containing only observation (and logical) terms which is a deductive consequence of the original theory is also a deductive consequence of its Ramsey sentence (see, for example, Rozeboom's article in this volume); thus, as far as any deductive systemization is concerned, any theory may be

---

[12] Carl G. Hempel, "The Theoretician's Dilemma," in *Minnesota Studies in Philosophy of Science*, Vol. II, Feigl, Scriven, and Maxwell, eds. Nagel, *loc. cit.*

[13] For an extended consideration of the Ramsey sentence see Professor William Rozeboom's essay in this volume.

eliminated and its Ramsey sentence used instead. However, it is also easy to prove (if indeed it is not obvious) that if a given theory (or a theory together with other considerations, theoretical or observational) entails that there exist certain kinds of unobservable entities, then the appropriate Ramsey sentence will also entail that there exist the same number of kinds of *unobservable* entities.[14] Although, insofar as deductive systemization is concerned, the Ramsey sentence can avoid the use of theoretical terms; it cannot, even in letter, much less in spirit (Hempel, *loc. cit.*, was too charitable), eliminate reference to unobservable (theoretical) entities.

The Craig result, like the Ramsey sentence, provides a "method" of reaxiomatizing a postulate set so that any arbitrarily selected class of terms may be eliminated, provided one is interested only in those theorems which contain none of these terms. Its "advantages" over the Ramsey sentence are that it does not quantify over predicates and class terms and that its final reaxiomatization eliminates reference both in spirit and in letter to unobservable entities. However, its shortcomings (for the purposes at hand) render it useless as an instrument of actual scientific practice and also preclude its having, even in principle, any implications for ontology. The resulting number of axioms will, in general, and particularly in the case of the empirical sciences, be infinite in number and practicably unmanageable.

But if the practical objections to the use of Craig's method as a means for elimination of theoretical terms are all but insurmountable, there are objections of principle which are even more formidable. Both Craig's method and the Ramsey device must operate upon theories (containing, of course, theoretical terms) which are "already there." They eliminate theoretical terms only after these terms have already been used in inter-

---

[14] The proof may be sketched as follows: Let 'T' designate the theory (conjoined, if necessary, with other statements in the accepted body of knowledge) which entails that the kinds of entities C, D, . . . are not observable, i.e., T entails that

$(\exists x)(\exists y)$ . . . $(Cx \cdot Dy$ . . . x is not observable $\cdot$ y is not observable . . .) which in turn entails
$(\exists f)(\exists g)$ . . . $(\exists x)(\exists y)$ . . . $(fx \cdot gy$ . . . x is not observable $\cdot$ y is not observable . . .).

Now the Ramsey result holds for any arbitrary division of nonlogical terms into two classes, so we may put 'observable' into the class with the observation terms, so that the latter formalized statement may be treated as an "observational" consequence of T (transitivity of entailment). But then it is also a consequence of the Ramsey sentence of T. Q.E.D.

mediary steps. Neither provides a method for axiomatization *ab initio* or a recipe or guide for invention of new theories. Consequently neither provides a method for the elimination of theoretical terms in the all-important "context of discovery." [15] It might be argued that this objection is not so telling, after all, for we also lack any recipe for the invention of theories themselves, and it is logically possible that we should discover, without the use of theories as intermediaries, Ramsey sentences or Craig end products which are just as useful for explaining and predicting observations as the theories which we happen to have (accidently) adduced. It might be added that it is also logically possible that we should discover just those observation statements (including predictions, etc.) which happen to be true without the use of any instrumental intermediaries.

We must reply that the accomplished fact that it *is* theories, referring to unobservables, which have been invented for this purpose and that many of them serve it so admirably—this fact, itself, cries out for explanation. To say that theories are *designed* to accomplish this task is no reply unless at least a schema of an instrumentalist recipe for such designing is provided. As far as I know this has not been done. The thesis that theoretical entities are "really" just "bundles" of observable objects or of sense data would, if true, provide an explanation; but it is not taken very seriously by most philosophers today—for the very good reason that it seems to be false. The only reasonable explanation for the success of theories of which I am aware is that well-confirmed theories are conjunctions of well-confirmed, genuine statements and that the entities to which they refer, in all probability, exist. That it is psychologically possible for us to invent such theories is explained by the fact that many of the entities to which they refer resemble in many respects (although

---

[15] The Ramsey sentence is intuitively tractable enough so that very simple "theories" might be invented as full-blown Ramsey sentences without the use of intermediary terms. However, Craig's theorem provides no means of operating *ab initio*. Craig points out (*loc. cit.*) that once the original theory is "there," reference, in letter, to theoretical entities in the application of his method may be avoided by using the names of theoretical terms rather than using the terms themselves (i.e. by mentioning theoretical terms rather than using them). But surely only a diehard instrumentalist can take more than very scant comfort from this. The question would still remain: Where did the theory come from in the first place, and why are the names of these particular terms arranged in this particular manner such admirable "instruments" for explanation and prediction of observations? Whatever ontological implications this modification of the Craig method may have, they seem to be exactly the same as those of instrumentalism proper.

they may differ radically from them in others) the entities which we have already observed.

It should also be remembered, at this point, that theories, even as instruments, are important not only for deductive systemization but also for inductive systemization (see Hempel, *loc. cit.*). We often reason theoretically using induction, and the conclusions may be either observational or theoretical. Thus we might infer from the facts that a certain substance was paramagnetic, that it catalyzed the recombination of free radicals, and that it *probably* contained a "one-electron" bond; and we might go on to infer, again inductively, that it would probably catalyze the conversion of orthohydrogen to parahydrogen. The Craig result applies only to deductive systemization and, thus, cannot, even in its Pickwickian fashion, eliminate theoretical terms where inductive theoretical reasoning is involved. Although Craig's theorem is of great interest for formal logic, we must conclude, to use Craig's (*loc. cit.*) own words, "[as far as] the meaning [and, I would add, the referents] of such expressions [auxiliary expressions (theoretical terms)] ... [is concerned] the method . . . fails to provide any . . . clarification."

We have seen that the elimination of theoretical *terms*, even by explicit definition, would not necessarily eliminate reference to theoretical (unobservable) *entities*. We have also seen that, even if reference to theoretical entities could be eliminated after the theories themselves have been used in such an elimination (for example, by a device such as Craig's), the reality (existence) of the theoretical entities is not thereby militated against. But the most crucial point follows. Even if we do come up with a gimmick—a prediction machine or "black box"—into which we can feed data and grind out all the completely veridical observational predictions which we may desire, the possibility—I should say the likelihood—of the existence of unobserved causes for the observed events would still remain. For unless an *explanation* of why any prediction machine or "calculating device" in terms of the established rules of explanation, confirmation, etc., were forthcoming, the task of science would still be incomplete.

This brings us to another mistaken assumption that has been responsible for much mischief in considerations concerning the cognitive status of theories—the assumption that science is concerned solely with the "fruitful" organization of observational data or, more specifically, with successful prediction. Surely the main concerns of, say, a theoretical physicist involve such things as the actual properties and varieties of subatomic particles rather than the mere predictions about where and how intense a certain spectral line will be. The instrumentalist has the picture entirely reversed; as far as pure science is concerned, most observational data—most predictions—are mere instruments and are of value only for their roles in confirming theoretical principles. Even if we obtain the prediction machine, many of the theories extant today are well confirmed enough to argue strongly for the reality of theoretical entities. And they are much more intellectually satisfactory, for they provide an explanation of the occurrence of the observational events which they predict. And—equally important—an explanation for the fact that theories "work" as well as they do is, as already noted, also forthcoming; it is simply that the entities to which they refer exist.

"Criteria" of Reality and Instrumentalism

It was pointed out in the beginning of this article that Professor Ernest Nagel considers the dispute between realists and instrumentalists to be merely a verbal one.[16] There follows here a brief and what I hope is a not too inaccurate summary of his argument. Various criteria of 'real' or 'exist' (runs the argument) are employed by scientists, philosophers, etc., in their considerations of the "reality problem." (Among these criteria—some of them competing, some compatible with each other—are public perceivability, being mentioned in a generally accepted law, being mentioned in more than one law, being mentioned in a "causal" law, and being invariant "under some stipulated set of transformation, projections, or perspectives."[17]) Since, then (it continues) any two disputants will, in all probability, be using 'real' or 'exist' in two different senses, such disputes are merely verbal. Now someone might anticipate the forthcoming objections to this argument by pointing out that the word 'criteria' is a troublesome one and that perhaps, for Nagel, the connection between criteria and reality or existence is a contingent one rather than one based on meaning. But a moment's reflection makes it obvious that for Nagel's argument to have force, 'criteria' must be taken in the latter sense; and, indeed, Nagel explicitly speaks for the connection between criteria and the "senses [sic!] of 'real' or 'exist.' "[18]

[16] *Op. cit.*, pp. 141–152.
[17] Nagel, *op. cit.*, pp. 145–150.
[18] *Op. cit.*, p. 151.

Before proceeding to a criticism of these arguments, let me point out that Professor Gustav Bergmann, completely independently, treats ontological questions in a similar manner. Rather than criteria, he speaks of "patterns," although he does say that he "could instead have spoken of criteria," and he makes explicit reference to various "uses" of 'exist.'[19]

There are two main points that I wish to make regarding this kind of approach to ontological issues. First, it seems to me that it commits the old mistake of confusing meaning with evidence. To be sure, the fact that a kind of entity is mentioned in well-confirmed laws or that such entities are publicly perceptible, etc.—such facts are evidence (very good evidence!) for the existence or "reality" of the entities in question. But I cannot see how a prima-facie—or any other kind of—case can be made for taking such conditions as defining characteristics of existence.

The second point is even more serious. One would hope that (Professor Norman Malcolm notwithstanding) over nine hundred years of debate and analysis have made it clear that existence is not a property. Now surely the characteristics of being mentioned in well-confirmed laws, being publicly perceptible, etc., are properties of sorts; and if these comprised part of the meaning of 'exists,' then 'existence' would be a predicate (and existence a property).

Thus, it is seen that the issue between instrumentalism and realism can be made into a merely verbal one only by twisting the meanings of 'existence' and 'reality,' not only beyond their "ordinary" meaning but, also, far beyond any reasonable meanings which these terms might be given. In fact, it seems not too much to say that such an interpretation of the "reality problem" commits a fallacy closely akin to that of the Ontological Argument.

What can be said about the meanings of 'real' and 'exists'? I submit that in "ordinary language," the most usual uses of these terms are such that

$$\Phi_s \text{ are real} =_{df} \Phi_s \text{ exist}$$

and that

$$\Phi_s \text{ exist} =_{df} \text{ there are } \Phi_s$$

and that the meanings of these definiens are clear enough so that no further explication is seriously needed. (In most "constructed languages," 'There are Φ_s' would, of course, be expressed by '$(\exists x)(\Phi x)$.') Thus, if

[19] "Physics and Ontology," *Philosophy of Science*, 28:1–14 (1961).

we have a well-confirmed set of statements (laws or theories plus initial conditions) which entail the statement 'There are Φ_s' (or '$(\exists x)(\Phi x)$'), then it is well confirmed that Φ_s are real—full stop!

In summary, let us recall three points concerning instrumentalism. First, as is shown above, it cannot be excused on the grounds that it differs from realism only in terminology. Second, it cannot provide an explanation as to why its "calculating devices" (theories) are so successful. Realism provides the very simple and cogent explanation that the entities referred to by well-confirmed theories exist. Third, it must be acutely embarrassing to instrumentalists when what was once a "purely" theoretical entity becomes, due to better instruments, etc., an observable one.[20]

The Ontological Status of Entities—Theoretical and Otherwise

As I have stated elsewhere (see the second reference in footnote 22), the key to the solution of all significant problems in ontology can be found in Carnap's classic article, "Empiricism, Semantics, and Ontology."[21] Taking this essay as our point of departure, we may say that in order to speak at all about any kind of entities whatever and thus, a fortiori, to consider their existence or nonexistence, one must first accept the "linguistic framework" which "introduces the entities."[22] This simply means that in order to understand considerations concerning the existence of any kind of entities one must understand the meanings of the linguistic expressions (sentences and terms) referring to them—and that such expressions have no meaning unless they are given a place in a linguistic framework which "talks about the world" and which has at least a minimum of comprehensiveness. (Since I am interested, here, primarily in empirical science, I neglect universes of discourse containing only "purely mathematical" or "purely logical" entities.)

Although wide latitude in choosing and constructing frameworks is permissible, any satisfactory framework will embody, at the very least,

[20] Although I cannot agree with all the conclusions of Professor Feyerabend's essay in this volume, the reader is referred to it for an interesting critique of instrumentalism.

[21] R. Carnap, *Meaning and Necessity*, 2nd ed. (Chicago: University of Chicago Press, 1959).

[22] For a more detailed discussion of linguistic frameworks as well as their relevance for ontological problems, see Carnap, *ibid.*; and G. Maxwell, "Theories, Frameworks, and Ontology," *Philosophy of Science*, vol. 28 (1961). For an elaboration of the linguistic theses presupposed by the latter article and, to some extent, by this essay,

the following features: (1) the usual L(ogical)-formation and L-transformation rules and the corresponding set of L-true sentences which they generate; (2) a set of confirmation rules, whose nature I shall not discuss here but which I shall assume are quite similar to those actually used in the sciences; (3) a set of sentences whose truth value is quickly decidable on other than purely linguistic grounds—these correspond to "singular observation statements," but, of course, as we have seen, it is neither necessary nor desirable that such statements be incorrigible or indubitable or that a sharp distinction between observation and theory be drawn; and (4) a set of lawlike sentences, which, among other things, provide that component of meaning which is nonostensive for every descriptive (nonlogical) term of the framework. (I have argued in the references given in footnote 22 that every descriptive term has a meaning component which is nonostensive.[23] Even a term such as 'red' has part of its meaning provided by, for example, the lawlike sentence 'No surface can be both red and green all over at the same time.' Such a view is sometimes stigmatized by the epithet 'holism.' But if there is any holism involved in the view I am advocating, it is completely conceptual or epistemological and not ontological. Just what relations are present, or absent, between the actual entities of the "real world" is an empirical question and must be decided by considerations *within* a descriptive linguistic framework rather than by consideration *about* such frameworks.)

At this point, two views may be mentioned. I will omit consideration of explicitly defined terms, since they are, in principle, always eliminable. According to one view, it is always a proper subset of the lawlike sentence containing a given term which contributes to the term's meaning. The sentences in this subset are A-true[24] (analytic in a broad sense) and are totally devoid of any factual content—their only function is to provide part of the meaning of the term in question. The situation is immensely complicated by the fact that when actual usage is considered,

see G. Maxwell and H. Feigl, "Why Ordinary Language Needs Reforming," *Journal of Philosophy*, 58:488–498 (1961); G. Maxwell, "Meaning Postulates in Scientific Theories," in *Current Issues in the Philosophy of Science*, Feigl and Maxwell, eds.; and my brief article, "The Necessary and the Contingent," in this volume.

[23] Cf. also the writings of Wilfrid Sellars, for example in "Some Reflections on Language Games," *Philosophy of Science*, 21:204–228 (1954).

[24] See R. Carnap, "Beobachtungsprache und theoretisch Sprache," *Dialectica*, 12:236–248 (1957); as well as the references in fn. 22.

a sentence which is A-true in one context may be contingent in another and that even in a given context it is, more often than not, not clear, unless the context is a rational reformation, whether a given sentence is being used as A-true or as contingent. This confusion can be avoided by engaging in rational reformation, i.e., by stipulating (subject to certain broad and very liberal limitations) which sentences are to be taken as A-true and which as contingent. Needless to say, this is the viewpoint which I prefer.

The complication just mentioned, however, has led many philosophers, including Professor Putnam[25]—to say nothing of W. V. Quine—to the other viewpoint. According to it, no segregation of the relevant lawlike sentences into A-true and contingent should be attempted; each lawlike sentence plays a dual role: (1) it contributes to the meanings of its descriptive terms and (2) it provides empirical information. Fortunately, we do not have to choose between these two viewpoints here, for the thesis of realism which I am advocating is (almost) equally well accommodated by either one.

Now when we engage in any considerations about any kinds of entities and, a fortiori, considerations about the existence of theoretical entities, it is to the lawlike sentences mentioning the entities—for theoretical entities, the theoretical postulates and the so-called correspondence rules—to which we turn. These sentences tell us, for example, how theoretical entities of a given kind resemble, on the one hand, and differ from, on the other, the entities with which we happen to be more familiar. And the fact that many theoretical entities, for example those of quantum theory, differ a great deal from our ordinary everyday physical objects is no reason whatever to ascribe a questionable ontological status to them or to contend that they are merely "calculating devices." After all, the very air we breathe as well as such things as shadows and mirror images are entities of quite different kinds from chairs and tables, but this provides no grounds for impugning their ontological status. The fact that molecules, atoms, etc., cannot be said in any non-Pickwickian sense to have a color has given some philosophers ontological qualms. But, of course, the air has no color (unless we invoke the color of the sky); and a transparent object whose refractive index was the same as that of air would be completely invisible, although it would have all

---

[25] See his essay in this volume.

the other properties of ordinary physical objects. Molecules, for example, are in about the same category; they are physical things which possess some but not all of the properties of everyday physical things.

A: Do molecules exist?

B: Certainly. We have an extremely well-confirmed theory, which when conjoined with other true sentences such as 'There are gases' entails that there are molecules.

A: But are they real?

B: What do you mean?

A: Well, I'm not sure. As a starter: Are they physical objects?

B: Certainly the large ones are. Take, for example, that diamond in your ring. As for those which are submicroscopic but still large enough to have large quantum numbers, it seems that in almost any reasonable reformation they would be classified as physical objects. It would seem unjustifiable to withhold from them this status simply because they cannot be said to have a color in any straightforward fashion. In fact, I would even be inclined to call the smallest, the molecule of hydrogen, a physical object. It has mass, a reasonably determinate diameter, and, usually, something which approximates simple location, etc.

A: How about electrons?

B: The decision here is more difficult. We might find it necessary to try several reformations, taking into account many facets of contemporary physical theory, before we arrived at the most satisfactory one. It would also be helpful to have a more specific problem in view than the one which we are now considering. At any rate, we might begin by pointing out that electrons do have mass, even rest mass. They can be simply located at the expense of refraining from ascribing to them a determinate momentum. They can be said to causally interact with "bona fide" physical objects, even by those who have a billiard-ball notion of causality. The important point is that the question 'Are electrons physical objects?' is a request for a rational reformation of a very thoroughgoing variety. For most purposes, a rational reformation would not need to answer it. For your purposes, why not be content to learn in what ways electrons are similar to, and in what ways they differ from, what you would call "ordinary physical objects"? This will enable you to avoid conceptual blunders.

A: Perhaps you are right. However, I am genuinely puzzled about fields, and even photons.

B: Take the last first. We would probably never call them physical objects. For example, they have no rest mass and it would be a conceptual mistake to ask, except in a Pickwickian sense, What is their color? However, it would be reasonable to say that they are a sort of physical continuant; and they can even interact with electrons in a billiard-ball manner. At any rate, we must agree, speaking loosely, that they are "every bit as real" as electrons. The concepts of field theories are so open textured that it is difficult to decide what kinds of reformations one should adopt here. And it is virtually impossible to find similar kinds of entities with which one is prescientifically familiar. Perhaps these theories will someday be enriched until decisions concerning the most appropriate rational reformations are easier to make—perhaps not. But even here, the meanings of the terms involved are usually sufficiently clear to avoid conceptual blunders and ontological anxieties. You might like to consider the "lines of force," which are often spoken of in connection with fields. These are often used as a paradigm of the "convenient fiction" by those who hold such a view of theories.[26] But though convenient, lines of force are not fictions. They "really exist." Let me try to make this a little more plausible. Consider the isobars of meteorology, or the isograms which connect points of equal elevation above sea level. Now at this very moment, the 1017 millibar isobar, i.e., the line along which the barometric pressure is 1017 millibars, exists right here in the United States. Its location can even be determined "operationally." And all of this is true whether anyone ever draws, or ever has drawn, a weather map. Since a well-confirmed theory (plus, perhaps, other

[26] Cf. B. Mayo, "The Existence of Theoretical Entities," Science News, 32:7–18 (1954), and "More about Theoretical Entities," ibid., 39:42–55 (1956). For a critique of these articles and for excellent constructive remarks concerning theoretical entities, see J. J. C. Smart, "The Reality of Theoretical Entities," Australasian Journal of Philosophy, 34:1–12 (1956).

In connection with convenient fictions, we might consider such entities as ideal gases and bodies uninfluenced by external forces. These actually are fictions. But no theory (or theory plus true sentences) entails that there are such things. To understand their function, we need only recourse to the notion of a limit, often used in mathematics. Roughly speaking, what we actually do when we use theories involving such "fictions" is to assume, for example, that the influence of external forces on the body in question is very, very small, or that the behavior of the gas with which we are concerned is approximately given by 'PV = nRT,' or, in early kinetic theory, that the diameter of a molecule is very, very small compared to the distance between molecules. Note that had van der Waals taken the calculating-device or convenient-fiction view, he probably would not have developed his equation which embodies a correction for the effect due to the finite (greater than zero) diameter of molecules.

well-confirmed sentences) entails that there are lines of force, lines of force exist. To be sure, they are very different from everyday physical objects. But as long as we are clear about this, what metaphysical—what ontological—problems remain?

One of the exciting aspects of the development of science has been the emergence of reference to strikingly new kinds of entities. This is particularly true in field theories and quantum theory. The great difference between these and the old, familiar categories seems to have caused many philosophers and philosophically inclined scientists to despair of effecting a satisfactory conceptual analysis of these powerful new conceptual tools. The attitude too often has been, "Let us proceed to use those new devices and, if necessary for heuristic reasons, even to behave as if they consisted of genuine statements about real entities. But let us remember that, in the last analysis, they are only meaningless calculating devices, or, at best, they talk only of convenient fictions, etc. The only real entities are the good old familiar ones which we sense directly everyday." To turn the purpose of a saying of Bertrand Russell's almost completely about-face: such a view has advantages—they are the same as those of theft over honest toil. The compulsion toward metaphysical asepsis which appears to have been the motivation for the espousal of many of these reductionistic philosophies seems, itself, to have arisen from a preoccupation with metaphysical pseudo problems, e.g., the conviction that there are very few ontologically legitimate kinds of entities, perhaps only one.

Explanation, Reduction, and Empiricism

The main contention of the present paper is that a formal account of reduction and explanation is impossible for general theories, or noninstantial theories,[1] as they have also been called. More especially, it will be asserted and shown that wherever such theories play a decisive role both Nagel's theory of reduction[2] and the theory of explanation associated with Hempel and Oppenheim[3] cease to be in accordance with actual scientific practice and with a reasonable empiricism. It is to be admitted that these two "orthodox" accounts fairly adequately represent the relation between sentences of the 'All-ravens-are-black' type, which abound in the more pedestrian parts of the scientific enterprise.[4] But if the attempt is made to extend these accounts to such comprehensive structures of thought as the Aristotelian theory of motion, the impetus theory, Newton's celestial mechanics, Maxwell's electrodynamics, the theory of relativity, and the quantum theory, then complete failure is the result. What happens *here* when transition is made from a theory T' to a wider theory T (which, we shall assume, is capable of covering all the phenomena

[1] In what follows, the usual distinction will be drawn between *empirical generalizations*, on the one side, and *theories*, on the other. Empirical generalizations are statements, such as 'All A's are B's' (the A's and B's are not necessarily observational entities), which are tested by inspection of instances (the A's). Universal theories, such as Newton's theory of gravitation, are not tested in this manner. Roughly speaking their test consists of two steps: (1) derivation, with the help of suitable boundary conditions, of empirical generalizations; and (2) tests, in the manner indicated above, of these generalizations. One should not be misled by the fact that universal theories, too, can be (and usually are) put in the form 'All A's are B's'; for, whereas, in the case of generalizations, this form reflects the test procedure in a very direct way, such an immediate relation between the form and the test procedures does not obtain in the case of theories. Many thinkers have been seduced by the similarity of form into thinking that the test procedures will be the same in both cases.

[2] Nagel has explained his theory in [60]. I shall quote from the reprint of the article in [20], pp. 288–312.

[3] For the theory of Hempel and Oppenheim see [47]. I shall quote from the reprint in [23], pp. 319–352.

[4] For important exceptions, see fn. 90.

that have been covered by T') is something much more radical than incorporation of the *unchanged* theory T' (unchanged, that is, with respect to the meanings of its main descriptive terms as well as to the meanings of the terms of its observation language) into the context of T. What does happen is, rather, a *complete replacement* of the ontology (and perhaps even of the formalism) of T' by the ontology (and the formalism) of T and a corresponding change of the meanings of the descriptive elements of the formalism of T' (provided these elements and this formalism are still used). This replacement affects not only the theoretical terms of T' but also at least some of the observational terms which occurred in its test statements. That is, not only will description of things and processes in the domain in which so far T' had been applied be infiltrated, either with the formalism and the *terms* of T, or if the terms of T' are still in use, with the *meanings* of the terms of T, but the sentences expressing what is accessible to direct observation inside this domain will now mean something different. In short: introducing a new theory involves changes of outlook both with respect to the observable and with respect to the unobservable features of the world, and corresponding changes in the meanings of even the most "fundamental" terms of the language employed. So far this is the position which will be defended in the present paper.

This position may be said to consist of two ideas. The first idea is that the influence, upon our thinking, of a comprehensive scientific theory, or of some other general point of view, goes much deeper than is admitted by those who would regard it as a convenient scheme for the ordering of facts only. According to this first idea scientific theories are ways of looking at the world; and their adoption affects our general beliefs and expectations, and thereby also our experiences and our conception of reality. We may even say that what is regarded as "nature" at a particular time is our *own product* in the sense that all the features ascribed to it have first been invented by us and then used for bringing order into our surroundings. As is well known, it was Kant who most forcefully stated and investigated this all-pervasive character of theoretical assumptions. However, Kant also thought that the very generality of such assumptions and their omnipresence would forever prevent them from being refuted. As opposed to this, the second idea implicit in the position to be defended here demands that our theories be *testable* and that they be abandoned as soon as a test does not produce the predicted result. It is this second idea which makes sciences proceed to better and better theories and which creates the changes described in the introductory paragraphs of the present paper.

Now, it is easily seen that the mere statement of the second idea will not do. What we need is a guarantee that despite the all-pervasive character of a scientific theory as it is asserted in the first idea, it is still possible to specify facts that are inconsistent with it. Such a possibility has been denied by some philosophers. These philosophers started out by reacting against the claim that scientific theories are nothing but predictive devices; they recognized that their influence goes much deeper; however, they then doubted that it would be possible ever to get outside any such theory; and they therefore either became apriorists (Poincaré, Eddington), or they returned to instrumentalism. For these thinkers there seemed to exist only a choice between two evils—instrumentalism or apriorism.

Now, a closer look at the arguments leading up to this dilemma shows that they all proceed from a test model in which a *single* theory is confronted with the facts. As soon as this model is replaced by a model in which we make use of at least two factually adequate but mutually inconsistent theories, the first idea becomes compatible with the demand for testability which must now be interpreted as a demand for crucial tests either between two explicitly formulated theories, or between a theory and our "background knowledge." In this form, however, the test model turns out to be inconsistent with the "orthodox" theory of explanation and reduction. It is one of the aims of the present paper to exhibit this inconsistency.

It will be necessary, for this purpose, to discuss two principles which underlie the orthodox approach: (A) the principle of deducibility; and (B) the principle of meaning invariance. According to the principle of deducibility, explanation is achieved by deduction in the strict logical sense. This principle leads to the demand, which is incompatible with the test model just outlined, that all successful theories in a given domain must be mutually consistent. According to the principle of meaning invariance, an explanation must not change the meanings of the main descriptive terms of the explanandum. This principle, too, will be found to be inconsistent with empiricism.

It is interesting to note that (A) and (B) play a role both within modern empiricism and within some very influential "school philosophies." Thus it is one of the basic assumptions of Platonism that the key terms of sentences expressing knowledge (epistēmē) refer to unchangeable entities and must therefore possess a stable meaning. Similarly, the key terms of Cartesian physics—i.e., the terms 'matter,' 'space,' 'motion'—and the terms of Cartesian metaphysics—such as the terms 'god' and 'mind'—are supposed to remain unchanged in any explanation involving them. Compared with these similarities[5] between the school philosophies, on the one side, and modern empiricism, on the other, the differences are of very minor importance. These differences lie in the *terms* of which stability of meaning is required. A Platonist will direct his attention to numbers and other "ideas," and he will demand that words referring to these entities retain their (Platonic) meanings. Modern empiricism, on the other hand, regards empirical terms as fundamental and demands that their meanings remain unchanged.

Now, it will turn out, in the course of this essay, that any form of meaning invariance is bound to lead to difficulties when the task arises either to give a proper account of the growth of knowledge, and of discoveries contributing to this growth, or to establish correlations between entities which are described with the help of what we shall later call incommensurable concepts. It will also turn out that these are exactly the difficulties we encounter in trying to solve such age-old problems as the mind-body problem, the problem of the reality of the external world, and the problem of other minds. That is, it will usually turn out that a solution of these problems is deemed satisfactory only if it leaves unchanged the meanings of certain key terms and that it is exactly *this* condition, i.e., the condition of meaning invariance, which makes them insoluble. It will also be shown that the demand for meaning invariance is incompatible with empiricism. Taking all this into account, we may hope that once contemporary empiricism has been freed from the elements which it still shares with its more dogmatic opponents, it will be able to make swift progress in the solution of the above problems. It is the purpose of the present paper to develop and to defend the outlines of such a disinfected empiricism.[6]

Popper's admirable *Logic of Scientific Discovery* and his paper "The

---

[5] Concerning these similarities, see Popper's discussion of essentialism in [65], Ch. III and *passim*, as well as Dewey's very different account in [21], especially Ch. II.

[6] As will be shown in Sec. 2, the empiricism of the thirties *was* disinfected in the sense desired here. However, later on modern empiricism readopted some very undesirable principles of traditional philosophy.

P. K. Feyerabend

Aim of Science"[7] have been both the starting point and the motive force of the investigations to follow. I have also profited a great deal from discussions with Professors Bohm (Bristol-Haifa), Feigl (Minneapolis), Körner (Bristol), Maxwell (Minneapolis), Putnam (Princeton), and Tranekjaer-Rasmussen (Copenhagen). Both Professor Körner and Professor Sellars (New Haven) seem to hold similar views with respect to the character of the observation language, and reading their publications has therefore been a great help.[8]

While the present paper was in progress I had an opportunity to consult various as yet unpublished papers by Professor T. S. Kuhn (Berkeley) in which the noncumulative character of scientific progress is illustrated very forcefully by historical examples. Despite some important and perhaps unalterable differences, the area of agreement between Professor Kuhn and myself seems to be quite considerable. One most important point of agreement is the emphasis which both of us put upon the need, in the process of the refutation of a theory, for at least another theory. As far as I am aware, this point has been made previously by K. R. Popper in his lectures on scientific method which I attended in 1948 and 1952. Popper has also pointed out[9] that the alternative theory used in the process of refutation need not be explicitly stated but can be part of our "background knowledge."

Bohm's theory of levels and Putnam's considerations in the present volume seem to lead in the same direction. What I regard as a most important feature of the situation—a feature, by the way, that has been emphasized by Bohm and Vigier—is that direct refutation of a fairly complicated theory may be impossible for *empirical* reasons. That this is so will be shown with the help of an example. Finally, I would like to thank Professor Popper and Mr. J. W. N. Watkins (London) for constructive criticism that has been utilized in the final version of the paper.

1. Two Assumptions of Contemporary Empiricism.

Nagel's theory of reduction is based upon two assumptions. The first assumption concerns the relation between the secondary science, i.e., the

---

[7] See [68] and [66]. The basic ideas of [66] can be found in [64], which was written earlier.

[8] I am referring here to Körner's [50] and to Sellars' admirable [70].

[9] In a lecture given at Stanford University in September 1960.

discipline to be reduced, on the one side, and the primary science, i.e., the discipline to which reduction is made, on the other. It is asserted that this relation is the relation of deducibility. Or, to quote Nagel,

(1)   "The objective of the reduction is to show that the laws, or the general principles of the secondary science, are simply logical consequences of the assumptions of the primary science."[10]

The second assumption concerns the relation between the meanings of the primitive descriptive terms of the secondary science and the meanings of the primitive descriptive terms of the primary science. It is asserted that the former will not be affected by the process of reduction. Of course, this second assumption is an immediate consequence of (1), since a derivation is not supposed to influence the meanings of the statements derived. However, for reasons which will become clear later, it is advisable to formulate this invariance of meaning as a separate principle. This is also done by Nagel, who says: "It is of the utmost importance to observe that the expressions peculiar to a science will possess meanings that are fixed by its own procedures, and are therefore intelligible in terms of its own rules of usage, *whether or not the science has been, or will be, reduced to some other discipline.*"[11] Or, to express it in a more concise manner:

(2)   Meanings are invariant with respect to the process of reduction.

(1) and (2) admit of two different interpretations, just as does any theory of reduction and explanation: such a theory may be regarded either as a *description* of actual scientific practice, or as a *prescription* which must be followed if the scientific character of the whole enterprise is to be guaranteed. Similarly, (1) and (2) may be interpreted as *assertions* concerning actual scientific practice, or as *demands* to be satisfied by the theoretician who wants to follow the scientific method. Both of these interpretations will be scrutinized in the present paper.

Two very similar assumptions, or demands, play a decisive role in the orthodox theory of explanation, which may be regarded as an elaboration of suggestions that were first made, in a less definite form by Popper.[12] The first assumption (demand) concerns again the relation

[10] [20], p. 301. A more elaborate form of this condition is called the "condition of derivability" on p. 354 of [61].
[11] [20], p. 301. My italics. See also [61], p. 345, 352.
[12] [68], Sec. 12.

between the explanandum, or the laws, or the facts to be explained, on the one side, and the explanans, or the discipline which functions as the basis of explanation, on the other. It is again asserted (required) that this relation is (be) the relation of deducibility. Or, to quote Hempel and Oppenheim

(3)   "The explanandum must be a logical consequence of the explanans; in other words, the explanandum must be logically deducible from the information contained in the explanans, for otherwise the explanans would not constitute adequate grounds for the explanation."[13]

Considering what has been said in the case of reduction one would expect the assumption (demand) concerning meanings to read as follows:

(4)   Meanings are invariant with respect to the process of explanation.

However, despite the fact that (4) is a trivial consequence of (3), this assumption has never been expressed in as clear and explicit a way as (2).[14] There was even a time when a consequence of (4), viz., the assertion that *observational* meanings are invariant with respect to the process of explanation, seemed to be in doubt. It is for this reason that I have separated (2) from (1), and (4) from (3).

It is not difficult to show that, with respect to observational terms, (4) or its implications, is consistent with the earlier positivism of the Vienna Circle. Their main thesis that all descriptive terms of a scientific theory can be explicitly defined on the basis of observation terms guarantees the stability of the meanings of observational terms (unless one assumes that an explicit definition changes the meaning of the definiens, a possibility that to my knowledge has never been considered by empiricists). And as the chain of definitions leaves unchanged terms already defined, (4) turns out to be correct as well.

However, since these happy and carefree days of the *Aufbau*, logical

[13] [47], p. 321.

[14] An exception is Nagel who, in [61], p. 338, defines reduction as "the explanation of a theory or a set of experimental laws established in one area of inquiry by a theory usually, though not invariably, formulated for some other domain." This implies that the condition of meaning invariance formulated by him for the process of reduction is supposed to be valid in the case of explanation also. On pp. 86–87, meaning invariance for observational terms is stated quite explicitly: an experimental law "retains a meaning that can be formulated independently of [any] theory . . . [It] has . . . a life of its own, not contingent on the continued life of any particular theory that may explain the law."

empiricism has been greatly modified. The changes that took place were mainly of two kinds. On the one side, new ideas were introduced concerning the relation between observational terms and theoretical terms. On the other side, the assumptions made about the observational language itself were modified. In both cases the changes were quite drastic. For our present purpose a brief outline must suffice: The early positivists assumed that observational terms refer to subjective impressions, sensations, and perceptions of some sentient being. Physicalism for some time retained the idea that a scientific theory should be based upon experiences, and that the ultimate constituents of experience were sensations, impressions, and perceptions. Later, however, a behavioristic account was given of these perceptions to make them accessible to intersubjective testing. Such a theory was held, for some time, by Carnap and Neurath.[15] Soon afterwards the idea that it is *experiences* to which we must refer when trying to interpret our observation statements was altogether abandoned.[16] According to Popper, who has been responsible for this decisive turn, we must *"distinguish sharply between objective science on the one hand, and 'our knowledge' on the other."* It is conceded "we can become aware of facts only by observation"; but it is denied that this implies an interpretation of observation sentences in terms of experiences, whether these experiences are explained subjectivistically or as features of objective behavior.[17] For example, we may admit that the sentence 'this is a raven' uttered by an observer who points at a bird in front of him is an observational sentence and that the observer has produced it because of the impressions, sensations, and perceptions he possesses. We may also admit that he would not have uttered the sentence had he not possessed these impressions. Yet, the sentence is not therefore about impressions; it is about a bird which is neither a sensation nor the behavior of some sentient being. Similarly, it may be admitted that the observation sentences which a scientific observer produces are prompted by his impressions. However, their content will again be determined, not by these impressions, but by the entities allegedly described. In the case of classical physics, therefore, "every basic statement must either be itself a statement about relative positions

---

[15] For this and the following, see Carnap's account in [13], especially the passages in small print on pp. 223–224.

[16] *Ibid.*, p. 223: "It is stipulated that under given circumstances any concrete statement may be regarded as a protocol statement."

[17] Popper [68], p. 98. Italics in the original.

of physical bodies . . . or it must be equivalent to some basic statement of this 'mechanistic' . . . kind."[18]

The descriptive terms of Carnap's "thing-language," too, no longer refer to experiences. They refer to properties of objects of medium size which are accessible to observation, i.e., which are such that a normal observer can quickly decide whether or not an object possesses such a property.[19] "What we have called observable predicates," says Carnap, "are predicates of the thing-language (they have to be clearly distinguished from what we have called perception terms . . . whether these are now interpreted subjectivistically, or behavioristically)."[20]

Now it is most important to realize that the characterization of observation statements implicit in the above quotations is a *causal* characterization, or if one wants to use more recent terminology, a *pragmatic* characterization:[21] an observation sentence is distinguished from other sentences of a theory, not, as was the case in earlier positivism, by its *content*; but by the *cause* of its production, or by the fact that its production conforms to certain *behavioral patterns.*[22] This being the case, the fact that a certain sentence belongs to the observation language does not allow us to infer anything about its content; more especially, it does not allow us to make any inference concerning the *kind* of entities described in it.

It is worthwhile to dwell a little longer on the features of this *pragmatic theory of observation*, as I shall call it. In the case of measuring instruments, the pragmatic theory degenerates into a triviality: nobody would ever dream of asserting that the way in which we interpret the movements of, say, the hand of a voltmeter is uniquely determined either by the character of this movement itself or by the processes inside the instrument; a person who can see and understand only these processes will be unable to infer that what is indicated is voltage, and he will be equally unable to understand what voltage is. Taken by themselves the indications of instruments do not mean anything unless we possess a *theory* which teaches us what situations we are to expect in

[18] Popper [68], p. 103. Popper himself does not restrict his characterization to the observation statements of classical physics.

[19] Carnap [14], p. 63, Explanation 1. Page references are to the reprint of this article in [22], pp. 47–92.

[20] *Ibid.*, p. 69.

[21] For this terminology see Morris [59], pp. 6ff.

[22] See again Explanation 1 of [14], as well as my elaboration of this explanation in [31].

the world, and which guarantees that there exists a reliable correlation between the indications of the instrument and such a particular situation. If a certain theory is replaced by a different theory with a different ontology, then we may have to revise the interpretation of *all* our measurements, however self-evident such a particular interpretation may have become in the course of time: according to the phlogiston theory, measurements of weight before and after combustion are measurements of the amount of phlogiston added or lost in the process. Today we must give a completely different interpretation of the results of these measurements. Again, Galileo's thermoscope was initially supposed to measure an intrinsic property of a heated body; however, with the discovery of the influence of atmospheric pressure, of the expansion of the substance of the thermoscope (which, of course, was known beforehand), and of other effects (nonideal character of the thermoscopic fluid), it was recognized that the property measured by the instrument was a very complicated function of such an intrinsic property, of the atmospheric pressure, of the properties of the particular enclosure used, of its shape, and so on.[23] Indeed, the point of view outlined in the beginning of the present paper gives an excellent account of the way in which results of measurement, or indications of instruments, are reinterpreted in the light of fresh theoretical insight. Nobody would dream of using the insight given by a new theory for the readjustment of some general beliefs only, leaving untouched the interpretation of the results of measurement. And nobody would dream of demanding that the meanings of observation statements *as obtained with the help of measuring instruments* remain invariant with respect to the change and progress of knowledge. Yet, precisely this is done when the measuring instrument is a human being, and the indication is the behavior of this human being, or the sensations he has, at a particular time.

It is not easy to set down in a few lines the reasons for this exceptional treatment of human observers. Nor is it possible to criticize them thoroughly and thereby fully pave the way for the acceptance of the pragmatic theory of observation. However, such a comprehensive criticism is not really necessary here. It was partly given by those very same philosophers who are responsible for the formulation of the pragmatic

[23] For historical references, see [18], especially the articles on the phlogiston theory (J. B. Conant) and on the early development of the concept of temperature (D. Roller).

theory[24] (which most of them dropped later on, their own excellent arguments in favor of it notwithstanding). I shall therefore content my-self with giving an outline of the idea leading to the assumption that human observers are something special and cannot be treated in the same manner as physical measuring instruments.

These ideas are connected with the (very old) belief that (a) some states of the mind (sensations or abstract ideas) can be known with certainty; that (b) it is exactly this knowledge that constitutes the foundation of whatever assertion we make about the world; and that (c) meaning invariance is obtained in the following manner: if it is in-deed the case that statements about, say, sensations are irrevocable once produced, then the same applies to the descriptive terms contained in them; their meaning is uniquely and irrevocably determined by the struc-ture of the statements in which they occur as well as by the circum-stances which lead to the certain production of these statements. (Simi-lar considerations apply if we are dealing, not with sensations, but with the 'clear and distinct' appearance of ideas.)

The theories behind meaning invariance are, of course, a little more complicated than I have just indicated, and they should perhaps be outlined in greater detail in order that their force be duly realized. Nevertheless, their most fundamental assumptions—viz., (a), (b), and (c)—can be eliminated on the basis of some very simple and almost trivial considerations. These considerations, which cannot be found in the above-mentioned writings of the original defenders of the pragmatic theory, proceed from the remark that in the argument leading up to (c) the distinction is obliterated between (psychological and sociological) *facts* and (linguistic) *conventions*.[25] It is assumed that the urge we feel under certain circumstances to say 'I am in pain' and the peculiar char-acter of this urge (it is different from the urge we feel when we say 'I am hungry') *already determine* the meaning of the main descriptive term of the sentence uttered, viz., the meaning of the term 'pain' or 'hunger').

Now, quite apart from leading to some very undesirable paradoxes[26] this procedure assumes that a *fact*, viz., the existence of either an urge

[24] Cf. Carnap [11] and [12].

[25] For a very clear presentation of this distinction, see Popper [65], Ch. V.

[26] For a more detailed discussion of these paradoxes, see my paper [31], especially Secs. 4 and 5.

to produce a sentence of a certain kind or the existence of psychological phenomenon, can without further ado transfer meaning upon a sentence, viz., the sentence 'I am in pain.' It is therefore unacceptable to any philosopher who takes seriously the distinction between facts and conventions. Conversely, the attempt to uphold this distinction leads at once to the separation, characteristic of the pragmatic theory, of the observational character of a statement from its meaning: according to the pragmatic theory, the fact that a statement belongs to the observational domain has no bearing upon its meaning. Even if its production is accompanied by very forceful sensations and related to them in a manner that makes substitution by a different sentence psychologically very difficult or perhaps even impossible, even then we are still free to interpret the sentence in whatever way we like. It is very important to point out that this freedom of interpretation obtains also in psychology, where our sentences are indeed about subjective events. Whatever restrictions of interpretation we accept are determined by the language we use, or by the theories or general points of view whose development has led to the formulation of this language.[27]

To repeat: strict adherence to the distinction between nature and convention at once eliminates the third of the three assumptions mentioned above and thereby introduces a very fundamental element of the pragmatic theory, namely, its emphasis upon the separation between observability and meaning. However, we cannot retain the first assumption either. The reason is that the sciences are the result of a decision to use only testable statements for the expression of laws and singular facts. This being the case we cannot admit any irrefutable statement,

[27] A pointed criticism of the idea that the interpretation of a statement is uniquely determined by the sensations that accompany its production has been given by Wittgenstein in his [74]. This book also emphasizes the dependence of interpretations upon the incorporation of the corresponding sentence into a language. It does not seem to me, however, that Wittgenstein possesses a clear idea of what we have called the pragmatic theory of observation. He fails to recognize that languages are not ends in themselves but are means for expressing *theories* and that they can and should be abandoned as soon as the corresponding theory has been refuted. Quite to the contrary—he dwells upon the difficulties one will encounter when trying to change a language in a very fundamental way, and he thereby insinuates that it may be altogether impossible to carry out decisive changes. The reason for this pessimism seems to be identical with the one I briefly discussed in the introductory part of this paper: it is assumed that the all-pervasive character of a language makes it impossible to specify grounds for abandoning it. For an application of this pessimism to more concrete problems, see Hanson [43], especially Chs. III and V. For a criticism, see my review of Hanson in [35].

however elevated and noble its source may seem to be.[28] Indeed, a whole theory may at some time turn out to be unsatisfactory, and the need may arise to replace it by a completely different idiom based upon a different point of view. Clearly, then, the interpretation of the observation sentences will have to follow suit, for again there is no way of conferring an interpretation upon them except by incorporating them into a new and better theory.

The pragmatic theory of observation thus turns out to be a presupposition of the feasibility of the point of view which has been outlined in my introductory remarks (and a consequence of the distinction between nature and convention). This position, this point of view, and especially the idea that our theories determine our entire conception of reality now emerge as a combination of (a) the demand to apply the terminology and the ontology of a given theory everywhere inside the domain of its validity with (b) the pragmatic theory of observation. It is in this form that I shall defend my position in the present paper.

The freedom of interpretation admitted by the pragmatic theory did not exist in the earlier positivism. Here sensations were thought to be the objects of observation. According to it, whether or not a statement is a sense-datum statement and, therefore, part of the observation language could be determined by logical analysis. Conversely, the assertion that a certain statement belongs to the observation language there implied an assertion about the kind of entities described (e.g., sense data). The ontology of the observational domain was therefore fixed independently of theorizing. This being the case, the demand for a unified ontology (which is still retained) could be met only by adopting the one or the other of the following two procedures: it could be met by either denying a descriptive function to the sentences of the theory and by declaring that these sentences are nothing but part of a complicated prediction machine (*instrumentalism*), or by conferring upon these sentences an interpretation that completely depends upon their connection with the observational language as well as upon the (fixed) interpretation of the latter (*reductionism*). It is important to realize that it is the clash between realism, on the one side, and the combination of the theory of sense data with the demand for a unified ontology, on the

---

[28] In [33] I discuss some of the consequences of the use of irrefutable statements of observation and thereby provide some reasons for their elimination from the body of the sciences and of knowledge in general.

other, which necessitates this transition to either instrumentalism or reductionism.

Now one of the most surprising features of the development of contemporary empiricism is that the very articulate formulation of the pragmatic account of observation was not at once followed by an equally articulate formulation of a realistic interpretation of scientific theories. After all, realism had been abandoned mainly because the theory of sense data had made it incompatible with the demand for a unified ontology. The arrival of the pragmatic theory of observation removed this incompatibility and thereby opened the way for a hypothetical realism of the kind outlined earlier. Yet, in spite of this possibility, the actual historical development was in a completely different direction. The pragmatic theory was retained for a while (and is still retained, in footnotes, by some empiricists[29]), but it was soon combined either with instrumentalism or with reductionism. As the reader can verify for himself, such a combination in effect amounts to abandoning the pragmatic theory, a more complicated language with a more complicated ontology now taking the place of the sense-data language of the earlier point of view. How close the most recent offspring of this development is to the old sense-data ideology may be seen in a recent paper by Professor Carnap.

In this paper Carnap analyzes scientific theories with the help of his well-known double-language model consisting of an observational language, L_O and a theoretical language, L_T, the latter containing a postulate system, T. The languages are connected to each other by correspondence rules, i.e., by sentences containing observational terms and theoretical terms. With respect to such a system, Carnap asserts that "there is no independent interpretation for L_T. The system T is itself an uninterpreted postulate system. The terms of [L_T] obtain only an indirect and incomplete interpretation by the fact that some of them are connected by correspondence rules with observational terms, and the remaining terms of [L_T] are connected with the first ones by the postulates of T."[30]

This procedure quite obviously presupposes that the meaning of the observational terms is fixed independently of their connection with theoretical systems. If the pragmatic theory of observation were still retained in this essay of Carnap's, then the interpretation of an observational

[29] See Hempel [46], especially fn. 10.
[30] See Carnap [15], p. 47.

statement would have to be independent of the behavioral pattern exhibited in the observational situation as well. It is not clear how, then, the observation sentence could be given any meaning at all. Now, Carnap is very emphatic about the fact that incorporation into a theoretical context is not sufficient for providing an interpretation, since no theoretical context possesses an "independent interpretation."[31] We must, therefore, suspect that, for Carnap, incorporation of a sentence into a complicated behavioral pattern has implications for its meaning, i.e., we must suspect that Carnap has silently dropped the pragmatic theory. This is indeed the case. He asserts that "a complete interpretation of L_O" is given since "L_O is used by a certain language community as a means of communication,"[32] adding in a later passage[33] that if people use a term in such a fashion that for some sentences containing the term "any possible observational result can never be absolutely conclusive evidence, but at best evidence yielding a high probability, then the appropriate place for [the term] in a dual language system . . . is in L_T rather than in L_O . . ." These two passages together seem to imply that the meaning of an observational statement is already fixed by the way in which the sentence expressing it is handled in the *immediate* observational situation (note the emphasis upon absolute confirmability for observational sentences!), i.e., they seem to imply the rejection of the pragmatic theory.

As I said above, this tacit withdrawal from the pragmatic theory of observation is one of the most surprising features of modern empiricism. It is responsible for the fact that this philosophy, despite the apparent progress that has been made since the thirties, is still in accordance with the assumption that observational meanings are invariant with respect to the process of explanation and perhaps even with full meaning invariance (if we only consider that the behavioristic criterion of observability will be satisfied by any language that has been used for a long time, a long history and observational plausibility brought about by it are the best preconditions for the petrification of meanings; this applies to Platonism as well as to modern empiricism).

This finishes a somewhat lengthy digression which started immediately after the pronouncement of (4). I will make only two points before

[31] For a detailed criticism of this assertion, cf. my [27] and [39].
[32] Carnap [15], p. 40.
[33] *Ibid.*, p. 69.

returning to the main argument of the present paper: first, that the unwitting and partial return to the ideology of sense data is responsible for many of the 'inner contradictions' which are so characteristic of contemporary empiricism as well as for the pronounced similarity of this philosophy to the "school philosophies" it has attacked; second, that (4) has been accepted, not only by philosophers, but also by many physicists who believe in the so-called Copenhagen interpretation of microphysics. It is one of Niels Bohr's most fundamental ideas that "however far the new phenomena" found on the microlevel "transcend the scope of classical physical *explanation*, the account of all evidence must be expressed in classical terms." [34] I shall not discuss, in the present section, the arguments which Bohr has developed in favor of this idea. Let me only say that this idea immediately leads to the invariance of the meanings of the descriptive terms of the observation language, the classical signs now playing the role of the observational vocabulary.

To sum up: two ideas which are common to both the modern empiricist's theory of reduction and to his theory of explanation are:

(A)   reduction or explanation is (or should be) by derivation;
(B)   the meanings of (observational) terms are invariant with respect to both reduction and explanation.

In the sections to follow it will be my task to scrutinize these two basic principles. I shall begin with (A).

2. Criticism of Reduction or Explanation by Derivation.

The task of science, so it is assumed by those who hold the theory about to be criticized, is the explanation, and the prediction, of known singular facts and regularities with the help of more general theories. In what follows we shall assume T′ to be the totality of facts and regularities to be explained, D′ the domain in which T′ makes correct predictions, and T (domain D′ ⊂ D) the theory which functions as the basis of explanation.[35] Considering (3) we shall have to demand that T be either strong enough to contain T′ as a logical consequence, or at least

[34] [6], pp. 209ff. For a more detailed account of Bohr's philosophy of science, see [32].
[35] In what follows it will not be necessary explicitly to distinguish between " "T′ " " and " "T," " and this distinction will therefore not be made. Also terms such as 'consistent,' 'incompatible,' and 'follows from' will be applied to pairs of theories, (T,T′), and they will then mean that T *taken together with the conditions of validity of T′*, or the boundary conditions characterizing D′, is compatible with, consistent with, or sufficient to derive, T′.

compatible with T′ (inside D′, that is). Only theories which satisfy one or the other of the two demands just stated are admissible as explanatia. Or, taking the demand for explanation for granted,

(5)   only such theories are admissible (for explanation and prediction) in a given domain which either *contain* the theories already used in this domain, or are at least *consistent* with them.

It is in this form that (A) will be discussed in the present section and in the sections to follow.

As has just been shown, condition (5) is an immediate consequence of the logical empiricist's theory of explanation and reduction, and it is therefore adopted—at least by implication—by all those who defend that theory. However, its correctness has been taken for granted by a much wider circle of thinkers, and it has also been adopted independently of the problem of explanation. Thus, in his essay "Studies in the Logic of Confirmation" C. G. Hempel demands that "every logically consistent observation report" be "logically compatible with the class of all the hypotheses which it confirms," and more especially, he has emphasized that observation reports do "not confirm any hypotheses which contradict each other."[36] If we adopt this principle, then a theory T (see the notation introduced at the beginning of the present section) will be confirmed by the observations confirming a more narrow theory T′ only if it is compatible with T′. Combining this with the principle that a theory is admissible only if it is confirmed to some degree by the evidence available, we at once arrive at (5).

Outside philosophy, (5) has been taken for granted by many physicists. Thus, in his *Waermelehre*, Ernst Mach makes the following remark: "Considering that there is, in a purely mechanical system of absolutely elastic atoms no real analogue for the *increase of entropy*, one can hardly suppress the idea that a violation of the second law . . . should be possible if such a mechanical system were the *real* basis of thermodynamic processes."[37] And he insinuates that, for this reason, the mechanical hypothesis must not be taken too seriously.[38] More recently, Max Born has based his arguments against the possibility of a return to

---

[36] [45], p. 105, condition (8.3). It was J. W. N. Watkins who drew my attention to this property of Hempel's theory.

[37] [53], p. 364.

[38] For a much more explicit statement of what appears in [53] only as an insinuation, see [54].

44

determinism upon (5) and the assumption, which we shall here take for granted,[39] that the theory of wave mechanics is incompatible with determinism. "If any future theory should be deterministic," says he, "it cannot be a modification of the present one, but must be entirely different. How this should be possible without sacrificing a whole treasure of well-established results I leave the determinist to worry about."[40]

The use of (5) is not restricted to such general remarks, however. A decisive part of the quantum theory itself, viz., the so-called quantum theory of measurement, is the immediate result of the postulate that the behavior of macroscopic objects, such as measuring instruments, must obey some classical laws *precisely* and not only *approximately*. For example, macroscopic objects must always dwell in a well-defined classical state, and this despite the fact that their microscopic constituents exhibit a very different behavior. It is this postulate which leads to the introduction of abrupt jumps in addition to the continuous changes that occur in accordance with Schrödinger's equation.[41] An account of measurement which very clearly exhibits this feature has been given by Landau and Lifshitz.[42] These authors point out that "the classical nature of the apparatus means that . . . the reading of the apparatus . . . has some definite value." "This," they continue, "enables us to say that the state of the system apparatus + electron after the measurement will in actual fact be described, not by the entire sum $[\Sigma A_n(q)\Phi_n(\mathfrak{s})$ where q is the coordinate of the electron, $\mathfrak{s}$ the apparatus coordinate] but by only the one term which corresponds to the 'reading' g_n of the apparatus, $A_n(q)\Phi_n(\mathfrak{s})$." Moreover, most of the arguments against suggestions such as those put forth by Bohm, de Broglie, and Vigier make more or less explicit use of (5).[43] A discussion of this condition is therefore very

---

[39] Born believes that this assumption has been established by von Neumann's proof. In this he is mistaken; see [29]. However, there exist different and quite plausible arguments for the incompatibility of determinism and wave mechanics, and it is for this reason that I take the assumption for granted. An outline of these plausible arguments is given in [37]. It should be noted that von Neumann himself did not share Born's inductivism. See [62], p. 327.

[40] [7], p. 109. In his treatment of the relation between Kepler's laws and Newton's theory, which, he thinks, applies to all pairs of theories which overlap in a certain domain and are adequate in this domain, Born explicitly accepts (5). For an analysis of Born's inductivism, see Popper [67].

[41] See [30].

[42] [52], p. 22. See also von Neumann's treatment of the Compton effect in [62], pp. 211–215.

[43] Cf. [32], [36], [38].

45

topical and leads right into the center of contemporary arguments about microphysics.

This discussion will be conducted in three steps. It will first be argued that most of the cases which have been used as shining examples of scientific explanation *do not* satisfy (5) and that it is not possible to adapt them to the deductive scheme. It will then be shown that (5) *cannot* be defended on empirical grounds and that it leads to very unreasonable consequences. Finally, it will turn out that once we have left the domain of empirical generalizations, (5) *should not* be satisfied either. In connection with this last, methodological step, the elements of a positive methodology for theories will be developed, and the historical, psychological, and semantical aspects of such a methodology will be discussed. Altogether the three steps will show that (A) is in disagreement both with actual scientific practice and with reasonable methodological demands. I start now with the discussion of the *actual* inadequacy of (5).

3. *The First Example.*

A favorite example of both reduction and explanation is the reduction of what Nagel calls the Galilean science to the physics of Newton,[44] or the explanation of the laws of the Galilean physics on the basis of the laws of the physics of Newton. By the Galilean science (or the Galilean physics) is meant, in this connection, the body of theory dealing with the motion of material objects (falling stone, penduli, balls on an inclined plane) near the surface of the earth. A basic assumption here is that the vertical accelerations involved are constant over any finite (vertical) interval. Using T' to express the laws of this theory, and T to express the laws of Newton's celestial mechanics, we may formulate Nagel's assertion to the effect that the one is reducible to the other (or explainable on the basis of the other) by saying that

(6) $\quad T \ \& \ d \mid\!- T'$

where d expresses, in terms of T, the conditions valid inside D'. In the case under discussion d will include description of the earth and its surroundings (supposed to be free from air; we shall also abstract from

[44] [20], p. 291. I am aware that, from a historical point of view, the discussion to follow is not adequate. However, I am here interested in the systematic aspect, and I have therefore allowed myself what could only be regarded as great liberties if the main interest were historical.

all those phenomena which are due to the rotation of the earth and whose inclusion would strengthen, rather than weaken our case), and reference will be made to the fact that the variation H of the height above ground level in the processes described is very small if compared with the radius R of the earth.

As is well known (6) cannot be correct: as long as H/R has some finite value, *however small*, T' will not follow (logically) from T and d. What will follow will rather be a law, T''', which, while being experimentally indistinguishable from T' (on the basis of the experiments which formed the inductive evidence for T' in the first place), is yet inconsistent with T'. If, on the other hand, we want to derive T' precisely, then we must replace d by a statement which is patently false, as it would have to describe the conditions in the close neighborhood of the earth as leading to a vertical acceleration that is constant over a finite interval of vertical distance. It is therefore impossible, *for quantitative reasons*, to establish a deductive relationship between T and T', or even to make T and T' compatible. This shows that the present example is not in agreement with (5) and is, therefore, also incompatible with (A), (1), and (3).

Now in this situation, we may adopt one or the other of the following two procedures. We may either declare that the Galilean science can neither be reduced to, nor explained in, terms of Newton's physics;[45] or we may admit that reduction and explanation are possible, but deny that deducibility, or even consistency (on the basis of suitable boundary conditions), is a necessary condition of either. It is clear that the question as to which of these two procedures is to be adopted is of subordinate importance (after all, it is purely a matter of terminology that is to be settled here!) if compared with the question whether newly invented theories should be consistent with, or contain, those of their predecessors with whom they overlap in empirical content. We shall therefore defer settlement of the terminological problem raised above and concentrate on the question of consistency, or derivability. And we shall use the terms 'explanation' and 'reduction' either in a vague and general sense, awaiting further explication, or in the manner suggested by Nagel and by Hempel and Oppenheim. The usage adopted should always be clear from the context.

[45] This suggestion was made to me by Professor Viktor Kraft.

P. K. Feyerabend

The objection which has just been developed—so it is frequently pointed out—cannot be said to endanger the correct theory of explanation, since everybody would admit that explanation may be by approximation only. This is a curious remark indeed! It criticizes us for taking seriously, and objecting to, a criterion which has either been universally stated as a necessary condition of explanation, or which plays a central role in some theories of confirmation, viz., condition (3). Now dropping (3) means altogether giving up the orthodox theory, for (3) formed the very core of this theory.[46] On the other hand, the remark that we explain "by approximation" is much too vague and general to be regarded as the statement of an alternative theory. As a matter of fact, it will turn out that the idea of approximation cannot any more be incorporated into a formal theory, since it contains elements which are essentially subjective. However, before dealing with this aspect of explanation we shall inquire a little more closely into the reasons for the failure of (3). Such an inquiry will lead to the result not only that (3) is false, but it is also very unreasonable to assume that it could be true.

4. Reasons for the Failure of (5) and (3).

The basic argument is really very simple, and it is very surprising that it has not been used earlier. It is based upon the fact that *one and the same set of observational data is compatible with very different and mutually inconsistent theories.* This is possible for two reasons: first, because theories, which are universal, always go beyond any set of observations that might be available at any particular time; second, because the truth of an observation statement can always be asserted within a certain margin of error only.[47] The first reason allows for theories to differ in domains where experimental results are not yet available. The second reason allows for such differences even in those domains where observations have been made, provided the differences are restricted to the margin of error connected with the observations.[48] Both reasons taken together sometimes allow considerable freedom in the construction of our theories.

[46] This has been emphasized, in private communication, by Professors Kraft (Vienna) and Rynin (Berkeley).
[47] As J. W. N. Watkins has pointed out to me, this invalidates Hempel's conditions 9.1 and 9.2 (in [45]). An attempt to bring logical order into the relation between observation statements and the more precise statements derived from a theory has been made by Professor S. Körner [50], p. 140.
[48] Even this condition is too strong, as will be shown below.

48

Now, it is very important to realize that this freedom which experi-
ace grants the theoretician is nearly always restricted by conditions of
a altogether different character. These additional conditions are neither
niversally valid, nor objective. They are connected partly with the tra-
ition in which the scientist works, with the beliefs and the prejudices
hich are characteristic of that tradition; and they are partly connected
ith his own personal idiosyncrasies. The formal apparatus available and
ae structure of the language he speaks will also strongly influence the
ctivity of the scientist. Whorff's assertion to the effect that the proper-
es of the Hopi language are not very favorable for the development
f a physics like the one with which we are acquainted may very well
e correct.[49] Of course, it must not be overlooked[50] that man is capable
ot only of applying, but also of inventing, languages. Still, the influence
f the language from which he starts should never be underestimated.
another factor which strongly influences theorizing is metaphysical be-
efs. The Neoplatonism of Copernicus was at least a contributing factor
a his acceptance of the system of Aristarchus.[51] Also, the contemporary
ssue between the followers of Niels Bohr and the realists, being still
ndecidable on the basis of contemporary experimentation, is mainly
metaphysical in character.[52] That the choice of theories may be influ-
nced even by aesthetic motives can be seen from Galileo's reluctance
o accept Kepler's ellipses.[53]

Taking all this into account we see that the theory which is suggested
oy a scientist will also depend, apart from the *facts* at his disposal, on
he *tradition* in which he participates, on the mathematical instruments
ae accidentally knows, on his preferences, on his aesthetic prejudices,
on the suggestions of his friends, and on other elements which are
ooted, not in facts, but in the mind of the theoretician and which are
herefore subjective. This being the case it is to be expected that theo-
eticians working in different traditions, in different countries, will ar-
ive at theories which, although in agreement with all the known facts,
are yet mutually inconsistent. Indeed, any consistency over a long period

[49] See [73].
[50] As is done by Bohr, Heisenberg, and von Weizsaecker in their philosophical
writings as well as by some Wittgensteinians. For the point of view of these physicists, see [34] and [38], as well as the end of Sec. 7 of the present paper.
[51] See T. S. Kuhn [51], pp. 128ff.
[52] See [36].
[53] See E. Panofsky [63].

of time would have to be regarded not, as is suggested by (3), (A), an(5), as a methodological virtue, but as an alarming sign that no neideas are being produced and that the activity of theorizing has conto an end. Only the inductivistic doctrine that theories are uniquedetermined by the facts could have persuaded people that lack of ideis praiseworthy and that its consequences are an essential feature of tldevelopment of our knowledge.[54]

At this point it is worth mentioning what will be explained in gredetail later: that the freedom of theorizing granted by the indetermnateness of facts is of great methodological importance. It will turn othat many test procedures presuppose the existence of a class of mutuaincompatible, but factually adequate, theories. Any attempt to reducthis class to a single theory would result in a decisive decrease of tlempirical content of this remaining theory and would therefore be udesirable from the point of view of empiricism. The freedom granted kthe indeterminateness of facts is therefore not only psychologically irportant (it allows scientists of different temperament to follow their diferent inclinations and thereby gives them satisfaction which goes byond the satisfaction derived from the exclusive consideration of factsit is also needed for methodological reasons.

The gist of the argument developed so far is that because of thlatitude which experience allows the theoretician, and because of thdifferent way in which this latitude will be exercised by thinkers odifferent tradition, temperament, and interests, it is to be expected thatwo different theories, and especially two theories of a different degreof generality, will be inconsistent with each other even in those casewhere both are confirmed by the set. In this argument it was assumethat the experimental evidence which inside D' confirms T and T' athe same in both cases. Although this may be so in the specific exampldiscussed, it is certainly not true in general. Experimental evidence doe

[54] This is true mainly of those more crude theories of induction which are held, bimplication, by many physicists. It would seem to me that discussion and criticism cthese theories is a much more effective way of advancing scientific knowledge thainvention of highly technical theories of confirmation which are of no interest to thscientist because they cannot be applied to a single noninstantial theory. Unfortunately, many philosophers of science consider it below their dignity to discuss succrude but effective theories, and they prefer the construction of sophisticated theoriewhich are totally ineffective and useless. The brief discussion of Hempel's paper seemto show that similar objections must be raised against some ideas held by contempcrary empiricists.

EXPLANATION, REDUCTION, AND EMPIRICISM

not consist of facts pure and simple, but of facts analyzed, modeled, and manufactured according to some theory.

The first indication of this manufactured character of the evidence is seen in the corrections which we apply to the readings of our measuring instruments, and in the selection which is made among those readings. Both the corrections and the selection made depend upon the theories held, and they may be different for the theoretical complex containing T, and for the theoretical complex containing T'. Usually T will be more general, more sophisticated, than T', and it will also be invented a considerable time after T'. New experimental techniques may have been introduced in the meantime. Hence, the 'facts,' within D', which count as evidence for T will be different from the 'facts,' within D', which counted as evidence for T' when the latter theory was first introduced. An example is the very different manner in which the apparent brightness of stars was determined in the seventeenth century and is determined now. This is another important reason why T usually will not satisfy (5) with respect to T': not only are T and T' connected with different theoretical ideas leading to different predictions even in the domain where they overlap and are both confirmed, but the better experimental techniques and the improved theories of measurement will usually provide evidence for T which is different from the evidence for T' even within the domain of common validity. In short: introducing T very often leads to recasting the evidence for T'. The demand that T should satisfy (5) with respect to T' would in this case imply the demand that new and refined measurements not be used, which is clearly inconsistent with empiricism.

Against the argument in the last paragraph it might be pointed out that results of measurement which are capable of improvement, and which therefore change, do not belong to the observational domain, but must be formulated with the help of singular statements of the theoretical language.[55] Observational statements proper are such qualitative statements as "pointer A coincides with mark B," or "A is greater than B"—and these statements will not change, or be eliminated, whatever the development of the theory, or of the methods of measurement. This point will be dealt with, and refuted, in Section 7 of this paper.

A further indication of the "manufactured" character of the experi-

[55] For this move, cf. Carnap [15], p. 40.

51

mental evidence is seen in the fact that observable results, and indeed anything conveyed with the help of a language, are always expressed in some theory or other. Because this fact will also be of importance in connection with my criticism of (B), and because it leads to a further criticism of (A), I shall discuss at length the example I have chosen for its elucidation.

5. Second Example: The Problem of Motion.[56]

From its very beginning, rational cosmology, this creation of the Ionian "physiologists," was faced with the problem of change and motion (in the general sense in which it includes locomotion, qualitative alteration, quantitative augmentation and diminution, as well as generation and corruption). The problem arose in two forms. The first was the possibility of change and motion. This form of the problem had to be solved by the invention of a cosmology which allowed for change, i.e., which was not such that the occurrence of change was (unwittingly) excluded from it by the very nature of the assumptions upon which it was based. The second form of the problem which arose, once the first had been solved in a satisfactory manner, was the cause of change. As was shown by Parmenides, the early monistic theories of Thales, Anaximander, and others could not solve the first form of the problem. For Parmenides himself, this did not refute monism; it refuted the existence of change.

The majority of thinkers went a different path, however. They regarded monism as refuted and started with pluralistic theories. In the case of the atomic theory, which was one of these pluralistic theories, this relation between Parmenides' arguments and pluralism is very clear. Leucippus, who "had associated with Parmenides in philosophy,"[57] "thought he had a theory which was in harmony with the senses, and did not do away with coming into being and passing away, nor motion, nor with the multiplicity of things."[58] This is how the atomic theory arose, as an attempt to solve problems created by the empirical inadequacy of the early monism of the Ionians.

However, the theory which was most influential in the Middle Ages

[56] For a more detailed account of the theories mentioned in this section, see M. Clagett [17]. Concerning the first part of the present section, see J. Burnet [10], as well as Clagett [16] and Popper [67].

[57] Aristotle [2], A, 8 324b35.

[58] Theophrastus quoted from Burnet [10], p. 333.

and which also tried to solve what I have above called the second form of the problem was Aristotle's theory of motion as the actualization of potentiality. According to Aristotle

(7)   "motion is a process arising from the continuous action of a source of motion, or a 'motor,' and a 'thing moving.' "[59]

This principle, according to which any motion (and not only accelerated motion) is due to the action of some kind of force, can be easily supported by such common observations as a cart drawn by a horse and a chair pushed around by an angry husband. It gets into difficulties when one considers the motion of things thrown: stones continue to move despite the fact that contact with the motor apparently ceases when they leave the hand. Various theories have been suggested to eliminate this difficulty. From the point of view of later developments, the most important one of these theories is the impetus theory. The impetus theory retains (7) and the general background of the Aristotelian theory of motion. Its distinction lies in the specific assumptions it makes concerning the causes that are responsible for the motion of the projectile. According to the impetus theory, the motor (for example the hand) transfers upon the projectile an inner moving force which is responsible for its continuation of path, and which is continually decreased by the resisting air and by the gravity of the projectile. A stone in empty space would therefore either remain at rest or move (along a straight line[60]) with constant speed, depending on whether its impetus is zero or possesses a finite value.

At this point a few words must be said about the characterization of locomotion. The question as to its proper characterization was a matter of dispute. To us it seems quite natural to characterize motion by space transversed, and, as a matter of fact, one of the suggested characterizations did just this: it defined motion kinematically by reference to space transversed. This apparently very simple characterization needs further specification if an account is to be given of nonuniform movements where the distinction becomes relevant between average velocity and instantaneous velocity. Compared with the actual space transversed by a given body, the instantaneous velocity is a rather abstract notion since

[59] Clagett [17], p. 425.
[60] The parentheses I have added because of the absence from the earlier forms of the impetus theory of an explicit consideration of direction.

it refers to the space that would be transversed if the velocity were to retain constancy over a finite interval of time.

Another characterization of motion is the dynamical. It defines motion in terms of the forces which bring it about in accordance with (7). Adopting the impetus theory the motion of a stone thrown would have to be characterized by its inherent impetus, which pushes it along until it is exhausted by the opposing forces of friction and gravity.

Which characterization is the better one to take? From an operationalistic point of view (and we shall adopt this point of view, since we want to follow the empiricist as far as possible), the dynamical characterization is definitely to be preferred: while it is fairly easy to observe the impetus enclosed in a moving body by bringing it to a stop in an appropriate medium (such as soft wax) and then noting the effect of such a maneuver, it is much more difficult, if not nearly impossible, to arrange matters in such a way that from a given moment on, a non-uniformly moving object assumes a constant speed with a value identical with the value of the instantaneous velocity of the object at that moment and then to watch the effect of this procedure.

With the use of the dynamical characterization, the "inertial law" pronounced above reads as follows:

(8)   The impetus of a body in empty space which is not under the influence of any outer force remains constant.

Now, in the case of inertial motions (8) gives correct predictions about the behavior of material objects. According to (3), explanation of this fact will involve derivation of (8) from a theory and suitable initial conditions. Disregarding the demand for explanation, we can also say, on the basis of (5), that any theory of motion that is more general than (8) will be adequate only if it contains (8) which, after all, is a very basic law. According to (2), the meanings of the key terms of (8) will be unaffected by such a derivation. Assuming Newton's mechanics to be the primary theory, we shall therefore have to demand that (8) be derivable from it *salva significatione*. Can this demand be satisfied?

At first sight it would seem that it is much easier to derive (8) from Newton's theory than it is to establish the correctness of (6): as opposed to Galileo's law (8) is not in quantitative disagreement with anything asserted by Newton's theory. Even better: (8) seems to be identical with Newton's first law so that the process of derivation seems to degenerate into a triviality.[61]

In the remainder of the present section, it will be shown that this is not so and that it is impossible to establish a deductive relationship between (8) and Newton's theory. Later on this will be the starting point of our criticism of (B).

Let me repeat, before beginning the argument, that (8), *taken by itself*, cannot be attacked on empirical grounds. Indeed, we have indicated a primitive method of measurement of impetus, and the attempt to confirm (8) by using this method will certainly show that within the domain of error connected with such crude measurements, (8) is perfectly all right. It is, therefore, quite in order to ask for the explanation, or the reduction, of (8), and the failure to arrive at a satisfactory solution of this task cannot be blamed upon the empirical inadequacy of (8).

We now turn to an analysis of the main terms of (8). According to Nagel the meaning of these terms is to be regarded as "fixed" by the procedures and assumption of the impetus theory, and any one of them is "therefore intelligible in terms of its own rules of usage." [62] What are these meanings, and what are the rules which establish them?

Take the term 'impetus.' According to the theory of which (8) is a part, the impetus is the force responsible for the movement of the object that has ceased to be in direct contact, by push, or by pull, with the material mover. If this force did not act, i.e., if the impetus were destroyed, then the object would cease to move and fall to the ground (or simply remain where it is, in case the movement were on a frictionless horizontal plane). A moving object which is situated in empty space and which is influenced neither by gravity nor by friction is not outside the reach of any force. It is pushed along by the impetus, which may be pictured as a kind of inner principle of motion (similar, perhaps, to the vital force of an organism which is the inner principle of *its* motion).

We now turn to Newton's celestial mechanics and the description, in terms of this theory, of the movement of an object in empty space. (Newton's theory still retains the notion of absolute space and allows therefore for such a description to be formed.) Quantitatively, the same

---

[61] There existed theories, among them the theory of *mail* by Abu'l-Barakat, where quantitative disagreement with Newton's laws was to be expected: in these theories, the impetus decreased with time in the same manner in which a hot poker that is removed from the fire gradually loses the heat stored in it. Cf. Clagett [17], p. 513.

[62] See Nagel [20], p. 301.

movement results. But can we discover in the description of this movement, or in the explanation given for it, anything resembling the impetus of (8)? It has been suggested that the momentum of the moving object is the perfect analogue of the impetus. It is correct that the measure of this magnitude (viz., mv) is identical with the measure that has been suggested for the impetus.[63] However, it would be very mistaken if we were, on that account, to identify impetus and momentum. For whereas the impetus is supposed to be something that pushes the body along,[64] the momentum is the result rather than the cause of its motion. Moreover, the inertial motion of classical mechanics is a motion which is supposed to occur by itself, and without the influence of any causes. After all, it is this feature which according to most historians, radical empiricists included, constitutes one of the main differences between the Aristotelian theory and the celestial mechanics of the seventeenth, eighteenth, and nineteenth centuries: in the Aristotelian theory, the natural state in which an object remains without the assistance of any causes is the state of rest. A body at rest (in its natural place, we should add) is not under the influence of any forces. In the Newtonian physics it is the state of being at rest or in uniform motion which is regarded as the natural state. This means, of course, the explicit denial of a force such as the impetus is supposed to represent.

Now this denial need not mean that the concept of such a force cannot be formed within Newton's mechanics. After all, we deny the existence of unicorns and use in this denial the very concept of a unicorn. Is it then perhaps possible to define a concept such as impetus in terms of the theoretical primitives of Newton's theory? The surprising fact is that any attempt to arrive at such a definition leads to disappointment (which shows, by the way, that theories such as Newton's are expressed in a language that is much more tightly knit than is the language of everyday life). I have already pointed out that the momentum, which would give us the correct mathematical value, is not what we want. What we want is a force that acts upon the isolated object and is responsible for its motion. The concept of such a force can of course be formed within Newton's theory. But considering (a) that the movement under review (the inertial movement) occurs with constant ve-

[63] See Clagett [17], p. 523.

[64] For an elaborate discussion of the difference between momentum and impetus, see Anneliese Maier [58]. For what follows, see also M. Bunge [9], Ch. 4.4.

locity, and (b) Newton's second law, we obtain in all relevant cases zero for the value of this force which is not the measure we want. A positive measure is obtained only if it is assumed that the movement occurs in a resisting medium (which is, of course, the original Aristotelian assumption), an assumption which is inconsistent with another feature of the case considered, i.e., with the fact that the inertial movement is supposed by Newton's theory to occur in empty space. I conclude from this that the concept of impetus, as fixed by the usage established in the impetus theory, cannot be defined in a reasonable way within Newton's theory. And this is not further surprising. For this usage involves laws, such as (7), which are inconsistent with the Newtonian physics.

In the last argument, the assumption that the concept force is the same in both theories played an essential role. This assumption was used in the transition from the assertion, made by the impetus theory, that inertial motions occur under the influence of forces to the calculation of the magnitude of these forces on the basis of Newton's second law. Its legitimacy may be derived from the fact that both the impetus theory and Newton's theory apply the concept force under similar circumstances (paradigm-case argument!). Still, meaning and application are not the same thing, and it might well be objected that the transition performed is not legitimate, since the different contexts of the impetus theory, on the one hand, and of Newton's theory, on the other, confer different meanings upon one and the same word 'force.' This being the case, our last argument is based upon a quaternio terminorum and is, therefore, invalid. In order to meet this objection, we may repeat our argument using the word 'cause' instead of the word 'force' (the latter has a somewhat more specific meaning). But if someone again retorts that 'cause' has a different meaning in Newton's theory from what it has in the impetus theory, then all I can say is that a consistent continuation of that kind of objection will in the end establish what I wanted to show in a more simple manner, viz., the impossibility of defining the notion of an impetus in terms of the descriptive terms of Newton's theory. To sum up: the concept impetus is not "explicable in terms of the theoretical primitives of the primary science."[65] And this is exactly as it should be, considering the inconsistency between some very basic principles of these two theories.

[65] Nagel [20], p. 302.

However, explication in terms of the primitives of the primary science is not the only method which was considered by Nagel in his discussion of the process of reduction. Another way to achieve reduction, which he mentions immediately after the above quotation, "is to adopt a material, or physical hypothesis according to which the occurrence of the properties designated by some expression in the premises of the primary science is a sufficient, or a necessary and sufficient condition for the occurrence of the properties designated by the expressions of the secondary discipline." Both procedures are in accordance with (4), or with (2), or at least Nagel thinks that they are: ". . . in this case" he says, referring to the procedure just outlined, "the meaning of the expressions of the secondary science *as fixed by the established usage of the latter*, is not declared to be analytically related to the meanings of the corresponding expressions of the primary science."[66] Let us now see what this second method achieves in the present case.

To start with, this method amounts to introducing a hypothesis of the form

(9)   impetus = momentum

where each side retains the meaning it possesses in its respective discipline. The hypothesis then simply asserts that wherever momentum is present, impetus will also be present (see the above quotation of Nagel's), and it also asserts that the measure will be the same in both cases. Now this hypothesis, although acceptable within the impetus theory (after all, this theory permits the incorporation of the concept of momentum), is incompatible with Newton's theory. It is therefore not possible to achieve reduction and explanation by the second method.

To sum up: a law such as (8) which, as I have argued, is empirically adequate, and in quantitative agreement with Newton's first law, is yet incapable of reduction to Newton's theory and therefore incapable of explanation in terms of the latter. Whereas the reasons we have so far found for irreducibility were of a quantitative nature, this time we met a qualitative reason, as it were, i.e., the incommensurable character of the conceptual apparatus of (8), on the one side, with that of Newton's theory, on the other.

Taking together the quantitative as well as the qualitative argument, we are now presented with the following situation: there exist pairs of

[66] *Ibid.* My italics.

theories, T and T', which overlap in a domain D' and which are incompatible (though experimentally indistinguishable) in this domain. Outside D', T has been confirmed, and it is also more coherent, more general, and less ad hoc than T'. The conceptual apparatus of T and T' is such that it is possible neither to define the primitive descriptive terms of T' on the basis of the primitive descriptive terms of T nor to establish correct empirical relations involving both these terms (correct, that is, from the point of view of T). This being the case, explanation of T' on the basis of T or reduction of T' to T is clearly impossible if both explanation and reduction are to satisfy (A) and (B). Altogether, the use of T will necessitate the elimination both of the conceptual apparatus of T' and of the laws of T'. The conceptual apparatus will have to be eliminated because its use involves principles, such as (7) in the example above, which are inconsistent with the principles of T; and the laws will have to be eliminated because they are inconsistent with what follows from T for events inside D'. (This would apply to the example above if the theory of mail had been used instead of the impetus theory.) This being the case the demand for explanation and reduction clearly cannot arise if this demand is interpreted as the demand for the explanation, or reduction, of T', rather than of a set of laws that is in some respect similar to T' but in other respects (meanings of fundamental terms included) very different from it. For such a demand would imply the demand to derive, from correct premises, what is false, and to incorporate what is incommensurable.

The effect of the transition from T' and T is rather to be described in the manner indicated in the introductory remarks of the present paper: where I said: What happens when transition is made from a restricted theory T' to a wider theory T (which is capable of covering all the phenomena which have been covered by T') is something much more radical than incorporation of the unchanged theory T' into the wider context of T. What happens is rather a complete replacement of the ontology of T' by the ontology of T, and a corresponding change in the meanings of all descriptive terms of T' (provided these terms are still employed). Let me add here that the not-too-well-known example of the impetus theory versus Newton's mechanical theory is not the only instance where this assertion holds. As I shall show a little later, more recent theories also correspond to it. Indeed, it will turn out that the principle correctly describes the relation between the elements of any

P. K. Feyerabend

pair of noninstantial theories satisfying the conditions which I have just enumerated.

This finishes step one of the argument against the assumption that reduction and explanation are by derivation. What I have shown (and shall show in later sections) is that some very important cases which have been, or could be used as examples of reduction (and explanation) are not in agreement with the condition of derivability. It will be left to the reader to verify that this holds in almost all cases of explanation by theories: assumption (A) does not give a correct account of actual scientific practice. It has also been shown that in this respect the thesis formulated in the beginning of this paper is much more adequate.

Now, as against this result it may be pointed out, with complete justification, that scientific method, as well as the rules for reduction and explanation connected with it, is not supposed to describe what scientists are actually doing. Rather, it is supposed to provide us with normative rules which should be followed, and to which actual scientific practice will correspond only more or less closely.[67] It is very important nowadays to defend such a normative interpretation of scientific method and to uphold reasonable demands even if actual scientific practice should proceed along completely different lines. It is important because many contemporary philosophers of science seem to see their task in a very different light. For them actual scientific practice is the material from which they start, and a methodology is considered reasonable only to the extent to which it mirrors such practice.[68] Looking at disciplines such as medicine, they discover (whether rightly or wrongly—this I do not want to discuss at the present moment) that what is called "explanation" here is not always the inverse of prediction, and they infer from this that the orthodox model which demands such an inverse relationship to hold between explanation and prediction[69] is unduly restrictive.

Two elements must be distinguished in this "discovery."[70] The first is of a purely linguistic character. It is connected with the problem as to what meaning should be given to the word 'explanation.' Clearly this

[67] This point has been made, most forcefully, by K. R. Popper [68], Sec. 10.

[68] If I understand him correctly, this is also the point of view held by my colleague, Professor T. S. Kuhn.

[69] Cf. Hempel and Oppenheim [47], p. 323.

[70] The "discovery" is due among others to Professor Barker. See his contribution to [24], my criticism in the same volume, and his reply.

element is without serious interest. It may be that the word 'explanation' sounds beautiful to some ears—but who would think it reasonable to start a war for, or against, its elimination? The second element, however, which usually remains hidden beneath the linguistic analysis, is much more serious. For what the suggested procedure amounts to is increased leniency with respect to questions of test: a certain medical hypothesis (which, let us say, is expressed by saying that a patient died *because* of tuberculosis) is accepted, and retained, despite the fact that independent tests (independent, that is, of the *past* histories of this case and of other cases) are not available, and its further use is defended by reference to the fact that it is in accordance with the "logic of medicine." Expressed in more pedestrian terms, this maneuver propagates the acceptance of unsatisfactory hypotheses on the grounds that this is what everybody is doing. It is conformism covered up with high-sounding language.[71] It is clear, however, that if this conformism had been propagated successfully in the Middle Ages, modern science with its so very different "logic" would never have come into existence. Modern science is the result of a conscious criticism of the theses propagated and the methods employed by the great majority of scholastic philosophers. For the thinker who demands that a subject be judged "according to its own standards," such criticism is of course impossible; he will be strongly inclined to reject any interference and to "leave everything as it

[71] It should be pointed out that almost all theses which terminate (if at all) the long-winded inquiries of linguistic philosophers possess this two-faced character. On the one side, they seem to be about the meanings of terms only, and therefore rather harmless and uninteresting (although there are enough enthusiastic buyers even for such products). However, on closer inspection, it often turns out that beneath these linguistic trappings there are hidden, and thereby removed from criticism, some highly questionable theories or methodological rules. Only consider the example discussed in the text: it may well be the case that in some disciplines the word 'explanation' is used in a manner that does not lead to the demand for additional predictions. This linguistic result may be expressed by saying that prediction is not essential to explanation (viz., to 'explanation' in the new sense that is characteristic for the disciplines under review). Now from this last result it is then very often inferred that the search for additional predictions is unnecessary and that all is well. Clearly this methodological consequence can be derived from the linguistic premise only if it is further assumed that all is well once an explanation has been given, and this regardless of the sense in which the word 'explanation' is being used in the discipline in question. This assumption, which, I submit, is the silent premise of many contemporary linguistic arguments about explanation, amounts to asserting that all is well as long as the word 'explanation' occurs somewhere in the description of the procedure to be analyzed. This is, of course, pure word magic. Curiously enough, it is the linguistic philosopher who is swayed by such word magic. Which only shows how little language is understood by some of its most verbose champions.

is."[72] It is somewhat puzzling to find that such demands are nowadays advertised under the title of philosophy of science.

Against such conformism it is of paramount importance to insist upon the normative character of scientific method. Adopting this point of view, one cannot regard the arguments of the last few sections as ultimately decisive. They are satisfactory insofar as they show that the "orthodox" are wrong when asserting that (A), (B), and (5) reflect actual scientific practice. But they do not dispose of these principles if they are interpreted as demands to be followed by the scientist (although, of course, they provide ample material for such disproof). I therefore proceed now to a methodological criticism of the demands of the orthodox. The first move in this criticism will be the examination of an argument which has sometimes been used to defend (5).

6. Methodological Considerations.

The argument runs as follows: (a) a good theory is a summary of facts; (β) the predictive success of T' (I will continue to use the notation introduced in Section 2) has shown T' to be good theory inside D'; hence (τ), if T, too, is to be successful inside D', then it must either give us all the facts contained in T', i.e., it must give us T', or at least it must be compatible with T'.

It is easily seen that this very popular argument[73] will not do. We can show this by considering its premises. Premise (a) is acceptable if it is not taken in too strict a sense (for example, if it is not interpreted as implying an ontology of mutually independent 'facts' as has been suggested by Mach and the early Wittgenstein). Interpreted in such a loose manner (a) simply says that a good theory not only will be able to answer many questions, but will also answer them correctly. Now if this is to be the interpretation of (a), then (β) cannot possibly be correct: in (β) the predictive success of T' is taken to indicate that T' will give a correct account of all the facts inside its domain. However, one must remember that because of the general character of statements expressing laws and theories, their predictive success can be established

---

[72] Wittgenstein [74], Sec. 124.

[73] A sloppy version of this argument occurs frequently in arguments by physicists. It ought to be mentioned, by the way, that Hempel's condition 8 leads to the very same result, viz., to the demand that new theories be consistent with their confirmed predecessors. A justification for discussing "crude" arguments of the kind outlined in the text above is given in fn. 54.

only with respect to part of their content. Only part of a theory can at any time be known to be in agreement with observation. From this limited knowledge nothing can be inferred (logically!) with respect to the remainder.[74]

We have also to consider the margin of error involved in every single test. Hence, from a purely logical point of view, new theories will be restricted only to the extent to which their predecessors have been tested and confirmed.[75] Only to this extent will it be necessary for them to agree with their predecessors. In domains where tests have not yet been carried out, or where only very crude tests have been made, we have complete freedom on how to proceed, and this quite independently of which theories were originally used here for the purpose of prediction. Clearly this last condition, which is in agreement with empiricism, is much less restrictive than either (3) or (5).

One might hope to arrive at more restrictive conditions by adding inductive argument to logical reasoning. True, from a logical point of view we can only say that *part* of T' has been found to be in agreement with observation and that T need agree only with that part and not, as is demanded in (5), with the whole of T'. However, if inductive reasoning is used as well, then we shall perhaps have to admit that this partial confirmation has established T', and that therefore the whole of T' should be covered by T. Does this help us to strengthen the condition mentioned at the end of the last paragraph and to demonstrate (5) after all?

It is clear that inductive reasoning cannot establish (5) either. For let us assume that T agrees with T' only where T' has been confirmed and is different from T' in all other instances without having as yet been refuted. In this case T will satisfy our own condition of the last paragraph, and it will not satisfy any stronger condition (except accidentally). Can inductive reasoning prompt us to eliminate T? It is not easily seen how this could be the case, since T shares all its confirming instances with T'. Hence, if T' is established by these instances, then so is T— unless we use formal considerations (which I shall discuss later). Again

[74] This point is indebted to Hume. That Hume's arguments are still not understood by many thinkers and are therefore still in need of repetition has been emphasized by Popper [68], Reichenbach [69], Goodman [42], and others.
[75] As was mentioned in Sec. 4, it is hardly ever the case that two theories which have been discussed in very different historical periods will be based upon exactly the same observations. The condition is therefore still too strict.

we arrive at the result that from the point of view of fact there is not much to choose between T and T', and that (5) cannot be defended on empirical grounds.

It is worthwhile to inquire a little more closely into the effects which adoption of (5), and, incidentally, also of Hempel's condition 8,[76] would have upon the development of scientific knowledge. Such adoption would lead to the elimination of a theory, not because it is inconsistent with the facts, but because it is inconsistent with another, and as yet unrefuted, theory whose confirming instances it shares. This is a strange procedure to be adopted by thinkers who, above anything, claim to be empiricists! However, the situation becomes even worse when we inquire why the one theory is retained and the other rejected. The answer (which is, of course, not the answer given by the empiricist) can only be that the theory which is retained was there first. This shows that in practice the allegedly empirical procedure (5) leads to the preservation of the old theories and to the rejection of the new theories even before these new theories have been confronted with the facts. That is, it leads to the same result as transcendental deduction, intuitive argumentation, and other forms of a priori reasoning, the only difference being that now it is in the name of experience that such results are obtained. This is not the only instance where, on closer scrutiny, a rather close relation emerges between some versions of modern empiricism and the 'school philosophies' it attacks.

We have now to consider the argument that formal criteria may provide a principle of choice between T and T' that is independent of fact. Such formal criteria can indeed be given.[77] However, while usually a more general and coherent theory is preferred to a less general collection of laws, and this on account of being less ad hoc, (5) tends to reverse this procedure. This is due to the fact that general theories of a high degree of coherence usually violate (5). Again this principle is seen to be incompatible with reasonable methodology.

Two things have been shown so far. First, the invalidity of an argument used for establishing (5). Second, the undesirability, from an empirical point of view, of some consequences of this argument. However, all this has little weight when compared with the following most important consideration.

[76] See [45], p. 105.
[77] See Popper [68], Ch. VI.

Within contemporary empiricism, discussions of test and of empirical content are usually carried out in the following manner: it is inquired how a theory is related to its empirical consequences and what these consequences are. True, in the derivation of these consequences reference will have to be made to principles or theorems which are borrowed from other disciplines and which then occur in the correspondence rules. However, these principles and these theorems play a subordinate role when compared with the theory under review; and it is, of course, also assumed that they are mutually consistent and consistent with the theory. One may therefore say that, for the orthodox procedure, the natural unit to which discussions of empirical content and of test methods are referred is always a single theory taken together with those of its consequences that belong to the observation language.

This manner of discussion does not allow us to give an adequate account of crucial experiments which involve more than one theory, none of which are expendable or of psychological importance only. A very good example of the structure of such crucial tests is provided by the more recent development of thermodynamics. As is well known, the Brownian particle is a perpetual motion machine of the second kind, and its existence refutes the (phenomenological) second law. However, could this fact have been discovered in a direct manner, i.e., by a direct investigation of the observational consequences of thermodynamics? Consider what such a refutation would have required! The proof that the Brownian particle is a perpetual motion machine of the second kind would have required (a) measurement of the exact motion of the particle in order to ascertain the changes of its kinetic energy plus the energy spent on the overcoming of the resistance of the fluid, and (b) precise measurements of temperature and heat transfer in the surrounding medium in order to ascertain that any loss occurring here was indeed compensated by the increase of the energy of the moving particle and the work done against the fluid as mentioned in (a). Such measurements, however, are beyond experimental possibilities.[78] Hence, a direct refutation of the second law, i.e., a refutation based upon an investigation of the testable consequences of thermodynamics alone, would have had to wait for one of those rare, not repeatable, and therefore, prima facié suspicious, large fluctuations in which the transferred heat is indeed

[78] Concerning the extreme difficulties of following the motion of the Brownian particle in all its details, see R. Fuerth [41].

accessible to measurement. This means that such a refutation would have never taken place, and, as is well known, the actual refutation of the second law was brought about in a very different manner. It was brought about via the kinetic theory and Einstein's utilization of it in the calculation of the statistical properties of the Brownian motion. In the course of this procedure the phenomenological theory (T') was incorporated into the wider context of statistical physics (T) in such a manner that (5) was violated; and then a crucial experiment was staged (Perrin's investigations).

Now it seems to me that the more general our knowledge gets the more important it will be to carry out tests in the manner indicated, i.e., not by comparing a single theory with experience, but by staging crucial experiments between theories which, although in accordance with all the known facts, are mutually inconsistent and give widely different answers in unexplored domains. This suggests that outside the domain of empirical generalizations the methodological unit to which we refer when discussing questions of test and empirical content consists of a whole set of partly overlapping, factually adequate, but mutually inconsistent theories. To the extent to which utilization of such a set provides additional tests which, for empirical reasons, could not have been carried out in a direct manner, the use of a set of this kind is demanded by empiricism. For the basic principle of empiricism is, after all, to increase the empirical content of whatever knowledge we claim to possess.[79]

On the other hand, the fact that (5) does not allow for the formation of such sets now proves this principle to be inconsistent with empiricism. By excluding valuable tests it decreases the empirical content of the theories that are permitted to remain (and which, as indicated above, will usually be the theories which were there first). This last result of a consistent application of (5) is of very topical interest: it may well be, as has been pointed out by Bohm and Vigier,[80] that the refutation of the quantum-mechanical uncertainties presupposes just an incorporation of the present theory into a wider context, which is not any more in

[79] With respect to this last demand, it might be objected that, given a certain theory, such an extension of content is not possible without changing the theory. This argument would be correct if it could be granted that the interpretation of a physical theory is wholly empirical. As I have shown elsewhere (see [39]), this is not the case. The demand to increase the empirical part of this interpretation is therefore a sensible demand.

[80] See the discussion remarks of these two physicists in [49], as well as those of Bohm [4].

accordance with the idea of complementarity and which therefore suggests new and decisive experiments. And it may also be that the insistence on part of the majority of contemporary physicists upon (5) will, if successful, forever protect these uncertainties from refutation. This is how modern empiricism may finally lead to a situation where a certain point of view petrifies into dogma by being, in the name of experience, completely removed from any conceivable criticism.

To sum up the arguments of the present section: it has been shown that neither (5) nor (A) can be defended on the basis of experience. Quite on the contrary, a strict empiricism will admit theories which are factually adequate and yet mutually inconsistent.[81] An analysis of the character of tests in the domain of theories has revealed, moreover, that the existence of sets of partly overlapping, mutually inconsistent, and yet empirically adequate theories is not only possible, but also required. I shall now conclude the present section by discussing a little more in detail the logical and psychological consequences of the use of such a set.

Increase of testability will not be the only result. The use of a set of theories with the properties indicated above will also improve our understanding of each of its members by making it very clear what is denied by the theory that happens to be accepted in the end. Thus, it seems to me that our understanding of Newton's somewhat obscure notion of absolute space and of its merits is greatly improved when we compare it with the relational ideas of Berkeley, Huyghens, Leibnitz, and Mach, and when we consider the failure of the latter ideas to give a satisfactory account of the phenomenon of inertial forces. Also, the study of general relativity will lead to a deeper understanding of this notion than could be obtained from a study of the *Principia* alone.[82] This is not meant to

[81] It is interesting to study in detail the dilemma of an inductivistic philosophy of science. To start with, a radical empiricist demands close adherence to the facts and is ultimately suspicious of any generalization; a summary of facts is all he will admit on empirical grounds. However, generalizations *do* play an important role in the sciences; scientific knowledge is, after all, a collection of theories. Hence, methods must be found to justify these theories on the basis of experience, and a "logic" must be constructed which allows us, again on the basis of experience, to confer some kind of certainty upon them. A "logic" of this kind is introduced, and theories are established with its help. But now the demand that the future theories should be consistent with the theories thus established turns out to be far too strict. The way out of this dilemma can only consist in abandoning the idea that theories can be "established" by experience and in admitting that insofar as they go beyond the facts we have no means whatever (except, perhaps, psychological ones) to guarantee their trustworthiness.

[82] This, by the way, is one of the reasons why an axiomatic exposition of physical

67

be understood in a psychological sense only. For just as the meaning of a term is not an intrinsic property of it but is dependent upon the way in which the term has been incorporated into a theory, in the very same manner the content of a whole theory (and thereby again the meaning of the descriptive terms which it contains) depends upon the way in which it is incorporated into both the set of its empirical consequences and the set of all the alternatives which are being discussed at a given time:[83] once the contextual theory of meaning has been adopted, there is no reason to confine its application to a single theory, or a single language, especially as the boundaries of such a theory or of such a language are almost never well defined. The considerations above have shown, moreover, that the unit involved in the test of a specific theory is not this theory taken together with its own consequences; they have shown

principles, such as Newton's, is inferior by far to a dialectical exposition where many ideas are considered, and the pros and cons discussed, until finally one theory is pronounced the most satisfactory one. Of course, if one holds that, concerning theories, the only relation of interest is the relation between a single theory and "the facts," and if one also believes that these facts single out a certain theory more or less uniquely, then one will be inclined to regard discussion of alternatives as a matter of history, or of psychology, and one will even wish to hide, with some embarrassment, the situation at the time when the clear message of the facts had not yet been grasped. However, as soon as it is recognized that the refutation (and thereby also the confirmation) of a theory necessitates its incorporation into a family of mutually inconsistent alternatives, in the very same moment, the discussion of these alternatives becomes of paramount importance for methodology and should be included in the presentation of the theory that is accepted in the end. For the same reason, adherence to either the distinction between context of discovery (where alternatives are considered, but given a psychological function only) and context of justification (where they are not mentioned any more), or strict adherence to the axiomatic approach must be regarded as an arbitrary and very misleading restriction of methodological discussion: much of what has been called "psychological," or "historical," in past discussions of method is a very relevant part of the theory of test procedures.

Considering all this, the increased attention paid to the historical aspects of a subject, and the attempts to break down the distinction between the synthetic and the analytic must be welcomed as steps in the right direction. However, even here there are drawbacks. Only very few of the enthusiastic proponents of an increased study of the history of a subject realize the methodological importance of their investigations. The justification they give for their interest is either sentimental or psychological ("it gives me ideas"), or based upon some very implausible (because Hegelian) notions concerning the "growth" of knowledge. What these thinkers need in order not to fall victims to all sorts of quasi-philosophies is a methodological backbone, and I hope that the theory of test which has been sketched above in its merest outlines will provide such a backbone.

[83] In the twentieth century, the contextual theory of meaning has been defended most forcefully by Wittgenstein; see [74] as well as my summary in [28]. However, it seems that Wittgenstein is inclined to restrict this theory to the inside of his language games: Platonism of concepts is replaced by Platonism of (theories or) games. For a brief criticism of this attitude see [35].

that this unit is a whole class of mutually incompatible and factually adequate theories. Hence, both consistency and methodological considerations suggest such a class as the context from which meanings are to be made clear.[84]

Also, the use of such a class rather than of a single theory is a most potent antidote against dogmatism. Psychologically speaking, dogmatism arises, among other things, from the inability to imagine alternatives to the point of view in which one believes. This inability may be due to the fact that such alternatives have been absent for a considerable time and that, therefore, certain ways of thinking have been left undeveloped; it may also be due to the conscious elimination of such alternatives. However that may be, persistence of a single point of view will lead to the gradual establishment of well-circumscribed methods of observation and measurement; it will lead to codification of the ways in which these results are interpreted; it will lead to a standardized terminology and to other developments of a similarly conservative kind. This being the case, the gradual acceptance of the theory by an ever-increasing number of people must finally bring about a transformation even of the most common idiom that is taught in very early youth. In the end, all the key terms will be fixed in an unambiguous manner, and the idea (which may have led to such a procedure in the first place) that they are copies of unchanging entities and that change of meaning, if it should happen, is due to human mistake—this idea will now be very plausible. Such plausibility reinforces all the maneuvers which may be used for the preservation of the theory (elimination of opponents included[85]).

The conceptual apparatus of the theory having penetrated nearly all means of communication, such methods as transcendental deduction and analysis of usage, which are further means of solidifying the theory, will be very successful. Altogether it will seem that at last an absolute and irrevocable truth has been arrived at. Disagreement with facts may of course occur, but, being now convinced of the truth of the existing

---

[84] Textbooks and historical presentations very often create the impression either that such classes never existed and that physicists (at least the "great" ones) at once arrived at the one and good theory or that their existence must not be taken too seriously. This is quite understandable. After all, historians have been just as much under the influence of inductivistic ideas as the physicists and the philosophers.

[85] Today, of course, the "elimination" takes the more "refined" form of a refusal to publish (or to read) what is not in agreement with the accepted doctrine. However, this "liberalism" applies to physical theories only. It does not seem to apply to political theories.

point of view, its proponents will try to save them with the help of ad hoc hypotheses. Experimental results that cannot be accommodated, even with the greatest ingenuity, will be put aside for later consideration. The result will be absolute truth, but, at the same time, it will decrease in empirical content to such an extent that all that remains will be no more than a verbal machinery which enables us to accompany any kind of event with noises (or written symbols) which are considered true statements by the theory.[86]

The picture painted above is by no means exaggerated. The way in which, for example, the theory of witchcraft and daemonic influence crept into the most common way of thinking, and could be preserved for quite a considerable time, offers a vivid illustration of each point mentioned in the last paragraph. Moreover, the story of its overthrow furnishes another illustration of our thesis that comprehensive theories cannot be eliminated by a direct confrontation with "the facts."

Now let us compare such a dogmatic procedure with the effects of the use of a class of theories rather than a single theory. First of all such a procedure will encourage the building of a great variety of measuring instruments. There will be no one way of interpreting the results, and the theoretician will be trained to switch quickly from one interpretation to another.[87] Intuitive appeal will lose its paralyzing effect, transcendental deduction which, after all, presupposes uniformity of usage, will be impossible; and the question of agreement with the facts will assume a very prominent position. Experimental results which are inconsistent with one theory may be consistent with a different theory; this elimi-

[86] The fact that all the features of absolute knowledge can be manufactured by exercising an absolute conformism I regard as the most important objection to any claim to finality: it would seem to show that whereas hypothetical knowledge, being the result of repeated corrections in the light of criticism, is at least in partial contact with the world, absolute knowledge is entirely man made and can therefore not raise any claim to factual content. Even the stability of a testable hypothesis cannot be regarded as a sign of its truth because it may be due to the fact that, owing to some particular astigmatism on our part, we have as yet overlooked some very decisive tests. There is no sign by which factual truth may be recognized.

It is interesting to note that the development toward dogmatism as it has been described in the text occurred both in the "school philosophies" and in empiricism, and, in the latter case, notably in physics: there is no indication that empiricism provides special protection against dogmatic petrification. The only difference is that its defense will be in terms of "experience" rather than in terms of "intuition" or "revelation." For this see Dewey [21], esp. Ch. II.

[87] As Professor Agassi has pointed out to me, this method was consciously used by Faraday in order to escape the influence of prejudice. Concerning its role in modern discussions about the microlevel, see [37].

nates the motives for ad hoc hypotheses, or at least reduces them considerably. Nor will it be necessary to use instrumentalism as a means of getting out of trouble, since a coherent account may be provided by an alternative to the theory considered. The likelihood that empirical results will be left lying around will also be smaller; if they do not fit one theory, they will fit another. It is not at all superfluous to mention the tremendous development of human capabilities encouraged by such a procedure and the antidotes it contains against the wish to set up, and to obey, all-powerful regimes, be they political, religious, or scientific. Taking all this into account, we are inclined to say that *whereas unanimity of opinion may be fitting for a church, or for the willing followers of a tyrant, or some other kind of 'great man,' variety of opinion is a methodological necessity for the sciences and, a fortiori, for philosophy.* Neither (A), nor (B), nor (5) allows for such variety. It follows that, to the extent to which both principles (and the philosophy behind them) delimit variety and demand future theories to be consistent with theories already in existence, they contain a theological element (which lies, of course, in the worship of "facts" so characteristic for nearly all empiricism).

The paralyzing effect of familiarity and intuitive appeal upon social reconstruction and the progress of knowledge has been understood and described in a most excellent manner by Bertolt Brecht. It is true that Brecht was mainly concerned with the function of the theater in the process of making familiar, and thereby creating the impression of the unchangeability of, social relations. But in his analysis, which was to establish the need for a new kind of theater, he hit upon a very important fact of a much more general character, namely, the paralyzing effect of familiarity and of the methods used to bring about such familiarity. In the theater these methods consist in trying to represent what is accidental and changeable (a particular social situation, for example) as essential to man or nature, and therefore unchangeable. "The theatre with which we are confronted shows the structure of society (represented on the stage) as being removed from the influence of society (which is situated in the audience). Oedipus, who has sinned against some principles which were supported by the society of his time, is executed; the gods take care of that—they cannot be criticised" ([8], p. 146). "What we need," Brecht continues, "is a theatre which does not only allow for those sensations, insights, and impulses that are permitted by

71

the field of human relations where the action takes place; what we need is a theatre which uses and creates thoughts and feelings that play a role in the change of the field itself" ([8], p. 147). *It is exactly the same thing that is needed in the domain of epistemology.* What is needed is a method which does not—in the name of either "universal principles," "revelation," or "experience"—put fetters on the scientist's imagination but which enables him to use alternatives to the point of view which is the one commonly accepted. What is needed is a method that also enables him to take a critical attitude with respect to any element of this point of view, be it a law, or a so-called empirical fact. I am afraid that only very few scientists have ever been aware of the need for such a method, and that most of them are to be compared to the transfixed audience of one of the familiar pieces of the "classical" repertoire which has been described so vividly by Brecht—textbooks and scientific journals replacing the images projected by the actors.

At all times, the existence, within a certain tradition, of a variety of opinions (or a variety of theories) has been regarded as proof of the unsoundness of the method adopted by the members of this tradition. It was assumed, as being nearly self-evident, that the proper method must lead to the truth, that the truth is one, and that the proper method must therefore result in the establishment of a single theory and the perennial elimination of all alternatives. Conversely, the existence of various points of view and of a community where discussion of alternatives was regarded as fundamental was always regarded as a sign of confusion. Curiously enough, this attitude is found in thinkers who otherwise have very little in common. This can be seen from an examination of various criticisms of the pre-Socratic philosophers that have been proffered in the course of history.

As has been shown by Popper [67], these early philosophers were not only the inventors of a theoretical science (as opposed to a science which is content with assembling empirical generalizations as was the physics, the mathematics, and the astronomy of the Egyptians), but they also invented the method characteristic for this kind of science, i.e., the method of test within a class of mutually inconsistent, partly overlapping, and to that extent empirically adequate theories. All this was thoroughly misunderstood. Thus the sophists are reported to have ridiculed the Ionians by pointing out that their motto seemed to be "To every philosopher his own principle." Plato made full use of this popular sentiment (*Sophist* 242ff), and so did the church fathers later on: "As of canonical authors," writes St. Augustine (quoted from [16], p. 132), "God forbid that they should differ . . . [But] let one look amongst all the multitude of philosophers' writings, and if he finds two that tell both one tale in all respects, it may be registered for a rarity." Soon after the rise of Baconian empiricism with its so very different message, the variety of the theories discussed by the pre-Socratics was used as an example of where one gets when leaving the solid ground, not of revelation, but of experience. The following quotation is very characteristic: "As to the particular tenets of Thales, and his successors of the *Ionian* school, the sum of what we learn from the imperfect accounts we have of them is that each overthrew what his predecessor had advanced; and met with the same treatment himself from his successor . . . So early did the passion for systems begin."[88] In fact, nearly all inventors of new methods in philosophy and in the sciences have been inspired by the hope that they would be able to put an end to the quarrel of the schools and to establish the one true body of knowledge (this inspiration is present even today in some of the defenders of the Copenhagen interpretation of the quantum theory). What emerges from our own considerations is that *once dispute has been made empirically testable* (and this was already the case with the early Ionians—see again Popper's article [67]), *it becomes an essential element of the development of knowledge;* and it also emerges that cessation of dispute is not to be regarded as a sign that now we have finally arrived at the truth, but rather as a sign of fatigue in those originally participating in the dispute (just as the more recent return to religious beliefs is a sign of fatigue and of despair in the capabilities of reason).

According to Popper this procedure of testing a theory by comparing it with experience, as seen in the light of alternatives, is identical with the scientific method. Professor Matson, who also emphasizes that "the key" to the method of the pre-Socratics lies in the fact that "they were not dogmatic about their predecessors, but rather . . . criticized them most acutely,"[89] differs from Popper insofar as he regards this method as relevant either to part of the sciences only (he regards it as relevant to contemporary cosmology), or even as a nondogmatic alternative to

[88] McLaurin [55], p. 28.
[89] [57], p. 445.

73

empiricism (for Popper the test procedure outlined *is* the empirical method). That the pre-Socratics were inventors of *method* as well as *theories* is emphasized in Professor Matson's admirable paper on Anaximander [56]. As far as I can make out, Popper's views on the matter were already developed at the time of the first publication of his *Open Society*. Both these thinkers represent a minority view which, in my opinion, is of the greatest importance for the understanding of the history of early Greek philosophy.

This finishes my criticism of (A) and (5). (A) has been shown to be in disagreement not only with actual scientific practice but also with the principles of a sound empiricism. The account of theorizing given in the introduction has been shown to be superior to the hierarchy of axioms and theorems which seems to be the favorite model of contemporary empiricism. The use of a set of mutually inconsistent and partially overlapping theories has been found to be of fundamental importance for methodology. The desideratum mentioned in connection with what has here been called the second idea has thereby been fulfilled. Serious doubt has been thrown upon the correctness and the desirability of (B). I now turn to the refutation of (B).

7. Criticism of the Assumption of Meaning Invariance.

In Section 5 it was shown that the "inertial law" (8) of the impetus theory is incommensurable with Newtonian physics in the sense that the main concept of the former, viz., the concept of impetus, can neither be defined on the basis of the primitive descriptive terms of the latter, nor related to them via a correct empirical statement. The reason for this incommensurability was also exhibited: although (8), *taken by itself*, is in quantitative agreement both with experience and with Newton's theory, the "rules of usage" to which we must refer in order to explain the meanings of its main descriptive terms contain the law (7) and, more especially, the law that constant forces bring about constant velocities. Both of these laws are inconsistent with Newton's theory. Seen from the point of view of this theory, any concept of a force whose content is dependent upon the two laws just mentioned will possess zero magnitude, or zero denotation, and will therefore be incapable of expressing features of actually existing situations. Conversely, it will be capable of being used in such a manner only if all connections with Newton's theory have first been severed. It is clear that this example refutes (B) if we interpret that thesis as the description of how science actually proceeds.

We may generalize this result in the following fashion: consider two theories, T' and T, which are both empirically adequate inside D', but which differ widely outside D'. In this case the demand may arise to explain T' on the basis of T, i.e., to derive T' from T and suitable initial conditions (for D'). Assuming T and T' to be in quantitative agreement inside D', such derivation will still be impossible if T' is part of a theoretical context whose "rules of usage" involve laws inconsistent with T.[90]

It is my contention that the conditions just enumerated apply to many pairs of theories which have been used as instances of explanation and reduction. Many (if not all) of such pairs on closer inspection turn out to consist of elements which are incommensurable and therefore incapable of mutual reduction and explanation. However, the above conditions admit of still wider application and then lead to very important consequences with regard to the structure and development both of our knowledge and of the language used for the expression of it. After all, the principles of the context of which T' is a part need not be explicitly formulated, and as a matter of fact they rarely are. To bring about the situation described above (sets of mutually incommensurable concepts), it is sufficient that they govern the use of the main terms of T'. In such a case T' is formulated in an idiom some of whose implicit rules of usage are inconsistent with T (or with some consequences of T in the domain where T' is successful). Such inconsistency will not be obvious at a glance; it will take considerable time before the incommensurability of T and T' can be demonstrated. However, as soon as this demonstration has been carried out, in the very same moment, the idiom of T' must be given up and must be replaced by the idiom of T. Of course, one need not go through the laborious and very uninteresting task of analyzing the context of which T' is part.[91] All that is needed is the adoption of the terminology and the 'grammar' of the most detailed and most

[90] Since this difficulty can arise even in the domain of empirical generalizations, the orthodox account may be inappropriate for them as well.

[91] There are many philosophers (including my friends in the Minnesota Center) who would admit that the importance of linguistic analysis is very limited. However, they would still hold that its application is necessary in order to find out to what extent the advent of a new theory modifies the customary idiom. The considerations above would show that even this is granting too much and that one travels best without any linguistic ballast.

successful theory throughout the domain of its application.[92] This auto matically takes care of whatever incommensurabilities may arise, and it does so without any linguistic detective work (which therefore turns out to be entirely unnecessary for the progress of knowledge).

What has just been said applies most emphatically to the relation between (theories formulated in) some commonly understood language and more abstract theories. That is, I assert that languages such as the "everyday language," this notorious abstraction of contemporary lin guistic philosophy, frequently contain (not explicitly formulated, that is, but implicit in the way in which its terms are used) principles which are inconsistent with newly introduced theories, and that they must therefore be either abandoned and replaced by the language of the new and better theories even in the most common situations, or they must be completely separated from these theories (which would lead to a situation where it is possible to believe in various kinds of "truth"): it is far from correct to assume that the everyday languages are so widely conceived, so tolerant, indefinite, and vague that they will be compatible with any scientific theory, that science can at most fill in details, and that a scientific theory will never run against the principles implicitly contained in them. *The very opposite is the case.* As will be shown later, even everyday languages, like languages of highly theoretical systems, have been introduced in order to give expression to some theory or point of view, and they therefore contain a well-developed and sometimes very abstract ontology. It is very surprising that the champions of the "ordi nary language" should have such a low opinion of its descriptive power.

However, before turning to this part of the argument, I shall briefly discuss another example where the questionable principles of T' have been explicitly formulated, or can at least be easily unearthed.

The example which is dealt with by Nagel is the relation between phenomenological thermodynamics and the kinetic theory. Employing his own theory of reduction and, more especially, the condition I have quoted in the text adjacent to my footnote 11, Nagel claims that the

[92] One hears frequently that a *complete* replacement of the grammar and the terminology of the "old language" is impossible because this old language will be needed for introducing the new language and will, therefore, infect at least part of the new language. This is curious reasoning indeed if we consider that children learn languages without the help of a previously known idiom. Is it really asserted that what is possible for a small child will be impossible for a philosopher, a linguistic philoso pher at that?

terms of the statements which have been derived from the kinetic theory (with the help of correlating hypotheses similar to (9)) will have the meanings they originally possessed within the phenomenological theory, and he repeatedly emphasizes that these meanings are fixed by "its own procedures" (i.e., by the procedures of the phenomenological theory) "whether or not [this theory] has been, or will be, reduced to some other discipline." [93]

As in the case of the impetus theory, we shall begin our study of the correctness of this assertion with an examination of these "procedures" and "usages"; more especially, we shall start with an examination of the usage of the term 'temperature,' "as fixed by the established procedures" of thermodynamics.

Within thermodynamics proper,[94] temperature ratios are defined by reference to reversible processes of operating between two levels, L' and L", each of these levels being characterized by one and the same temperature throughout. The definition, viz.,

(10) $T':T'' = Q':Q''$

identifies (after a certain arbitrary choice of units) the ratio of the temperature with the ratio between the amount of heat absorbed at the higher level and the amount of heat rejected at the lower level. Closer inspection of the "established usage" of the temperature thus defined shows that it is supposed to be

(11)   independent of the material of the substance chosen for the cycle, and unique.

This property can be inferred from the extension of the concept of temperature thus defined to radiation fields and from the fact that the constants of the main laws in this domain are universal, rather than dependent upon either the thermometric substance or the substance of the system investigated.

Now, it can be shown by an argument not to be presented here that (10) and (11) taken together imply the second law of thermodynamics in its strict (phenomenological) form: the concept of temperature as "fixed by the established usages" of thermodynamics is such that its application to concrete situations entails the strict (i.e., nonstatistical) second law.

[93] Nagel [20], p. 301.
[94] See Fermi [22], Sec. 9.

77

Now whatever procedure is adopted, the kinetic theory does not give us such a concept. First of all, there does not exist any dynamical concept that possesses the required property.[95] The statistical account, on the other hand, allows for fluctuations of heat back and forth between two levels of temperature and, therefore, again contradicts one of the laws implicit in the "established usage" of the thermodynamic temperature. The relation between the thermodynamic concept of temperature and what can be defined in the kinetic theory, therefore, can be seen to conform to the pattern that has been described at the beginning of the present section: we are again dealing with two incommensurable concepts. The same applies to the relation between the purely thermodynamic entropy and its statistical counterpart; whereas the latter admits of very general application, the former can be measured by infinitely slow reversible processes only. Taking all this into consideration we must admit that it is impossible to relate the kinetic theory and the phenomenological theory in the manner described by Nagel, or to explain all the laws of the phenomenological theory in the manner demanded by Hempel and Oppenheim on the basis of the statistical theory. Again replacement rather than incorporation, or derivation (with the help, perhaps, of premises containing statistical as well as phenomenological concepts), is seen to be the process that characterizes the transition from a less general theory to a more general one.

It ought to be pointed out that the discussion is very idealized. The reason is that a purely kinetic account of the phenomena of heat does not yet seem to exist. What exists is a curious mixture of phenomenological and statistical elements, and it is this mixture which has received the name 'statistical thermodynamics.' However, even if this is admitted, it remains that the concept of temperature as it is used in this new and mixed theory is different from the original, purely phenomenological concept. To our point of view, according to which terms change their meanings with the progress of science, Nagel raises the following objection: "The redefinition of expressions with the development of inquiry [so it is noted], is a recurrent feature in the history of science. Accordingly, though it must be admitted that in an earlier use the word 'temperature' had a meaning specified exclusively by the rules and procedures of thermometry and classical thermodynamics, it is now so used

---

[95] I shall not discuss, in the present paper, the somewhat different situation with respect to the first law.

that temperature is 'identical by definition' with molecular energy. The deduction of Boyle-Charles' law does not therefore require the introduction of a further postulate, whether in the form of a coordinating definition or a special empirical hypothesis, but simply makes use of this definitional identity. This objection illustrates the unwitting double talk into which it is so easy to fall. It is certainly possible to redefine the word 'temperature' so that it becomes synonymous with 'mean kinetic energy.' But it is equally certain that on this redefined usage the word has a different meaning from the one associated with it in the classical science of heat, and therefore a meaning different from the one associated with the word in the statement of the Boyle-Charles law. However, if thermodynamics is to be reduced to mechanics, it is temperature in the sense of the term in the classical science of heat which must be asserted to be proportional to the mean kinetic energy of gas molecules. Accordingly, if the word 'temperature' is redefined as suggested by the objection, the hypothesis must be invoked that the state of bodies described as 'temperature' (in the classical thermodynamic sense) is also characterized by 'temperature' in the redefined sense of the term. This hypothesis, however, will then be one that does not hold as a matter of definition . . . Unless this hypothesis is adopted, it is not the Boyle-Charles law which can be derived from the assumptions of the kinetic theory of gases. What is derivable without the hypothesis is a sentence similar in syntactical structure to the standard formulation of the law, but possessing a sense that is unmistakably different from what the law asserts."[96] So far Nagel.

Commencing my criticism, I shall at once admit the correctness of the last assertion. After all, it has been my contention all through this paper that extension of knowledge leads to a decisive modification of the previous theories both as regards the quantitative assertions made and as regards the meanings of the main descriptive terms used. Applying this to the present case I shall therefore at once admit that incorporation into the context of the statistical theory is bound to change the meanings of the main descriptive terms of the phenomenological theory. The difference between Nagel and myself lies in the following. For me, such a change to new meanings and new quantitative assertions is a natural occurrence which is also desirable for methodological reasons

[96] [61], pp. 357–358.

(the last point will be established later in the present section). For Nagel such a change is an indication that reduction has not been achieved, for reduction in Nagel's sense is supposed to leave untouched the meanings of the main descriptive terms of the discipline to be reduced (cf. his "if thermodynamics is to be reduced to mechanics, it is temperature in the sense of the term in the classical science of heat which must be asserted to be proportional to the mean kinetic energy of gas-molecules"). "Accordingly," he continues, quite obviously assuming that reduction in his sense can be carried through, "if the word 'temperature' is redefined as suggested by the objection, the hypothesis must be invoked that the state of bodies described as 'temperature' (in the classical thermodynamic sense) is also characterized by 'temperature' in the redefined sense of the term. This hypothesis . . . will then be one that does not hold as a matter of definition." It will also be a false hypothesis because the conditions for the definition of the phenomenological temperature are never satisfied in nature (see the arguments above in the text and compare also the arguments in connection with formula (9)), which is only another sign of the fact that reduction, in the sense of Nagel, of the phenomenological theory to the statistical theory is not possible (obviously the additional premises used in the reduction are not supposed to be false). Once more arguments of meaning have led to quite unnecessary complications.

Further examples exhibiting the same features can be easily provided. Thus in classical, prerelativistic physics the concept of mass (and, for that matter, the concept of length and the concept of time duration) was absolute in the sense that the mass of a system was not influenced (except, perhaps, causally) by its motion in the coordinate system chosen. Within relativity, however, mass has become a relational concept whose specification is incomplete without indication of the coordinate system to which the spatiotemporal descriptions are all to be referred. Of course, the values obtained on measurement of the classical mass and of the relativistic mass will agree in the domain D', in which the classical concepts were first found to be useful. This does not mean that what is measured is the same in both cases: what is measured in the classical case is an *intrinsic property* of the system under consideration; what is measured in the case of relativity is a *relation* between the system and certain characteristics of D'. It is also impossible to define the exact classical concepts in relativistic terms or to relate them with the help of an empirical generalization. Any such procedure would imply the false assertion that the velocity of light is infinitely large. It is therefore again necessary to abandon completely the classical conceptual scheme once the theory of relativity has been introduced; and this means that it is imperative to use relativity in the theoretical considerations put forth for the explanation of a certain phenomenon as well as in the observation language in which tests for these considerations are to be formulated; after all, the empirically untenable consequences of the attempts above to give a reduction of classical terms to relativistic terms emerges whether or not the elements of the definition belong to the observation language.

Many more examples can be added to those discussed in the present paper (viz., the impetus theory, phenomenological thermodynamics, and the classical conception of mass). All these examples show that the postulate of meaning invariance is incompatible with actual scientific practice. That is, it has been shown that in most cases it is impossible to relate successive scientific theories in such a manner that the key terms they provide for the description of a domain D′, where they overlap and are empirically adequate, either possess the same meanings or can at least be connected by empirical generalizations. It is also clear that the methodological arguments against meaning invariance will be the same as the arguments against the derivability condition and the consistency condition. After all, the demand for meaning invariance implies the demand that the laws of later theories be compatible with the principles of the context of which the earlier theories are part, and this demand is, therefore, seen to be a special case of condition (5). Using our earlier arguments against (5) we may now infer the untenability, on methodological grounds, of meaning invariance as well. And as our argument is quite general we may also infer that it is undesirable that the "ordinary" usage of terms be preserved in the course of the progress of knowledge. Wherever such preservation is observed, we shall feel inclined to think that the suggested new theories are not as revolutionary as they perhaps ought to be, and we shall have the suspicion that some ad hoc procedures have perhaps been adopted. Violation of ordinary usage, and of other "established" usages, on the other hand, is a sign that real progress has been made, and it is to be welcomed by anybody interested in such progress (provided of course that this violation is connected with the suggestion of a new point of view or a new theory and is not just the result of linguistic arbitrariness).

81

P. K. Feyerabend

Our argument against meaning invariance is simple and clear. It proceeds from the fact that usually some of the principles involved in the determination of the meanings of older theories or points of view are inconsistent with the new, and better, theories. It points out that it is natural to resolve this contradiction by eliminating the troublesome and unsatisfactory older principles and to replace them by principles, or theorems, of the new and better theory. And it concludes by showing that such a procedure will also lead to the elimination of the old meanings and thereby to the violation of meaning invariance.

The most important method used for escaping the force of this clear and simple argument is the transition to instrumentalism. Instrumentalism maintains that the new theory must not be interpreted as a series of statements but that it is rather to be understood as a predictive machine whose elements are tools rather than statements and therefore cannot be incompatible with any principle already in existence. This very popular move (popular, that is, because used also by scientists) admittedly cuts the ground from beneath our argument and makes it inapplicable. However, it has never been explained why a new and satisfactory theory should be interpreted as an instrument, whereas the principles behind the established usage, which can easily be shown to be empirically inadequate, are not so interpreted. After all, the only advantage of the latter is that they are *familiar*—an advantage which is a psychological and historical accident and which should therefore not have any influence upon questions of interpretation and of reality. One may try to answer this criticism by ascribing an instrumental function to all principles, old or new, and not only to those contained in the most recent theory. Such a procedure means acceptance of a sense-data account of knowledge. Having shown elsewhere[97] that such an account is impossible, I can now say that this consequence of a universal instrumentalism is tantamount to its refutation. Result: neither a restricted nor a universal instrumentalism can be carried through in a satisfactory manner. This disposes of the instrumentalistic move.

While instrumentalism possesses at least a semblance of plausibility, the arguments to be discussed now are devoid even of this feature. Indeed, I am very hesitant to apply the word 'arguments' to these expressions of confused thinking, their wide acceptance and asserted self-evi-

[97] [38].

82

dence notwithstanding. Consider for example the following question (which is supposed to be a criticism of our suggestion that after the acceptance of the kinetic theory the word 'temperature' will be in need of reinterpretation): [98] "If the meaning of 'temperature' is [now] the same as that of 'mean kinetic energy of molecular motion,' what are we talking about when milk is said to have a temperature of 10° Cels? Surely not the kinetic energy of the molecular constituents of the liquid, for the uninstructed layman is able to understand what is here said without possessing any notion about the molecular composition of the milk."

Now it may be quite correct that the "uninstructed layman" [99] does not think of molecules when speaking about the temperature of his milk and that he has not the slightest notion of the molecular constitution of the liquid either. However, what has the reference to him got to do with our argument according to which a person who has already accepted *and understood* the theory of the molecular constitution of gases, liquids, and solids cannot at the same time demand that the premolecular concept of temperature be retained? It is not at all denied by our argument that the "uninstructed layman" may possess a concept of temperature that is very different from the one connected with the molecular theory (after all, some "uninstructed laymen," intelligent clergymen included, still believe in ghosts and in the devil). What is denied is that anybody can consistently continue using this more primitive concept and at the same time believe in the molecular theory. Again, this does not mean that a person may not, on different occasions, use concepts which belong to different and incommensurable frameworks. The only thing that is forbidden for him is the use of both kinds of concepts *in the same argument*; for example, he may not use the one kind of concept in his observation language and the other kind in his theoretical language. Any such combination—and this is the gist of our considerations in the pres-

[98] The argument in connection with this question can be found in Nagel [20], p. 293. It is not clear to me whether or not Nagel would be prepared to support the argument.

[99] By the way, who *is* this uninstructed layman? From the purpose for which he is being employed in many arguments, it would seem to emerge that he is not supposed to know much science, or much politics, or much religion, or much of anything. This means that in these times of mass communication and mass education he must be very careful not to read the wrong parts of his newspaper, he must be careful not to leave his television set on for too long a time, and he must also not allow himself to converse too much with his friends, his children, etc. That is, he must be either a savage or an idiot. I really wonder what are the motives which lead to a philosophy where the most interesting language is the language of savages or of idiots.

ent section—would introduce principles which are mutually inconsistent and thereby destroy the argument in which it is supposed to occur. It is evident that this position is not at all endangered by the objection implied by the question above.

However, quite apart from being so obviously irrelevant to our thesis, the objection reflects an attitude that must appear quite incredible to anybody who possesses even the slightest acquaintance with the history of knowledge. The question insinuates that the layman's ability to handle the word 'temperature' according to the rules prescribed for it in some simple idiom indicates his understanding of the thermal properties of bodies. It insinuates that the existence of an idiom allows us to infer the truth of the principles which underlie this idiom. Or, to be more specific, it insinuates that *what is being used is, on that account alone, already exhibited as adequate, useful, and perhaps irreplaceable.* After all, the reference to the layman's understanding of the word 'temperature' is not made without purpose. It is made with the purpose of preserving the common meaning of this word since, it is alleged, this common meaning can be understood and is not in need of replacement. The discussion of a specific example will at once show the detrimental effect of any such procedure.

The example chosen now brings us to the second part of the present section where the relation is investigated, not between explicitly formulated theories, but between a theory and the implicit principles that govern the usage of the descriptive terms of some idiom. As has been said a little earlier, it is our conviction that "everyday languages," far from being so widely and generally conceived that they can be made compatible with any scientific theory, contain principles that may be inconsistent with some very basic laws. It was also pointed out that these principles are rarely expressed in an explicit manner (except, perhaps, in those cases where there is an attempt to defend the corresponding idiom against replacement or change) but that they are implicit in the rules that govern the use of its main descriptive terms. And our point was that, once these principles are found to be empirically inadequate, they must be given up and with them the concepts that are obtained by using terms in accordance with them. Conversely, the attempt to retain these concepts will lead to the conservation of false laws and to a situation where every connection between concepts or facts is severed.

84

The example which I have chosen to show this involves the pair 'up-down.' There existed a time when this pair was used in an absolute fashion, i.e., without reference to a specified center, such as the center of the earth. That it was used in such a manner can be easily seen from the "vulgar" remark that the antipodes would "fall off" the earth if the earth were spherical,[100] as well as from the more sophisticated attempts of Thales, Xenophanes, and others to find support for the earth as a whole, assuming that it would otherwise fall "down."[101] These attempts, as well as that remark about the antipodes employ two assumptions: first, that any material object is under the influence of a force; second, that this force acts in a privileged direction in space and must therefore be regarded as anisotropic. It is this privileged direction to which the pair 'up-down' refers. The second assumption is not explicitly made; it can only be derived from the way in which the pair 'up-down' is used in arguments such as those mentioned above.[102] We have here an example of a cosmological assumption (anisotropic character of space) implicit in the common idiom.

This example refutes the thesis which has been defended by some philosophers that "everyday languages" are fairly free from hypothetical elements and therefore ideally suited as observational languages.[103] It refutes the thesis by showing that even the most harmless part of a common idiom may rest upon very far-reaching hypotheses and must therefore be regarded as hypothetical to a very high degree.

Another remark concerns the changes of meaning needed once the Newtonian (or perhaps even the Aristotelian) explanation of the fall of heavy bodies is adopted. Newtonian space is isotropic and homogeneous. Hence, accepting this theory, one cannot anymore use the pair 'up-down' in the previous fashion and at the same time assume that one is describing actual features of physical situations. More especially, one cannot retain the absolute use of this pair for the description of observable features, since such features are quite obviously assumed to exist. Any per-

---

[100] For a discussion of this remark and of a related "vulgar" remark concerning shape and arrangement of the terrestrial waters, see Pliny, *Natural History*, II, 161–166, quoted in Cohen and Drabkin [19], pp. 159–161.

[101] For a description and criticism of these attempts, see [1], 294a12ff; also quoted in [19], pp. 143–148.

[102] For the atomist's conception of space, which, at least since Epicurus, seems to be influenced by the popular ideas discussed above, cf. M. Jammer [48], p. 11.

[103] This thesis was introduced by Professor Herbert Feigl in discussions with me. For my own position, see also Philipp Frank [40].

son accepting Newton's physics and the conception of space it contain must, therefore, give a new meaning even to such a familiar pair of term as is the pair 'up-down,' and he must now interpret it as a relation be tween the direction of a motion and a center that has been fixed in ad vance. And as Newton's theory is preferable, on empirical grounds, to the older and "absolute" cosmology, it follows that the relational usage of the pair 'up-down' will be preferable, too. Conversely, the attempt to retain the old usage amounts to retaining the old cosmology, and this despite the discoveries which have shown it to be obsolete.

To this argument it may be, and has been, objected that the "vulgar" usage of the pair 'up-down' was never supposed to be so general as to be applicable to the universe as a whole. This may be the case (although I do not see any reason for assuming that "ordinary" people are so very cautious as to apply the pair to the surface of the earth only; all the passages referred to in the above quotations contradict this assumption and so does the fact that at all times *real* ordinary people—and not only their Oxford substitutes—were very much interested in celestial phe nomena[104]). However, even such a restriction would not invalidate our argument. It would rather show that the pair was used for singling out an absolute direction near the surface of the earth and that it did not assume such a direction to exist throughout the universe. It is clear that even this modest position is incompatible with the ideas implicit in the Newtonian point of view, which does not allow for local anisotropies either.

Consider now, after this example, the following argument in favor of the thesis that what is being used is, on that account alone, already ex hibited as adequate, useful, and perhaps irreplaceable. The argument is the late Professor Austin's, and it has been repeated by G. J. War nock.[105] "Language," writes Warnock, "is to be used for a vast number of highly important purposes; and it is at the very least unlikely that it should contain either much more, or much less, than those purposes re quire. If so, the existence of a number of different ways of speaking is very likely indeed to be an indication that there is a number of different things to be said . . . Where the topic at issue really is one that does

[104] The reason why Oxford philosophers so rarely discuss the influence of astronomy upon everyday languages may perhaps be found in the weather of their favorite dis cussion place. However, this reason unfortunately does not explain their ignorance in physics, theology, mythology, biology, and even linguistics.
[105] [71], pp. 150–151.

constantly concern most people in some practical way—as for example perception, the ascription of responsibility, or the assessment of human character or conduct—then it is certain that everyday language is as it is for some extremely good reasons; its verbal variety is certain to provide clues to important distinctions."[106]

If I understand this passage correctly, it means that the existence of certain distinctions in a language may be taken as an indication of similar distinctions in the nature of things, situations, and the like. And the reason for this is that people who are in constant contact with things and situations will soon develop the correct linguistic means for describing their properties. In short: human beings are good inductive machines in domains of concentrated interest, and their inductive ability will be the better the greater their concern, or the greater the practical value of the topic treated. Consequently, languages containing distinctions of practical interest are very likely to be adequate and irreplaceable.

There are many objections against this train of reasoning. First of all, it would seem to be somewhat arbitrary to restrict interests to those which can be derived from the immediate necessities of the physical life of the human race. From history we learn that the motives emerging from abstract considerations such as those found in a myth, or in a theo-astronomical system, are at least as strong as the more pedestrian motives connected with the immediate fulfillment of material needs (after all, people have died for their convictions!). Now if a language can be trusted because of the commitment of those who use it and if commitment is found to range over a much wider area than has first been imagined, if it is found to range over physics, astronomy (think of Giordano Bruno!), and biology, then the result will be that the principle we are discussing at the present moment (viz., the principle that what is being used for a purpose is on that account alone already useful and irreplaceable) must be applied to any language and any theory that has ever been developed and seriously tested. However—and this is the second point—there exist many theories and languages which have been found to be inadequate, and this despite their usefulness and despite the zeal of those who had developed them. This applies to the language of

[106] Astronomy is again omitted. It would seem to me that problems of astronomy had a much greater influence upon the formation of our language than problems of perception, which are of a very ephemeral nature and also are very technical. The skies and the stars (which, after all, were assumed to be gods) were everyone's concern.

87

the Aristotelian physics, which had to be introduced into medieval thinking under very great difficulties and whose influence went much further than is sometimes realized; it applies to the language of the physics of Newton (mechanicism); and it applies to many other languages. Of course, this is the result one would expect: success under even very severe tests does not guarantee infallibility; no amount of commitment and no amount of success can guarantee the perennial reliability of inductions.

The principle which we have been discussing just now does not occur only in philosophy. Bohr's contention[107] that the account of all quantum mechanical evidence must forever "be expressed in classical terms" has been defended in a very similar manner. According to Bohr, we need our classical concepts not only if we want to give a summary of facts; without these concepts the facts to be summarized could not even be stated. As Kant before him, Bohr observes that our experimental statements are always formulated with the help of certain theoretical terms and that the elimination of these terms would lead, not to the "foundations of knowledge" as a positivist would have it, but to complete chaos. "Any experience," he asserts, "makes its appearance within the frame of our customary points of view and forms of perception"[108]— and at the present moment the forms of perception are those of classical physics.

Now does it follow, as is asserted by Bohr, that we can never go beyond the classical framework and that therefore all our future microscopic theories will have to use the notion of complementarity as a fundamental notion?

It is quite obvious that the actual use of classical concepts for the description of experiments within contemporary physics can never justify such an assumption, even if these concepts happen to have been very successful in the past (Hume's problem). For a theory may be found whose conceptual apparatus, when applied to the domain of validity of classical physics, would be just as comprehensive and useful as the classical apparatus without coinciding with it. Such a situation is by no means uncommon. The behavior of the planets, of the sun, and of the satellites can be described both by the Newtonian concepts and by the concepts

---

[107] See p. 43 above.

[108] [5], p. 1. For a more detailed account of what follows, see [32], [34], [36], [37], [38].

of general relativity. The order introduced into our experiences by Newton's theory is retained and improved upon by relativity. This means that the concepts of relativity are sufficiently rich for the formulation of all the facts which were stated before with the help of Newtonian physics. Yet the two sets of concepts are completely different and bear no logical relation to each other.

Other examples of the same kind can be provided very easily. What we are dealing with here is, of course, again the old problem of induction. No number of examples of usefulness of an idiom is ever sufficient to show that the idiom will have to be retained forever. And if it is objected, as it has been in the case of the quantum theory, that the language of classical physics is the only actual language in existence for the description of experiments,[109] then the reply must be that man is not only capable of using theories and languages but that he is also capable of inventing them.[110] How else could it have been possible, to mention only one example, to replace the Aristotelian physics and the Aristotelian cosmology with the new physics of Galileo and Newton? The only conceptual apparatus then available was the Aristotelian theory of change with its opposition of actual and potential properties, the four causes, and the like. Within this conceptual scheme, which was also used for the description of experimental results, Galileo's (or rather Descartes') law of inertia does not make sense, nor can it be formulated. Should, then, Galileo have followed Heisenberg's advice and have tried to get along with the Aristotelian concepts as well as possible, since his "actual situation . . . [was] such that [he did] use the Aristotelian concepts"[111] and since "there is no use discussing what could be done if we were other beings than we are"? By no means. What was needed was not improvement or delimitation of the Aristotelian concepts; what was needed was an entirely new theory. This concludes our argument against the principle that a useful language is to be regarded as adequate and irreplaceable and, thereby, fully restores the force of our attack against meaning invariance, as well as reinforces the positive suggestions made in connection with this attack and especially the idea that conceptual changes may occur anywhere in the system that is employed at a certain time for the explanation of the properties of the world we live in.

[109] See Heisenberg [44], p. 56, and von Weizsaecker [72], p. 110.
[110] See also fn. 92.
[111] This is a paraphrase of a passage in Heisenberg [44], p. 56.

P. K. Feyerabend

As I indicated in the introductory discussion, this transition from a point of view which demands that certain 'basic' terms retain their meaning, come what may, to a more liberal point of view which allows for changes anywhere in the system employed is bound to influence profoundly our attitude with respect to many philosophical problems and will also facilitate their solution. Let me take the mind-body problem as an example. It seems to me that the difficulties of this problem are to be sought precisely in the fact that meaning invariance is regarded as a necessary condition of its satisfactory solution. That is, it is assumed, or even demanded, that the meanings of at least some terms of the problem must remain constant throughout the discussion of the problem and further that these terms must retain their meanings in the solution as well.

Of course, different schools will apply the demand for meaning invariance to different concepts. A Platonist will demand that terms such as 'mind' and 'matter' remain unchanged, whereas an empiricist will require that some observational terms, such as the term 'pain,' or the more abstract term 'sensation,' retain their (common) meaning. Now a closer analysis of these key terms will, I think, reveal that they are incommensurable in exactly the sense in which this term has been defined at the beginning of the present section. This being the case, it is of course completely impossible either to reduce them to each other, or to relate them to each other with the help of an empirical hypothesis, or to find entities which belong to the extension of both kinds of terms. That is, the conditions under which the mind-body problem has been set up as well as the particular character of its key terms are such that a solution is forever impossible: a solution of the problem would require relating what is incommensurable without allowing for a modification of meanings which would eliminate this incommensurability.

All these difficulties disappear if we are prepared to admit that, in the course of the progress of knowledge, we may have to abandon altogether a certain point of view and the meanings connected with it—for example, if we are prepared to admit that the mental connotation of mental terms may be spurious and in need of replacement by a physical connotation according to which mental events, such as pains, states of awareness, and thoughts, are complex physical states of either the brain or the central nervous system, or perhaps the whole organism. I personally happen to avor this idea that at some time sensations will turn out to be fairly complex central states which therefore possess a definite location inside the human body (which need not coincide with the place where the sensation is *felt* to be). I also hope that it will be possible to carry out a similar analysis of all so-called mental states.

Now whatever the merit of this belief of mine, it cannot be refuted by reference to the fact that what we "mean" by a sensation, or by a thought, is nothing that could have a location,[112] an internal structure, or physical ingredients. For if my belief is correct, and if it is indeed possible to develop a "materialistic" theory of human beings, then we shall of course be forced to abandon the "mental" connotations of the mental terms, and we shall have to replace them by physical connotations. According to the point of view which I am defending in the present paper, the only legitimate way of criticizing such a procedure would be to criticize this new materialistic theory by either showing that it is not in agreement with experimental findings or pointing out that it possesses some undesirable formal features (for example, by pointing out that it is *ad hoc*). Linguistic counter arguments have, I hope, been shown to be completely irrelevant.

The considerations in these last paragraphs are of course very sketchy. Still I hope that they give the reader an indication of the tremendous changes implied by the renunciation of the principle of meaning invariance as well as of the nefarious influence this principle has had upon traditional philosophy (modern empiricism included).

8. *Summary and Conclusion.*

Two basic assumptions of the orthodox theory of reduction and explanation have been found to be in disagreement with actual scientific practice and with reasonable methodology. The first assumption was that the explanandum is *derivable* from the explanans. The second assumption was that *meanings are invariant* with respect to the process of reduction and explanation. We may sum up the results of our investigation in the following manner:

Let us assume that T and T' are two theories satisfying the conditions outlined at the beginning of Section 3. Then, from the point of view of scientific method, T will be most satisfactory if it is (*a*) *inconsistent*

---

[112] See Wittgenstein's investigation of the "grammar" of mental terms as presented in [74].

with T' in the domain where they both overlap;[113] and if it is (β) *incommensurable* with T'.

Now it is clear that a theory which is satisfactory according to the criterion just pronounced will not be capable of functioning as an explanans in any explanation or reduction that satisfies the principles put forth by Hempel and Oppenheim or Nagel. Paradoxically speaking: *Hempel Oppenheim explanations cannot use satisfactory theories as explanantia And satisfactory theories cannot function as explanantia in Hempel-Oppenheim explanations.* How is the theory of explanation and reduction to be changed in order to eliminate this very undesirable paradox?

It seems to me that the changes that are necessary will make it impossible to retain a formal theory of explanation, because these changes will introduce pragmatic or "subjective" considerations into the theory of explanation. This being the case, it seems perhaps advisable to eliminate altogether considerations of explanation from the domain of scientific method and to concentrate upon those rules which enable us to compare two theories with respect to their formal character and their predictive success and which guarantee the constant modification of our theories in the direction of greater generality, coherence, and comprehensiveness. I shall now give a more detailed outline of the reasons which have prompted me to adopt this pragmatic point of view.

Consider again T and T' as described above. Under these circumstances, the set of laws T'' following from T inside D' will either be inconsistent with T' or incommensurable with it. In what sense, then, can T be said to explain T'? This question has been answered by Popper for the case of the inconsistency of T' and T''. "Newton's theory," he says, "unifies Galileo's and Kepler's. But far from being a mere conjunction of these two theories—which play the part of *explicanda* for Newton— *it corrects them while explaining them.* The original explanatory task was the deduction of the earlier results. It is solved, not by deducing them, but by deducing something better in their place: new results which, under the special conditions of the older results, come numerically very close to these older results, and at the same time correct them. Thus the empirical success of the old theory may be said to corroborate the new theory; and in addition, the corrections may be tested in their

---

[113] This condition has been discussed with great clarity in [66]. It was this discussion (as well as dissatisfaction with [60]) that was the starting point of the present analysis of the problem of explanation.

turn . . . What is brought out strongly by [this] . . . situation . . . is the fact that the new theory cannot possibly be *ad hoc* . . . Far from repeating its *explicandum*, the new theory contradicts it and corrects it. In this way, even the evidence of the *explicandum* itself becomes independent evidence for the new theory." [114]

In a letter to me, J. W. N. Watkins has suggested that this theory may be summarized as follows: Explanation consists of two steps. The first step is derivation, from T, of those laws which obtain under the conditions characterizing D'. The second step is comparison of T'' and T' and realization that both are empirically adequate, i.e., fall within the domain of uncertainty of the observational results. Or, to express it in a more concise manner: T explains T' satisfactorily only if T is true and there exists a consequence T'' of T for the conditions of validity of T' such that T'' and T' are at least equally strong and also experimentally indistinguishable.

The first question that arises in connection with Dr. Watkins' formulation is this: experimentally indistinguishable on the basis of which observations? T' and T'' may be indistinguishable by the crude methods used at the time when T was first suggested, but they may well be distinguishable on the basis of later and more refined methods. Reference to a certain observational method will therefore have to be included in the clause of experimental indistinguishability. The notion of explanation will be relative to this observational material. It will not make sense any longer to ask whether or not T explains T'. The proper question will be whether T explains T' *given the observational material, or the observational methods O*. Using this new mode of speech we are forced to deny that Kepler's laws are explained by Newton's theory *relative to the present observations*—and this is perfectly in order; for these present observations in fact refute Kepler's laws and thereby eliminate the demand for explanation. It seems to me that this theory can well deal with all the problems that arise when T and T' are commensurable, but inconsistent inside D'. It does not seem to me that it can deal with the case where T' and T are incommensurable. The reason is as follows.

As soon as reference to certain observational material has been included in the characterization of what counts as a satisfactory explanation, in the very same moment the question arises as to how this observational

[114] Popper [66], p. 33.

material is to be presented. If it is correct, as has been argued all the way through the present paper, that the meanings of observational terms depend on the theory on behalf of which the observations have been made, then the observational material referred to in this modified sketch of explanation must be presented in terms of this theory also. Now incommensurable theories may not possess any comparable consequences, observational or otherwise. Hence, there may not exist any possibility of finding a characterization of the observations which are supposed to confirm two incommensurable theories. How, then, is the above account of explanation to be modified to cover the case of incommensurable theories also? [115]

It seems to me that the only possible way lies in closest adherence to the pragmatic theory of observation. According to this theory, it will be remembered, we must carefully distinguish between the *causes* of the production of a certain observational sentence, or the features of the process of production, on the one side, and the *meaning* of the sentence produced in this manner on the other. More especially, a sentient being must distinguish between the fact that he possesses a certain sensation, or disposition to verbal behavior, and the interpretation of the sentence being uttered in the presence of this sensation, or terminating this verbal behavior. Now our theories, apart from being pictures of the world, are also instruments of prediction. And they are good instruments if the information they provide, taken together with information about initial conditions characterizing a certain observational domain D_o, would enable a robot, who has no sense organs, but who has this information built into himself (or herself), to react in this domain in exactly the same manner as sentient beings who, without knowledge of the theory, have been trained to find their way about D_o and who are able to answer, 'on the basis of observation,' many questions concerning their surroundings.[116] This is the criterion of predictive success, and it is seen not at all to involve reference to the *meanings* of the reactions carried out either by the robot or by the sentient beings (which latter need not be humans, but can also be other robots). All it involves is agreement of behavior.

Now this criterion involves "subjective" elements. Agreement is de-

---

[115] As Professor Feigl has pointed out to me, this difficulty also arises in the case of crucial experiments.

[116] Of course, the *motivations* of the robot and of the sentient being must also be the same.

manded between the behavior of (nonsentient, but theory-fed) robots and that of sentient beings, and it is thereby assumed that the latter possesses a privileged position. Considering that perceptions are influenced by belief in theories and that behavior, too, is influenced by belief in theories, this criterion would seem to be somewhat arbitrary. It is easily seen, however, that it cannot be replaced by a less arbitrary and more "objective" criterion. What would such an objective criterion be? It would be a criterion which is either based upon behavior that is not connected with any theoretical element—and this is impossible (cf. my criticism of the theory of sense data above)—or it would be behavior that is tied up with an irrefutable and firmly established theory—which is equally impossible. We have to conclude, therefore, that a formal and "objective" account of explanation cannot be given.

REFERENCES

1. Aristotle. *De Coelo*.
2. Aristotle. *De Generatione et Corruptione*.
3. Barker, S. "The Role of Simplicity in Explanation," in *Current Issues in the Philosophy of Science*, H. Feigl and G. Maxwell, eds. New York: Holt, Rinehart, and Winston, 1961. Pp. 265–274.
4. Bohm, D. *Causality and Chance in Modern Physics*. London: Routledge and Kegan Paul, 1957.
5. Bohr, N. *Atomic Theory and the Description of Nature*. Cambridge: Cambridge University Press, 1932.
6. Bohr, N. "Discussions with Einstein," in *Albert Einstein, Philosopher-Scientist*, P. A. Schilpp, ed. Evanston, Ill.: Library of Living Philosophers, 1948. Pp. 201–241.
7. Born, M. *Natural Philosophy of Cause and Chance*. Oxford: Oxford University Press, 1948.
8. Brecht, B. *Schriften zum Theater*. Berlin and Frankfurt/Main: Suhrkamp Verlag, 1957.
9. Bunge, M. *Causality*. Cambridge, Mass.: Harvard University Press, 1959.
10. Burnet, J. *Early Greek Philosophy*. London: Adam and Charles Black, 1930.
11. Carnap, R. "Die Physikalische Sprache als Universalsprache der Wissenschaft," *Erkenntnis*, 2:432–465 (1932).
12. Carnap, R. "Psychologie in Physikalischer Sprache," *Erkenntnis*, 3:107–142 (1933).
13. Carnap, R. "Über Protokollsaetze," *Erkenntnis*, 3:215–228 (1933).
14. Carnap, R. "Testability and Meaning," *Philosophy of Science*, 3:419–471 (1936) and 4:1–40 (1937).
15. Carnap, R. "The Methodological Character of Theoretical Concepts," in *Minnesota Studies in the Philosophy of Science*, Vol. I, H. Feigl and M. Scriven, eds. Minneapolis: University of Minnesota Press, 1956. Pp. 38–76.
16. Clagett, M. *Greek Science in Antiquity*. London: Abelard-Schuman, 1957.
17. Clagett, M. *The Science of Mechanics in the Middle Ages*. Madison: University of Wisconsin Press, 1959.
18. Conant, J. B. *Case Histories in the Experimental Sciences*, Vol. I. Cambridge, Mass.: Harvard University Press, 1957.

19. Cohen, M. R., and I. E. Drabkin, eds. *A Source Book in Greek Science.* New York: McGraw-Hill, 1948.
20. Danto, A., and S. Morgenbesser, eds. *Philosophy of Science.* New York: Meridian Books, 1960.
21. Dewey, John. *The Quest for Certainty.* New York: Capricorn Books, 1960.
22. Fermi, E. *Thermodynamics.* New York: Dover Publications, 1956.
23. Feigl, H., and M. Brodbeck, eds. *Readings in the Philosophy of Science.* New York: Appleton-Century-Crofts, 1953.
24. Feigl, H., and G. Maxwell, eds. *Current Issues in the Philosophy of Science.* New York: Holt, Rinehart, and Winston, 1961.
25. Feigl, H., and M. Scriven, eds. *Minnesota Studies in the Philosophy of Science,* Vol. I. Minneapolis: University of Minnesota Press, 1956.
26. Feigl, H., M. Scriven, and G. Maxwell, eds. *Minnesota Studies in the Philosophy of Science,* Vol. II. Minneapolis: University of Minnesota Press, 1958.
27. Feyerabend, P. K. "Carnap's Theorie der Interpretation Theoretischer Systeme," *Theoria,* 21:55–62 (1955).
28. Feyerabend, P. K. "Wittgenstein's 'Philosophical Investigations,' " *Philosophical Review,* 54:449–483 (1955).
29. Feyerabend, P. K. "Eine Bemerkung zum Neumannschen Beweis," *Zeitschrift für Physik,* 145:421–423 (1956).
30. Feyerabend, P. K. "On the Quantum Theory of Measurement," in *Observation and Interpretation,* S. Körner, ed. London: Butterworth, 1957. Pp. 121–130.
31. Feyerabend, P. K. "An Attempt at a Realistic Interpretation of Experience," *Proceedings of the Aristotelian Society,* New Series, 58:143–170 (1958).
32. Feyerabend, P. K. "Complementarity," *Proceedings of the Aristotelian Society,* Supplementary Vol., 32:75–104 (1958).
33. Feyerabend, P. K. "Das Problem der Existenz Theoretischer Entitaeten," *Probleme der Wissenschaftstheorie.* Vienna: Springer, 1960. Pp. 35–72.
34. Feyerabend, P. K. "O Interpretacji Relacyj Nieokreslonosci," *Studia Filozoficzne,* 19:23–78 (1960).
35. Feyerabend, P. K. "Patterns of Discovery," *Philosophical Review,* 59:247–252 (1960).
36. Feyerabend, P. K. "Professor Bohm's Philosophy of Nature," *British Journal for the Philosophy of Science,* 10:321–338 (1960).
37. Feyerabend, P. K. "Bohr's Interpretation of the Quantum Theory," in *Current Issues in the Philosophy of Science,* H. Feigl and G. Maxwell, eds. New York: Holt, Rinehart, and Winston, 1961. Pp. 371–390.
38. Feyerabend, P. K. "On the Interpretation of Microphysical Theories," to appear in *Minnesota Studies in the Philosophy of Science,* Vol. IV, H. Feigl and G. Maxwell, eds.
39. Feyerabend, P. K. "On the Interpretation of Scientific Theories," to appear in *Proceedings of the XIIth International Congress of Philosophy,* Milan.
40. Frank, P. *Relativity, a Richer Truth.* Boston: Beacon Press. 1950.
41. Fuerth, R. "Über einige Beziehungen Zwischen Klassischer Statistik und Quantenmechanik," *Zeitschrift für Physik,* 81:143–162 (1933).
42. Goodman, N. *Fact, Fiction, and Forecast.* Cambridge, Mass: Harvard University Press, 1955.
43. Hanson, N. R. *Patterns of Discovery.* Cambridge: Cambridge University Press, 1958.
44. Heisenberg, W. *Physics and Philosophy.* New York: Harper and Brothers, 1958.
45. Hempel, C. G. "Studies in the Logic of Confirmation," *Mind,* 54:1–26, 97–121 (1945).
46. Hempel, C. G. "A Logical Appraisal of Operationism," in *Validation of Scientific Theories,* P. Frank, ed. Boston: Beacon Press, 1954. Pp. 52–67.

47. Hempel, C. G., and P. Oppenheim. "Studies in the Logic of Explanation," *Philosophy of Science*, 15:135–175 (1948).
48. Jammer, M. *Concepts of Space*. Cambridge, Mass.: Harvard University Press, 1957.
49. Körner, S., ed. *Observation and Interpretation*. London: Butterworth, 1957.
50. Körner, S. *Conceptual Thinking*. New York: Dover Publications, 1960.
51. Kuhn, T. S. *The Copernican Revolution*. New York: Random House, 1959.
52. Landau, L. D., and E. M. Lifschitz. *Quantum Mechanics*. Reading, Mass.: Addison-Wesley, 1958.
53. Mach, E. *Waermelehre*. Leipzig: Johann Ambrosius Barth, 1897.
54. Mach, E. *Zwei Aufsaetze*. Leipzig: Johann Ambrosius Barth, 1912.
55. McLaurin, C. *An Account of Sir Isaak Newton's Philosophical Discoveries*. London: Buchanan's Head, 1750.
56. Matson, W. I. "The Naturalism of Anaximander," *Review of Metaphysics*, 6: 387–395 (1953).
57. Matson, W. I. "Cornford and the Birth of Metaphysics," *Review of Metaphysics*, 8:443–454 (1955).
58. Maier, A. *Die Vorlaeufer Galilei's im 14, Jahrhundert*. Rome: Edizioni di Storiae Litteratura, 1949.
59. Morris, E. "Foundation of the Theory of Signs," *International Encyclopaedia of Unified Science*, Sec. II/7. Chicago: University of Chicago Press, 1942.
60. Nagel, E. "The Meaning of Reduction in the Natural Sciences," in *Science and Civilization*, R. C. Stauffer, ed. Madison: University of Wisconsin Press, 1949. Pp. 99–145.
61. Nagel, E. *The Structure of Science*. New York: Harcourt, Brace, and Company, 1961.
62. Neumann, J. von. *Mathematical Foundations of Quantum Mechanics*. Princeton: Princeton University Press, 1957.
63. Panofsky, E. "Galileo as a Critic of the Arts," *Isis*, 47:3–15 (1956).
64. Popper, K. R. "Naturgesetze und Theoretische Systeme," in *Gesetz und Wirklichkeit*, S. Moser, ed. Innsbruck: Hochschulverlag, 1948. Pp. 65–84.
65. Popper, K. R. *The Open Society and Its Enemies*. Princeton: Princeton University Press, 1950.
66. Popper, K. R. "The Aim of Science," *Ratio*, 1:24–35 (1957).
67. Popper, K. R. "Back to the Pre-Socratics," *Proceedings of the Aristotelian Society*, New Series, 54:1–24 (1959).
68. Popper, K. R. *The Logic of Scientific Discovery*. New York: Basic Books, 1959.
69. Reichenbach, H. *Experience and Prediction*. Chicago: University of Chicago Press, 1948.
70. Sellars, W. "The Language of Theories," in *Current Issues in the Philosophy of Science*, H. Feigl and G. Maxwell, eds. New York: Holt, Rinehart, and Winston, 1961. Pp. 57–77.
71. Warnock, J. *British Philosophy in 1900*. Oxford: Oxford University Press, 1956.
72. Weizsaecker, C. F. von. *Zum Weltbild der Physik*. Leipzig: Verlag Hirzel, 1954.
73. Whorff, B. L. *Language, Thought, and Reality: Selected Writings*, John B. Carroll, ed. Cambridge, Mass.: Technology Press of Massachusetts Institute of Technology, 1956.
74. Wittgenstein, L. *Philosophical Investigations*. Oxford: Basil Blackwell, 1953.

Deductive-Nomological vs. Statistical Explanation

1. Objectives of This Essay.

This essay is concerned with the form and function of explanation in the sense in which it is sought, and often achieved, by empirical science. It does not propose to examine all aspects of scientific explanation; in particular, a closer study of historical explanation falls outside the purview of the present investigation. My main object is to propose, and to elaborate to some extent, a distinction of two basic modes of explanation—and similarly of prediction and retrodiction—which will be called the deductive and the inductive mode.[1]

The structure of deductive explanation and prediction conforms to what is now often called the covering-law model: it consists in the deduction of whatever is being explained or predicted from general laws in conjunction with information about particular facts. The logic of this procedure was examined in some earlier articles of mine, and especially in a study carried out in collaboration with P. Oppenheim.[2]

Since then, various critical comments and constructive suggestions concerning those earlier efforts have appeared in print, and these as well

---

[1] This distinction was developed briefly in Hempel [25], Sec. 2.

[2] See Hempel [24], especially Secs. 1–4; Hempel [25]; and Hempel and Oppenheim [26]. This latter article will henceforth be referred to as SLE. The point of these discussions was to give a more precise and explicit statement of the deductive model of scientific explanation and to exhibit and analyze some of the logical and methodological problems to which it gives rise: the general conception of explanation as deductive subsumption under more general principles had been set forth much earlier by a variety of authors, some of whom are listed in SLE, fn. 4. In fact, in 1934 that conception was explicitly presented in the following passage of an introductory textbook: "Scientific explanation consists in subsuming under some rule or law which expresses an invariant character of a group of events, the particular event it is said to explain. Laws themselves may be explained, and in the same manner, by showing that they are consequences of more comprehensive theories." (Cohen and Nagel [10], p. 397.) The conception of the explanation of laws by deduction from theories was

s discussions with interested friends and with my students have led me
to reconsider the basic issues concerning the deductive model of scientific explanation and prediction. In the first of the two principal parts of
this essay, I propose to give a brief survey of those issues, to modify in
certain respects the ideas set forth in the earlier articles, and to examine
some new questions concerning deductive explanation, deductive prediction, and related procedures.

The second major part of the present study is an attempt to point out,
and to shed some light on, certain fundamental problems in the logic of
inductive explanation and prediction.

Part I. Deductive-Nomological Systematization

2. *The Covering-Law Model of Explanation.*

The deductive conception of explanation is suggested by cases such as
the following: The metal screwtop on a glass jar is tightly stuck; after
being placed in warm water for a short while, it can be readily removed.
The familiar explanation of this phenomenon is, briefly, to the effect
that the metal has a higher coefficient of thermal expansion than glass,
so that a given rise in temperature will produce a larger expansion of
the lid than of the neck of the glass jar; and that, in addition, though
the metal is a good conductor of heat, the temperature of the lid will
temporarily be higher than that of the glass—a fact which further increases the difference between the two perimeters. Thus, the loosening
of the lid is here explained by showing that it came about, by virtue of
certain antecedent circumstances, in accordance with certain physical
laws. The explanation may be construed as an argument in which the
occurrence of the event in question is inferred from information expressed by statements of two kinds: (a) general laws, such as those concerning the thermal conductivity of metal and the coefficients of expansion for metal and for glass, as well as the law that heat will be
transferred from one body to another of lower temperature with which
it is in contact; (b) statements describing particular circumstances, such
as that the jar is made of glass, the lid of metal; that initially, at room

developed in great detail by N. R. Campbell; for an elementary account see his book
[4], which was first published in 1921. K. R. Popper, too, has set forth this deductive
conception of explanation in several of his publications (cf. fn. 4 in *SLE*); his earliest
statement appears in Sec. 12 of his book [38], which has at long last been published
in a considerably expanded English version [40].

temperature, the lid fitted very tightly on the top of the jar; and that then the top with the lid on it was immersed in hot water. To show that the loosening of the lid occurred "by virtue of" the circumstances in question, and "in accordance with" those laws, is then to show that the statement describing the result can be validly inferred from the specified set of premises.

Thus construed, the explanation at hand is a deductive argument of this form:

$$
(2.1) \quad \frac{\begin{array}{l} L_1, L_2, \ldots, L_r \\ C_1, C_2, \ldots, C_k \end{array}}{E}
$$

Here, $L_1, L_2, \ldots, L_r$ are general laws and $C_1, C_2, \ldots, C_k$ are statements of particular occurrences, facts, or events; jointly, these premises form the explanans. The conclusion E is the explanandum statement; it describes the phenomenon (or event, etc.) to be explained, which will also be called the explanandum phenomenon (or event, etc.); thus, the word 'explanandum' will be used to refer ambiguously either to the explanandum statement or to the explanandum phenomenon. Inasmuch as the sentence E is assumed to be a logical consequence of the premises, an explanatory argument of form (2.1) deductively subsumes the explanandum under "covering laws." [3] I will say, therefore, that (2.1) represents the *covering-law model of explanation*. More specifically, I will refer to explanatory arguments of the form (2.1) as *deductive-nomological*, or briefly as deductive, explanations: as will be shown later, there are other explanations invoking general laws that will have to be construed as inductive rather than as deductive arguments.

In my illustration, the explanandum is a particular event, the loosening of a certain lid, which occurs at a definite place and time. But deductive subsumption under general laws can serve also to explain general uniformities, such as those asserted by laws of nature. For example, the

[3] The suggestive terms 'covering law' and 'covering-law model' are borrowed from Dray, who, in his book [13], presents a lucid and stimulating critical discussion of the question whether, or to what extent, historical explanation conforms to the deductive pattern here considered. To counter a misunderstanding that might be suggested by some passages in Ch. II, Sec. 1 of Dray's book, I would like to emphasize that the covering-law model must be understood as permitting reference to *any* number of laws in the explanation of a given phenomenon: there should be no restriction to just *one* "covering law" in each case.

uniformity expressed by Galileo's law for free fall can be explained by deduction from the general laws of mechanics and Newton's law of gravitation, in conjunction with statements specifying the mass and radius of the earth. Similarly, the uniformities expressed by the law of geometrical optics can be explained by deductive subsumption under the principles of the wave theory of light.[4]

3. Truth and Confirmation of Deductive Explanations.

In *SLE* (Section 3) two basic requirements are imposed upon a scientific explanation of the deductive-nomological variety:[5] (i) It must be a deductively valid argument of the form (2.1), whose premises include at least one general law essentially, i.e., in such a way that if the law were deleted, the argument would no longer be valid. Intuitively, this means that reliance on general laws is essential to this type of explanation; a given phenomenon is here explained, or accounted for, by showing that it conforms to a general nomic pattern. (ii) The sentences constituting the explanans must be true, and hence so must the explanandum sentence. This second requirement was defended by the following consideration: suppose we required instead that the explanans be highly

[4] More accurately, the explanation of a general law by means of a theory will usually show (1) that the law holds only within a certain range of application, which may not have been made explicit in its standard formulation; (2) that even within that range, the law holds only in close approximation, but not strictly. This point is well illustrated by Duhem's emphatic reminder that Newton's law of gravitation, far from being an inductive generalization of Kepler's laws, is actually incompatible with them, and that the credentials of Newton's theory lie rather in its enabling us to compute the perturbations of the planets, and thus their deviations from the orbits assigned to them by Kepler. (See Duhem [14], pp. 312ff, and especially p. 317. The passages referred to here are included in the excerpts from P. P. Wiener's translation of Duhem's work that are reprinted in Feigl and Brodbeck [15], under the title "Physical Theory and Experiment.")

Analogously, Newtonian theory implies that the acceleration of a body falling freely in a vacuum toward the earth will increase steadily, though over short distances it will be very nearly constant. Thus, strictly speaking, the theory contradicts Galileo's law, but shows the latter to hold true in very close approximation within a certain range of application. A similar relation obtains between the principles of wave optics and those of geometrical optics.

[5] No claim was made that this is the only kind of scientific explanation; on the contrary, at the end of Sec. 3, it was emphasized that "Certain cases of scientific explanation involve 'subsumption' of the explanandum under a set of laws of which at least some are statistical in character. Analysis of the peculiar logical structure of that type of subsumption involves difficult special problems. The present essay will be restricted to an examination of the causal type of explanation . . ." A similar explicit statement is included in the final paragraph of Sec. 7 and in Sec. 5.3 of the earlier article, Hempel [24]. These passages seem to have been overlooked by some critics of the covering-law model.

Carl G. Hempel

confirmed by all the relevant evidence available, though it need not necessarily be true. Now it might happen that the explanans of a given argument of the form (2.1) was well confirmed at a certain earlier stage of scientific research, but strongly disconfirmed by the more comprehensive evidence available at a later time, say, the present. In this event, we would have to say that the explanandum was correctly explained by the given argument at the earlier stage, but not at the later one. And this seemed counterintuitive, for common usage appeared to construe the correctness of a given explanation as no more time dependent than, say, the truth of a given statement. But this justification, with its reliance on a notion of correctness that does not appear in the proposed definition of explanation, is surely of questionable merit. For in reference to explanations as well as in reference to statements, the vague idea of correctness can be construed in two different ways, both of which are of interest and importance for the logical analysis of science: namely, as truth in the semantical sense, which is independent of any reference to time or to evidence; or as confirmation by the available relevant evidence—a concept which is clearly time dependent. We will therefore distinguish between *true explanations*, which meet the requirement of truth for their explanans, and *explanations that are more or less well confirmed* by a given body of evidence (e.g., by the total evidence available). These two concepts can be introduced as follows:

First, we define a *potential explanation* (of deductive-nomological form)[6] as an argument of the form (2.1) which meets all the requirements indicated earlier, except that the statements forming its explanans and explanandum need not be true. But the explanans must still contain a set of sentences, $L_1, L_2, \ldots, L_r$, which are lawlike, i.e., which are like laws except for possibly being false.[7] Sentences of this kind will also be called *nomic*, or *nomological*, *statements*. It is this notion of potential

[6] This was done already in *SLE*, Sec. 7.

[7] The term 'lawlike sentence' and the general characterization given here of its intended meaning are from Goodman [20]. The difficult problem of giving an adequate general characterization of those sentences which if true would constitute laws will not be dealt with in the present essay. For a discussion of the issues involved, see, for example, *SLE*, Secs. 6–7; Braithwaite [3], Ch. IX, where the central question is described as concerning "the nature of the difference, if any, between 'nomic laws' and 'mere generalizations'"; and the new inquiry into the subject by Goodman [20, 21]. All the sentences occurring in a potential explanation are assumed, of course, to be empirical in the broad sense of belonging to some language adequate to the purposes of empirical science. On the problem of characterizing such systems more explicitly, see especially Scheffler's stimulating essay [46].

explanation which is involved, for example, when we ask whether a tentatively proposed but as yet untried theory would be able to explain certain puzzling empirical findings.)

Next, we say that a given potential explanation is more or less highly confirmed by a given body of evidence according as its explanans is more or less highly confirmed by the evidence in question. If the explanation is formulated in a formalized language for which an adequate quantitative concept of degree of confirmation or of inductive probability is available, we might identify the probability of the explanation relative to e with the probability of the explanans relative to e.

Finally, by a *true explanation* we understand a potential explanation with true explanans—and hence also with true explanandum.

4. *Causal Explanation and the Covering-Law Model.*

One of the various modes of explanation to which the covering-law model is relevant is the familiar procedure of accounting for an event by pointing out its "cause." In our first illustration, for example, the expansion of the lid might be said to have been caused by its immersion in hot water. Causal attributions of this sort presuppose appropriate laws, such as that whenever metal is heated under constant pressure, it expands. It is by reason of this implicit presupposition of laws that the covering-law model is relevant to the analysis of causal explanation. Let us consider this point more closely.

We will first examine general statements of causal connections, i.e., statements to the effect that an event of a given kind A—for example, motion of a magnet near a closed wire loop—will cause an event of some specified kind B—for example, flow of a current in the wire. Thereafter, we will consider statements concerning causal relations among individual events.

In the simplest case, a general statement asserting a causal connection between two kinds of events, A and B, is tantamount to the statement of the general law that whenever and wherever an instance of A occurs, it is accompanied by an instance of B. This analysis fits, for example, the statement that motion of a magnet causes a current in a neighboring wire loop. Many general statements of causal connection call for a more complex analysis, however. Thus, the statement that in a mammal, stoppage of the heart will cause death presupposes that certain "normal" conditions prevail, which are not explicitly stated, but which are surely meant to preclude, for example, the use of a heart-lung machine. "To say that X causes Y is to say that under proper conditions, an X will be followed by a Y," as Scriven[8] puts it. But unless the "proper conditions" can be specified, at least to some extent, this analysis tells us nothing about the meaning of 'X causes Y.' Now, when this kind of causal locution is used in a given context, there usually is at least some general understanding of the kind of background conditions that have to be assumed; but still, to the extent that those conditions remain indeterminate, a general statement of causal connection falls short of making a definite assertion and has at best the character of a promissory note to the effect that there are further background factors whose proper recognition would yield a truly general connection between the "cause" and "effect" under consideration.

Sentences concerning causal connections among individual events show similar characteristics. For example, the statement that the death of a certain person was caused by an overdose of phenobarbital surely presupposes a generalization, namely, a statement of a general causal connection between one kind of event, a person's taking an overdose of phenobarbital, and another, the death of that person.

Here again, the range of application for the general causal statement is not precisely stated, but a sharper specification can be given by indicating what constitutes an overdose of phenobarbital for a person—this will depend, among other things, on his weight and on his habituation to the drug—and by adding the proviso that death will result from taking such an overdose if the organism is left to itself, which implies, in particular, that no countermeasures are taken. To explain the death in question as having been caused by the antecedent taking of phenobarbital is therefore to claim that the explanandum event followed according to law upon certain antecedent circumstances. And this argument, when stated explicitly, conforms to the covering-law model.

Generally, the assertion of a causal connection between individual events seems to me unintelligible unless it is taken to make, at least implicitly, a nomological claim to the effect that there are laws which provide the basis for the causal connection asserted. When an individual event, say b, is said to have been caused by a certain antecedent event, or configuration of events, a, then surely the claim is intended that

[8] [49], p. 185.

whenever "the same cause" is realized, "the same effect" will recur. This claim cannot be taken to mean that whenever a recurs then so does b; for a and b are individual events at particular spatio-temporal locations and thus occur only once. Rather, a and b are, in this context, viewed as particular events of certain *kinds*—e.g., the expansion of a piece of metal or the death of a person—of which there may be many further instances. And the law tacitly implied by the assertion that b, as an event of kind B, was caused by a, as an event of kind A, is a general statement of causal connection to the effect that, under suitable circumstances, an instance of A is invariably accompanied by an instance of B. In most causal explanations offered in other than advanced scientific contexts, the requisite circumstances are not fully stated; for these cases, the import of the claim that b, as an instance of B, was caused by a may be suggested by the following approximate formulation: event b was in fact preceded by an event a of kind A, and by certain further circumstances which, though not fully specified or specifiable, were of such a kind that an occurrence of an event of kind A under such circumstances is universally followed by an event of kind B. For example, the statement that the burning (event of kind B) of a particular haystack was caused by a lighted cigarette carelessly dropped into the hay (particular event of kind A) asserts, first of all, that the latter event did take place; but a burning cigarette will set a haystack on fire only if certain further conditions are satisfied, which cannot at present be fully stated; and thus, the causal attribution at hand implies, second, that further conditions of a not fully specifiable kind were realized, under which an event of kind A will invariably be followed by an event of kind B.

To the extent that a statement of individual causation leaves the relevant antecedent conditions—and thus also the requisite explanatory laws—indefinite, it is like a note saying that there is a treasure hidden somewhere. Its significance and utility will increase as the location of the treasure is narrowed down, as the revelant conditions and the corresponding covering laws are made increasingly explicit. In some cases, such as that of the barbiturate poisoning, this can be done quite satisfactorily; the covering-law structure then emerges, and the statement of individual causal connection becomes amenable to test. When, on the other hand, the relevant conditions or laws remain largely indefinite, a statement of causal connection is rather in the nature of a program, or of a sketch, for an explanation in terms of causal laws; it might also be viewed as a "working hypothesis" which may prove its worth by giving new, and fruitful, direction to further research.

I would like to add here a brief comment on Scriven's observation that "when one asserts that X causes Y one is certainly committed to the generalization that an identical cause would produce an identical effect, but this in no way commits one to any necessity for producing laws not involving the term 'identical,' which justify this claim. Producing laws is one way, not necessarily more conclusive, and usually less easy than other ways of supporting the causal statement."[9] I think we have to distinguish here two questions, namely (i) what is being claimed by the statement that X causes Y, and in particular, whether asserting it commits one to a generalization, and (ii) what kind of evidence would support the causal statement, and in particular, whether such support can be provided only by producing generalizations in the form of laws.

As for the first question, I think the causal statement does imply the claim that an appropriate law or set of laws holds by virtue of which X causes Y; but, for reasons suggested above, the law or laws in question cannot be expressed by saying that an identical cause would produce an identical effect. Rather, the general claim implied by the causal statement is to the effect that there are certain "relevant" conditions of such a kind that whenever they occur in conjunction with an event of kind X, they are invariably followed by an event of kind Y.

In certain cases, some of the laws that are claimed to connect X and Y may be explicitly statable—as, for example, in our first illustration, the law that metals expand upon heating; and then, it will be possible to provide evidential support (or else disconfirmation) for them by the examination of particular instances; thus, while laws are implicitly claimed to underlie the causal connection in question, the claim can be supported by producing appropriate empirical evidence consisting of particular cases rather than of general laws. When, on the other hand, a nomological claim made by a causal statement has merely the character of an existential statement to the effect that there are relevant factors and suitable laws connecting X and Y, then it may be possible to lend some credibility to this claim by showing that under certain conditions an event of kind X is at least very frequently accompanied by an event of kind Y. This might justify the working hypothesis that the background

[9] *Ibid.*, p. 194.

conditions could be further narrowed down in a way that would eventually yield a strictly causal connection. It is this kind of statistical evidence, for example, that is adduced in support of such claims as that cigarette smoking is "a cause of" or "a causative factor in" cancer of the lung. In this case, the supposed causal laws cannot at present be explicitly stated. Thus, the nomological claim implied by this causal conjecture is of the existential type; it has the character of a working hypothesis that gives direction to further research. The statistical evidence adduced lends support to the hypothesis and justifies the program, which clearly is the aim of further research, of determining more precisely the conditions under which smoking will lead to cancer of the lung.

The most perfect examples of explanations conforming to the covering-law model are those provided by physical theories of deterministic character. A theory of this kind deals with certain specified kinds of physical systems, and limits itself to certain aspects of these, which it represents by means of suitable parameters; the values of these parameters at a given time specify the state of the system at that time; and a deterministic theory provides a system of laws which, given the state of an isolated system at one time, determine its state at any other time. In the classical mechanics of systems of mass points, for example, the state of a system at a given time is specified by the positions and momenta of the component particles at that time; and the principles of the theory— essentially the Newtonian laws of motion and of gravitation—determine the state of an isolated system of mass points at any time provided that its state at some one moment is given; in particular, the state at a specified moment may be fully explained, with the help of the theoretical principles in question, by reference to its state at some earlier time. In this theoretical scheme, the notion of a cause as a more or less narrowly circumscribed antecedent event has been replaced by that of some antecedent state of the total system, which provides the "initial conditions" for the computation, by means of the theory, of the later state that is to be explained; if the system is not isolated, i.e., if relevant outside influences act upon the system during the period of time from the initial state invoked to the state to be explained, then the particular circumstances that must be stated in the explanans include also those "outside influences"; and it is these "boundary conditions" in conjunction with the "initial" conditions which replace the everyday notion of cause, and which have to be thought of as being specified by the statements C_1,

$C_2, \ldots, C_k$ in the schematic representation (2.1) of the covering-law model.

Causal explanation in its various degrees of explicitness and precision is not the only type of explanation, however, to which the covering-law model is relevant. For example, as was noted earlier, certain empirical regularities, such as that represented by Galileo's law, can be explained by deductive subsumption under more comprehensive laws or theoretical principles; frequently, as in the case of the explanation of Kepler's laws by means of the law of gravitation and the laws of mechanics, the deduction yields a conclusion of which the generalization to be explained is only an approximation. Then the explanatory principles not only show why the presumptive general law holds, at least in approximation, but also provide an explanation for the deviations.

Another noncausal species of explanation by covering laws is illustrated by the explanation of the period of swing of a given pendulum by reference to its length and to the law that the period of a mathematical pendulum is proportional to the square root of its length. This law expresses a mathematical relation between the length and the period (a dispositional characteristic) of a pendulum *at the same time*; laws of this kind are sometimes referred to as *laws of coexistence*, in contradistinction to *laws of succession*, which concern the changes that certain systems undergo in the course of time. Boyle's, Charles's, and Van der Waals's laws for gases, which concern concurrent values of pressure, volume, and temperature of a gas; Ohm's law; and the law of Wiedemann and Franz (according to which, in metals, electric conductivity is proportional to thermal conductivity) are examples of laws of coexistence. Causal explanation in terms of antecedent events clearly calls for laws of succession in the explanans; in the case of the pendulum, where only a law of coexistence is invoked, we would not say that the pendulum's having such and such a length at a given time *caused* it to have such and such a period.[10]

It is of interest to note that in the example at hand, a statement of

[10] Note, however, that from a law of coexistence connecting certain parameters it is possible to derive laws of succession concerning the rates of change of those parameters. For example, the law expressing the period of a mathematical pendulum as a function of its length permits the derivation, by means of the calculus, of a further law to the effect that if the length of the pendulum changes in the course of time, then the rate of change of its period at any moment is proportional to the rate of change of its length, divided by the square root of its length, at that moment.

he length of a given pendulum in conjunction with the law just referred
o will much more readily be accepted as explaining the pendulum's
)eriod, than a statement of the period in conjunction with the same law
vould be considered as explaining the length of the pendulum; and this
s true even though the second argument has the same logical structure
.s the first: both are cases of deductive subsumption, in accordance with
he schema (2.1), under a law of coexistence. The distinction made here
eems to me to result from the consideration that we might change the
ength of the pendulum at will and thus control its period as a "depend-
:nt variable," whereas the reverse procedure does not seem possible. This
dea is open to serious objections, however; for clearly, we can also
:hange the period of a given pendulum at will, namely, by changing its
ength; and in doing so, we will change its length. It is not possible to
retort that in the first case we have a change of length independently
)f a change of the period; for if the location of the pendulum, and thus
the gravitational force acting on the pendulum bob, remains unchanged,
then the length cannot be changed without also changing the period. In
cases such as this, the common-sense conception of explanation appears
to provide no clear and reasonably defensible grounds on which to decide whether a given argument that deductively subsumes an occurrence
under laws is to qualify as an explanation.

The point that an argument of the form (2.1), even if its premises
are assumed to be true, would not always be considered as constituting
an explanation is illustrated even more clearly by the following example,
which I owe to my colleague Mr. S. Bromberger. Suppose that a flagpole stands vertically on level ground and subtends an angle of 45 degrees when viewed from the ground level at a distance of 80 feet. This
information, in conjunction with some elementary theorems of geometry, implies deductively that the pole is 80 feet high. The theorems in
question must here be understood as belonging to physical geometry
and thus as having the status of general laws, or, better, general theoretical principles, of physics. Hence, the deductive argument is of the
type (2.1). And yet, we would not say that its premises *explained* the
fact that the pole is 80 feet high, in the sense of showing why it is that
the pole has a height of 80 feet. Depending on the context in which it
is raised, the request for an explanation might call here for some kind of
causal account of how it came about that the pole was given this height,
or perhaps for a statement of the purpose for which this height was

109

chosen. An account of the latter kind would again be a special case of causal explanation, invoking among the antecedent conditions certain dispositions (roughly speaking, intentions, preferences, and beliefs) on the part of the agents involved in erecting the flagpole.

The geometrical argument under consideration is not of a causal kind; in fact, it might be held that if the particular facts and the geometrical laws here invoked can be put into an explanatory connection at all, then at best we might say that the height of the pole—in conjunction with the other particulars and the laws—explains the size of the substended angle, rather than vice versa. The consideration underlying this view would be similar to that mentioned in the case of the pendulum: It might be said that by changing the height of the pole, a change in the angle can be effected, but not vice versa. But here as in the previous case this contention is highly questionable. Suppose that the other factors involved, especially the distance from which the pole is viewed, are kept constant; then the angle can be changed, namely by changing the length of the pole; and thus, if the angle is made to change, then, trivially, the length of the pole changes. The notion that somehow we can "independ ently" control the length and thus make the angle a dependent variable but not conversely, does not seem to stand up under closer scrutiny.

In sum then, we have seen that among those arguments of the form (2.1) which are not causal in character there are some which would not ordinarily be considered as even potential explanations; but ordinary usage appears to provide no clear general criterion for those arguments which are to be qualified as explanatory. This is not surprising, for our everyday conception of explanation is strongly influenced by preanalytic causal and teleological ideas; and these can hardly be expected to provide unequivocal guidance for a more general and precise analysis of scientific explanation and prediction.

5. Covering Laws: Premises or Rules?

Even if it be granted that causal explanations presuppose general laws, it might still be argued that many explanations of particular occurrences as formulated in everyday contexts or even in scientific discourse limit themselves to adducing certain particular facts as the presumptive causes of the explanandum event, and that therefore a formal model should construe these explanations as accounting for the explanandum by means of suitable statements of particular fact, C_1, C_2, . . ., C_k, alone. Laws would have to be cited, not in the context by *giving* such an explanation, but in the context of *justifying* it; they would serve to show that the antecedent circumstances specified in the explanans are indeed connected by causal laws with the explanandum event. Explanation would thus be comparable to proof by logical deduction, where explicit reference to the rules or laws of logic is called for, not in stating the successive steps of the proof, but only in justifying them, i.e., in showing that they conform to the principles of deductive inference. This conception would construe general laws and theoretical principles, not as scientific statements, but rather as extralogical rules of scientific inference. These rules, in conjunction with those of formal logic, would govern inferences—explanatory, predictive, retrodictive, etc.—that lead from given statements of particular fact to other statements of particular fact.

The conception of scientific laws and theories as rules of inference has been advocated by various writers in the philosophy of science.[11] In particular, it may be preferred by those who hesitate, on philosophic grounds, to accord the status of bona fide statements, which are either

---

[11] Among these is Schlick [48], who gives credit to Wittgenstein for the idea that a law of nature does not have the character of a statement, but rather that of an instruction for the formation of statements. Schlick's position in this article is prompted largely by the view that a genuine statement must be definitively verifiable—a condition obviously not met by general laws. But this severe verifiability condition cannot be considered as an acceptable standard for scientific statements.

More recently, Ryle—see, for example, [44], pp. 121–123—has described law statements as statements which are true or false, but one of whose jobs is to serve as inference tickets: they license their possessors to move from the assertion of some factual statements to the assertion of others.

Toulmin [53], has taken the view, more closely akin to Schlick's, that laws of nature and physical theories do not function as premises in inferences leading to observational statements, but serve as modes of representation and as rules of inference according to which statements of empirical fact may be inferred from other such statements. An illuminating discussion of this view will be found in E. Nagel's review of Toulmin's book, in *Mind*, 63:403–412 (1954); it is reprinted in Nagel [35], pp. 303–315.

Carnap [5], par. 51, makes explicit provision for the construction of languages with extralogical rules of inferences. He calls the latter physical rules, or P-rules, and emphasizes that whether, or to what extent, P-rules are to be countenanced in constructing a language is a question of expedience. For example, adoption of P-rules may oblige us to alter the rules—and thus the entire formal structure—of the language of science in order to account for some new empirical findings which, in a language without P-rules, would prompt only modification or rejection of certain statements previously accepted in scientific theory.

The admission of material rules of inference has been advocated by W. Sellars in connection with his analysis of subjunctive conditionals; see [51, 52]. A lucid general account and critical appraisal of various reasons that have been adduced in support of construing general laws as inference rules will be found in Alexander [1].

true or false, to sentences which purport to express either laws covering an infinity of potential instances or theoretical principles about unobservable "hypothetical" entities and processes.[12]

On the other hand, it is well known that in rigorous scientific studies in which laws or theories are employed to explain or predict empirical phenomena, the formulas expressing laws and theoretical principles are used, not as rules of inference, but as statements—especially as premises—quite on a par with those sentences which presumably describe particular empirical facts or events. Similarly, the formulas expressing laws also occur as conclusions in deductive arguments; for example, when the laws governing the motion of the components of a double star about their common center of gravity are derived from broader laws of mechanics and of gravitation.

It might also be noted here that a certain arbitrariness is involved in any method of drawing a line between those formulations of empirical science which are to count as statements of particular fact and those which purport to express general laws, and which accordingly are to be construed as rules of inference. For any term representing an empirical characteristic can be construed as dispositional, in which case a sentence containing it acquires the status of a generalization. Take, for example, sentences which state the boiling point of helium at atmospheric pressure, or the electric conductivity of copper: are these to be construed as empirical statements or rather as rules? The latter status could be urged on the grounds that (i) terms such as 'helium' and 'copper' are dispositional, so that their application even to one particular object involves a universal assertion, and that (ii) each of the two statements attributes a specific disposition to any body of helium or of copper at any spatio-temporal location, which again gives them the character of general statements.

The two conceptions of laws and theories—as statements or as rules of inference—correspond to two different formal reconstructions, or models, of the language of empirical science; and a model incorporating laws and theoretical principles as rules can always be replaced by one which includes them instead as scientific statements.[13] And what matters for our present purposes is simply that in either mode of representation, ex-

---

[12] For detailed discussions of these issues, see Barker [2], especially Ch. 7; Scheffler [46], especially Secs. 13–18; Hempel [25], especially Sec. 10.

[13] On this point, see the review by Nagel mentioned in fn. 11.

planations of the kind here considered "presuppose" general theoretical principles essentially: either as indispensable premises or as indispensable rules of inference.

Of the two alternative construals of laws and theories, the one which gives them the status of statements seems to me simpler and more perspicuous for the analysis of the issues under investigation here; I will therefore continue to construe deductive-nomological explanations as having the form (2.1).

5. *Explanation, Prediction, Retrodiction, and Deductive Systematization—a Puzzle about 'About.'*

In a deductive-nomological explanation of a particular past event, the explanans logically implies the occurrence of the explanandum event; hence we may say of the explanatory argument that it could also have served as a predictive one in the sense that it could have been used to predict the explanandum event if the laws and particular circumstances adduced in its explanans had been taken into account at a suitable earlier time.[14] Predictive arguments of the form (2.1) will be called *deductive-nomological predictions,* and will be said to conform to the covering-law model of prediction. There are other important types of scientific prediction; among these, statistical prediction, along with statistical explanation, will be considered later.

Deductive-nomological explanation in its relation to prediction is instructively illustrated in the fourth part of the *Dialogues Concerning Two New Sciences.* Here, Galileo develops his laws for the motion of projectiles and deduces from them the corollary that if projectiles are fired from the same point with equal initial velocity, but different elevations, the maximum range will be attained when the elevation is 45°. Then, Galileo has Sagredo remark: "From accounts given by gunners, I was already aware of the fact that in the use of cannon and mortars, the maximum range . . . is obtained when the elevation is 45° . . . but to understand why this happens far outweighs the mere information obtained by the testimony of others or even by repeated experiment."[15] The reasoning that affords such understanding can readily be put into

---

[14] This remark does not hold, however, when all the laws invoked in the explanans are laws of coexistence (see Sec. 4) and all the particular statements adduced in the explanans pertain to events that are simultaneous with the explanandum event. I am indebted to Mr. S. Bromberger for having pointed out to me this oversight in my formulation.

[15] [18], p. 265.

the form (2.1); it amounts to a deduction, by logical and mathematical means, of the corollary from a set of premises which contains (i) the fundamental laws of Galileo's theory for the motion of projectiles and (ii) particular statements specifying that all the missiles considered are fired from the same place with the same initial velocity. Clearly then, the phenomenon previously noted by the gunners is here *explained*, and thus *understood*, by showing that its occurrence was to be expected under the specified circumstances, in view of certain general laws set forth in Galileo's theory. And Galileo himself points with obvious pride to the predictions that may in like fashion be obtained by deduction from his laws; for the latter imply "what has perhaps never been observed in experience, namely, that of other shots those which exceed or fall short of 45° by equal amounts have equal ranges." Thus, the explanation afforded by Galileo's theory "prepares the mind to understand and ascertain other facts without need of recourse to experiment," [16] namely, by deductive subsumption under the laws on which the explanation is based.

We noted above that if a deductive argument of the form (2.1) explains a past event, then it could have served to predict it if the information provided by the explanans had been available earlier. This remark makes a purely logical point; it does not depend on any empirical assumptions. Yet it has been argued, by Rescher, that the thesis in question "rests upon a tacit but unwarranted assumption as to the nature of the physical universe." [17]

The basic reason adduced for this contention is that "the explanation of events is oriented (in the main) towards the past, while prediction is oriented towards the future," [18] and that, therefore, before we can decide whether (deductive-nomological) explanation and prediction have the same logical structure, we have to ascertain whether the natural laws of our world do in fact permit inferences from the present to the future as well as from the present to the past. Rescher stresses that a given system might well be governed by laws which permit deductive inferences concerning the future, but not concerning the past, or conversely; and on this point he is quite right. As a schematic illustration, consider a model "world" which consists simply of a sequence of colors, namely, Blue (B),

[16] *Ibid.*
[17] [42], p. 282.
[18] *Ibid.*, p. 286.

114

Green (G), Red (R), and Yellow (Y), which appear on a screen during successive one-second intervals i_1, i_2, i_3, . . . Let the succession of colors be governed by three laws:

(L_1) B is always followed by G.
(L_2) G and R are always followed by Y.
(L_3) Y is always followed by R.

Then, given the color of the screen for a certain interval, say i_3, these laws unequivocally determine the "state of the world," i.e., the screen color, for all later intervals, but not for all earlier ones. For example, given the information that during i_3 the screen is Y, the laws predict the colors for the subsequent intervals uniquely as RYRYRY . . .; but for the preceding states i_1 and i_2, they yield no unique information, since they allow here two possibilities: BG and YR.

Thus, it is possible that a set of laws governing a given system should permit unique deductive *predictions* of later states from a given one, and yet not yield unique deductive *retrodictions* concerning earlier states; conversely, a set of laws may permit unique retrodiction, but no unique prediction. But—and here lies the flaw in Rescher's argument—this is by no means the same thing as to say that such laws, while permitting deductive prediction of later states from a given one, do not permit explanation; or, in the converse case, that while permitting explanation, they do not permit prediction. To illustrate by reference to our simple model world: Suppose that during i_3 we find the screen to be Y, and that we seek to explain this fact. This can be done if we can ascertain, for example, that the color for i_1 had been B; for from the statement of this particular antecedent fact we can infer, by means of L_1, that the color for i_2 must have been G and hence, by L_2, that the color for i_3 had to be Y. Evidently, the same argument, used before i_3, could serve to predict uniquely the color for i_3 on the basis of that for i_1. Indeed, quite generally, any predictive argument made possible by the laws for our model world can also be used for explanatory purposes and vice versa. And this is so although those laws, while permitting unique predictions, do not always permit unique retrodictions. Thus, the objection under consideration misses its point because it tacitly confounds explanation with retrodiction.[19]

[19] In Sec. 3 of *SLE*, to which Rescher refers in his critique, an explanation of a past event is explicitly construed as a deductive argument inferring the occurrence of the event from "antecedent conditions" and laws; so that the temporal direction of

Carl G. Hempel

The notion of scientific retrodiction, however, is of interest in its own right; and, as in the case of explanation and prediction, one important variety of it is the deductive-nomological one. It has the form (2.1), but with the statements $C_1, C_2, \ldots, C_k$ referring to circumstances which occur later than the event specified in the conclusion E. In astronomy an inference leading, by means of the laws of celestial mechanics, from data concerning the present positions and movements of the sun, the earth, and Mars to a statement of the distance between earth and Mars a year later or a year earlier illustrates deductive-nomological prediction and retrodiction, respectively; in this case, the same laws can be used for both purposes because the processes involved are reversible.

It is of interest to observe here that in their predictive and retrodictive as well as in their explanatory use, the laws of classical mechanics, or other sets of deterministic laws for physical systems, require among the premises not only a specification of the state of the system for some time t_0, earlier or later than the time, say t_1, for which the state of the system is to be inferred, but also a statement of the boundary conditions prevailing between t_0 and t_1; these specify the external influences acting upon the system during the time in question. For certain purposes in astronomy, for example, the disturbing influence of celestial objects other than those explicitly considered may be neglected as insignificant, and the system under consideration may then be treated as "isolated"; but this should not lead us to overlook the fact that even those laws and theories of the physical sciences which provide the exemplars of deductively nomological prediction do not enable us to forecast certain future events strictly on the basis of information about the present: the predictive argument also requires certain premises concerning the future— e.g., absence of disturbing influences, such as a collision of Mars with an unexpected comet—and the temporal scope of these boundary conditions

the inference underlying explanation is the same as that of a predictive nomological argument, namely, from statements concerning certain initial (and boundary) conditions to a statement concerning the *subsequent* occurrence of the explanandum event

I should add, however, that although all this is said unequivocally in *SLE*, there is a footnote in *SLE*, Sec. 3, which is certainly confusing, and which, though not referred to by Rescher, might have encouraged him in his misunderstanding. The footnote, numbered 2a, reads: "The logical similarity of explanation and prediction and the fact that one is directed towards past occurrences, the other towards future ones, is well expressed in the terms 'postdictability' and 'predictability' used by Reichenbach [in *Philosophic Foundations of Quantum Mechanics*, p. 13]." To reemphasize the point at issue: postdiction, or retrodiction, is not the same thing as explanation must extend up to the very time at which the predicted event is to occur. The assertion therefore that laws and theories of deterministic form enable us to predict certain aspects of the future from information about the present has to be taken with a considerable grain of salt. Analogous remarks apply to deductive-nomological retrodiction and explanation.

I will use the term 'deductive-nomological systematization' to refer to any argument of the type (2.1), irrespective of the temporal relations between the particular facts specified by C_1, C_2, . . ., C_k and the particular events, if any, described by E. And, in obvious extension of the concepts introduced in Section 3 above, I will speak of *potential* (deductive-nomological) *systematizations*, of *true systematizations*, and of *systematizations* whose joint premises are more or less well confirmed by a given body of evidence.

To return now to the characterization of an explanation as a potential prediction: Scriven[20] bases one of his objections to this view on the observation that in the causal explanation of a given event (e.g., the collapse of a bridge) by reference to certain antecedent circumstances (e.g., excessive metal fatigue in one of the beams) it may well happen that the only good reasons we have for assuming that the specified circumstances were actually present lie in our knowledge that the explanandum event did take place. In this situation, we surely could not have used the explanans predictively since it was not available to us before the occurrence of the event to be predicted. This is an interesting and important point in its own right; but in regard to our conditional thesis that an explanation could have served as a prediction *if* its explanans had been taken account of in time, the argument shows only that the thesis is sometimes counterfactual (i.e., has a false antecedent), but not that it is false.

In a recent article, Scheffler[21] has subjected the idea of the structural equality of explanation and prediction to a critical scrutiny; and I would like to comment here briefly on at least some of his illuminating observations.

Scheffler points out that a prediction is usually understood to be an assertion rather than an argument. This is certainly the case; and we might add that, similarly, an explanation is often formulated, not as an argument, but as a statement, which will typically take the form 'q be-

[20] [50].
[21] [47].

117

cause p.' But predictive statements in empirical science are normally established by inferential procedures (which may be deductive or inductive in character) on the basis of available evidence; thus, there arises the question as to the logic of predictive arguments in analogy to the problem of the logic of explanatory arguments; and the idea of structural equality should be understood as pertaining to explanatory, predictive, retrodictive, and related arguments in science.

Scheffler also notes that a scientific prediction statement may be false, whereas, under the requirement of truth for explanations as laid down in Section 3 of *SLE*, no explanation can be false. This remark is quite correct; however, I consider it to indicate, not that there is a basic discrepancy between explanation and prediction, but that the requirement of truth for scientific explanations is unduly restrictive. The restriction is avoided by the approach that was proposed above in Section 3, and again in the present section in connection with the general characterization of scientific systematization; this approach enables us to speak of explanations no less than of predictions as being possibly false, and as being more or less well confirmed by the empirical evidence at hand.

Another critical observation Scheffler puts forth concerns the view, presented in *SLE*, that the difference between an explanatory and a predictive argument does not lie in its logical structure, but is "of a pragmatic character. If . . . we know that the phenomenon described by E has occurred, and a suitable set of statements C_1, C_2, . . ., C_k, L_1, L_2, . . ., L_r is provided afterwards, we speak of an explanation of the phenomenon in question. If the latter statements are given and E is derived prior to the occurrence of the phenomenon it describes, we speak of a prediction."[22] This characterization would make explanation and prediction mutually exclusive procedures, and Scheffler rightly suggests that they may sometimes coincide, since, for example, one may reasonably be said to be both predicting and explaining the sun's rising when, in reply to the question 'Why will the sun rise tomorrow?' one offers the appropriate astronomical information.[23]

I would be inclined to say, therefore, that in an explanation of the deductive-nomological variety, the explanandum event—which may be past, present, or future—is taken to be "given," and a set of laws and particular statements is then adduced which provides premises in an

[22] *SLE*, Sec. 3
[23] Scheffler [47], p. 300.

appropriate argument of type (2.1); whereas in the case of prediction, it is the premises which are taken to be "given," and the argument then yields a conclusion about an event to occur after the presentation of the predictive inference. Retrodiction may be construed analogously. The argument referred to by Scheffler about tomorrow's sunrise may thus be regarded, first of all, as predicting the event on the basis of suitable laws and presently available information about antecedent circumstances; then, taking the predicted event as "given," the premises of the same argument constitute an explanans for it.

Thus far, I have dealt with the view that an explanatory argument is also a (potentially) predictive one. Can it be held equally that a predictive argument always offers a potential explanation? In the case of deductive-nomological predictions, an affirmative answer might be defended, though as was illustrated at the end of Section 4, there are some deductive systematizations which one would readily accept as predictions while one would find it at least awkward to qualify them as explanations. Construing the question at hand more broadly, Scheffler, and similarly Scriven,[24] have rightly pointed out, in effect, that certain sound predictive arguments of the nondeductive type cannot be regarded as affording potential explanations. For example, from suitable statistical data on past occurrences, it may be possible to "infer" quite soundly certain predictions concerning the number of male births, marriages, or traffic deaths in the United States during the next month; but none of these arguments would be regarded as affording even a low-level explanation of the occurrences they serve to predict. Now, the inferences here involved are inductive rather than deductive in character; they lead from information about observed finite samples to predictions concerning as yet unobserved samples of a given population. However, what bars them from the role of potential explanations is not their inductive character (later I will deal with certain explanatory arguments of inductive form) but the fact that they do not invoke any general laws either of strictly universal or of statistical form: it appears to be characteristic of an explanation, though not necessarily of a prediction, that it present the inferred phenomena as occurring in conformity with general laws.

In concluding this section, I would like briefly to call attention to a puzzle concerning a concept that was taken for granted in the preceding

[24] See *ibid.*, p. 296; Scriven [49].

discussion, for example, in distinguishing between prediction and retro diction. In drawing that distinction, I referred to whether a particula given statement, the conclusion of an argument of form (2.1), wa "about" occurrences at a time earlier or later than some specified time such as the time of presentation of that argument. The meaning of thi latter criterion appears at first to be reasonably clear and unproblematic If pressed for further elucidation, one might be inclined to say, by way of a partial analysis, that if a sentence explicitly mentions a certain mo ment or period of time then the sentence is about something occurring at that time. It seems reasonable, therefore, to say that the sentence 'The sun rises on July 17, 1958,' says something about July 17, 1958, and that therefore, an utterance of this sentence on July 16, 1958, constitutes a pre diction.

Now the puzzle in question, which might be called the puzzle of 'about,' shows that this criterion does not even offer a partially satis factory explication of the idea of what time a given statement is about For example, the statement just considered can be equivalently restated in such a way that, by the proposed criterion, it is about July 15 and thus, if uttered on July 16, is about the past rather than about the future. The following rephrasing will do: 'The sun plus-two-rises on July 15,' where plus-two-rising on a given date is understood to be the same thing as ris ing two days after that date. By means of linguistic devices of this sort, statements about the future could be reformulated as statements about the past, or conversely; we could even replace all statements with tem poral reference by statements which are, all of them, ostensibly "about" one and the same time.

The puzzle is not limited to temporal reference, but arises for spatial reference as well. For example, a statement giving the mean temperature at the North Pole can readily be restated in a form in which it speaks ostensibly about the South Pole; one way of doing this is to attribute to the South Pole the property of having, in such and such a spatial relation to it, a place where the mean temperature is such and such; another de vice would be to use a functor, say 'm,' which, for the South Pole, takes as its value the mean temperature at the North Pole. Even more gen erally there is a method which, given any particular object *o*, will re formulate any statement in such a way that it is ostensibly about *o*. If, for example, the given statement is 'The moon is spherical,' we intro duce a property term, 'moon-spherical,' with the understanding that it is to apply to o just in case the moon is spherical; the given statement then is equivalent to 'o is moon-spherical.'

The puzzle is mentioned here in order to call attention to the difficulties that face an attempt to explicate the idea of what a statement is "about," and in particular, what time it refers to; and that idea seems essential for the characterization of prediction, retrodiction, and similar concepts.[25]

Part II. Statistical Systematization
7. Laws of Strictly General and Statistical Form.

The nomological statements adduced in the explanans of a deductive-nomological explanation are all of a strictly general form: they purport to express strictly unexceptionable laws or theoretical principles interconnecting certain characteristics (i.e., qualitative or quantitative properties or relations) of things or events. One of the simplest forms a statement of this kind can take is that of a universal conditional: 'All (instances of) F are (instances of) G.' When the attributes in question are quantities, their interconnections are usually expressed in terms of mathematical functions, as is illustrated by many of the laws and theoretical principles of the physical sciences and of mathematical economics.

On the other hand, there are important scientific hypotheses and theoretical principles which assert that certain characters are associated, not unexceptionally or universally, but with a specified long-range frequency; we will call them statistical generalizations, or laws (or theoretical principles) of statistical form, or (statistical) probability statements. The laws of radioactive decay, the fundamental principles of quantum mechanics, and the basic laws of genetics are examples of such probability statements. These statistical generalizations, too, are used in science for the systematization of various empirical phenomena. This is illustrated, for example, by the explanatory and predictive applications of quantum theory and of the basic laws of genetics as well as by the postdictive use of the laws of radioactive decay in dating archeological relics by means of the radio-carbon method.

The rest of this essay deals with some basic problems in the logic of

[25] Professor Nelson Goodman, to whom I had mentioned my difficulties with the notion of a statement being "about" a certain subject, showed me a draft of an article entitled "About," which has now appeared in *Mind*, 70:1–24 (1961); in it, he proposes an analysis of the notion of aboutness which will no doubt prove helpful in dealing with the puzzle outlined here, and which may even entirely resolve it.

Carl G. Hempel

statistical systematizations, i.e., of explanatory, predictive, or similar arguments which make essential use of statistical generalizations.

Just as in the case of deductive-nomological systematization, arguments of this kind may be used to account not only for particular facts or events, but also for general regularities, which, in this case, will be of a statistical character. For example, from statistical generalizations stating that the six different results obtainable by rolling a given die are equiprobable and statistically independent of each other, it is possible to deduce the statistical generalization that the probability of rolling two aces in succession is 1/36; thus the latter statistical regularity is accounted for by subsumption (in this case purely deductive) under broader statistical hypotheses.

But the peculiar logical problems concerning statistical systematization concern the role of probability statements in the explanation, prediction, and postdiction of individual events or finite sets of such events. In preparation for a study of these problems, I shall now consider briefly the form and function of statistical generalizations.

Statistical probability hypotheses, or statistical generalizations, as understood here, bear an important resemblance to nomic statements of strictly general form: they make a universal claim, as is suggested by the term 'statistical law,' or 'law of statistical form.' Snell's law of refraction, which is of strictly general form, is not simply a descriptive report to the effect that a certain quantitative relationship has so far been found to hold, in all cases of optical refraction, between the angle of incidence and that of refraction: it asserts that that functional relationship obtains universally, in all cases of refraction, no matter when and where they occur.[26] Analogously, the statistical generalizations of genetic theory or the probability statements specifying the half lives of various radioactive substances are not just reports on the frequencies with which certain phenomena have been found to occur in some set of past instances;

[26] It is sometimes argued that a statement asserting such a universal connection rests, after all, only on a finite, and necessarily incomplete, body of evidence; that, therefore, it may well have exceptions which have so far gone undiscovered, and that, consequently, it should be qualified as probabilistic, too. But this argument fails to distinguish between the claim made by a given statement and the strength of its evidential support. On the latter score, all empirical statements have to count as only more or less well supported by the available evidence; but the distinction between laws of strictly universal form and those of statistical form refers to the claim made by the statements in question: roughly speaking, the former attribute a certain character to all members of a specified class; the latter, to a fixed proportion of its members.

122

rather, they serve to assert certain peculiar but universal modes of connection between certain attributes of things or events.

A statistical generalization of the simplest kind asserts that the probability for an instance of F to be an instance of G is r, or briefly that p(G,F) = r; this is intended to express, roughly speaking, that the proportion of those instances of F which are also instances of G is r. This idea requires clarification, however, for the notion of the proportion of the (instances of) G among the (instances of) F has no clear meaning when the instances of F do not form a finite class. And it is characteristic of probability hypotheses with their universal character, as distinguished from statements of relative frequencies in some finite set, that the reference class—F in this case—is not assumed to be finite; in fact, we might underscore their peculiar character by saying that the probability r does not refer to the class of all actual instances of F but, so to speak, to the class of all its potential instances.

Suppose, for example, that we are given a homogeneous regular tetrahedron whose faces are marked 'I,' 'II,' 'III,' 'IV.' We might then be willing to assert that the probability of obtaining a III, i.e., of the tetrahedran's coming to rest on that face, upon tossing it out of a dice box is ¼; but while this assertion would be meant to say something about the frequency with which a III is obtained as a result of rolling the tetrahedron, it could not be construed as simply specifying that frequency for the class of all tosses which are in fact ever performed with the tetrahedron. For we might well maintain our probability hypothesis even if the given tetrahedron were tossed only a few times throughout its existence, and in this case, our probability statement would certainly not be meant to imply that exactly or even nearly, one fourth of those tosses yielded the result III. In fact, we might clearly maintain the probability statement even if the tetrahedron happened to be destroyed without ever having been tossed at all. We might say, then, that the probability hypothesis ascribes to the tetrahedron a certain disposition, namely, that of yielding a III in about one out of four cases in the long run. That disposition may also be described by a subjunctive or counterfactual statement: If the tetrahedron were to be tossed (or had been tossed) a large number of times, it would yield (would have yielded) the result III in about one fourth of the cases.[27]

[27] The characterization given here of the concept of statistical probability seems to me to be in agreement with the general tenor of the "propensity interpretation" advo-

Carl G. Hempel

Let us recall here in passing that nomological statements of strictly general form, too, are closely related to corresponding subjunctive and counterfactual statements. For example, the lawlike statement 'All pieces of copper expand when heated' implies the subjunctive conditional 'If this copper key were heated it would expand' and the counterfactual statement, referring to a copper key that was kept at constant temperature during the past hour, 'If this copper key had been heated half an hour ago, it would have expanded.'[28]

To obtain a more precise account of the form and function of probability statements, I will examine briefly the elaboration of the concept of statistical probability in contemporary mathematical theory. This examination will lead to the conclusion that the logic of statistical systematization differs fundamentally from that of deductive-nomological systematization. One striking symptom of the difference is what will be called here *the ambiguity of statistical systematization.*

In Section 8, I will describe and illustrate this ambiguity in a general manner that presupposes no special theory of probability; then in Section 9, I will show how it reappears in the explanatory and predictive

cated by Popper in recent years. This interpretation "differs from the purely statistical or frequency interpretation only in this—that it considers the probability as a characteristic property of the experimental arrangement rather than as a property of a sequence"; the property in question is explicitly construed as *dispositional.* (Popper [39], pp. 67–68. See also the discussion of this paper at the Ninth Symposium of the Colston Research Society, in Körner [30], pp. 78–89 *passim.*) However, the currently available statements of the propensity interpretation are all rather brief (for further references, see Popper [40]); a fuller presentation is to be given in a forthcoming book by Popper.

[28] In fact, Goodman [20], has argued very plausibly that one symptomatic difference between lawlike and nonlawlike generalizations is precisely that the former are able to lend support to corresponding subjunctive or counterfactual conditionals; thus the statement 'If this copper key were to be heated it would expand' can be supported by the law mentioned above. By contrast, the general statement 'All objects ever placed on this table weigh less than one pound' is nonlawlike, i.e., even if true, it does not count as a law. And indeed, even if we knew it to be true, we would not adduce it in support of corresponding counterfactuals; we would not say, for example, that if a volume of Merriam-Webster's Unabridged Dictionary had been placed on the table, it would have weighed less than a pound. Similarly, it might be added, general statements of this latter kind possess no explanatory power: this is why the sentences $L_1 L_2, \ldots, L_n$ in the explanans of any deductive-nomological explanation are required to be lawlike.

The preceding considerations suggest the question whether there is a category of statistical probability statements whose status is comparable to that of accidental generalizations. It would seem clear, however, that insofar as statistical probability statements are construed as dispositional in the sense suggested above, they have to be considered as being analogous to lawlike statements.

124

use of probability hypotheses as characterized by the mathematical theory of statistical probability.

8. *The Ambiguity of Statistical Systematization.*

Consider the following argument which represents, in a nutshell, an attempt at a statistical explanation of a particular event: "John Jones was almost certain to recover quickly from his streptococcus infection, for he was given penicillin, and almost all cases of streptococcus infection clear up quickly upon administration of penicillin." The second statement in the explanans is evidently a statistical generalization, and while the probability value is not specified numerically, the words 'almost all cases' indicate that it is very high.

At first glance, this argument appears to bear a close resemblance to deductive-nomological explanations of the simplest form, such as the following: This crystal of rock salt, when put into a Bunsen flame, turns the flame yellow, for it is a sodium salt, and all sodium salts impart a yellow color to a Bunsen flame. This argument is basically of the form:

(8.1)
$$\frac{\text{All F are G.}}{\begin{array}{c}\text{x is F.}\\\hline\text{x is G.}\end{array}}$$

The form of the statistical explanation, on the other hand, appears to be expressible as follows:

(8.2)
$$\frac{\text{Almost all F are G.}}{\begin{array}{c}\text{x is F.}\\\hline\text{x is almost certain to be G.}\end{array}}$$

Despite this appearance of similarity, however, there is a fundamental difference between these two kinds of argument: A nomological explanation of the type (8.1) accounts for the fact that x is G by stating that x has another character, F, which is uniformly accompanied by G, in virtue of a general law. If in a given case these explanatory assumptions are in fact true, then it follows logically that x must be G; hence x cannot possibly possess a character, say H, in whose presence G is uniformly absent; for otherwise, x would have to be both G and non-G. In the argument (8.2), on the other hand, x is said to be almost certain to have G because it has a character, F, which is accompanied by G in almost all instances. But even if in a given case the explanatory statements are both true, x

125

may possess, in addition to F, some other attribute, say H, which is almost always accompanied by non-G. But by the very logic underlying (8.2), this attribute would make it almost certain that x is not G.

Suppose, for example, that almost all, but not quite all, penicillin-treated, streptococcal infections result in quick recovery, or briefly, that almost all P are R; and suppose also that the particular case of illness of patient John Jones which is under discussion—let us call it j—is an instance of P. Our original statistical explanation may then be expressed in the following manner, which exhibits the form (8.2):

(8.3a)

Almost all P are R.

$$\frac{\text{j is P.}}{\text{j is almost certain to be R.}}$$

Next, let us say that an event has the property P* if it is either the event j itself or one of those infrequent cases of penicillin-treated streptococcal infection which do not result in quick recovery. Then clearly j is P*, whether or not j is one of the cases resulting in recovery, i.e., whether or not j is R. Furthermore, almost every instance of P* is an instance of non-R (the only possible exception being j itself). Hence, the argument (8.3a) in which, on our assumption, the premises are true can be matched with another one whose premises are equally true, but which by the very logic underlying (8.3a), leads to a conclusion that appears to contradict that of (8.3a):

(8.3b)

Almost all P* are non-R.

$$\frac{\text{j is P*.}}{\text{j is almost certain to be non-R.}}$$

If it should be objected that the property P* is a highly artificial property and that, in particular, an explanatory statistical law should not involve essential reference to particular individuals (such as j in our case), then another illustration can be given which leads to the same result and meets the contemplated requirement. For this purpose, consider a number of characteristics of John Jones at the onset of his illness, such as his age, height, weight, blood pressure, temperature, basal metabolic rate, and IQ. These can be specified in terms of numbers; let $n_1, n_2, n_3, \ldots$ be the specific numerical values in question. We will say that an event has the property S if it is a case of streptococcal infection in a patient who at the onset of his illness has the height n_1, age n_2, weight n_3, blood pressure n_4, and so forth. Clearly, this definition of S in terms of numerical characteristics no longer makes reference to j. Finally, let us say that an event has the property P^{**} if it is either an instance of S or one of those infrequent cases of streptococcal infection treated with penicillin which do not result in quick recovery. Then evidently j is P^{**} because j is S; and furthermore, since S is a very rare characteristic, almost every instance of P^{**} is an instance of non-R. Hence, (8.3a) can be matched with the following argument, in which the explanatory probability hypothesis involves no essential reference to particular cases:

(8.3c)

Almost all P^{**} are non-R.

$$\frac{\text{j is } P^{**}.}{\text{j is almost certain to be non-R.}}$$

The premises of this argument are true if those of (8.3a) are, and the conclusion again appears to be incompatible with that of (8.3a).

The peculiar phenomenon here illustrated will be called the *ambiguity of statistical explanation*. Briefly, it consists in the fact that if the explanatory use of a statistical generalization is construed in the manner of (8.2), then a statistical explanation of a particular event can, in general, be matched by another one, equally of the form (8.2), with equally true premises, which statistically explains the nonoccurrence of the same event. The same difficulty arises, of course, when statistical arguments of the type (8.2) are used for predictive purposes. Thus, in the case of our illustration, we might use either of the two arguments (8.3a) and (8.3c) in an attempt to predict the effect of penicillin treatment in a fresh case, j, of streptococcal infection; and even though both followed the same logical pattern—that exhibited in (8.2)—and both had true premises, one argument would yield a favorable, the other an unfavorable forecast. We will, therefore, also speak of the *ambiguity of statistical prediction* and, more inclusively, of the *ambiguity of statistical systematization*.

This difficulty is entirely absent in nomological systematization, as we noted above; and it evidently throws into doubt the explanatory and predictive relevance of statistical generalizations for particular occurrences. Yet there can be no question that statistical generalizations are widely invoked for explanatory and predictive purposes in such diverse fields as physics, genetics, and sociology. It will be necessary, therefore, to examine more carefully the logic of the arguments involved and, in particular, to reconsider the adequacy of the analysis suggested in (8.2). And while for a general characterization of the ambiguity of statistical explanation it was sufficient to use an illustration of statistical generalization of the vague form 'Almost all F are G,' we must now consider the explanatory and predictive use of statistical generalizations in the precise form of quantitative probability statements: 'The probability for an F to be a G is r.' This brings us to the question of the theoretical status of the statistical concept of probability.

9. The Theoretical Concept of Statistical Probability and the Problem of Ambiguity.

The mathematical theory of statistical probability[29] seeks to give a theoretical systematization of the statistical aspects of random experiments. Roughly speaking, a random experiment is a repeatable process which yields in each case a particular finite or infinite set of "results," in such a way that while the results vary from repetition to repetition in an irregular and practically unpredictable manner, the relative frequencies with which the different results occur tend to become more or less constant for large numbers of repetitions. The theory of probability is intended to provide a "mathematical model," in the form of a deductive system, for the properties and interrelations of such long-run frequencies, the latter being represented in the model by probabilities.

In the mathematical theory of probability, each of the different outcomes of a random experiment which have probabilities assigned to them is represented by a set of what might be called elementary possibilities. For example, if the experiment is that of rolling a die, then get-

[29] The mathematical theory of statistical probability has been developed in two major forms. One of these is based on an explicit definition of probabilities as limits of relative frequencies in certain infinite reference sequences. The use of this limit definition is an ingenious attempt to permit the development of a simple and elegant theory of probability by means of the apparatus of mathematical analysis, and to reflect at the same time the intended statistical application of the abstract theory. The second approach, which offers certain theoretical advantages and is now almost generally adopted, develops the formal theory of probability as an abstract theory of certain set-functions and then specifies rules for its application to empirical subject matter. The brief characterization of the theory of statistical probability given in this section follows the second approach. However, the problem posed by the ambiguity of statistical systematization arises as well when the limit definition of probability is adopted.

ing an ace, a deuce, and so forth, would normally be chosen as elementary possibilities; let us refer to them briefly as I, II, . . ., VI, and let F be the set of these six elements. Then any of those results of rolling a die to which probabilities are usually assigned can be represented by a subset of F: getting an even number, by the set (II, IV, VI); getting a prime number, by the set (II, III, V); rolling an ace, by the unit set (I); and so forth. Generally, a random experiment is represented in the theory by a set F and a certain set, F*, of its subsets, which represent the possible outcomes that have definite probabilities assigned to them. F* will sometimes, but not always, contain all the subsets of F. The mathematical theory also requires F* to contain, for each of its member sets, its complement in F; and also for any two of its member sets, say G_1 and G_2, their sum, $G_1 \vee G_2$, and their products, $G_1 \cdot G_2$. As a consequence, F* contains F as a member set.[30] The probabilities associated with the different outcomes of a random experiment then are represented by a real-valued function $p_F(G)$ which ranges over the sets in F*.

The postulates of the theory specify that p_F is a nonnegative additive set function such that $p_F(F) = 1$; i.e., for all G in F*, $p_F(G) \geqq 0$; if G_1 and G_2 are mutually exclusive sets in F* then $p_F(G_1 \vee G_2) = p_F(G_1) + p_F(G_2)$. These stipulations permit the proof of the theorems of elementary probability theory; to deal with experiments that permit infinitely many different outcomes, the requirement of additivity is suitably extended to infinite sequences of mutually exclusive member sets of F*.

The abstract theory is made applicable to empirical subject matter by means of an interpretation which connects probability statements with sentences about long-run relative frequencies associated with random experiments. I will state the interpretation in a form which is essentially that given by Cramér,[31] whose book *Mathematical Methods of Statistics* includes a detailed discussion of the foundations of mathematical probability theory and its applications. For convenience, the notation '$p_F(G)$' for the probability of G relative to F will now be replaced by '$p(G, F)$.'

(9.1)　　*Frequency interpretation of statistical probability*: Let F be a given kind of random experiment and G a possible result of it; then the statement that $p(G, F) = r$ means that in a long series

[30] See, for example, Kolmogoroff [31], Sec 2.

[31] [11], pp. 148–149. Similar formulations have been given by other representatives of this measure-theoretical conception of statistical probability, for example, by Kolmogoroff [31], p. 4.

of repetitions of F, it is practically certain that the relative frequency of the result G will be approximately equal to r.

Evidently, this interpretation does not offer a precise definition of probability in statistical terms: the vague phrases 'a long series,' 'practically certain,' and 'approximately equal' preclude that. But those phrases are chosen deliberately to enable formulas stating precisely fixed numerical probability values to function as theoretical representations of near-constant relative frequencies of certain results in extended repetitions of a random experiment.

Cramér also formulates two corollaries of the above rule of interpretation; they refer to those cases where r differs very little from 0 or from 1. These corollaries will be of special interest for an examination of the question of ambiguity in the explanatory and predictive use of probability statements, and I will therefore note them here (in a form very similar to that chosen by Cramér):

(9.2a) If $0 \leqq p(G, F) < \epsilon$, where ϵ is some very small number, then, if a random experiment of kind F is performed one single time, it can be considered as practically certain that the result G will not occur.[32]

(9.2b) If $1 - \epsilon < p(G, F) \leqq 1$, where ϵ is some very small number, then if a random experiment of kind F is performed one single time, it can be considered as practically certain that the result G will occur.[33]

I now turn to the explanatory use of probability statements. Consider the experiment, D, of drawing, with subsequent replacement and thorough mixing, a ball from an urn containing one white ball and 99 black ones of the same size and material. Let us suppose that the probability, $p(W, D)$, of obtaining a white ball as a result of a performance of D is .99. According to the statistical interpretation, this is an empirical hypothesis susceptible of test by reference to finite statistical samples, but for the moment, we need not enter into the question how the given hypothesis might be established. Now, rule (9.2b) would seem to indicate that this hypothesis might be used in statistically explaining or predicting the results of certain individual drawings from the urn. Suppose, for example, that a particular drawing, d, produces a white ball. Since $p(W, D)$ differs from 1 by less than, say, .015, which is a rather small

---

[32] Cf. Cramer [11], p. 149; see also the very similar formulation in Kolmogoroff [31], p. 4.

[33] Cf. Cramér [11], p. 150.

number, (9.2b) yields the following argument, which we might be in-clined to consider as a statistical explanation of the fact that d is W:

(9.3)

$1 - .015 < p(W, D) \leqq 1$; and .015 is a very small number.

d is an instance of D.

It is practically certain that d is W.

This type of reasoning is closely reminiscent of our earlier argument (8.3a), and it leads into a similar difficulty, as will now be shown. Suppose that besides the urn just referred to, which we will assume to be marked '1,' there are 999 additional urns of the same kind, each containing 100 balls, all of which are black. Let these urns be marked '2,' '3' . . . '1000.' Consider now the experiment E which consists in first drawing a ticket from a bag containing 1000 tickets of equal size, shape, etc., bearing the numerals '1,' '2' . . . '1000,' and then drawing a ball from the urn marked with the same numeral as the ticket drawn. In accordance with standard theoretical considerations, we will assume that $p(W, E)$ = .00099. (This hypothesis again is capable of confirmation by statistical test in view of the interpretation (9.1).) Now, let e be a particular performance of E in which the first step happens to yield the ticket numbered 1. Then, since e is an instance of E, the interpretative rule (9.2a) permits the following argument:

(9.4a)

$0 \leqq p(W, E) < .001$; and .001 is a very small number.

e is an instance of E.

It is practically certain that e is not W.

But on our assumption, the event e also happens to be an instance of the experiment D of drawing a ball from the first urn; we may therefore apply to it the following argument:

(9.4b)

$1 - .015 < p(W, D) \leqq 1$; and .015 is a very small number.

e is an instance of D.

It is practically certain that e is W.

Thus, in certain cases the interpretative rules (9.2a) and (9.2b) yield arguments which again exhibit what was called above the ambiguity of statistical systematization.

This ambiguity clearly springs from the fact that (a) the probability of obtaining an occurrence of some specified kind G depends on the random experiment whose result G is being considered, and that (b) a particular instance of G can normally be construed as an outcome of different kinds of random experiment, with different probabilities for the outcome in question; as a result, under the frequency interpretation given in (9.2a) and (9.2b), an occurrence of G in a particular given case may be shown to be both practically certain and practically impossible. This ambiguity does not represent a flaw in the formal theory of probability: it arises only when the empirical interpretation of that theory is brought into play.

It might be suspected that the trouble arises only when an attempt is made to apply probability statements to individual events, such as one particular drawing in our illustration: statistical probabilities, it might be held, have significance only for reasonably large samples. But surely this is unconvincing since there is only a difference in degree between a sample consisting of just one case and a sample consisting of many cases. And indeed, the problem of ambiguity recurs when probability statements are used to account for the frequency with which a specified kind G of result occurs in finite samples, no matter how large.

For example, let the probability of obtaining recovery (R) as the result of the "random experiment" P of treating cases of streptococcus infection with penicillin be $p(R, P) = .75$. Then, assuming statistical independence of the individual cases, the frequency interpretation yields the following consequence, which refers to more or less extensive samples: For any positive deviation d, however small, there exists a specifiable sample size n_d such that it is practically certain that in one single series of n_d repetitions of the experiment P, the proportion of cases of R will deviate from .75 by less than d.[34] It would seem therefore that a recovery rate of close to 75 per cent in a sufficiently large number of instances of P could be statistically explained or predicted by means of the probability statement that $p(R, P) = .75$. But any such series of instances can also be construed as a set of cases of another random experiment for which it is practically certain that almost all the cases in the sample recover; alternatively, the given cases can be construed as a set of instances of yet another random experiment for which it is practically certain that none of the cases in a sample of the given size will recover. The arguments leading to this conclusion are basically similar to those

[34] *Ibid.*, pp. 197–198.

presented in connection with the preceding illustrations of ambiguity; he details will therefore be omitted.

In its essentials, the ambiguity of statistical systematization can be characterized as follows: If a given object or set of objects has an attribute A which with high statistical probability is associated with another attribute C, then the same object or set of objects will, in general, also have an attribute B which, with high statistical probability, is associated with non-C. Hence, if the occurrence of A in the particular given case, together with the probability statement which links A with C, is regarded as constituting adequate grounds for the predictive or explanatory conclusion that C will almost certainly occur in the given case, then there exists, apart from trivial exceptions, always a competing argument which in the same manner, from equally true premises, leads to the predictive or explanatory conclusion that C will not occur in that same case. This peculiarity has no counterpart in nomological explanation: If an object or set of objects has a character A which is invariably accompanied by C then it cannot have a character B which is invariably accompanied by non-C.[35]

The ambiguity of statistical explanation should not, of course, be taken to indicate that statistical probability hypotheses have no explanatory or predictive significance, but rather that the above analysis of the logic of statistical systematization is inadequate. That analysis was suggested by a seemingly plausible analogy between the systematizing use of statistical generalizations and that of nomic ones—an analogy which seems to receive strong support from the interpretation of statistical generalizations which is offered in current statistical theory. Nevertheless, that analogy is deceptive, as will now be shown.

10. *The Inductive Character of Statistical Systematization and the Requirement of Total Evidence.*

It is typical of the statistical systematizations considered in this study that their "conclusion" begins with the phrase 'It is almost certain that,'

[35] My manuscript here originally contained the phrase 'is invariably (or even in some cases) accompanied by non-C.' By reading the critique of this passage as given in the manuscript of Professor Scriven's contribution to the present volume, I became aware that the claim made in parentheses is indeed incorrect. Since the point is entirely inessential to my argument, I deleted the parenthetical remark after having secured Professor Scriven's concurrence. However, Professor Scriven informed me that he would not have time to remove whatever references to this lapse his manuscript might contain: I therefore add this note for clarification.

which never occurs in the conclusion of a nomological explanation or prediction. The two schemata (8.1) and (8.2) above exhibit this difference in its simplest form. A nomological systematization of the form (8.1) is a deductively valid argument: if its premises are true then so is its conclusion. For arguments of the form (8.2), this is evidently not the case. Could the two types of argument be assimilated more closely to each other by giving the conclusion of (8.1) the form 'It is certain that x is G'? This suggestion involves a misconception which is one of the roots of the puzzle presented by the ambiguity of statistical systematization. For what the statement 'It is certain that x is G' expresses here can be restated by saying that the conclusion of an argument of form (8.1) cannot be false if the premises are true, i.e., that the conclusion is a logical consequence of the premises. Hence, the certainty here in question represents not a property of the conclusion that x is G, but rather a relation which that conclusion bears to the premises of (8.1). Generally, a sentence is certain, in this sense, relative to some class of sentences just in case it is a logical consequence of the latter. The contemplated reformulation of the conclusion of (8.1) would therefore be an elliptic way of saying that

(10.1)   'x is G' is certain relative to, i.e., is a logical consequence of, the two sentences 'All F are G' and 'x is F.'[36]

But clearly this is not equivalent to the original conclusion of (8.1); rather, it is another way of stating that the entire schema (8.1) is a deductively valid form of inference.

Now, the basic error in the formulation of (8.2) is clear: near certainty, like certainty, must be understood here not as a property but as a relation; thus, the "conclusion" of (8.2) is not a complete statement but an elliptical formulation of what might be more adequately expressed as follows:

(10.2)   'x is G' is almost certain relative to the two sentences 'Almost all F are G' and 'x is F.'

The near certainty here invoked is sometimes referred to as (high)

---

[36] A sentence of the form 'It is certain that x is G' ostensibly attributes the modality of certainty to the proposition expressed by the conclusion in relation to the propositions expressed by the premises. For the purposes of the present study, involvement with propositions can be avoided by construing the given modal sentence as expressing a logical relation that the conclusion, taken as a sentence, bears to the premise sentences. Concepts such as near certainty and probability can, and will here, equally be treated as applying to pairs of sentences rather than to pairs of propositions.

probability; the conclusion of arguments like (8.2) is then expressed by such phrases as '(very) probably, x is G,' or 'it is (highly) probable that x is G'; a nonelliptic restatement would then be given by saying that the sentences 'Almost all F are G' and 'x is F' taken jointly lend strong support to, or confer a high probability or a high degree of rational credibility upon, 'x is G.' The probabilities referred to here are logical or inductive probabilities, in contradistinction to the statistical probabilities mentioned in the premises of the statistical systematization under examination. The notion of logical probability will be discussed more fully a little later in the present section.

As soon as it is realized that the ostensible "conclusions" of arguments such as (8.2) and their quantitative counterparts, such as (9.3), are elliptic formulations of relational statements, one puzzling aspect of the ambiguity of statistical systematization vanishes: the apparently conflicting claims of matched argument pairs such as (8.3a) and (8.3b) or (9.4a) and (9.4b) do not conflict at all. For what the matched arguments in a pair claim is only that each of two contradictory sentences, such as 'j is R' and 'j is not R' in the pair (8.3), is strongly supported by certain other statements, which, however, are quite different for the first and for the second sentence in question. Thus far then, no more of a "conflict" is established by a pair of matched statistical systematizations than, say, by the following pair of deductive arguments, which show that each of two contradictory sentences is even conclusively supported, or made certain, by other suitable statements which, however, are quite different for the first and for the second sentence in question:

(10.3a)
$$\frac{\text{All F are G.}}{\text{a is F.}}$$
$$\text{a is G.}$$

(10.3b)
$$\frac{\text{No H is G.}}{\text{a is H.}}$$
$$\text{a is not G.}$$

The misconception thus dispelled arises from a misguided attempt to construe arguments containing probability statements among their premises in analogy to deductive arguments such as (8.1)—an attempt which prompts the construal of formulations such as 'j is almost certain to be

R' or 'probably, j is R' as self-contained complete statements rather than as elliptically formulated statements of a relational character.[37]

The idea, repeatedly invoked in the preceding discussion, of a statement or set of statements e (the evidence) providing strong grounds for asserting a certain statement h (the hypothesis), or of e lending strong support to h, or making h nearly certain is, of course, the central concept of the theory of inductive inference. It might be conceived in purely qualitative fashion as a relation S which h bears to e if e lends strong support to h; or it may be construed in quantitative terms, as a relation capable of gradations which represents the extent to which h is supported by e. Some recent theories of inductive inference have aimed at developing rigorous quantitative conceptions of inductive support: this

[37] These remarks seem to me to be relevant, for example, to C. I. Lewis's notion of categorical, as contradistinguished from hypothetical, probability statements. For in [32], p. 319, Lewis argues as follows: "Just as 'If D then (certainly) P, and D is the fact,' leads to the categorical consequence, 'Therefore (certainly) P'; so too, 'If D then probably P, and D is the fact,' leads to a categorical consequence expressed by 'It is probable that P'. And this conclusion is not merely the statement over again of the probability relation between 'P' and 'D'; any more than 'Therefore (certainly) P' is the statement over again of 'If D then (certainly) P'. 'If the barometer is high tomorrow will probably be fair; and the barometer *is* high,' categorically assures something expressed by 'Tomorrow will probably be fair'. This probability is still relative to the grounds of judgment; but if these grounds are actual, and contain all the available evidence which is pertinent, then it is not only categorical but may fairly be called *the* probability of the event in question."

This position seems to me to be open to just those objections which have been suggested in the main text. If 'P' is a statement, then the expressions 'certainly P' and 'probably P' as envisaged in the quoted passage are not statements: if we ask how one would go about trying to ascertain whether they were true, we realize that we are entirely at a loss unless and until a reference set of statements or assumptions is specified relative to which P may then be found to be certain, or to be highly probable, or neither. The expressions in question, then, are essentially incomplete; they are elliptic formulations of relational statements; neither of them can be the conclusion of an inference. However plausible Lewis's suggestion may seem, there is no analogue in inductive logic to *modus ponens*, or the "rule of detachment" of deductive logic, which, given the information that 'D,' and also 'if D then P,' are true statements, authorizes us to detach the consequent 'P' in the conditional premise and to assert it as a self-contained statement which must then be true as well.

At the end of the quoted passage, Lewis suggests the important idea that 'probably P' might be taken to mean that the total relevant evidence available at the time confers high probability upon P; but even this statement is relational in that it tacitly refers to some unspecified time; and besides, his general notion of a categorical probability statement as a conclusion of an argument is not made dependent on the assumption that the premises of the argument include all the relevant evidence available.

It must be stressed, however, that elsewhere in his discussion, Lewis emphasizes the relativity of (logical) probability, and thus the very characteristic which rules out the conception of categorical probability statements.

s true especially of the systems of inductive logic constructed by Keynes and others and recently, in a particularly impressive form, by Carnap.[38] If—as in Carnap's system—the concept is construed so as to possess the formal characteristics of a probability, it will be referred to as the logical (or inductive) probability, or as the degree of confirmation, c(h, e), of h relative to e. (This inductive probability, which is a function of statements, must be sharply distinguished from statistical probability, which is a function of classes of events.) As a general phrase referring to a quantitative notion of inductive support, but not tied to any one particular theory of inductive support or confirmation, let us use the expression '(degree of) inductive support of h relative to e.'[39]

An explanation, prediction, or retrodiction of a particular event or set of events by means of principles which include statistical generalizations has then to be conceived as an inductive argument. I will accordingly speak of *inductive systematization* (in contradistinction to *deductive systematization*, where whatever is explained, predicted, or retrodicted is a deductive consequence of the premises adduced in the argument).

When it is understood that a statistical systematization is an inductive argument, and that the high probability or near certainty mentioned in the conclusions of such arguments as (8.3a) and (8.3b) is relative to the premises, then, as shown, one puzzle raised by the ambiguity of statistical explanation is resolved, namely the impression of a conflict, indeed a near incompatibility, of the claims of two equally sound inductive systematizations.

But the same ambiguity raises another, more serious, problem, which now calls for consideration. It is very well to point out that in (8.3a) and (8.3b) the contradictory statements 'j is R' and 'j is not R' are shown to be almost certain by referring to different sets of "premises": it still remains the case that both of these sets are true. Here, the analogy to (10.3a) and (10.3b) breaks down: in these deductive arguments with contradictory conclusions the two sets of premises cannot both be true. Thus, it would seem that by statistical systematizations based on suitably

---

[38] See especially [7, 8], and, for a very useful survey [6].

[39] In a recent study, Kemeny and Oppenheim [29], have proposed, and theoretically developed, an interesting concept of "degree of factual support" (of a hypothesis by given evidence), which differs from Carnap's concept of degree of confirmation, or inductive probability, in important respects; for example, it does not have the formal character of a probability function. For a suggestive distinction and comparison of different concepts of evidence, see Rescher [43].

Carl G. Hempel

chosen bodies of true information, we may lend equally strong suppor
to two assertions which are incompatible with each other. But then—an
this is the new problem—which of such alternative bodies of evidence i
to be relied on for the purposes of statistical explanation or prediction

An answer is suggested by a principle which Carnap calls *the require*
ment of total evidence. It lays down a general maxim for all application
of inductive reasoning, as follows: "in the application of inductive logi
to a given knowledge situation, the total evidence available must be take
as basis for determining the degree of confirmation."[40] Instead of th
total evidence, a smaller body, e_1, of evidence may be used on conditio
that the remaining part, e_2, of the total evidence is inductively irrelevan
to the hypothesis h whose confirmation is to be determined. If, as i
Carnap's system, the degree of confirmation is construed as an inductiv
probability, the irrelevance of e_2 for h relative to e_1 can be expressed b
the condition that $c(h, e_1 \cdot e_2) = c(h, e_1)$.[41]

The general consideration underlying the requirement of total evidenc
is obviously this: If an investigator wishes to decide what credence to giv
to an empirical hypothesis or to what extent to rely on it in planning hi
actions, then rationality demands that he take into account all the rele
vant evidence available to him; if he were to consider only part of that evi
dence, he might arrive at a much more favorable, or a much less favorable,
appraisal, but it would surely not be rational for him to base his decision
on evidence he knew to be selectively biased. In terms of the concept o
degree of confirmation, the point might be stated by saying that the de
gree of confirmation assigned to a hypothesis by the principles of inductive

[40] Carnap [7], p. 211. In his comments, pp. 211–213, Carnap points out that in
less explicit form, the requirement of total evidence has been recognized by various
authors at least since Bernoulli. The idea also is suggested in the passage from Lewis
[32], quoted in fn. 36. Similarly, Williams, whose book *The Ground of Induction*
centers about various arguments that have the character of statistical systematizations,
speaks of "the most fundamental of all rules of probability logic, that 'the' probability of any proposition is its probability in relation to the known premises and
them only." (Williams [55], p. 72.)

I wish to acknowledge here my indebtedness to Professor Carnap, to whom I
turned in 1945, when I first noticed the ambiguity of statistical explanation, and who
promptly pointed out to me in a letter that this was but one of several apparent
paradoxes of inductive logic which result from violations of the requirement of total
evidence.

In his recent book, Barker [2], pp. 70–78, concisely and lucidly presents the gist
of the puzzle under consideration here and examines the relevance to it of the principle of total evidence.

[41] Cf. Carnap [7], pp. 211, 494.

ogic will represent the rational credibility of the hypothesis for a given investigator only if the argument takes into account all the relevant evidence available to the investigator.

The requirement of total evidence is not a principle of inductive logic, which is concerned with relations of potential evidential support among statements, i.e., with whether, or to what degree, a given set of statements supports a given hypothesis. Rather, the requirement is a maxim for the application of inductive logic; it might be said to state a necessary condition of rationality in forming beliefs and making decisions on the basis of available evidence. The requirement is not limited to arguments of the particular form of statistical systematizations, where the evidence, represented by the "premises," includes statistical generalizations: it is a necessary condition of rationality in the application of any mode of inductive reasoning, including, for example, those cases in which the evidence contains no generalizations, statistical or universal, but only data on particular occurrences.

Let me note here that in the case of deductive systematization, the requirement is automatically satisfied and thus presents no special problem.[42] For in a deductively valid argument whose premises constitute only part of the total evidence available at the time, that part provides conclusive grounds for asserting the conclusion; and the balance of the total evidence is irrelevant to the conclusion in the strict sense that if it were added to the premises, the resulting premises would still constitute conclusive grounds for the conclusion. To state this in the language of inductive logic: the logical probability of the conclusion relative to the premises of a deductive systematization is 1, and it remains 1 no matter what other parts of the total evidence may be added to the premises.

The residual problem raised by the ambiguity of probabilistic explanation can now be resolved by requiring that if a statistical systematization is to qualify as a rationally acceptable explanation or prediction (and not just as a formally sound *potential* explanation or prediction), it must satisfy the requirement of total evidence. For under this requirement, the "premises" of an acceptable statistical systematization whose "conclusion" is a hypothesis h must consist either of the total evidence e or of some subset of it which confers on h the same inductive probability as e; and

[42] Carnap [7], p. 211, says "There is no analogue to this requirement [of total evidence] in deductive logic"; but it seems more accurate to say that the requirement is automatically met here.

Carl G. Hempel

the same condition applies to an acceptable systematization which has th
negation of h as its "conclusion." But one and the same evidence, e, can
not—if it is logically self-consistent—confer a high probability on h as we
as on its negation, since the sum of the two probabilities is unity. Henc
of two statistical systematizations whose premises confer high probabil
ties on h and on the negation of h, respectively, at least one violates th
requirement of total evidence and is thus ruled out as unacceptable.

The preceding considerations suggest that a statistical systematizatio
may be construed generally as an inductive argument showing that a ce
tain statement or finite set of statements, e, which includes at least on
statistical law, gives strong but not logically conclusive support to a stat
ment h, which expresses whatever is being explained, predicted, retr
dicted, etc. And if an argument of this kind is to be acceptable in scienc
as an empirically sound explanation, prediction, or the like—rather tha
only a formally adequate, or potential one—then it will also have to me
the requirement of total evidence.

But an attempt to apply the requirement of total evidence to statistic
systematizations of the simple kind considered so far encounters a seriou
obstacle. This was noted, among others, by S. Barker with special refer
ence to "statistical syllogisms," which are inductive arguments with tw
premises, very similar in character to the arguments (9.4a) and (9.4b
above. Barker points out, in effect, that the statistical syllogism is subjec
to what has been called here the ambiguity of statistical systematizatior
and he goes on to argue that the principle of total evidence will be of n
avail as a way to circumvent this shortcoming because generally our tota
evidence will consist of far more than just two statements, which would
moreover have to be of the particular form required for the premises of
statistical syllogism.[43] This observation would not raise a serious difficulty
at least theoretically speaking, if an appropriate general system of induc
tive logic were available: the rules of this system might enable us to shov
that that part of our total evidence which goes beyond the premises of ou
simple statistical argument is inductively irrelevant to the conclusion i
the sense specified earlier in this section. Since no inductive logic of th
requisite scope is presently at hand, however, it is a question of great inter
est whether some more manageable substitute for the requirement of tota
evidence might not be formulated which would not presuppose a full sys

[43] See Barker [2], pp. 76–78. The point is made in a more general form by Carnap
[7], p. 404.

140

em of inductive logic and would be applicable to simple statistical sys-
ematizations. This question will be examined in the next section on the
asis of a closer analysis of simple statistical systematizations offered by
mpirical science.

1. *The Logical Form of Simple Statistical Systematizations: A Rough
Criterion of Evidential Adequacy.*

Let us note, first of all, that empirical science offers many statistical
ystematizations which accord quite well with the general characterization
o which we were led in the preceding section.

For example, by means of Mendelian genetic principles it can be shown
hat in a random sample taken from a population of pea plants each of
whose parent plants represents a cross of a pure white-flowered and a pure
ed-flowered strain, approximately 75 per cent of the plants will have red
lowers and the rest white ones. This argument, which may be used for
explanatory or for predictive purposes, is a statistical systematization; what
t explains or predicts are the approximate percentages of red- and white-
lowered plants in the sample; the "premises" by reference to which the
pecified percentages are shown to be highly probable include (1) the
pertinent laws of genetics, some of which are statistical generalizations,
whereas others are of strictly universal form; and (2) particular informa-
ion of the kind mentioned above about the genetic make-up of the parent generation of the plants from which the sample is taken. (The genetic
principles of strictly universal form include the laws that the colors in
question are tied to specific genes; that the red gene is dominant over the
white one; and various other general laws concerning the transmission, by
genes, of the colors in question—or, perhaps, of a broader set of gene-
linked traits. Among the statistical generalizations invoked is the hypothesis that the four possible combinations of color-determining genes—
WW, WR, RW, RR—are statistically equiprobable in their occurrence
in the offspring of two plants of the hybrid generation.) These premises
may fairly be regarded as exhausting that part of the total available evidence that is relevant to the hypothesis about the composition of the sample. Similar considerations apply to the kind of argument that serves retro-
dictively to establish the time of manufacture of a wooden implement
found at an archeological site when the estimate is based on the amount
of radioactive carbon the implement contains. Again, in addition to statements of particular fact, the argument invokes hypotheses of strictly uni-

141

versal form as well as a statement, crucial to the argument at hand, concerning the rate of decay of radioactive carbon; this statement has the form of a statistical probability hypothesis.

Let us now examine one further example somewhat more closely. The statistical law that the half life of radon is 3.82 days may be invoked for a statistical explanation of the fact that within 7.64 days, a particular sample consisting of 10 milligrams of radon was reduced, by radioactive decay, to a residual amount falling somewhere within the interval from 2 to 3 milligrams; it could similarly be used for predicting a particular outcome of this kind. The gist of the explanatory and predictive argument is, briefly, this: The statement giving the half life of radon conveys two statistical laws, (i) that the statistical probability for an atom of radon to undergo radioactive decay within a period of 3.82 days is $\frac{1}{2}$, and (ii) that the decaying of different radon atoms constitutes statistically independent events. One further premise needed is the statement that the number of atoms in 10 milligrams of radon is enormously large (in excess of 10^{19}). As mathematical probability theory shows, the two laws in conjunction with this latter statement imply deductively that the statistical probability is exceedingly high that the mass of the radon atoms surviving after 7.64 days will not deviate from 2.5 milligrams by more than .5, i.e., that it will fall within the specified interval. More explicitly, the consequence deducible from the two statistical laws in conjunction with the information on the large number of atoms involved is another statistical law to this effect: The statistical probability is very high that the random experiment F of letting 10 milligrams of radon decay for 7.68 days will yield an outcome of kind G, namely a residual amount of radon whose mass falls within the interval from 2 to 3 milligrams. Indeed, the probability is so high that, according to the interpretation (9.2b), if the experiment F is performed just once, it is "practically certain" that the outcome will be of kind G. In this sense, it is rational on the basis of the given information to expect the outcome G to occur as the result of a single performance of F; and also in this sense, the information concerning the half life of radon and the large number of atoms involved in an experiment of kind F affords a statistical explanation or prediction of the occurrence of G in a particular performance of the experiment.[44]

[44] By reference to a physical theory that makes essential use of statistical systematization, Hanson [23], has recently advanced an interesting argument against the view

In the statistical systematization here outlined, the requirement of total evidence is satisfied at least in the broad sense that according to the total body of present scientific knowledge, the rate of radioactive decay of an element is independent of all other factors, such as temperature and pressure, ordinary magnetic and electric influences, and chemical interactions; none of these need be taken into consideration in appraising the probability of the specified outcome.

Other statistical explanations offered in science for particular phenomena follow the same general pattern: To account for the occurrence of a certain kind of event under specified (e.g., experimental) conditions, certain laws or theories of statistical form are adduced, and it is shown that as a consequence of these, the statistical probability for an outcome of the specified kind under circumstances of the specified kind is extremely high, so that that outcome may be expected with practical certainty in any one case where the specified conditions occur. (For example, the probabilistic

that any explanation constitutes a potential prediction. According to Hanson, that view fits the character of the explanations and predictions made possible by the laws of Newtonian classical mechanics, which are deterministic in character; but it is entirely inappropriate for quantum theory, which is fundamentally nondeterministic. More specifically, Hanson holds that the laws of quantum theory do not permit the *prediction* of any individual quantum phenomenon P, such as the emission of a beta particle from a radioactive substance, but that "P can be completely *explained* ex post facto; one can understand fully just what kind of event occurred, in terms of the well-established laws of . . . quantum theory . . . These laws give the *meaning* of explaining single microevents'." (Hanson [23], p. 354; the italics are the quoted author's.) I quite agree that by reason of their statistical character, the laws of quantum theory permit the prediction of events such as the emission of beta particles by a radioactive substance only statistically and not with deductive-nomological certainty for an individual case. But for the same reason it is quite puzzling in what sense those laws could be held to permit a complete explanation ex post facto of the single event P. For if the explanans contains the statement that P has occurred, then the explanation is unilluminatingly circular; it might be said, at best, to provide a description of what in fact took place, but surely not an understanding of why it did; and to answer the question 'why?' is an essential task of explanation in the characteristic sense with which we have been, and will be, concerned throughout this essay. If, on the other hand, the explanans does not contain the statement that P has occurred, but only statements referring to antecedent facts plus the laws of quantum theory, then the information thus provided can at best show that an event of the kind illustrated by P—namely, emission of a beta particle—was highly probable under the circumstances; this might then be construed, in the sense outlined in the text, as constituting a probabilistic explanation for the occurrence of the particular event P. Thus, it still seems correct to say that an explanation in terms of statistical laws is also a potential prediction, and that both the explanation and the prediction are statistical-probabilistic in character, and provide no complete accounts of individual events in the manner in which deductive-nomological systematization permits a complete account of individual occurrences.

143

explanation provided by wave mechanics for the diffraction of an electron beam by a narrow slit is essentially of this type.)

Let us examine the logic of the argument by reference to a simple model case: Suppose that a statistical explanation is to be given of the fact that a specified particular sequence S of 10 successive flippings of a given coin yielded heads at least once—let this fact be expressed by the sentence h and suppose furthermore we are given the statements that the statistical probabilities of heads and of tails for a flipping of the given coin both equal $\frac{1}{2}$, and that the results of different flippings are statistically independent of each other. These statements might then be invoked to achieve the desired explanation; for jointly they imply that the probability for a set of 10 successive flippings of the given coin to yield heads at least once is $1 - (\frac{1}{2})^{10}$, which is greater than .999. But this probability is still statistical in character; it applies to a certain kind of event (heads at least once) relative to a certain other kind of event (10 flippings of the given coin), but not to any individual event, such as the appearance of heads at least once in the particular unique set S of 10 flippings. If the statistical probability statement is to be used in explaining this latter event, then an additional principle is needed which makes statistical probabilities relevant to rational expectations concerning the occurrence of particular events.

One such principle is provided by the interpretation of a very high statistical probability as making it practically certain that the kind of outcome in question will occur in any one particular case (see (9.2b) above). This idea can be expressed in the following rule:

(11.1) On the information that the statistical probability $p(G, F)$ exceeds $1 - \epsilon$ (where ϵ is some very small positive number) and that b is a particular instance of F, it is practically certain that b is an instance of F.

Another way of giving statistical probability statements relevance for rational expectations concerning individual events would be to develop a system of inductive logic for languages in which statistical probability statements can be expressed. Such a system would assign, to any "hypothesis" h expressible in the language, a logical probability $c(h, e)$ with respect to any logically consistent evidence sentence e in that language. Choosing as evidence the sentence e_1, 'The statistical probability of obtaining heads at least once in a set of 10 flippings of this coin is $1 - (\frac{1}{2})^{10}$, and S is a articular set of such flippings,' and as hypothesis the sentence h_1, 'S yields eads at least once,' we would then obtain the logical probability conerred by e_1 on h_1. Now the systems of inductive logic presently available— y far the most advanced of which is Carnap's—do not cover languages ich enough to permit the formulation of statistical probability statements.[45] However, for the simple kind of argument under consideration ere, it is clear that the value of the logical probability should equal that f the corresponding statistical probability, i.e., that we should have $c(h_1,$ $_1) = 1 - (\frac{1}{2})^{10}$. Somewhat more generally, the idea may be expressed n the following rule:

11.2) If e is the statement '$(p(G, F) = r) \cdot Fb$' and h is 'Gb,' then $c(h, e) = r$.

This rule is in keeping with the conception, set forth by Carnap, of ogical probability as a fair betting quotient for a bet on h on the basis of ; and it accords equally with Carnap's view that the logical probability n evidence e of the hypothesis that a particular case b will have a speciied property M may be regarded as an estimate, based on e, of the relative requency of M in any class K of cases on which the evidence e does not eport. Indeed, Carnap adds that the logical probability of 'Mb' on e may n certain cases be considered as an estimate of the statistical probability f M.[46] If, therefore, e actually contains the information that the statistial probability of M is r, then it seems clear that the estimate, on e, of that tatistical probability, and thus the logical probability of 'Mb' on e, should e r as well.

The rules (11.1) and (11.2) may be regarded as schematizing at least imple kinds of statistical systematization. But, arguments conforming to hose rules will constitute acceptable explanations or predictions only if hey satisfy the principle of total evidence. For example, suppose that the otal evidence e contains the information e_1 that F_1b and $p(G, F_1) =$ 9999; then e_1 makes it practically certain that Gb; and yet it would not e acceptable as the premise of a statistical explanation or prediction of Gb' if e also contained the information, e_2, and F_2b and $p(G, F_2) =$ 0001. By itself, e_2 makes it practically certain that b is not G; and if e onsists of just e_1 and e_2, then the simple rule (11.2) does not enable us

[45] I learned from Professor Carnap, however, that in as yet unpublished work, his ystem of inductive logic has been extended to cover also statistical probability statements.

[46] See Carnap [7], pp. 168–175.

to assign a logical probability to 'Gb.' But suppose that, besides e_1 and e e also contains e_3: '$p_3(G, F_1 \cdot F_2) = .9997$, and nothing else (i.e., nothin that is not logically implied by e_1, e_2, and e_3 in conjunction). Then seems reasonable to say that the probability of 'Gb' on e should be equa or at least close, to .9997. Similarly, if e contains just the further inform: tion that F_3b and $p(G, F_1 \cdot F_2 \cdot F_3) = .00002$ then the probability of 'G on e should be close to .00002, and so on.

This consideration suggests the possibility of meeting the desideratur expressed at the end of Section 10 by the following rough substitute fc the requirement of total evidence:

(11.3) *Rough criterion of evidential adequacy for simple statistical sy tematizations*: A statistical systematization of the simple type in dicated in rules (11.1) and (11.2) may be regarded as satisfyin the requirement of total evidence if it is based on the statistic: probability of G within the narrowest class, if there is one, fc which the total evidence e available provides the requisite statist cal probability.[47] More explicitly, a statistical systematization wit the premises 'Fb' and '$p(G, F) = r$' may be regarded as rough satisfying the requirement of total evidence if the following cor ditions are met: (i) the total evidence e contains (i.e., explicitl states or deductively implies) those two premises; (ii) e implies* that F is a subclass of any class F^* for which e contains the state ment that F^*b and in addition a statistical law (which must nc

[47] This idea is closely related to one used by Reichenbach (see [41], Sec. 72) i an attempt to show that it is possible to assign probabilities to individual events withi the framework of a strictly statistical conception of probability. Reichenbach propose that the probability of a single event, such as the safe and successful completion of particular scheduled flight of a given commercial plane, be construed as the statistic: probability which the *kind* of event considered (safe and successful completion of flight) possesses within the narrowest reference class to which the given case (th specified flight of the given plane) belongs, and for which reliable statistical informa tion is available (this might be, for example, the class of scheduled flights undertaken so far by planes of the line to which the given plane belongs, and under weathe conditions similar to those prevailing at the time of the flight in question). Our work ing rule, however, assigns a probability to (a statement describing) a single even only if the total evidence specifies the value of the pertinent statistical probability whereas Reichenbach's interpretation refers to the case where the total evidence pro vides a statistical report on a finite sample from the specified reference class (in ou illustration, a report on the frequencies of safe completion in the finite class of simila flights undertaken so far); note that such a sample report is by no means equivalen to a statistical probability statement, though it may well suggest such a statemen and may serve as supporting evidence for it. (On this point, cf. also Sec. 7 of th present essay.)

[48] This requirement of implication serves to express the idea that F is the narrow est class of which b is *known* (namely, as a consequence of the total evidence) to b an element.

be simply a theorem of formal probability theory)[49] stating the value of the probability $p(G, F^*)$. The classes F, F^*, etc., are of course understood here simply as the classes of those elements which have the characteristics F, F^*, etc.

Condition (ii) might be liberalized by the following qualification: F need not be the narrowest class of the kind just specified; it suffices if e implies that within any subclass of F to which e assigns b, the statistical probability of G is the same as in F. For example, in the prediction, considered above, of the residual mass of radon, the total information available may well include data on temperature, pressure, and other characteristics of the given sample s: In this case, e assigns the particular event under study to a considerably narrower class than the class F of cases where a 10 milligram sample of radon is allowed to decay for 7.64 days. But the theory of radioactivity, likewise included in e, implies that those other characteristics do not affect the probability invoked in the prediction; in other words, the statistical probability of decay in the corresponding subclasses of F is the same as in F itself.

The working rule suggested here would also avoid an embarrassment which the general requirement of total evidence creates for the explanatory use of statistical systematizations. Suppose, for example, that an individual case b has been found to have the characteristic G (or to belong to the class G); and consider a proposed explanation of Gb by reference to the statements 'Fb' and '$p(G, F) = .9999$.' Even assuming that nothing else is known, the total evidence then includes, in addition to these latter two statements, the sentence 'Gb.' Hence if we were strictly to enforce the requirement of total evidence, then the explanans, by virtue of containing the explanandum, would trivially imply the later without benefit of any statistical law, and would confer upon it the logical probability 1. Thus, no nontrivial inductive explanation would be possible for any facts or events that are known (reported by e) to have occurred. This consequence cannot be avoided by the convention that e with the explanandum statement omitted is to count as total evidence for the statistical explanation of an event known to have occurred; for despite its apparent clarity, the notion of omitting the explanandum statement from e does not admit of a precise logical explication. It is surely not a matter of just deleting the

---

[49] Statistical probability statements which are theorems of mathematical probability theory cannot properly be regarded as affording an explanation of empirical subject matter. The condition will prove significant in a context to be discussed a little later in this section.

Carl G. Hempel

explanandum sentence from e, for the total evidence can always be so fo[r]
mulated as not to contain that sentence explicitly; for example, 'Gb' ma[y]
be replaced by the two sentences 'Gb v Fb' and 'Gb v — Fb.'

On the other hand, the working rule would circumvent the difficult[y]
For even though, in the illustration, e contains 'Gb,' the rule qualifies th[e]
statistical explanation of 'Gb' by means of 'Fb' and 'p(G, F) = .9999[']
alone as satisfying the requirement of total evidence. For the statistic[al]
law invoked here specifies the probability of G for the narrowest referenc[e]
class to which e assigns b, namely the class F. (To be sure, e also assigns [b]
to the narrower reference class F · G, for which clearly p(G, F · G) = [1.]
It will be reasonable to say that e (trivially) contains this latter statemen[t]
since it is simply a logical consequence of the measure-theoretical postu[-]
lates for statistical probability. But precisely for this reason, the statemen[t]
'p(G, F · G) = 1' is not an empirical law; hence, under the working rul[e]
this part of the content of e need not be taken into consideration.)[50]

But while a rule such as (11.3) does seem in accord with the rational[e]
of scientific arguments intended to explain or to predict individual occu[r]
rences by means of statistical laws, it offers no more than a rough workin[g]
principle, which must be used with caution and discretion. Suppose, fo[r]
example, that the total evidence e consists of the statements 'Fb,' 'Hb[,]'
'p(G, F) = .9999,' and a report on 10,000 individual cases other than [b]
to the effect that all of them were H and non-G. Then the statistical argu[-]
ment with 'Fb' and 'p(G, F) = .9999' as its premises and 'Gb' as its co[n]

[50] While here our rule permits us to disregard, as it were, the occurrence of th[e]
explanandum in the total evidence, this is not so in all cases. Suppose, for exampl[e]
that the total evidence e consists of the following sentences: e_1: 'p(G, F) = .4[']
e_2: 'p(G,(G v H) · F) = .9999'; e_3: 'Fb'; e_4: 'Gb.' Here again, e assigns b to the cla[ss]
F · G, for which p(G, F · G) = 1; as before, we may disregard this narrowest referenc[e]
class. But e implies as well that b belongs to the class (G v H) · F, which is the narrow[-]
est reference class relative to which e also specifies an empirical probability for G. Henc[e]
under our rule the statistical systematization with the premises e_2 and '(Gb v Hb)
Fb' and with the conclusion 'Gb' satisfies the requirement of total evidence (wherea[s]
the argument with e_1 and e_3 as premises and 'Gb' as conclusion does not). Thus, w[e]
have here an argument that statistically explains b's being G by reference to b's bein[g]
(G v H) · F, though to establish this latter fact, we made use of the sentence 'Gb[.']
In this case, then, our rule does not allow us to disregard the occurrence of the e[x]
planandum in the total evidence.

The logical situation illustrated here seems to be analogous to that described b[y]
Scriven [50], in reference to causal explanation. Scriven points out that when w[e]
causally explain a certain event by reference to certain antecedent circumstances, [it]
may happen that practically the only ground we have for assuming that those e[x]
planatory antecedents were in fact present is the information that the explanandu[m]
event did occur. Similarly, in our illustration, the information that b is G provide[s]
the ground for the assertion that b has the explanatory characteristic (G v H) · [F]

148

clusion would qualify, under the rule, as meeting the requirement of total evidence; but even though e does not state the statistical probability of G relative to H, its sample statistics on 10,000 cases of H, in conjunction with the statement that b is H, must surely cast serious doubt upon the acceptability of the proposed statistical argument as an explanation or prediction of Gb. Hence, the information relevant to 'Gb' that is provided by e cannot generally and strictly be identified with the information provided by e concerning the statistical probability of G in the narrowest available reference class; nor, of course, can the logical probability of 'Gb' on e be strictly equated with the statistical probability of G in that narrowest reference class. Thus, as a general condition for a statistical systematization that is to be not only a formally correct argument (a potential systematization) but a scientifically acceptable one, the requirement of total evidence remains indispensable.

12. On Criteria of Rational Credibility.

Besides the requirement of total evidence, there is a further condition which it might seem any statistical systematization ought to satisfy if it is to qualify as an adequate explanation, prediction, or retrodiction; namely, that the information contained in its "premises" e should provide so strong a presumption in favor of the "conclusion" h as to make it rational, for someone whose total evidence is e, to believe h to be true, or, as I will also say, to include h in the set of statements accepted by him as presumably true. In a deductive-nomological systematization, the premises afford such presumption in an extreme form: they logically imply the conclusion; hence someone whose system of accepted statements includes those premises has the strongest possible reason to accept the conclusion as well.

Thus, the study of inductive generalization gives rise to the question whether it is possible to formulate criteria for the rational acceptability of hypotheses on the basis of information that provides strong, but not conclusive, evidence for them.

I will first construe this question in a quite general fashion without limiting it specifically to the case where the supporting information provides the premises of a statistical systematization. Toward the end of this section I will return to this latter, special case.

Let us assume that the total body of scientific knowledge at a given time t can be represented by a set K_t, or briefly K, whose elements are all the statements accepted as presumably true by the scientists at time t. The class K will contain statements describing particular events as well as assertions of statistical and universal law and in addition various theoretical statements. The membership of K will change in the course of time; for as a result of continuing research, additional statements come to be established, and thus accepted into K; while others, formerly included in K, may come to be disconfirmed and then eliminated from the system.

We can distinguish two major ways in which a statement may be accepted into K: *direct acceptance*, on the basis of suitable experiences or observations, and *inferential acceptance*, by reference to previously accepted statements. An observer who records the color of a bird or notes the reading of an instrument accepts the corresponding statements directly, as reporting what he immediately observes, rather than as hypotheses whose acceptability is warranted by the fact that they can be inferred from other statements, which have been antecedently accepted and thus are already contained in K. Inferential acceptance may be either deductive or (strictly) inductive, depending on whether the statement in question is logically implied or only more or less highly supported by the previously accepted statements.

This schematic model does not require, then, that the statements representing scientific knowledge at a given time be true; rather, it construes scientific knowledge as the totality of beliefs that are accepted at a given time as warranted by appropriate scientific procedures. I will refer to this schematization as the *accepted-information model of scientific knowledge*.

Now, we have to consider the rules of acceptance or rejection which regulate membership in K. In its full generality, this question calls for a comprehensive set of principles for the formulation, test, and validation of scientific hypotheses and theories. In the context of our investigation, however, it will suffice to concentrate on some general rules for indirect acceptance; the question of criteria for direct acceptance, which would bear on standards for observational and experimental procedures, is not relevant to the central topic of this essay.

The rules to be discussed here may be considered as stating certain necessary conditions of rationality in the formation of beliefs. One very obvious condition of this kind is the following:

(CR1) Any logical consequence of a set of accepted statements is likewise an accepted statement; or, K contains all logical consequences of any of its subclasses.

The reason for this requirement is clear: If an investigator believes a certain set of statements, and thus accepts them as presumably true, then, to be rational, he has to accept also their logical consequences because any logical consequence of a set of true statements is true.

Note that (CR1) does not express a rule or principle of logic but rather a maxim for the rational *application* of the rules of deductive logic. These rules, such as *modus ponens* or the rules of the syllogism, simply indicate that if sentences of a specified kind are true, then so is a certain other sentence; but they say nothing at all about what it is rational to believe. Another rule is the following:

(CR2) The set K of accepted statements is logically consistent.

Otherwise, by reason of (CR1), K would also contain, for every one of its statements, its contradictory. This would defeat the objective of science of arriving at a set of presumably true beliefs (if a statement is presumably true, its contradictory is not); and K could provide no guidance for expectations about empirical phenomena since whatever K asserted to be the case it would also assert not to be the case.

(CR3) The inferential acceptance of any statement h into K is decided on by reference to the total system K (or by reference to a subset K' of it whose complement is irrelevant to h relative to K').

This is simply a restatement of the requirement of total evidence. As noted earlier, it is automatically satisfied in the case of deductive acceptance.

Now we must look for more specific rules of inferential acceptance. The case of deductive acceptance is completely settled by (CR1), which makes it obligatory for rational procedure to accept all statements that are deductively implied by those already accepted. Can analogous rules be specified for rational inductive acceptance? Recent developments in the theory of inductive procedures suggest that this question might best be considered as a special case of the general problem of establishing criteria of rationality for choices between several alternatives; in the case at hand, the choice would be that of accepting a proposed new statement h into K, rejecting it (in the strong sense of accepting its contradictory), or leaving the decision in suspense (i.e., accepting neither h nor its contradictory).

I will consider the problem first on the assumption that a system of inductive logic is available which, for any hypothesis h and for any logically

Carl G. Hempel

consistent "evidence" sentence e, determines the logical probability, or the degree of confirmation, $c(h, e)$, of h relative to e.

The problem of specifying rational rules of decision may now be construed in the following schematic fashion: An agent X has to choose one from among n courses of action, $A_1, A_2, \ldots, A_n$, which, on the total evidence e available to him, are mutually exclusive and jointly exhaust all the possibilities open to him. Each of these may eventuate, with certain probabilities (some of which may be zero), in any one of m outcomes, $O_1, O_2, \ldots, O_m$, which, on the evidence e, are mutually exclusive and exhaustive. The agent's decision to choose a particular course of action, say A_k, will be rational only if it is based on a comparison of its probable consequences with those of the alternative choices that are open to him. For such a comparison, inductive logic would provide one important tool. Let $a_1, a_2, \ldots, a_n$ be statements to the effect that X follows course of action $A_1, A_2, \ldots, A_n$, respectively; and let $o_1, o_2, \ldots, o_m$ be statements asserting the occurrence of $O_1, O_2, \ldots, O_m$, respectively. Then the probability, relative to e, for a proposed course of action, say A_j, to yield a specified outcome, say O_k, is given by $c(o_k, e \cdot a_j)$. The principles of the given system of inductive logic would determine all these probabilities, but they would not be sufficient to determine a rational course of action for X. Indeed, rationality is a relative concept; a certain decision or procedure can be qualified as rational only relative to some objective, namely by showing, generally speaking, that the given decision or procedure offers the optimal prospect of attaining the stated objective.

One theoretically attractive way of specifying such objectives is to assume that for X each of the outcomes $O_1, O_2, \ldots, O_n$ has a definite value or disvalue, which is capable of being represented in quantitative terms by a function assigning to any given outcome, say O_k, a real number u_k, the utility of O_k for X at the time in question. The idea of such a utility function raises a variety of problems which cannot be dealt with here, but which have been the object of intensive discussion and of much theoretical as well as experimental research.[51] The utility function, together with the probabilities just mentioned, determines the expectation value, or the probability estimate, based on e, of the utility attached to A_j for X:

$$(12.1) \quad u'(A_j, e) = c(o_1, e \cdot a_j) \cdot u_1 + \ldots + c(o_m, e \cdot a_j) \cdot u_m.$$

[51] For details and further bibliographic references, see, for example, Neumann and Morgenstern [36]; Savage [45]; Luce and Raiffa [33]; Carnap [7], par. 51.

In the context of our schematization, the conception of rationality of decision as relative to some objective can now be taken into account in a more precise form; this is done, for example, in the following rule for rational choice, which was proposed by Carnap:

Rule of maximizing the estimated utility: In the specified circumstances, X acts rationally if he chooses a course of action, A_j, for which the expectation value of the utility is maximized, i.e., is not exceeded by that associated with any of the alternative courses of action.[52]

I will now attempt to apply these considerations to the problem of establishing criteria of rational inductive acceptance. The decision to accept, or to reject, a given hypothesis, or to leave it in suspense might be considered as a special kind of choice required of the scientific investigator. This conception invites an attempt to obtain criteria of rational inductive belief by applying the rule of maximizing the expected utility to this purely scientific kind of choice with its three possible "outcomes": K enlarged by the contemplated hypothesis h; K enlarged by the contradictory of h; K unchanged. But what could determine the utilities of such outcomes?

The pursuit of knowledge as exemplified by pure scientific inquiry, by "basic research" not directly aimed at any practical applications with corresponding utilities, is often said to be concerned with the discovery of truth. This suggests that the acceptance of a hypothesis might be considered a choice as a result of which either a truth or a falsehood is added to the previously established system of knowledge. The problem then is to find a measure of the purely scientific utility, or, as I will say, the *epistemic utility*, of such an addition.

It seems reasonable to say that the epistemic utility of adding h to K depends not only on whether h is true or false but also on how much of what h asserts is new, i.e., goes beyond the information already contained in K. Let k be a sentence which is logically implied by K, and which in turn implies every sentence in K, just as the conjunction of the postulates in a finite axiomatization of geometry implies all the postulates and theorems of geometry. Then k has the same informational content as K. Now,

---

[52] Cf. Carnap [7], p. 269; the formulation given there is "Among the possible actions choose that one for which the estimate of the resulting utility is a maximum." Carnap proposes this rule after a critical examination, by reference to instructive illustrations, of several other rules for rational decision that might seem plausible (*ibid.*, Secs. 50, 51).

153

Carl G. Hempel

the common content of two statements is expressed by their disjunction, which is the strongest statement logically implied by each of them. Hence, the common content of h and K is given by h ∨ k. But h is equivalent to $(h ∨ k) \cdot (h ∨ -k)$, where the two component sentences in parentheses have no common content: their disjunction is a logical truth. Hence that part of the content of h which goes beyond the information contained in K is expressed by $(h ∨ -k)$. To indicate *how much* is being asserted by this statement, we make use of the concept of a content measure for sentences in a (formalized) language L. By a *content measure function* for a language L we will understand a function m which assigns, to every sentence s of L, a number $m(s)$ in such a way that (i) $0 \le m(s) \le 1$; (ii) $m(s) = o$ if and only if s is a logical truth of L; (iii) if the contents of s_1 and s_2 are mutually exclusive—i.e., if the disjunction $s_1 ∨ s_2$ is a logical truth of L—then $m(s_1 \cdot s_2) = m(s_1) + m(s_2)$.[53] (If these requirements are met, then m can readily be seen to satisfy also the following conditions: (iv) $m(s) = 1 - m(-s)$; (v) if s_1 logically implies s_2, then $m(s_1) \ge m(s_2)$; (vi) logically equivalent sentences have equal measures.)

Let m be a content measure function for an appropriately formalized language suited to the purposes of empirical science. Then, in accordance with the idea suggested above, it might seem plausible to accept the following:

(12.2) *Tentative measure of epistemic utility*: The epistemic utility of accepting a hypothesis h into the set K of previously accepted scientific statements is $m(h ∨ -k)$ if h is true, and $-m(h ∨ -k)$ if h is false; the utility of leaving h in suspense, and thus leaving K unchanged, is 0.

The rule of maximizing the estimated utility now qualifies the decision to accept a proposed hypothesis as epistemically rational if the probability estimate of the corresponding utility is at least as great as the estimates attached to the alternative choices. The three estimates can readily be computed. The probability, on the basis of K, that the proposed hypothesis h is true is $c(h, k)$, and that it is false, $1 - c(h, k)$. Denoting the three alternative actions of accepting h, rejecting h, and leaving h in suspense

[53] Content measures satisfying the specified conditions can readily be constructed for various kinds of formalized languages. For a specific measure function applicable to any first-order functional calculus with a finite number of predicates of any degrees, and a finite universe of discourse, see *SLE*, par. 9, or Carnap and Bar-Hillel [9], Sec. 6.

by 'A,' 'R,' 'S,' respectively, we obtain the following formulas for the estimated utilities attached to these three courses of action:

$$(12.3a) \quad u'(A, k) = c(h, k) \cdot m(h \vee -k) - (1 - c(h, k)) \cdot m(h \vee -k)$$
$$= m(h \vee -k) \cdot (2c(h, k) - 1).$$

Analogously, considering that rejecting h is tantamount to accepting $-h$, which goes beyond K by the assertion $-h \vee -k$, we find

$$(12.3b) \quad u'(R, k) = m(-h \vee -k) \cdot (1 - 2c(h, k)).$$

Finally, we have

$$(12.3c) \quad u'(S, k) = 0.$$

Now the following can be readily verified:[54]

(i) If $c(h, k) = \frac{1}{2}$, then all three estimates are zero;
(ii) If $c(h, k) > \frac{1}{2}$, then $u'(A, k)$ exceeds the other two estimates;
(iii) If $c(h, k) < \frac{1}{2}$, then $u'(R, k)$ exceeds the other two estimates.

Hence, the principle of maximizing the estimated utility leads to the following rule:

(12.4) *Tentative rule for inductive acceptance:* Accept or reject h, given K, according as $c(h, k) > \frac{1}{2}$ or $c(h, k) < \frac{1}{2}$; when $c(h, k) = \frac{1}{2}$, h may be accepted, rejected, or left in suspense.

It is of interest to note that the principle of maximizing the estimated utility, in conjunction with the measure of epistemic utility specified in (12.2), implies this rule of acceptance quite independently of whatever particular inductive probability function c and whatever particular measure function m might be adopted.

Unfortunately, the criteria specified by this rule are far too liberal to be acceptable as general standards governing the acceptance of hypotheses in pure science. But this does not necessarily mean that the kind of approach attempted here is basically inadequate: the fault may well lie with the oversimplified construal of epistemic utility. It would therefore seem a problem definitely worth further investigation whether a modified version of the concept of epistemic utility cannot be construed which, via

[54] We have
$$u'(A, k) - u'(R, k) = (2c(h, k) - 1) \cdot (m(h \vee -k) + m(-h \vee -k)).$$

Since m is nonnegative, the second factor on the right could be 0 only if both of its terms were 0. But this would require $h \vee -k$ as well as $-h \vee -k$ to be logically true, in which case k would logically imply both h and $-h$; and this is precluded by the consistency requirement, (CR2), for K. Hence, $u'(A, k)$ exceeds $u'(R, k)$ or is exceeded by it according as $c(h, k)$ is greater or less than $\frac{1}{2}$; and whichever of the two estimates is the greater will also be positive and thus greater than $u'(S, k)$.

the principle of maximizing estimated utility, will yield a more satisfactory rule for the inductive acceptance or rejection of hypotheses in pure science. Such an improved measure of epistemic utility might plausibly be expected to depend, not only on the change of informational content, but also on other changes in the total system of accepted statements which the inductive acceptance of a proposed hypothesis h would bring about. These would presumably include the change in the simplicity of the total system, or, what may be a closely related characteristic, the change in the extent to which the theoretical statements of the system would account for, or systematize, the other statements in the system, in particular those which have been directly accepted as reports of previous observational or experimental findings. As yet, no fully satisfactory general explications of these concepts are available, although certain partial results have been obtained.[55] And even assuming that the concepts of simplicity and degree of systematization can be made explicit and precise, it is yet another question whether the notion of epistemic utility permits a satisfactory explication, which can serve as a basis for the construction of rules of inductive acceptance.

We will now consider briefly an alternative construal of scientific knowledge, which would avoid the difficulties just outlined: it will be called the *pragmatist* or *instrumentalist model*. Let us note, first of all, that the epistemic utilities associated with the decision inductively to accept (or to reject, or to leave in suspense) a certain hypothesis would have to represent "gains" or "losses" as judged by reference to the objectives of "pure" or "basic" scientific research; in contradistinction to what will be called here *pragmatic utilities*, which would represent gains or losses in income, prestige, intellectual or moral satisfaction, security, and so forth, that may accrue to an individual or to a group as a result of "accepting" a proposed hypothesis in the practical sense of basing some course of action on it. Theories of rational decision making have usually been illustrated by, and applied to, problems in which the utilities are of this pragmatic kind, as for example, in the context of quality control. The hypotheses that have to be considered in that case concern the items produced by a certain technological process during a specified time; e.g., vitamin capsules which

---

[55] For an illuminating discussion of the concept of simplicity of a total system of statements, see Barker [2] (especially Chs. 5 and 9); also see the critical survey by Goodman [22]. One definition (applicable only to formalized languages of rather simple structure) of the systematizing power of a given theory with respect to a given class of data has been proposed in *SLE*, Secs. 8 and 9.

must meet certain standards, or tablets containing a closely specified amount of a certain toxic ingredient, or ball bearings for whose diameter a certain maximum tolerance has been fixed, or light bulbs which must meet various specifications. The hypothesis under test will assert, in the simplest case, that the members of the population (e.g., the output produced by a given industrial plant in a week) meet certain specified standards (e.g., that certain of their quantitative characteristics fall within specified numerical intervals). The hypothesis is tested by selecting a random sample from the total population and examining its members in the relevant respects. The problem then arises of formulating a general decision rule which will indicate, for every possible outcome of the test, whether on the evidence afforded by that outcome the hypothesis is to be accepted or rejected. But what is here referred to as acceptance or rejection of a hypothesis clearly amounts to adopting or rejecting a certain practical course of action (e.g., to ship the ball bearings to the distributors, or to reprocess them). In this kind of situation, we may distinguish four possible "outcomes": the hypothesis may be accepted and in fact true, rejected though actually true, accepted though actually false, or rejected and in fact false. To each of these outcomes there will be attached a certain positive or negative utility, which in cases of the kind considered might be represented, at least approximately, in monetary terms. Once such utilities have been specified, it is possible to formulate decision rules which will indicate for every possible outcome of the proposed testing procedure whether, on the evidence provided by the outcome and in consideration of the utilities involved, the hypothesis is to be accepted or to be rejected. For example, the principle of maximizing estimated utilities affords such a rule, which presupposes, however, that a suitable inductive logic is available which assigns to any proposed hypothesis h, relative to any consistent "evidence" statement e, a definite logical probability, $c(h, e)$.

Alternatively, there have been developed, in mathematical statistics and in the theory of games, certain methods of arriving at decision rules which do not require any such general concept of inductive or logical probability. These methods are limited to certain special types of hypotheses and evidence sentences; normally, their application is to hypotheses in the form of probability statements (statistical generalizations), and to evidence sentences in the form of reports on statistical findings in finite samples. One of the best known methods of this kind is based on the minimax principle. This method uses the concept of probability only in its statistical form. It is intended to select the most rational from among various possible rules that might be followed in deciding on the acceptance or rejection of a proposed hypothesis h in consideration of (i) the results of a specified kind of test and (ii) the utilities assigned to the possible "outcomes" of accepting or rejecting the hypothesis. Briefly, the minimax principle directs that we adopt, from among the various possible decision rules, one that minimizes the maximum risk, i.e., one for which the largest of the (statistically defined) probability estimates of the losses that might be incurred in the given context as a result of following this rule is no greater than the largest of the corresponding risks (loss estimates) attached to any of the alternative decision rules.[56]

Clearly, the minimax principle is not itself a decision rule, but rather a metarule specifying a standard of adequacy, or of rationality, for decision rules pertaining to a suitably characterized set of alternative hypotheses, plus testing procedure, and a given set of utilities.[57]

But whatever decision rules, or whatever general standards for the choice of decision rules, may be adopted in situations of the kind referred to here, the crucial point remains that the pragmatic utilities involved, and thus the decision dictated by the rule once the test results are given, will depend on, and normally vary with, the kind of action that is to be based upon the hypothesis. Consider, for example, the hypothesis that all of the vials of vaccine produced during a given period of time by a pharma-

[56] The minimax principle was proposed and theoretically developed by A. Wald; see especially his book [54]. A lucid and stimulating less technical account and appraisal of the minimax method, of special interest from a philosophical point of view, is given in Braithwaite [3], Ch. VII. Recent very clear presentations of the fundamentals of minimax theory, plus critical comments and further developments, may be found, for example, in Savage [45], and in Luce and Raiffa [33]. Carnap [7], par. 98, gives an instructive brief comparison of those methods of estimation which are based on inductive logic with those which, like the minimax method, have been developed within the framework of statistical probability theory, without reliance on a general inductive logic.

[57] The standard set up by the minimax principle is by no means the only possible standard of rationality that can be proposed for decision rules in problem situations of the kind referred to here; and indeed, the minimax standard has been criticized in certain respects, and alternatives to it have been suggested by recent investigators. For details, see, for example, Savage [45], Ch. 13; Luce and Raiffa [33], Ch. 13. In an article which includes a lucid examination of the basic ideas of the minimax principle, R. C. Jeffrey points out that in applying this principle the experimenter acts on the assumption that this is the worst of all possible worlds for him; thus "the minimax criterion is at the pessimistic end of a continuum of criteria. At the other end of this continuum is the 'minimin' criterion, which advises each experimenter to minimize his minimum risk. Here each experimenter acts as if this were the best of all possible worlds for him." (Jeffrey [28], p. 244.)

ceutical firm meet certain standards of purity; and suppose that a test has been performed by analyzing the vials in a random sample. Then the gains or losses to be expected from correct or incorrect assumptions as to the truth of the hypothesis will depend on the action that is intended, for example, on whether the vaccine is to be administered to humans or to animals. By reason of the different utilities involved, a given decision rule—be it the rule of maximizing estimated utility or a rule selected in accordance with the minimax, or a similar, standard—may then well specify, on one and the same evidence, that the hypothesis is to be rejected in the case of application to human subjects, but accepted if the application is to be to animals.

Clearly then, in cases of this kind we cannot properly speak of a decision to accept or reject a hypothesis per se; the decision is rather to adopt one of two (or more) alternative courses of action. Moreover, it is not even clear on what grounds the acceptance or rejection of this hypothesis per se, on the given evidence, could be justified—unless it is possible to specify a satisfactory concept of epistemic utility, whose role for the decisions of pure science would be analogous to that of pragmatic utility in decisions concerning actions based on scientific hypotheses.

Some writers on the problems of rational decision have therefore argued that one cannot strictly speak of a decision to accept a scientific hypothesis, and that the decisions in question have to be construed as concerning choices of certain courses of action.[58] A lucid presentation and defense of

[58] See, for example, De Finetti [12], p. 219; Neyman [37], pp. 259–260. Savage [45], Ch. 9, Sec. 2, strongly advocates a "behavioralistic" as opposed to a "verbalistic" outlook on statistical decision problems; he argues that these problems are concerned with acts rather than with "assertions" (i.e., of scientific hypotheses). However, he grants the possibility of considering an "assertion" as a special kind of behavioral act and thus does not rule out explicitly the possibility of speaking of the acceptance—as presumably the same thing as "assertion"—of hypotheses in science. Savage here also makes some suggestive though all too brief remarks on the subtle practical consequences resulting from the assertion of a hypothesis in pure science (such as that the velocity of light is between 2.99×10^{10} and 3.01×10^{10} cm/sec); those consequences would presumably have to be taken into account, from his behavioralistic point of view, in appraising the utilities attached to the acceptance or rejection of purely scientific hypotheses. But Savage stresses that "many problems described according to the verbalistic outlook as calling for decisions between assertions really call only for decisions between much more down-to-earth acts, such as whether to issue single—or double—edged razors to an army . . ." (loc. cit., p. 161). A distinction similar to that drawn by Savage is considered by Luce and Raiffa [33], who contrast "classical statistical inference" with "modern statistical decision theory" (Ch. 13, Sec. 10). In this context, the authors briefly consider the question of how to appraise the losses from falsely rejecting or accepting a scientific research hypothesis. They suggest that no such evaluation appears possible, but conclude with a remark that seems to

Carl G. Hempel

this point of view has been given by R. C. Jeffrey, who accordingly arrives at the conclusion that the scientist's proper role is to provide the rational agents of his society with probabilities for hypotheses which, on the more customary account, he would be described as simply accepting or rejecting.[59]

This view, then, implies a rejection of the accepted-information model of scientific knowledge and suggests an alternative which might be called a tool-for-optimal-action model, or, as I said earlier, an instrumentalist model of scientific knowledge. This label is meant to suggest the idea that whether a hypothesis is to be accepted or not will depend upon the sort of action to be based on it, and on the rewards and penalties attached to the possible outcomes of such action. An instrumentalist model might be formulated in different degrees of refinement. A very simple version would represent the state of scientific knowledge at a given time t by a set D, or more explicitly, D_t, of directly accepted statements, plus a theory of inductive support which assigns to each proposed hypothesis, or to at least some of them, a certain degree of support relative to D_t. Like K in the accepted-information model, D_t would be assumed to be logically consistent and to contain any statement logically implied by any of its own subsets. But no statement other than those in D_t, however strongly confirmed by D_t, would count as accepted, or as belonging to the scientific knowledge at the given time. Rather, acceptance would be understood pragmatically in the context of some contemplated action, and a decision would then depend on the utilities involved.

If, in particular, the theory of inductive support assumed here is an inductive logic in Carnap's sense, then it will assign a degree of confirmation $c(h, e)$ to any statement h relative to any logically consistent statement e in the language of science, which we assume to be suitably chosen and formalized. In this case, scientific knowledge at a given time t might be represented by a functional k_t assigning to every sentence S that is expressible in the language of science a real-number value, $k_t(S)$, which lies between 0 and 1 inclusive. The value $k_t(S)$ would simply be the logical probability of S relative to D_t; in particular, for any S included in or logi-

hint at what I have called the concept of epistemic utility: ". . . if information is what is desired, then this requirement should be formalized and attempts should be made to introduce the appropriate information measures as a part of the loss structure. This hardly ends the controversy, however, for decision theorists are only too aware that such a program is easier suggested than executed!" (*Loc. cit.*, p. 324.)

[59] Jeffrey [28], p. 245.

cally implied by D_t, $k_t(S)$ would be 1; for any S logically incompatible with D_t, $k_t(S)$ would be 0. Temporal changes of scientific knowledge would be reflected by changes of k_t, and thus by changes in the numbers assigned to some of the sentences in the language of science.

But clearly, this version of a pragmatist model is inadequate: It construes scientific knowledge as consisting essentially of reports on what has been directly observed, for the formal theory of inductive probability which it presupposes for the appraisal of other statements would presumably be a branch of logic rather than of empirical science. This account of science disregards the central importance of theoretical concepts and principles for organizing empirical data into patterns that permit explanation and prediction. So important is this aspect of science that theoretical considerations will often strongly influence the decision as to whether a proposed report on some directly observed phenomenon is to be accepted: What is a fact is to some extent determined by theory. In this respect the notion of a system D_t of statements which are accepted directly and independently of theoretical considerations, and by reference to which the rational credibility of all other scientific statements is adjudged, is a decided oversimplification. And it is an oversimplification in yet another respect: In theoretically advanced disciplines, many of the terms that the experimenter would use to record his observations, and thus to formulate his directly accepted statements, belong to the theoretical vocabulary rather than to that of everyday observation and description; and the appropriate theoretical framework has to be presupposed if those statements are to make sense.

Another inadequacy of the model lies in the assumption that any individual hypothesis that may be proposed in the language of science can be assigned a reasonable degree of confirmation by checking it against the total set D_t of statements that have been directly accepted, for the test of any even moderately advanced scientific hypothesis will require the assumption of other hypotheses in addition to observational findings. As Duhem emphasized so strongly, what can be tested experimentally is never a single theoretical statement, but always a comprehensive and complexly interconnected body of statements.

If we were to try to construct a pragmatist model on the basis of statistical decision theory, the difficulties would become even greater; for this theory, as noted earlier, eschews the assignment of degrees of support to statements relative to other statements. Hence, here, the scientist

161

would have to assume the role of a consultant who, in a limited class of experimental contexts, provides decisions concerning the acceptance or rejection of certain statistical hypotheses for the guidance of action, provided that the pertinent utilities have been furnished to him.

At present, I do not know of a satisfactory *general* way of resolving the issue between the two conceptions of science which are schematized by our two models. But the preceding discussion of these models does seem to suggest an answer to the question raised at the beginning of this section, namely, whether it should be required of a statistical explanation, prediction, etc. in science that its premises make its conclusion rationally acceptable.

The preceding considerations seem to indicate that it would be pointless to formulate criteria of acceptability by reference to pragmatic utilities; for we are concerned here with purely theoretical (in contrast to applied) explanatory and predictive statistical arguments. We might just add the remark that criteria of rational acceptability based on pragmatic utilities might direct us to accept a certain predictive hypothesis, even though it was exceedingly improbable on the available evidence, on the ground that, if it were true, the utility associated with its adoption would be exceedingly large. In other words, if a decision rule of this kind, which is based on statistical probabilities and on an assignment of utilities, singles out, on the basis of evidence e, a certain hypothesis h from among several alternatives, then what is qualified as rational is, properly speaking, not the decision to believe h to be true, but the decision to act in the given context as *if* one believed h to be true even though e may offer very little support for that belief.

The rational credibility of the conclusion, in a sense appropriate to the purely theoretical, rather than the applied, use of statistical systematizations will thus have to be thought of as represented by a suitable concept of inductive support (perhaps in conjunction with a concept of epistemic utility). And at least for the types of statistical systemization covered by rule (11.1) or (11.2), the statistical or logical probability specified in the argument itself may serve as an indicator of inductive support; the requirement of high credibility for the conclusion can then be met by requiring, in the case of (11.1), that ϵ be sufficiently small, and in the case of (11.2), that r be sufficiently large.

But the notions "sufficiently small" and "sufficiently large" invoked here cannot well be construed as implying the existence of some fixed probability value, say r^*, such that a statistical systematization will meet the requirement of rational credibility just in case the probability associated with it exceeds r^*: The standards of rational credibility will vary with the context in which a statistical systematization is used.[60] It will therefore be more satisfactory, for an explication of the logic of statistical explanation, prediction, and similar arguments, explicitly to construe *statistical systematization* as *admitting of degrees*: The evidence e adduced in an argument of this kind may then be said to explain, or predict, or retrodict, or generally to systematize its "conclusion" h to degree r, where r is the inductive support that e gives to h. In this respect, statistical systematization differs fundamentally from its deductive-nomological counterpart: In a deductive-nomological systematization, the explanandum follows logically from the explanans and thus is *certain* relative to the latter; no higher degree of rational credibility (relative to the information provided by the premises) is possible, and anything less than it would vitiate the claim of a proposed argument to constitute a deductive systematization.

13. *The Nonconjunctiveness of Statistical Systematization.*

Another fundamental difference between deductive and statistical systematization is this: Whenever a given explanans e deductively explains each of n different explananda, say h_1, h_2, . . ., h_n, then e also deductively explains their conjunction; but if an explanans e statistically explains each of n explananda, h_1, h_2, . . ., h_n to a positive degree, however high, it may still attribute a probability of zero to their conjunction. Thus, e may statistically explain (or analogously, predict, retrodict, etc.) very strongly whatever is asserted by each of n hypotheses, but not at all what is asserted by them conjointly: statistical systematization is, in this sense, nonadditive, or nonconjunctive (whereas deductive systematization is additive, or conjunctive). This point can be stated more precisely as follows:

(13.1) *Nonconjunctiveness of statistical systematization:* For any probability value p^*, however close to 1, there exists a set of statisti-

[60] Even decision rules of the kind discussed earlier, which are formulated by reference to certain probabilities and utilities, provide only a comparative, not an absolute (classificatory) concept of rationality, i.e., they permit, basically, a *comparison* of any two in the proposed set of alternative choices and determine which of them, if any, is more rational than the other; thus, they make it possible to single out a most rational choice from among a set of available alternatives. But they do not yield a classificatory criterion which would characterize any one of the alternatives, either as rational or as nonrational in the given context.

cal systematizations which have the same "premise" e, but different "conclusions," h_1, h_2, . . ., h_n, such that e confers a probability of at least p* on every one of these conclusions but the probability zero on their conjunction.

The proof can readily be outlined by reference to a specific example. Let us assume that p* has been chosen as .999 (and similarly, that ϵ, for use of the rule (11.1), has been chosen as .001). Then consider the case, mentioned earlier, of ten successive flippings of a given coin. Choose as "premise" the statement e: 'The statistical probabilities of getting heads and of getting tails by flipping this coin are both ½; the results of different flippings are statistically independent; and S is a particular sequence of 10 flippings of this coin'; furthermore, let h_1 be 'S does not yield tails 10 times in succession'; h_2: 'S does not yield 9 tails followed by 1 head'; h_3: 'S does not yield 8 tails followed by 2 heads'; and so on to h_{1024}: 'S does not yield heads 10 times in succession.' Each of these hypotheses h_j ascribes to S a certain kind of outcome O_j; and as is readily seen, the probability statements included in e imply logically that for each of these 2^{10} different possible outcomes, the statistical probability of obtaining it as a result of 10 successive flippings of the given coin is $1 - (½)^{10}$. But according to rule (11.1), this makes it practically certain, for any one of the O_j that the particular sequence S will have O_j as its outcome; in other words, this makes it practically certain, for each one of our hypotheses h_j, that h_j is true. Rule (11.2) more specifically ascribes the logical probability $1 - (½)^{10}$ to each of the h_j on the basis of the statistical probability for O_j which is implied by e.[61]

On the other hand, the conjunction of the h_j is tantamount to the assertion that none of the 10 particular flippings that constitute the indi-

---

[61] Thus the basis for the assignment, under rule (11.1), of the probability $1 - (½)^{10}$ to each h_j is, strictly speaking, not e, but the sentence e_j: 'The statistical probability of obtaining O_j as a result of 10 successive flippings of the coin is $1 - (½)^{10}$, and S is a particular set of n such successive flippings'; this e_j is a logical consequence of, but not equivalent to, e. Now, in general, if $c(h^*, e^*) = q$ and e^{**} is a logical consequence of e^*, then $c(h^*, e^{**})$ need not equal q at all; but in our case, it is extremely plausible to assume that whatever information e contains beyond e_j is inductively irrelevant to h_j; and on this assumption, we then have $c(h_j, e) = 1 - (½)^{10}$ for each j. The requisite assumption may also be expressed more generally in the following rule, which is a plausible extension of (11.2):

Let e be a sentence which (i) specifies, for various outcomes G_k of a random experiment F, their statistical probabilities $p(G_k, F) = r_k$, (ii) states that the outcomes of different performances of F are statistically independent of each other, and (iii) asserts that a certain particular event b is a case of n successive performances of F; and let e assert nothing else. Next, let h be a statement ascribing to each of the idual sequence S will yield either heads or tails—a kind of outcome, say O*, for which e implies the statistical probability zero. This, under rule (11.1), makes it practically certain that this outcome will not occur in S, i.e., that the conjunction of the h_j is false—even though each of the conjoined hypotheses is practically certain to be true. And if (11.2) is invoked, then the statement that the statistical probability of O* is zero confers upon the conjunction of the h_j the logical probability zero even though, on the basis of statistical information also provided by e, each of the h_j has a logical probability exceeding .999.[62]

A similar argument can be presented for the case, considered earlier, of the radioactive decay of a particular sample S of 10 milligrams of radon over a period of 7.64 days. For the interval from 2 to 3 milligrams referred to in our previous discussion can be exhaustively divided into mutually exclusive subintervals i_1, i_2, . . ., i_n, which are so small that for each i_j there is a statistical probability exceeding .999999, let us say, that the residual mass of radon left of an initial 10 milligrams after 7.64 days will not lie within i_j. Hence, given the information that the half life of radon is 3.82 days, it will be practically certain, according to rule (11.1) that if the experiment is performed just with the one particular sample S, the residual mass of radon will not lie within the interval i_1; it will also be practically certain that the residual mass will not lie within i_2; and so forth. But conjointly these hypotheses, each of which is qualified as practically certain, assert that the residual mass will not lie within the interval from 2 to 3 milligrams; and as was noted earlier, the law stating the half life of radon makes it practically certain that precisely the contradictory of this assertion is true! Thus, the statistical information about the half life of radon statistically explains (or predicts, etc., depending on the context) to a very high degree each of the individual hypotheses referring to the subintervals; but it does not thus explain (or predict, etc.) their conjunction.

Though superficially reminiscent of the ambiguity of statistical systematization, which was examined earlier, this nonadditivity is a logically quite different characteristic of statistical systematization. In reference to statistical systematizations of the simple kind suggested by rule (11.1), ambiguity can be characterized as follows: If the fact that b is G can be sta-

particular performances of F that constitute b some particular one of the various outcomes G_x. Then c(h, e) equals the product of the statistical probabilities of those n outcomes. (For example, if b consists of three performances of F and h asserts that the first and third of these yield G_2, and the third G_4, then c(h, e) $= r_2 \cdot r_4 \cdot r_2$.)

[62] The observation made in the preceding note applies here in an analogous manner.

tistically explained (predicted) by a true explanans stating that b is F and
that $p(G, F) > 1 - \epsilon$, then there is in general another true statement to
the effect that b is F' and that $p(-G, F') > 1 - \epsilon$, which in the same
sense statistically explains (predicts) that b is non-G. This ambiguity can
be prevented by requiring that a statistical systematization, to be scientifically acceptable, must satisfy the principle of total evidence; for one and
the same body of evidence cannot highly confirm both 'Gb' and '—Gb'.

But the principle of total evidence does not affect at all the nonconjunctiveness of statistical systematization, which lies precisely in the fact
that one and the same set of inductive "premises" (one and the same
body of evidence) e may confirm to within $1 - \epsilon$ each of n alternative
"conclusions" (hypotheses), while confirming with equal strength also
the negation of their conjunction. This fact is rooted in the general multiplication theorem of elementary probability theory, which implies that
the probability of the conjunction of two items (characteristics or statements, according as statistical or logical probabilities are concerned) is, in
general, less than the probability of either of the items taken alone. Hence
once the connection between "premises" and "conclusion" in a statistical
systematization is viewed as probabilistic in character, nonconjunctiveness appears as inevitable, and as one of the fundamental characteristics
that distinguish statistical systematization from its deductive-nomological
counterpart.

14. Concluding Remarks.

Commenting on the changes that the notion of causality has undergone as a result of the transition from deterministic to statistical forms of
physical theory, R. von Mises holds that "people will gradually come to
be satisfied by causal statements of this kind: It is *because* the die was
loaded that the 'six' shows more frequently (but we do not know what
the next number will be); or, *Because* the vacuum was heightened and
the voltage increased, the radiation became more intense (but we do not
know the precise number of scintillations that will occur in the next min-
ute)." [63] This passage clearly refers to statistical explanation in the sense
considered in the present essay; it sets forth what might be called a statistical-probabilistic concept of "because," in contradistinction to a strictly deterministic one, which would correspond to deductive-nomological
explanation. Each of the two concepts refers to a certain kind of subsump-

[63] Mises [34], p. 188; italics in original text.

ion under laws—statistical in one case, strictly universal in the other; but, as has been argued in the second part of this study, they differ in a number of fundamental logical characteristics: The deterministic "because" is deductive in character, the statistical one is inductive; the deterministic "because" is an either-or relation, the statistical one permits degrees; the deterministic "because" is unambiguous, while the statistical one exhibits an ambiguity which calls for relativization with respect to the total evidence available; and finally, the deterministic "because" is conjunctive whereas the statistical one is not.

The establishment of these fundamental logical differences is at best just a small contribution toward a general analytic theory of statistical modes of explanation and prediction. The fuller development of such a theory raises a variety of other important issues, some of which have been touched upon in these pages; it is hoped that those issues will be further clarified by future investigations.

REFERENCES

1. Alexander, H. Gavin. "General Statements as Rules of Inference?" in *Minnesota Studies in the Philosophy of Science*, Vol. II, H. Feigl, M. Scriven, and G. Maxwell, eds. Minneapolis: University of Minnesota Press, 1958. Pp. 309–329.
2. Barker, S. F. *Induction and Hypothesis*. Ithaca: Cornell University Press, 1957.
3. Braithwaite, R. B. *Scientific Explanation*. Cambridge: Cambridge University Press, 1953.
4. Campbell, Norman. *What Is Science?* New York: Dover Press, 1952.
5. Carnap, R. *The Logical Syntax of Language*. New York: Harcourt, Brace, and Co., 1937.
6. Carnap, R. "On Inductive Logic," *Philosophy of Science*, 12:72–97 (1945).
7. Carnap, R. *Logical Foundations of Probability*. Chicago: University of Chicago Press, 1950.
8. Carnap, R. *The Continuum of Inductive Methods*. Chicago: University of Chicago Press, 1952.
9. Carnap, R., and Y. Bar-Hillel. *An Outline of a Theory of Semantic Information*. Massachusetts Institute of Technology, Research Laboratory of Electronics. Technical Report No. 247. 1952.
10. Cohen, M. R., and E. Nagel. *An Introduction to Logic and Scientific Method*. New York: Harcourt, Brace, and Co., 1934.
11. Cramér, H. *Mathematical Methods of Statistics*. Princeton: Princeton University Press, 1946.
12. De Finetti, Bruno. "Recent Suggestions for the Reconciliations of Theories of Probability," in *Proceedings of the Second Berkeley Symposium on Mathematical Statistics and Probability*, J. Neyman, ed. Berkeley: University of California Press, 1951. Pp. 217–226.
13. Dray, W. *Laws and Explanation in History*. London: Oxford University Press, 1957.
14. Duhem, Pierre. *La Théorie physique, son objet et sa structure*. Paris: Chevalier et Rivière, 1906.

Carl G. Hempel

15. Feigl, H., and May Brodbeck, eds. *Readings in the Philosophy of Science.* New York: Appleton-Century-Crofts, 1953.
16. Feigl, H., and W. Sellars, eds. *Readings in Philosophical Analysis.* New York Appleton-Century-Crofts, 1949.
17. Feigl, H., M. Scriven, and G. Maxwell, eds. *Minnesota Studies in the Philosoph of Science,* Vol. II. Minneapolis: University of Minnesota Press, 1958.
18. Galilei, Galileo. *Dialogues Concerning Two New Sciences.* Transl. by H. Crev and A. de Salvio. Evanston, Ill.: Northwestern University, 1946.
19. Gardiner, Patrick, ed. *Theories of History.* Glencoe, Ill.: Free Press, 1959.
20. Goodman, Nelson. "The Problem of Counterfactual Conditionals," *Journal o Philosophy,* 44:113–128 (1947). Reprinted, with minor changes, as the firs chapter of Goodman [20].
21. Goodman, Nelson. *Fact, Fiction, and Forecast.* Cambridge, Mass.: Harvard Uni versity Press, 1955.
22. Goodman, Nelson. "Recent Developments in the Theory of Simplicity," *Philoso phy and Phenomenological Research,* 19:429–446 (1959).
23. Hanson, N. R. "On the Symmetry between Explanation and Prediction," *Philo sophical Review,* 68:349–358 (1959).
24. Hempel, C. G. "The Function of General Laws in History," *Journal of Philoso phy,* 39:35–48 (1942). Reprinted in Feigl and Sellars [16], and in Jarrett an McMurrin [27].
25. Hempel, C. G. "The Theoretician's Dilemma," in *Minnesota Studies in the Phi losophy of Science,* Vol. II, H. Feigl, M. Scriven, and G. Maxwell, eds. Minne apolis: University of Minnesota Press, 1958. Pp. 37–98.
26. Hempel, C. G., and P. Oppenheim. "Studies in the Logic of Explanation," *Phi losophy of Science,* 15:135–175 (1948). Secs. 1–7 of this article are reprinted in Feigl and Brodbeck [15].
27. Jarrett, J. L., and S. M. McMurrin, eds. *Contemporary Philosophy.* New York Henry Holt, 1954.
28. Jeffrey, R. C. "Valuation and Acceptance of Scientific Hypotheses," *Philosophy of Science,* 23:237–246 (1956).
29. Kemeny, J. G., and P. Oppenheim. "Degree of Factual Support," *Philosophy of Science,* 19:307–324 (1952).
30. Körner, S., ed. *Observation and Interpretation.* Proceedings of the Ninth Sym posium of the Colston Research Society. New York: Academic Press Inc., 1957 London: Butterworth, 1957.
31. Kolmogoroff, A. *Grundbegriffe der Wahrscheinlichkeitrechnung.* Berlin: Springer 1933.
32. Lewis, C. I. *An Analysis of Knowledge and Valuation.* La Salle, Ill.: Open Court Publishing Co., 1946.
33. Luce, R. Duncan, and Howard Raiffa. *Games and Decisions.* New York: Wiley, 1957.
34. Mises, Richard von. *Positivism. A Study in Human Understanding.* Cambridge, Mass.: Harvard University Press, 1951.
35. Nagel, E. *Logic without Metaphysics.* Glencoe, Ill.: The Free Press, 1956.
36. Neumann, John von, and Oskar Morgenstern. *Theory of Games and Economic Behavior.* Princeton: Princeton University Press, 2d ed., 1947.
37. Neyman, J. *First Course in Probability and Statistics.* New York: Henry Holt, 1950.
38. Popper, K. R. *Logik der Forschung.* Vienna: Springer, 1935.
39. Popper, K. R. "The Propensity Interpretation of the Calculus of Probability, and the Quantum Theory," in *Observation and Interpretation,* S. Körner, ed. Proceedings of the Ninth Symposium of the Colston Research Society. New York: Academic Press Inc., 1957. London: Butterworth, 1957. Pp. 65–70.

40. Popper, K. R. The Logic of Scientific Discovery. London: Hutchinson, 1959.
41. Reichenbach, H. The Theory of Probability. Berkeley and Los Angeles: University of California Press, 1949.
42. Rescher, N. "On Prediction and Explanation," British Journal for the Philosophy of Science, 8:281–290 (1958).
43. Rescher, N. "A Theory of Evidence," Philosophy of Science, 25:83–94 (1958).
44. Ryle, G. The Concept of Mind. London: Hutchinson, 1949.
45. Savage, L. J. The Foundations of Statistics. New York: Wiley, 1954.
46. Scheffler, I. "Prospects of a Modest Empiricism," Review of Metaphysics, 10: 383–400, 602–625 (1957).
47. Scheffler, I. "Explanation, Prediction, and Abstraction," British Journal for the Philosophy of Science, 7:293–309 (1957).
48. Schlick, M. "Die Kausalität in der gegenwärtigen Physik," Die Naturwissenschaften, 19:145–162 (1931).
49. Scriven, M. "Definitions, Explanations, and Theories," in Minnesota Studies in the Philosophy of Science, Vol. II, H. Feigl, M. Scriven, and G. Maxwell, eds. Minneapolis: University of Minnesota Press, 1958. Pp. 99–195.
50. Scriven, M. "Explanations, Predictions, and Laws," in this volume of Minnesota Studies in the Philosophy of Science, pp. 170–230.
51. Sellars, W. "Inference and Meaning," Mind, 62:313–338 (1953).
52. Sellars, W. "Conterfactuals, Dispositions, and the Causal Modalities," in Minnesota Studies in the Philosophy of Science, Vol. II, H. Feigl, M. Scriven, and G. Maxwell, eds. Minneapolis: University of Minnesota Press, 1958. Pp. 225–308.
53. Toulmin, S. The Philosophy of Science. London: Hutchinson, 1953.
54. Wald, A. Statistical Decision Functions. New York: Wiley, 1950.
55. Williams, D. C. The Ground of Induction. Cambridge, Mass.: Harvard University Press, 1947.

Explanations, Predictions, and Laws

1. Preface

There are two distinct kinds of problem encountered by the logician in studying the physical sciences. In certain areas of theoretical physics the current problems for the physicist are of an essentially logical kind, so that the logician's technical skills are immediately relevant. The contribution he makes, if successful, is to the solution of the exact difficulty the physicist faces. To do this he requires considerable ability in both fields; and when successful he earns a double accolade. There is another kind of problem, however, of a less spectacular but no less interesting kind. Rather than selecting the logical difficulties out of the physicist's collection, one may consider those difficulties in the logician's bundle of problems which have some relevance to physics. The discussion of these with detailed reference to physics is often viewed with suspicion by the physicist; but it may prove enlightening to the philosopher, and the path between fundamental physics and such logical problems has proved startlingly short.

Examples of the first kind are the relativity of simultaneity, the idea of identity for fundamental particles, the existence of subquantum particles, the reversibility of temporal processes, the measurement of the dimensions of the universe, the existence of temperatures below absolute zero, the conventionality of the (one-way) velocity of light, etc. These are problems concerned with the substance of physics, and I shall refer to them as problems of applied logic in the physical sciences or problems in the substantial logic of the physical sciences. If this has an honorific connotation, it is well deserved. It would require a form of theory-be-damned

NOTE: The present paper has benefited greatly from helpful criticisms of my fellow contributors, and my students and colleagues at Indiana, and is based in part on work done with support from the National Science Foundation and the Hill Foundation; to them I am most grateful.

igotry virtually unknown among physicists today to deny the importance f these issues. A more common position is to deny that these are problems of "an essentially logical kind." By this phrase I mean to exclude issues which are entirely "decided by the facts," or by calculation, or which ould be so decided, and by "the facts" I mean undisputed (not indisputable) observations. I believe these definitions make the issues mentioned inexceptionable examples of logical issues; put positively, they are issues where the very criteria for interpreting observational evidence and the ery meaning of the fundamental concepts are in dispute. Such disputes annot be settled entirely within the subject of physics as it is usually conceived; they are necessarily in the province of logic itself, where modes of easoning and the presuppositions of concepts are themselves the object f study. And to the extent that a physicist is concerned with such issues, he is playing logician, just as in other cases he must perforce play mathematician or writer or recording instrument. He is no less a theoretical physicist for this; a theoretical physicist is one who combines these roles n treating a certain range of problem. Some of these problems are essentially logical, others essentially experimental or mathematical; and some are mixtures of all these. But as I have previously said, even the ones that are 'essentially' of one of these kinds usually require considerable knowledge of physics itself for their solution. A student of comparative linguistics may have an essentially logical problem on his hands but it does not follow that a logician can solve it without knowing the languages under consideration, for translating the problem into the appropriate logical terms is very often the really difficult step, and requires understanding both the terms of the original problem and those of logic.

The second kind of problem is exemplified by discussions of the frequency (and other) analyses of probability, the testability of indeterminism, the role of simplicity in evaluating theories, the status of unobservable entities, the definitional element in fundamental laws, the notions of 'significant results" and "crucial experiments," the relation between explanation and prediction, the connection between dispositional properties, states, and laws, the utility of idealization, etc. These are problems about the structure of physical theories, language, and experiments, and I shall refer to them as those problems of *pure logic with a bearing on the physical sciences*, or as *problems in the structural logic of the physical sciences*. This volume contains papers dealing with topics of both kinds, examples of the first and second kind respectively being those by Professor

Michael Scriven

Grünbaum on geometry and Professor Putnam on analyticity. This paper represents an attempt to handle some of the closely connected problems of the second kind. It is in no sense a comprehensive study, being chiefly concerned with explanations and predictions and to a lesser extent with causes, laws, determinism, and probability. The general approach can be very crudely indicated as claiming that problems of structural logic can only be solved by reference to concepts previously condemned by many logicians as "psychological, not logical," e.g., understanding, belief, and judgment.

2. Outline

I shall pay special attention to an account of scientific explanation and prediction given by Hempel and Braithwaite, and attempt to improve on it. In so doing I shall be led to comment on the notion of physical law (since this is said to be involved in all explanations), and this will introduce the problematic notion of universally law-governed behavior, i.e., determinism. Comments on reductionism and causation follow from these considerations. A second necessity in studying explanations is some reference to the notion of probability (for, in the absence of general law, we may find that statistical laws provide a basis for a weaker notion of "probability explanation"). As the discussion of law brings in determinism, so the discussion of probability will lead me to say something about the nature of physical knowledge and understanding.

It is perhaps worth mentioning that this paper has a twin, devoted to the corresponding problems in the social sciences and, especially, history; and they both stand on the shoulders of the treatment of certain logical problems in "Definitions, Explanations, and Theories," in Volume II of this series.

3. Preliminary Issues

3.1. Explanations: Introduction.

I am going to take a series of suggested analytical claims about the logic of explanation and gradually develop a general idea of what is lacking in them, or too restrictive about them. I shall at each stage try to formulate criteria which will survive the difficulties while retaining the virtues of the current candidate. Eventually I shall try to draw the surviving criteria together into an outline of a new account of both explanation and under-

[1] "Truisms as the Grounds for Historical Explanations" in *Theories of History*, P. L. Gardiner, ed. (Glencoe, Ill.: Free Press, 1959).

standing; but this will not be possible until I have encompassed the whole field of topics envisioned above. The questions with which I begin will seem quite unimportant; but they are in fact more significant than they appear because of the cumulative error in the standard answers to them.

3.2. Explanations as Answers to "Why" Questions.

"To explain the phenomena in the world of our experience, to answer the question 'Why?' rather than only the question 'What?' . . ." With these words, Hempel and Oppenheim begin their monograph on scientific explanation.[2] Braithwaite says, "Any proper answer to a 'Why?' question may be said to be an explanation of a sort. So the different kinds of explanation can best be appreciated by considering the different sorts of answers that are appropriate to the same or to different 'Why?' questions."[3]

This happens to be a non sequitur, but it is the conclusion that I particularly wish to consider. The "answer-to-a-why-question" criterion could have been proposed in the absence of a serious attempt to think of counterexamples. "How can a neutrino be detected, when it has zero mass and zero charge?" is a perfectly good request for a perfectly standard scientific explanation. "What is it about cepheid variables that makes them so useful for the determination of interstellar distances?" Likewise, and the same can be said of suitable Which, Whither, and When questions. Not all, perhaps not even most, of the answers to such questions involve explanations, whereas it is perhaps true that most answers to Why questions are explanations. (But not all, as for example an answer that rebuts a presupposition of the question "Why do you persist in lying?" "I have never lied about this affair.") But explanations are also given when no questions are asked at all, as in the course of a lecture, or in correcting or supporting an assertion. The identifying feature of an explanation cannot, therefore, be the grammatical form of the question which (sometimes) produces it. Indeed, it is fairly clear that one does not teach a foreigner or a child the word "explanation" simply by reference to Why questions; so the authors quoted presumably had some prior (or at least alternative) notion of explanation in mind which enabled them to identify answers to Why

[2] "The Logic of Explanation" in Readings in the Philosophy of Science, H. Feigl and M. Brodbeck, eds. (New York: Appleton-Century-Crofts, 1953), p. 319. This is an abridgment of "Studies in the Logic of Explanation," Philosophical of Science, 15: 135–175 (1948). All page references hereafter are to the later version.

[3] Scientific Explanation (Cambridge: Cambridge University Press, 1953), p. 319.

questions as explanations. Should we not look for the meaning of tha
notion?

It is sometimes replied that our common notion of explanation is ex
cessively vague, and it is therefore quite unrewarding to seek its exac
meaning; far better to concentrate on some substantial concept which
clearly does occur. This is a very good reply and represents a sensible ap
proach, if only it can be shown to be true. This requires showing (a) tha
the ordinary notion *is* excessively vague, and (b) that the 'substantial
alternative occurs often enough to justify any general conclusions about
explanations which are inferred from studying it. I shall be arguing tha
neither of these seemingly innocuous premises can be established, and ir
consequence the analysis suggested by the reconstructionist authors i
fundamentally unsatisfactory.

Explanation is undoubtedly a notion whose analysis must be sought in
the practical foundation of language; but it is too much to hope one can
identify explanations by such a simple linguistic device as the one suggested. Nor will it do to suppose that explanations are such that they are
answers to *potential* Why questions; for then they are also potentially
answers to What-about questions, How-possibly questions, etc. Thus, to
take an example quoted by Hempel and Oppenheim, the question "Why
does a stick half-immersed in water appear bent?" can readily be rephrased
as "What makes a stick (in such circumstances) seem to be bent?" Indeed, such a question as "How can the sun possibly continue to produce
so much energy with a negligible loss of mass?" is only with some difficulty
rephrased as a Why question.[4] In sum, the grammatical indicators of
explanations are complicated and none of them are necessary; some more
illuminating and reliable criteria must be sought.

3.3. Explanations as "More Than" Descriptions.

Another common remark in the literature is that explanations are more
than descriptions. This is put by Hempel and Oppenheim in the following words: ". . . especially, scientific research in its various branches
strives to go beyond a mere description of its subject matter by providing
an explanation of the phenomena it investigates."[5] But if one goes on
to examine their own examples of explanations one finds what seem to

[4] The discussion of How-possibly questions has been initiated and sustained by William Dray. See his *Laws and Explanation in History* (London: Oxford University Press,
1957), especially pp. 164ff.
[5] *Op. cit.*, p. 319.

e simply complex descriptions.[6] Thus they offer an explanation of the
act that when "a mercury thermometer is rapidly immersed in hot water,
there occurs a temporary drop of the mercury column, which is then fol-
owed by a swift rise." And the explanation consists of the following account: "The increase in temperature affects at first only the glass tube
of the thermometer; it expands and thus provides a larger space for the
mercury inside, whose surface therefore drops. As soon as by heat conduction the rise in temperature reaches the mercury, however, the latter
xpands, and as its coefficient of expansion is considerably larger than that
of glass, a rise of the mercury level results."[7]

This is surely intended to be a narrative description of exactly what
happens. The one feature which might suggest a difference from a 'mere
description' is the occurrence of such words as "thus," "however," and
"results." These are *reminiscent* of an argument or demonstration, and I
think partially explain the analysis proposed by Hempel and Oppenheim,
and others. But they are not part of an argument or demonstration here,
simply of an explanation; and they or their equivalents occur in some of
the simplest descriptions. "The curtains knocked over the vase" is a description which includes a causal claim and it could equally well be put,
style aside, as "The curtains brushed against the vase, *thus* knocking it
over" (or ". . . *resulting* in it being knocked over"). The fact that it is
an explanatory account is therefore not in any way a ground for saying it
s not a descriptive account (cf. "historical narrative"). Indeed, if it was
not descriptive of what happens, it could hardly be explanatory. The question we have to answer is how and when certain descriptions count as
explanations. Explaining how fusion processes enable the sun to maintain
its heat output consists exactly in describing these processes and their
products. Explaining therefore sometimes consists simply in giving the
right description. What counts as the right description? Tentatively we
can consider the vague hypothesis that the right description is the one
which fills in a particular gap in the understanding of the person or people
to whom the explanation is directed. That there is a gap in understanding, or a misunderstanding, seems plausible since whatever an explanation
actually does, in order to be called an explanation at all it must be *capable*
of making clear something not previously clear, i.e., of increasing or producing understanding of something. The difference between explaining

[6] For an alternative and acceptable interpretation of their remark, see 3.4 below.
[7] *Op. cit.*, p. 320.

and "merely" informing, like the difference between explaining and de scribing, does not, I shall argue, consist in explaining being something "more than" or even something intrinsically different from informing o describing, but in its being the appropriate piece of informing or describ ing, the appropriateness being a matter of its *relation to a particular con text*. Thus, what would in one context be "a mere description" can in another be "a full explanation." The distinguishing features will be found not in the verbal form of the question or answer, but in the known or in ferred state of understanding and the proposed explanation's relation t it. To these, of course, the form of the question and answer are often im portant clues, though not the only clues. But this is only a rough indica tion of the *direction* of the solution to be proposed in this paper, and i may be that the notion of understanding will present us with substantia difficulties, quite apart from the problem of identifying the criteria fo "closing the gap" in understanding (or rectifying the misunderstanding) However, let me remind the reader that understanding is *not* a subjec tively appraised state any more than knowing is; both are objectively test able and are, in fact, tested in examinations. We may first benefit from examining the relation between explanation and another important scien tific activity.

3.4. Explanations as "Essentially Similar" to Predictions.

The next suggestion to be considered is a much more penetrating one and although it cannot be regarded as satisfactory, the reasons for dis satisfaction are more involved. Quoting from Hempel and Oppenhein once more: ". . . the same formal analysis . . . applies to scientific pre diction as well as to explanation. The difference between the two is of a pragmatic character . . . It may be said, therefore, that an explanation i not fully adequate unless . . . if taken account of in time, [it] could have served as a basis for predicting the phenomenon under consideration."

(3.41) The full treatment of this view will require some points that wil only be made later in the paper; but we can begin with several rather weighty objections. First, there certainly seem to be occasions when we can predict some phenomenon with the greatest success, but cannot pro vide any explanation of it. For example, we may discover that wheneve cows lie down in the open fields by day, it always rains within a few hours We are in an excellent position for prediction, but we could scarcely offe

[8] *Op. cit.*, pp. 322–323.

the earlier event as an explanation of the latter. It appears that explanation requires something "more than" prediction; and my suggestion would be that, whereas an understanding of a phenomenon often enables us to forecast it, the ability to forecast it does not constitute an understanding of a phenomenon.

(3.42) Indeed, the forecast is simply a description of an event (or condition, etc.) given prior to its occurrence and identified as referring to a future time; whereas an explanation will have to do more than merely describe *those features of the thing to be explained that identify it*. (In this sense, it is more than a (particular) description.[9]) At the very least some other features of it must be mentioned, and often some reference is made to previous or (other) concurrent events and/or laws. Since none of this is required of a prediction, it seems rather extraordinary to suppose that the *contents* of a prediction are logically identical to those of an explanation. And our first point showed that the *grounds* for the two are often quite different, in that one can be inferred from a mere correlation and the other not.

Such cases also demonstrate that explaining something is by no means the same as showing it was to be expected, since the latter task can be accomplished without any explanation being given.[10] For our purpose, however, the crucial point is that, however achieved, a prediction is what it is, simply because it is produced in advance of the event it predicts; it is *intrinsically* nothing but a bare description of that event. Whereas an explanation of the event *must* be more than the *identifying* description of it, else to request an explanation of X (where "X" is a description, not a name) is to give an explanation of X. Of course, there is usually a difference of tense, but we could agree to this as a pragmatic difference. However, it is the *least* and not the *only* difference between explanations and predictions.

(3.43) There also seem to be cases where explanation is not in terms of temporally ordered and causally related events, and we are consequently never able to make predictions. These cases are common enough outside the physical sciences, e.g., in explaining the rules of succession in an Egyptian dynasty, or the symbolism of a tribal dance. Within science

---

[9] And this salvages the theme which forms the title of 3.3—but at the expense of making that of 3.4 untenable. I am thus uncertain which interpretation to accept.

[10] See also "Explanation, Prediction, and Abstraction" by Israel Scheffler, *British Journal for the Philosophy of Science*, 7: 293–309 (1957), where several of the points in this section are discussed.

Michael Scriven

there are of course all the cases of explaining a theory or mechanism o proof; these are normally dismissed by supporters of the Hempel and Oppenheim position, on the grounds that they are clearly a different kind of explanation, the explanation of *meaning*, not at all related to *scientific* explanation. While there is no doubt about the difference in procedure between explaining a theory and explaining some phenomenon in terms of the theory, it is not enough to appeal to intuition for support of the claim that they are not 'fundamentally,' i.e., for all logical purposes, the same except for subject matter, much as definition in mathematics might be said to be fundamentally the same as definition in the empirical sci ences except for subject matter. In fact, it seems clear enough that one important element is held in common between the two 'kinds' of ex planation, viz., the provision of understanding. But is there not a great deal of difference between the kinds of understanding provided in the two cases?

Now, *not* understanding a theory may be due to not understanding what its assumptions are, to not understanding the meaning of some of its terms, or to not understanding how the derivations said to be possible from it are to be made. One might suppose this to be quite unlike not understanding why a stick half-immersed in water appears bent.

But instead of asking how we go about explaining a natural phenome non, let us ask how we come to ask for an explanation, i.e., what it is that we think *needs* explanation. It may seem that science is committed to the view that *everything* needs to be explained. Now it is clear that every *thing* cannot be explained *every time* we give an explanation of some par ticular thing (or set of things) which is all we ever do in a given context So we can rephrase our question as, What is it that needs explanation in a given context? It seems clear that it is those things which are not prop erly understood (by whomever the explanation is addressed to). Now, lack of understanding of a natural phenomenon may be due to the ab sence of certain information about the situation, to the presence of false beliefs about it, or to an inability to see the connections between what is understood and what is not understood. These are much the same kinds of difficulty as occur in not understanding a theory, although the informa tion will be in one case about a verbal construction out of our knowledge and in the other about, for example, a mechanical construction out of our raw materials. However important the differences of subject may be, it is not obvious that the notion of understanding or explanation involved is

178

in any important way different; and it *is* quite obvious that no predictions are possible on the basis of an explanation of the meaning of a theory (except, irrelevantly, those which the theory makes possible, if any—and it may be a theory whose advantages lie solely in its unifying powers). Certainly one should feel uneasy about any general claims of common logical structure for explanation and prediction which have to be defended by rejecting clear cases of explanation as "essentially different," without detailed examination. I shall argue that the differences are much less important than the similarities: in effect we are in both cases providing a series of comprehensible statements that have some of a wide range of logical relations to other statements. Lest this seem to be a proof of similarity by simply weakening the definition of 'similar,' I also try to show that the narrower definition is independently unsatisfactory.

(3.44) Again, we often talk of *explaining* laws: indeed half of Hempel and Oppenheim's examples are of this kind. Now, when we offer an explanation of Newton's Law of Cooling (that a body cools at a rate proportional to the difference between its temperature and that of its surroundings), we do so—according to Hempel and Oppenheim—by deriving this law from more general laws.[11] What predictions could be made which would have "the same formal structure" as this kind of explanation? The 'pragmatic' difference between the two as they see it is essentially that explanation occurs after the phenomenon, and prediction before. But in the case of laws, which are presumably believed to hold at all times, what does it mean to talk of predicting the phenomenon? It is surely the case that the truth of Newton's law is *simultaneous* with that of the more general laws from which it is derivable. We cannot speak of being able to predict the inclusion of the class of A's in the class of C's if we already know that A's are B's and B's are C's.

It may seem that this argument can only be countered by saying that a law is a generalization about a number of events and that to 'predict a law' is to predict the outcome of experiments done to determine the pattern of these events. It is true that this is different from predicting an

[11] The reader may be worried by the fact that his law is known to be only an approximation. This is true of almost all 'laws,' but we do give explanations of them, e.g., of Kepler's laws, Snell's law. Now, for any such law, deducing *it* from any premises would simply show the premises to be inaccurate. Hence, explanation cannot require deduction from true premises. We must substitute a weaker requirement; there are several possibilities which appear to retain much of the Hempel analysis (but see 3.5, 4.5, and 6 below).

Michael Scriven

eclipse, where the actual event to which the prediction refers is in the future, not merely the discovery of its nature. However, (i) it is certainly true that some—but not all—events governed by most laws lie entirely in the future, and its truth depends on these and these are predictable in the usual sense. (ii) Certainly, too, we want to say that inferences about the past, which *generate* predictions about what archaeologists or geologists will discover, have exactly the same logical structure as inferences about the future, a fact well brought out by the practice of calling them post-dictions or retrodictions. So, if *explaining a law* consists in explaining the over-all pattern of events, past, present, and future, *predicting a law*, as it seems we might interpret it, could be regarded as compounded out of such predictions and postdictions. This interpretation represents at the very least an *extension of meaning* since we cannot in the usual sense call inferences about the activities of the earth's crust in Jurassic times (which will be covered by geological laws) predictions. Although we could quite properly apply this term to inferences about what will be found by geologists upon searching in certain areas, this is *not* what the law is about (for else we must say that the law asserts something different every time something new is discovered by the geologists, there being that much less for them still to discover). Indeed, we land in a well-known swamp if we make this move; for the same argument makes all historical statements into statements about contemporary evidence and all statements about distant places into statements about local evidence, etc. It is the argument which confuses the *reference of* a statement with the *evidence for* a statement. So explanations of laws only have a correlate among predictions if we *extend* the meaning of the notion of prediction to include postdiction.

Now this extension may not seem very significant until one reminds oneself that the whole significance of the term "prediction" resides in the temporal relation of its utterance to the event it mentions. "Prediction" is a term defining a category of sentences, in the same way as "command," "argument," "description," etc.; it defines descriptive sentences in the future tense made when the tense is appropriate. The sentence uttered or written in making any given prediction can be repeated after the event has occurred as a perfectly good historical description, provided only that the tense of the verbs (or the corresponding construction) is changed, i.e. apart from tense, predictions are not identifiable. So this extension of meaning amounts to an *elimination* of the meaning with which one began. We may agree that one procedure of inferring past events is essentially the same as one procedure of inferring future events, but we cannot possibly conclude that the *results* in the first case are essentially the same as predictions. This is like saying that analytic statements are essentially the same as synthetic statements since both can be inferred syllogistically. If the *only* way of inferring to such different kinds of statement was syllogistic, one *might* be more inclined to call them logically identical. One would still not be very impressed, since their logical character is written on their face, and appealing to their common ancestry cannot prove that all siblings are twins; it cannot eliminate the obvious differences. But it is clear that there is nothing unique about the type of inference suggested here; predictions and postdictions can be obtained from arguments of virtually any logical form and also without any argument at all, as in the case of the expert but inarticulate diagnostician or the precognitive. I conclude that the explanation of laws has no proper counterpart among predictions, since there is no general concept of predicting laws; for (i) if what can be predicted is said to be the *discovery* of a law, this fails because the counterpart to explaining an *event* is predicting *it*, not its *discovery* (which would require laws about discovering laws); (ii) in the only other possible interpretation, a large number of the conclusions inferred are simply not predictions at all; (iii) even ignoring the first two points, nothing is more obvious than the difference in logical structure between the "prediction" "All A's are (or even, will be found to be) C's" and anything that might conceivably be said to be an explanation of it.

(3.45) I think little can be salvaged from the impact of this set of four points (3.41 through 3.44) against the 3.4 thesis, but I wish to indicate another series of difficulties which will help us to develop a constructive alternative position. The first involves a rather lengthy example, but the same example is of some assistance in dealing with the notion of cause as well as those of explanation and prediction. Suppose we are in a position to explain the collapse of a bridge as due to the fatigue of the metal in one of the thrust members. This is not an unusual kind of situation, and it is, of course, one where no prediction was *in fact* made. According to Hempel and Oppenheim, if this is a satisfactory explanation, then, if taken account of in time, it could have formed the basis for a prediction. We can abandon the idea—presumably but incorrectly taken to be equivalent by Hempel and Oppenheim—that it would actually have the same

181

logical structure as the prediction.) Let us examine in a little detail how this could be so.

We begin our search for the explanation with an eyewitness account that locates a particular girder as the first to go. We already know that there is a substantial deterioration in the elastic properties of carbon steel as it ages and is subjected to repeated compressions; we also know that the amount of this deterioration is not predictable with great reliability since it depends on the conditions in the original welding, casting, and annealing processes, the size and frequency of subsequent temperature changes to which the formed metal is subjected, the special stresses to which it may have been subjected, e.g., by lightning discharges which put heavy currents through it, and, of course, irregularities in and perhaps violations of the design load. The only way to deal with these sources of error is to 'over-design,' i.e., to make an allowance for the unpredictables and provide a safety factor on top of that. But the cost of materials and the pressure of competitive tenders puts limits on the size of such safety margins, and every now and then, as in the spectacular case of the Launceston Bridge, where the wind set up resonant vibrations, a failure occurs. In the present case, where internal rather than external circumstances are the significant factor in the failure, we obtain samples of the metal from the girder in question and discover that its elastic properties have substantially deteriorated. But as we do not have any exact data about the load at the time of failure, we cannot immediately prove that such a load would definitely have produced failure.

Now we go over the rest of the bridge carefully, searching for other possible causes of failure and find none. The bridge is of standard design, sited on good bedrock, and well built. We do have good reason to suppose that the load-causing failure was no greater than the bridge had withstood on many previous occasions, though greater than the static load (assume standard traffic and moderate wind); so we are forced to look for the cause in the structural changes. In the light of all this information, we can have great confidence in our explanation of the failure as due to fatigue in the particular beam; but we simply do not have the data required for a prediction that the failure would take place on a certain date.

It is perfectly true that if we also had exact data about the load when the failure occurred, could obtain some exact and reliable elastic coefficients from the fatigued sample, were in no doubt about our theory, and found on calculation with the revised elastic coefficients that the load exceeded

182

he residual strength, we could be *even more confident* of our explana-
ion. But I have described a much more realistic situation in which we
an still have a very high level of rational confidence in our explanation,
ι level which places it beyond reasonable doubt.

Now, in both cases—with and without the exact details—we can make
ome kind of a *conditional* prediction—that the bridge will fail *if* the load
goes over a certain point, for obviously we can give some load which exceeds any known bridge's capacity. It must be noticed first that such a prediction has no practical interest at all except in so far as we can predict
he occurrence of such loads. It is a conditional prediction not a categorial prediction, and if the only kind of prediction which is associated with
explanations is conditional prediction, especially if they are of this "upper
imit" kind, this is of very little interest indeed for scientists or engineers
vho cannot predict when the conditions are met, or who know they are
very rarely met.

These considerations make us realize that the crucial element in the
duality thesis' about explanations and predictions is the existence of a
specific correlative prediction for each explanation. Naturally, we can
make a number of conditional predictions as soon as, or indeed *before*,
we have any data about the material and form of the bridge; but these
are independent of the particular circumstances of the failure. The 'dual-
ty' claim presumably implies that to every different good explanation
here corresponds a different prediction relating to precisely those circumstances to which the explanation applies. But it is easy enough to see that
we can attain all reasonable certainty about an explanation with less evidence than is required to justify even a conditional prediction with the
ame *specific* reference.

In the bridge example, we have so far been much too profligate of our
investigator's time. In fact, he knows very well that the only causes of
failure other than a load in excess of anything for which the structure was
originally designed are metal fatigue or external damage by, for example,
corrosion, abrasion, or explosion. It is easy to check for the symptoms
of external damage, it is relatively easy to judge that the load was not beyond the design limits. Consequently, he can almost certainly identify
the cause of failure immediately as fatigue. In suitable circumstances, i.e.,
with suitable evidence for the above statements, it is only a formality to
go through with testing samples of the structural steel. Suppose, however,
as a final check, we do a rough computation of Young's modulus for ma-

Michael Scriven

terial in the beam and find it has substantially decreased and by muc
more than for the other beams; but we take no exact measurements of i
and none at all of the other elastic coefficients (which normally vary i
the same direction as Young's modulus). Our hypothesized explanatio
has been very strongly confirmed. It is now beyond reasonable doubt. No
what conditional prediction can we make? It seems there is only the e
tremely weak one that if sufficient substantial fatiguing takes place, an
a somewhat higher load than normal is imposed, failure will take place.
Not only does such a 'prediction' correspond to an indefinitely large num
ber of explanations and hence fail to meet the uniqueness condition pr
viously mentioned, but it is couched in such vague terms as to be almos
wholly uninformative. Yet, I wish to maintain that such a prediction i
all that can be said to be correlated with some very well-established e
planations.

Let us examine the most natural counterargument. This would consis
in saying that no such explanation could ever be regarded as certain i
view of its lack of precise support. Imagine an attorney for the steel con
pany attacking this explanation in court. "How can you be sure that th
metal fatigue was enough to produce a failure? You made no calculatio
and no measurements from which you could in any way infer that a bridg
built of that same steel, in the same condition you found it, would fa
even under twice the load impressed on it that windy night. Hence, I sut
mit that no evidence for blaming the steel has been produced, and henc
no evidence that the steel was at fault."

The weakness of this argument as far as our considerations are cor
cerned is twofold. First, there is not the least difference between direc
and indirect evidence for establishing a conclusion beyond reasonab
doubt; indirect evidence is often more reliable and the distinction betwee
the two is largely arbitrary. If only A, B, or C can cause X, and A and I
are ruled out, it is unnecessary to show C is present; in this case, howeve
it was also shown that C was present, and the only debate is over whethe
sufficient C was present. The reply to such doubts is simply "What els
then?" A redoubtable prima-facie case has been made, and if not rebutte
it must be accepted.

Second, there are some grounds for doubting the significance of an
'direct' test, which do not apply to the indirect evidence. Suppose we tak

---

[12] Hempel, replying to this point in the present volume, suggests another, more sp
cific candidate. I comment on this later.

all the exact measurements we can and make all the calculations we can, and they indicate that a bridge made of the metal tested would *not* collapse under any stress that seems likely to have occurred in the circumstances at the time of failure. Here we have *two conflicting* indications. Far from it being the case that the 'direct' test affords the crucial test, we find it substantially less reliable than the other evidence. First the sample of metal tested is not known to be identical with that which failed: we take a sample adjacent to the *fracture*, but it is a very difficult matter to determine where the fracture *begins*, i.e., where the *failure* occurs. It is quite certain that different spots on the same girder—and along the same fracture—are under very different stresses and hence at very different stages of fatigue. Second, the steps involved in going from the data on materials to conclusions about bridge strength involve a vast number of assumptions of various kinds, few of them more than approximations whose errors may in sum be fatal to the argument. For example, it is *possible* that exceptional conditions did prevail in local areas around the bridge structure, producing strains such as would not normally be associated with a moderate storm—a typical example is provided by the random development of wind resonance, which can build up a considerable, though not precisely known, extra force from a mild breeze that happens to be blowing in the right direction at the right velocity. Hence our 'direct' calculations by no means settle the matter; and the recent examples of wing failure in the Electra airliner show that fatigue can be identified as the cause of failure even when exact theory is wholly inapplicable. The moral of this example is that explanations can be supported by assertions about qualitative *necessary* conditions whereas even a conditional prediction requires quantitative *sufficient* conditions. (This point would of course be completely lost if one proceeds on the common assumption that causes are simply sufficient conditions: see 4.9.)

We have thus discovered that the 'direct test' of the indirectly supported hypothesis is by no means immune to rejection. But the general issues about confirmation are not important here; it is the existence of *some* cases where we can have every confidence in an explanation and yet be in no position to make a prediction, even an applicable conditional prediction. This counterexample to the 'duality' view is the analogue of the counterexamples already mentioned where we are in an excellent position to make a prediction but cannot produce an explanation. A simple and somewhat rough way of putting the point of the last example would

185

be to say that a prediction has to say *when* something will happen, or *what* will (sometime) happen, a causal explanation only *what made it* happen. The first requires either attaching a time or range of times (unconditional prediction), or a value of some other variable (conditional prediction), to the description of an event, whereas the second often requires only giving a cause, i.e., picking out (not estimating the size of) a variable, or another event.

Naturally there are some cases where more than these minimum requirements are available. Sometimes the nature of the problem is such that when the explanation is certain the prediction *is* possible: the Farnborough research into the fuselage fatigue of the De Havilland Comet airliners is a case in point. This was possible only because they had excellent data on the circumstances of the failure (from service records plus recovered instruments). The first type of case we have described is of central importance in the social sciences, because most of our knowledge of human behavior can be expressed only in necessary condition propositions or judgment propositions. Hence it enables us to explain but not predict with equal accuracy. We can confidently explain the migration of the Okies to California in terms of the drought in Oklahoma, though we could not have predicted it with any reliability. For we know (i) that there must have been a reason for migrating and (ii) that drought produces economic conditions which can provide such a reason, and (iii) that nothing else with such effects was present. But we do not know how much of a drought is required to produce a migration and hence could not have predicted this with any confidence. Hempel mistakenly regards this as grounds for doubting the explanation.[13] We must insist on making a distinction between a *dubious* explanation and one for which further confirmation—in the technical sense—is still possible: every empirical claim has the latter property.

(3.46) To summarize, in part. The idea that a causal explanation can only be justified by direct test of the conditions from which a prediction could be made is a root notion in the Hempel and Oppenheim treatment of explanation, and they try to give it a precise formulation. It is said that an explanation must have the form of a deduction from (a) causal laws (L_v) connecting certain antecedent conditions (C_v) to certain consequent conditions (E_v), plus (b) assertions that the conditions C_v ob-

[13] "The Function of General Laws in History," *Journal of Philosophy*, 39:35–48 (1942).

tained in the case under consideration, where we are trying to explain X, which is the sum of the conditions Ev. In the bridge example, we would have to show (by appeal to connections involving Lv) that material with the properties of the sample taken $(C_1, C_2, \ldots, C_{n-1})$ under the ambient conditions of the failure $(C_n, C_{n+1}, \ldots, C_{m-1})$ would lead to the behavior described, i.e., X, the collapse of the bridge (the bridge's design and state prior to failure being described in terms of $C_m, C_{m+1}, \ldots, C_p$). I have been arguing that an indirect approach may be just as effective, i.e., one showing fatigue (C_1) to be a necessary condition for X under the circumstances $(C_2, \ldots, C_p)$. This would involve appealing to a proposition of the form "If X occurs, then either A_1, B_1, or C_1 caused it" and showing that $C_2, \ldots, C_p$ rule out A_1 and B_1. The main trouble with such laws for the thesis of Hempel and Oppenheim is that they do not permit any predictions of X, since the occurrence of X is required for their application. Nor can such laws be reformulated for predictive use, for they are quite different from "If A_1 or B_1 or C_1 occur, they will produce X," not just because one states necessary and the other sufficient conditions, but because the first does not and the second does require quantitative formulation if it is to be true—for it is obviously false that any degree of fatigue produces failure. These laws incidentally demonstrate that the duality thesis about explanation and prediction was actually a separate, fifth condition and not a consequence of the four conditions R_1–R_4.

(3.47) In concluding this discussion of the prediction criterion for explanations, I think it is worth mentioning some points which are neither wholly independent of those discussed above nor, it seems to me, quite so strong. First, it is a consequence of Hempel and Oppenheim's analysis that whatever we explain must be a true statement, since they explicitly require all statements in an explanation to be true. Now it is certainly not the case that all the predictions we make must be true: we often err in predicting the behavior of the stock market, the weather, and the ponies. This point is thought by Scheffler to show that explanations and predictions are different in this respect; but I take it to be mainly a difficulty with Hempel and Oppenheim's analysis of explanation. For one can talk of explaining things that do not happen, just as one can talk of the consequences of things that do not happen. "If you hadn't got here on time, I know who would have been responsible," the irate parent says to the almost-wayward daughter; "If the fourth stage had failed to (ever fails to)

187

fire, you may be sure it would have been (will be) because of a valve failure in the fuel-line," the missile technician may say. This use is *derivative*, i.e., it can be explained by reference to the commonest use; and in the commonest use I think we can agree that to say something is an explanation of X is to presuppose (in Strawson's sense) that X occurred. But this is not to say that in *all proper uses*, this can be inferred.

Apart from the case just cited, where it is known that X did not occur, known even by the giver of the explanation, and apart from explanations of events in fiction, there are other cases where this condition does not hold. In the modified phlogiston theory of about 1785, the explanation of the limited phlogistication of air when calcination occurs in a closed vessel was in terms of the finite capacity of a finite volume of air for absorbing phlogiston. The very phenomenon here explained does not exist, although even Cavendish thought it did. The explanation given is within the theory, of something described in *theoretical* terms. This is not to be confused with the case where we quite commonly put single quotation marks around the term "explanation," meaning that the term is not properly applicable, as when Conant says "an 'explanation' of metallurgy was at hand: Metallic ore (an oxide) + Phlogiston from charcoal → Metal." [14] For here we are referring to an incorrect explanation of something we know *does* occur, viz., smelting, and we know this *not* to be the correct explanation: compare the previous cases discussed where we know the phenomenon *does not* occur. I am therefore unwilling to agree that all proper uses of the term "explanation" presuppose that the phenomenon explained occurs; though I would agree that in the primary use this is so.[15] I shall say something about the necessity for the truth of the body of the explanation itself in the next section.

In the primary use of explanation, then, we know something when we are called on for an explanation that we do not know when called on for a prediction, viz., that the event referred to has occurred. This is sometimes a priceless item of information since it may demonstrate the existence or absence of a hitherto unknown strength of a certain power. Thus, to take a simpler example than the bridge case, a man in charge of an

[14] *Harvard Case Histories in Experimental Science*, Vol. I (Cambridge, Mass.: Harvard University Press, 1958), pp. 70, 110.

[15] There are of course a number of terms besides "explanation" (e.g., description (see 3.6), observation, insight) that are used in such a way that the description "incorrect (explanation)" can be synonymous with "not an (explanation) at all." The points just made do not, however, depend on this ambiguity.

open-hearth furnace may be suspiciously watching a roil on the surface of the liquid steel, wondering if it is a sign of a "boil" (an occasionally serious destructive reaction) on the furnace lining down below or just due to some normal oxidizing of the additives in the mixture. Suddenly, a catastrophe: the whole charge drops through the furnace lining into the basement. It is now absolutely clear that there was a boil which has eaten through the lining: apart from sabotage (easily disproved by examination) there's no other possibility. But no prediction is possible to the event, using the data then available. This renders almost empty Hempel and Oppenheim's (and even Scheffler's) conclusion that explanations provide a basis for predictions. For "Had we known what was going to happen, we could have predicted it" is a vacuous claim. One might mutter something about "If the furnace was in exactly the same state *again* we could predict it would dump," but I have already pointed out that this is a virtually empty remark since we usually can't identify "exactly the same state"; it is simply a dubious determinist slogan, not even a genuine conditional prediction. Since it is technically entirely impossible to rebuild the furnace to the point where it is identical down to the temperature distribution in the mixture (a crucial factor) and the shape of the irregularities in the floor (also crucial), even if we knew these specifications, it will be pure chance if the conditions ever recur and when they do they won't be identifiable. Thus our grounds for thinking the determinist's slogan to be true—if we do—are entirely indirect, and the explanation certainly does not rest on subsumption under the slogan since we cannot even tell when the latter applies, whereas we can be sure the explanation is correct.

The problem of direct vs. indirect confirmation which arises here is of great importance throughout structural logic. To say that "same cause, same effect" is a determinist's slogan is not to say it has *no* empirical content. It has, and it is actually false, as far as present evidence goes, though only to a small extent for macroscopic observations. (It is also not equivalent to the idea of determinism as universal law-governed behavior.) What it lacks is single-case applicability and hence direct confirmability when complex systems are involved—for it is often impossible to specify what counts as the 'same conditions.' It may still be felt on general grounds that unless we *do* know what counts as the 'same conditions' in a given case, we cannot be sure of the proposed causal explanation in that case. The opposite thesis will be defended in a later section of this paper, and to prepare for it I shall need to make several further distinctions and points. At this stage, however, let me summarize by saying that any prediction specifically associated with an explanation is (i) often conditional, and (ii) either so general as to be almost empty or so specific as to refer to no other case, and (iii) often not assertible until it is known the event occurred, i.e., not a true prediction.

3.5. Explanations as Sets of True Statements.

It is not possible to claim that explanations can only be offered for events that actually occur or have occurred. They can be given for events in the future (Scheffler), for events in fiction, for events known not to occur, and for events wrongly believed to occur—and also for some laws, states, and relationships which are timeless (see above). Assuming Hempel and Oppenheim's analysis to be in other respects correct, it follows that in such cases some of the propositions comprising the explanation itself cannot be true, contrary to one of their explicit conditions.[16] The reason they give for this condition is a very plausible one, however, and it is of interest to see if a more general account can be given which will contain allowance for their point. They say: ". . . it might seem more appropriate to stipulate that the (explanation) has to be highly confirmed by all the relevant evidence rather than that it should be true. This stipulation, however, leads to awkward consequences. Suppose that a certain phenomenon was explained at an earlier stage of science by means of an (explanation) which was well supported by the evidence then at hand, but which had been highly disconfirmed by more recent empirical findings. In such a case, we would have to say that originally the explanatory account was a correct explanation, but that it ceased to be one later, when unfavorable evidence was discovered. This does not appear to accord with sound common usage, which directs us to say that . . . the account in question was not—and had never been—a correct explanation."[17]

It is roughly on these grounds that Conant puts the term "explanation" in quotes when he is referring to the phlogiston theory's account of calcination. For much the same reason we refer to an astrologer's remarks as an 'explanation' of Henry Ford's successful business career.

But notice we can talk perfectly well about "two competing explanations" of some phenomenon in contemporary physics without feeling it improper to refer to both as explanations although only one can be true.

[16] Op. cit., pp. 321–322.
[17] Op. cit., p. 322.

And there certainly seem to be cases where we want to say, for example, that the Babylonian explanation of the origin of the universe was basically naturalistic, without using inverted commas. The best treatment of these cases, it seems to me, is to regard them as secondary uses which have become fairly standard, the notion of a secondary use being defined in terms of the fact that understanding it depends logically on understanding the primary use. But these are definitely proper uses and the term "explanation" is hence perhaps less a 'success word' or 'achievement word' than, for example, "knowledge" and "perception." We cannot say of two contradictory claims that both are known since this implies both are true. And this suggests a solution to our present problem.

The proper way of avoiding Hempel and Oppenheim's powerful argument is, I think, very simple; the secondary uses of "explanation" are legitimate but there are no such secondary uses of "correct explanation," the term which they substitute halfway through the argument. Remove the qualifying adjective "correct" and you will see that the argument is no longer persuasive. For consistency, this term must be and can be added to the occurrences of "explanation" in the premises. Overwhelming counter-evidence does not necessarily lead us to abandon or even to put quotes around "explanation," but, as the argument rightly says, it does lead us to abandon the application of the term "correct explanation" (or "the explanation" which is often used equivalently). Hence we should regard Hempel and Oppenheim's analysis as an analysis of "correct explanation" rather than of "explanation," or "an explanation," and this is surely what they were most interested in. "Explanations," or "an explanation," or "his explanation," or "a possible explanation," do not always have to be true (or of the appropriate type, or adequate); they only need high confirmation, at some stage.

Doesn't the notion of confirmation come into the analysis of "correct explanation" at all? It is not part of the analysis, which only involves truth; but it is our only means of access to the truth. We have not got the correct explanation unless it contains only true assertions, but if we want to know which explanation is most likely to meet that condition, we must select the one with the highest degree of confirmation. Good evidence does not guarantee true conclusions but it is the best indicator, so we need no excuse for appealing to degree of confirmation. Moreover, we have no need to adopt the skeptic's position that all possibility of knowing when we have a correct explanation is by now beyond reasonable doubt, and to restrict "knowing" to cases of absolute logical necessity is to mistake the empty glitter of definitional truth for the fallible flame of knowledge. The notion of reasonable doubt is highly dependent on context, but highly unambiguous in a given context, and it sets the threshold level which distinguishes knowledge from likelihood. Anything that is to be called the "correct explanation" of something that is known to have happened must contain only statements from the domain of knowledge.

Now among the things we know are some statements about the probability of certain events under certain circumstances, e.g., about the probability of throwing a six with a die that passes various specifiable tests. Could we not use such propositions as part of an explanation? Hempel and Oppenheim—in the papers cited—countenance the possibility that what they call statistical explanations may be of great importance but they neither undertake to discuss them nor, more significantly, restrict most of their conclusions about explanation in general in the way that would be appropriate if we do take seriously the claims of statistical explanations (which I include in the broader class of explanations based on probability statements). In the present volume, Hempel sets out an account of statistical explanation on which I comment later. Such explanations cannot be subsumed under Hempel and Oppenheim's original analysis as it stands, because no *deduction* of a nonprobability statement from them is possible, and it is hence impossible for them to explain any actual occurrence, since actual occurrences have to be described by nonprobability statements. In particular we could make a (probable) prediction from such 'laws,' but could not—using the same premises—be said to explain the event predicted, if it does come about.

(3.51) It is of some importance to notice that Hempel and Oppenheim's analysis of explanation absolutely presupposes a descriptive language. For them there can be nothing to explain if there is no language, since the thing-to-be-explained is dealt with *only* via the 'explanandum' which is its description in the relevant language. One suspects such a restriction immediately because there are clearly cases where we can explain without language, e.g., when we explain to the mechanic in a Yugoslav garage what has gone wrong with the car. Now this is hardly a scientific explanation, but it seems reasonable to suppose that the scientific explanation represents a refinement on, rather than a totally different kind of entity from, the ordinary explanation. In our terms, it is the *understand-*

ing which is the essential part of an explanation and the *language* which is a useful accessory for the process of communicating the understanding. By completely eliminating consideration of the step from the phenomenon to the description of the phenomenon, Hempel and Oppenheim make it much easier to convince us that deducibility is a criterion of explanation. In fact, within the language there is only one other relation possible, viz., inducibility. We shall argue that good inductive inferribility is the only required relation involved in explanations, deduction being a dispensable and overrestrictive requirement which may of course sometimes be met. But a source of both error and understanding is left out of account in such a debate; for unexplained things are sometimes such that we do not describe them in asking for an explanation and such that they are explained *merely* by being described in the correct way regardless of deduction from laws.[18] (And on the other hand, sometimes a *true* description *and* deduction is not enough.)

3.6. *Explanations as Involving Descriptions of What Is to Be Explained.*

Once we have realized the extraordinary difficulty there is in supposing that explanations and predictions have a common structure, it is natural to ask what the structure of an explanation really is. The 'structure' of a prediction, we noticed, is simply that of a declarative statement using an appropriate future tense and any kind of descriptive language. It may indeed be of the form "C will bring about X" but is more usually of the form "X will occur" or "X, at time t." The structure of an argument, to take a further example, is such as to involve several statements which are put forward as bearing upon each other or upon some other statements in the relation of premises to conclusions. Now what is the structure of an explanation? A bridge's failure may have as its explanation the fatigue of the metal in a particular member or the overload due to a bomb blast. These appear to be a state and an event which could be held to be the cause of the event to be explained. A different account will have to be

---

[18] A common case is that when someone, greatly puzzled, asks What on earth is this? or What's going on here? and is told, for example, that it is an initiation ceremonial on which he has stumbled. Analogous cases in particle physics, engineering, and astronomy are obvious. The point of these examples is that understanding is roughly the perception of relationships and hence may be conveyed by any process which locates the puzzling phenomenon in a system of relations. When we supply a law, we supply part of the system; but a description may enable us to supply a whole framework which we already understand, but of whose *relevance* we had been unaware. We deduce nothing; our understanding comes because we *see* the phenomenon for what it is, and are *in a position* to make other inferences from this realization.

given of the explanation of laws, but for the moment we can profitably concentrate on the explanation of events, to which Hempel and Oppenheim devote a good deal of space, and which has some claim to be epistemologically prior to the explanation of laws.

Now there is a further apparent ambiguity about "explanation": it can either refer to the linguistic structure which describes certain states or events or to the states (or events) themselves. This kind of ambiguity even occurs in connection with such terms as "consequence," "concept," "cause," "inference," and "argument"; it is common throughout logic and best illustrated, perhaps, by the very term "fact." We shall usually be referring to the linguistic entity when we use the term "explanation," but clearly *neither* this entity nor its referents include whatever it is that is to be explained. In the simple but standard examples just given, the explanation, in this sense, is an assertion about a state or event that is entirely different from (assertions about) the state or event to be explained. But Hempel and Oppenheim say, "We divide an explanation into two major constituents . . . the sentence describing the phenomenon to be explained . . . [and] the class of those sentences which are adduced to account for the phenomenon."[19] The former is plainly not a constituent of the explanation at all (except where it is all of the explanation—see 3.5). Only if we find its consequences very confusing or inconvenient should we abandon such a clear distinction as this.

The first difficulty that strikes us about this version of the Hempel and Oppenheim account, then, is that it asserts all explanations of a phenomenon X consist in a deductive argumentlike structure, with a statement about X as the conclusion, whereas our simple examples above are merely statements about something or other that is held to be the *cause* of whatever is to be explained. And are there not occasions, on which one is going over—demonstrating—an explanation, when one does finish off by giving as the last step the description of what is to be explained? It is certainly not a common practice, scientifically or ordinarily, and even when it occurs, it only shows that part of a proof *that* something is an explanation of X may involve a description of X. The *explanation* of the photoelectric effect does not involve the description of the effect—this is presupposed by the explanation. The point may be minor, but it puts us on our guard, for we cannot be sure whether it may not have unfortunate consequences,

[19] *Op. cit.*, p. 321.

analogous to those involved in saying that predictions have the same logical structure as explanations. In fact, we have already seen one error that results from this incautious amalgamation of (i) phenomena, (ii) their description, and (iii) their explanation, in 3.5. We could state part of it by saying that a sixth requirement is actually implicit in their account, viz., the requirement of accuracy and relevance of the description of X, which for them is part of the explanation of X.

In fact, the most serious error of all those I believe to be involved in Hempel and Oppenheim's analysis also springs from the very same innocuous-seeming oversimplification: the requirement of deducibility itself, plausible only if we forget that our concern is fundamentally with a phenomenon, not a statement. It may seem unjust to suggest that Hempel and Oppenheim amalgamate the phenomenon and its description (though certainly they do amalgamate the description and the explanation) when they make clear that the 'conclusion' of the explanation is "a sentence describing the phenomenon (not that phenomenon itself)."[20] The justice of my complaint rests, not on their failure ever to make this distinction, but on their failure to be consistent in dealing with its consequences. For it is a consequence of this distinction that a nondeductive step is involved between the statements in an explanation and the phenomenon explained. And we may then ask why they should suppose deducibility to be the only logical relation in a good explanation. They never address themselves to this question directly, chiefly, I think, because they do not realize the consequences of the distinction they do once make. Attention to it would surely have led them to notice (i) cases of explanatory description (see 3.5), (ii) cases where the completeness or (iii) the uniqueness of the description are crucial in assessing the explanation (see 6.2). Only if we assume that getting as far as the description is getting to the phenomenon, i.e., doing what an explanation is supposed to do, could we overlook such interesting cases. (I think the fact that "description" can be taken to mean "accurate description" also led them to overlook the independent importance of this requirement.)

3.7. The Last Two Conditions and a Summary of Difficulties.

It is stated that the explanation "must contain general laws, and these must actually be required for the derivation . . ." And finally, it is said that the derivation must be deductive, ". . . for otherwise the (explana-

---

[20] Op. cit., p. 321.

tion) would not constitute adequate grounds for (the proposition describing the phenomenon)." [21] We now have a general idea of Hempel and Oppenheim's model of explanation, which I have elsewhere christened, for obvious reasons, 'the deductive model.' [22] I wish to maintain against it the following criticisms in particular, and some others incidentally;

1. It fails to make the crucial logical distinctions between explanations, grounds for explanations, predictions, things to be explained, and the description of these things.
2. It is too restrictive in that it excludes their own examples and almost every ordinary scientific one.
3. It is too inclusive and admits entirely nonexplanatory schema.
4. It requires an account of cause, law, and probability which are basically unsound.
5. It leaves out of account three notions that are in fact essential for an account of scientific explanation: context, judgment, and understanding.

These objections are not wholly independent, and I have already dealt with some of them.

4. Fundamental Issues

4.1. The Distinction between Explanations and the Grounds for Explanations.

It is certainly not the case that our grounds for thinking a plain descriptive statement to be true are part of the statement itself; no one thinks that a more complete analysis of "Gandhi died at an assassin's hand in 1953" would include "I read about Gandhi's death in a somewhat unreliable newspaper" or "I was there at the time and saw it happen, the only time I've been there, and it was my last sabbatical leave so I couldn't be mistaken about the date," etc. Why, then, should one suppose that our grounds for (believing ourselves justified in putting forward) [23] a particular explanation of a bridge collapsing, e.g., the results of our tests on samples of the metal, our knowledge about the behavior of metals, eyewitness accounts, are part of the explanation? They might indeed be produced as part of a *justification* of (the claim that what has been produced is) the explanation. But surely an explanation does not have to contain

[21] *Op. cit.*, p. 321.

[22] "Certain Weaknesses in the Deductive Model of Explanation," paper read at the Midwestern Division of the American Philosophical Association, May 1955.

[23] I shall abbreviate some more precise formulations by omitting the words in parentheses where I think they are not essential.

its own justification any more than a statement about Gandhi's death has to contain the evidence on which it is based. Yet, the deductive model of explanation requires that an explanation include what are often nothing but the grounds for the explanation.

Not only linguistic impropriety but absolute impossibility is involved in the attempt to market the joint package as the "whole explanation" or "complete explanation." The linguistic impropriety is twofold: first, perfectly proper explanations would be rejected for the quite unjust reason that they did not contain the grounds on which they were asserted; second, the indefinite number of possible grounds for an explanation makes absurd the idea of a single correct explanation since there is, in terms of the model, nothing more or less correct about any one of the wide range of possible sets of deductively adequate true grounds. And clearly these are circumstances in which we do identify a particular account as "The correct explanation." The impossibility derives from the second impropriety. There is no sense in which one could ever provide a complete justification of an explanation, out of context; for a justification is a defense against some specific doubt or complaint, and there is an indefinite number of possible doubts.

The deductive model apparently provides an answer to the latter objection in an interesting way. It prescribes that the only kind of justification required is deduction from general laws and specific antecedent conditions. Once this is given, a complete explanation has been given; until this has been done, only (at best) an 'explanation sketch' has been given.

When we say that a perfectly good explanation of one event, e.g., a bridge collapsing, may be no more than an assertion about another event, e.g., a bomb exploding, might it not plausibly be said that this can only be an explanation if some laws are assumed to be true, which connect the two events? After all, the one is an explanation of the other, not because it came before it, but because it caused it. In which case, a full statement of the explanation would make explicit these essential, presupposed laws.

The major weakness in this argument is the last sentence; we can put the difficulty again by saying that, if completeness requires not merely the existence but the quoting of all necessary grounds, there are no complete explanations at all. For just as the statement about the bomb couldn't be an explanation of the bridge collapsing unless there was some connection between the two events, it couldn't be an explanation unless it was true. So, if we must include a statement of the relevant laws to justify our be-

197

Michael Scriven

lief in the connection, i.e., in the soundness of the explanation, then we must include a statement of the relevant data to justify our belief in the claim that a bomb burst, on which the soundness of the explanation also depends."[24]

Certainly in putting forward one event as an explanation of another in the usual cause-seeking contexts, we are committed to the view that the first event caused the second, and we are also committed to the view that the first took place. Of course, we may be wrong about either view and then we are wrong in thinking we have given the explanation. But it is a mistake to suppose this error can be eliminated by quoting further evidence (whether laws or data); it is merely that the error may be then located in a more precise way—as due to a mistaken belief in such and such a datum or law. The function of deduction is only to shift the grounds for doubt, though doubts sometimes get tired and give up after a certain amount of this treatment.

Perhaps the most important reason that Hempel and Oppenheim have for insisting on the inclusion of laws in the explanation is what I take to be their belief (at the time of writing the paper in question) that only if one had such laws in mind could one have any rational grounds for putting forward one's explanation. This is simply false as can be seen immediately by considering an example of a simple physical explanation of which we can be quite certain. If you reach for a cigarette and in doing so knock over an ink bottle which then spills onto the floor, you are in an excellent position to explain to your wife how that stain appeared on the carpet, i.e., why the carpet is stained (if you cannot clean it off fast enough). You knocked the ink bottle over. This is the explanation of the state of affairs in question, and there is no nonsense about it being in doubt because you cannot quote the laws that are involved, Newton's and all the others; in fact, it appears one cannot here quote any unambiguous true general statements, such as would meet the requirements of the deductive model.

The fact you cannot quote them does not show they are not somehow *involved*, but the catch lies in the term "involved." Some kind of connection must hold, and if we say this *means* that laws are involved, then of

[24] Their model requires the truth of the asserted explanation, but it doesn't require the inclusion of evidence for this. Instead of similarly requiring a causal connection, it actually requires the inclusion of one *special kind* of evidence for this. If it treated both requirements equitably, the model would be either trivial (causal explanations must be true and causally relevant) or deviously arbitrary (. . . must include *deductively* adequate grounds for the truth of any assertions and for the causal connection).

course the point is won. The suggestion is debatable, but even if true, it does not follow that we will be able to state a law that guarantees the connection. The explanation requires that there be a connection, but not any particular one—just one of a wide range of alternatives. Certainly it would not be the explanation if the world was governed by *antigravity*. But then it would not be the explanation if you *had not* knocked over the ink bottle—and you have just as good reasons for believing that you did knock it over as you have for believing that knocking it over led to (caused) the stain. Having reasons for causal claims thus does not always mean being able to quote laws. We shall return to this example later. For the moment, it is useful mainly to indicate that (i) there is a reply to the claim that one cannot have good reasons for a causal ascription unless one can quote intersubjectively verifiable general statements and (ii) there is an important similarity between the way in which the production of an appropriate law supports the claim that one event explains another, and the way in which the production of further data (plus laws) to confirm the claim that the prior event occurred supports the same claim. They are defenses against two entirely different kinds of error or doubt, indeed, but they are also both support for the same kind of claim, viz., the claim that one event (state, etc.) explains another.

This is perhaps obscured by the fact that when we make an assertion our claim is in full view, so to speak, whereas when we put forward an assertion as an explanation, its further role is entirely derived from the context, e.g., that it is produced in answer to a request for an explanation, and so its further obligations seem to require explicit statement. This is a superficial view. All that we actually identify in the linguistic entity of a 'declarative statement' is the subject, predicate, tense, etc. We have no reason at all, apart from the context of its utterance, for supposing it to be *asserted*, rather than proposed for consideration, pronounced for a grammatical exercise, mouthed by an actor, produced as an absurdity, etc.[25] *That* it is asserted to be true we infer from the context just as we infer that it is proffered as the explanation of something else; and for both these tasks it may need support. We may concede that assertion is the primary role of indicative sentences without weakening this point.

It is in fact the case that considerations of context, seen to be necessary even at the level of identifying assertions and explanations themselves,

[25] See Max Black's "Definition, Presupposition, and Assertion," in his *Problems of Analysis* (London: Routledge and Kegan Paul, 1954).

not only open up another dimension of error for an explanation, that of pragmatic inappropriateness, but simultaneously offer a possible way of identifying *the* explanation of something, where this notion is applicable

A particular context—such as a discussion between organic chemists working on the same problem—may make one of many deductively acceptable explanations of a biochemical phenomenon entirely inappropriate, and make another of exactly the right type. (Of course, I also wish to reject the criteria of the deductive model; but even if one accepted it, the consideration of context turns out to be *also* necessary. So its importance is not only apparent in dealing with alternative analyses.)

We may generalize our observations in the following terms. An explanation is sometimes said to be incorrect or incomplete or improper. I suggest we pin down these somewhat general terms along with their slightly more specific siblings as follows. If an explanation explicitly contains false propositions, we can call it *incorrect* or *inaccurate*. If it fails to explain what it is supposed to explain because it cannot be 'brought to bear' on it, e.g., because no causal connection exists between the phenomenon as so far specified and its alleged effect, we can call it *incomplete* or *inadequate*. If it is satisfactory in the previous respects but is clearly not the explanation required in the given context, either because of its difficulty or its field of reference, we can call it *irrelevant*, *improper*, or *inappropriate*.

Corresponding to these possible failings there are types of defense which may be relevant. Against the charge of inaccuracy, we produce what I shall call *truth-justifying grounds*. Against the charge of inadequacy, we produce *role-justifying grounds*, and against the complaint of inappropriateness, we invoke *type-justifying grounds*. To put forward an explanation is to commit oneself on truth, role, and type, though it is certainly not to have explicitly considered grounds of these kinds in advance, any more than to speak English in England implies language-type consideration for a life-long but polylingual resident Englishman.

The mere production of, for example, truth-justifying grounds does not guarantee their acceptance, of course. They may be questioned, and they may be defended further by appeal to further evidence; we defend our claim that a bomb damaged a bridge by producing witnesses or even photographs taken at the time; and we may defend the accuracy of the latter by producing the unretouched negatives and so on. The second line of defense involves *second-level grounds*, and they may be of the same

EXPLANATIONS, PREDICTIONS, AND LAWS

three kinds. That they can be of these kinds is partly fortuitous (since they are not explanations of anything) and due to the fact that the relation of being-evidence-for is in certain ways logically similar to being-an-explanation-of. In each case, truth, role, and type may be in doubt; in fact, this coincidence of logical character is extremely important. We notice, however, that there is no similarity of any importance between these two and being-a-prediction-of, where truth is not relevant in the same way, role is wholly determined by time of utterance and syntax, and only the type can be—in some sense—challenged.

4.2. Completeness in Explanations.

The possibility of indefinitely challenging the successive grounds of an explanation has suggested to some people—not Hempel and Oppenheim—that a complete explanation cannot be given within science. Such people are adopting another use of "complete"—even less satisfactory than Hempel and Oppenheim's—according to which the idea of a complete explanation becomes not only foreign to science but in fact either wholly empty, essentially teleological, or capable of completion by appeal to a self-caused cause. Interesting though this move is in certain respects, it essentially requires saying that we can better understand something in the world by ultimately ascribing its existence and nature to the activities of a mysterious entity whose existence and nature cannot be explained in the same way, than by relating it to its proximate causes or arguing that the world has existed indefinitely. I shall only add that we are supposed to be studying scientific explanations, and if none of them are complete in this sense, we may as well drop this sense while making a note of the point—which is equivalent to the point that the causal relation is irreflexive and hence rather unexciting—for there is an important and standard use of "complete" which does apply to some suggested scientific explanation, and not to others, and is well worth analyzing.

Now, if some scientific explanations are complete—and think how a question in a physics exam may ask for a complete explanation of, for example, the effects noticed by Hertz in his experiments to determine whether electromagnetic waves existed—it cannot be because there is a last step in the process of challenging grounds, for there is no stage at which a request for further proof could not make sense. But *in any given context* such requests eventually become absurd, because in any scientific context certain kinds of data are taken as beyond question, and there is

Michael Scriven

no meaning to the notions of explanation and justification which is not directly or indirectly, dependent on a context. This situation is of a very familiar kind of logic. It makes perfectly good sense to ask for the spatial location of any physical object; and perfectly proper and complete answers will involve a reference to the location of some other physical objects. Naturally we can, and often do, go on to ask the further question concerning where these other objects are ("Where's Carleton College?" "In Northfield." "But where's Northfield?"). And no question in a series of this kind is meaningless, unless one includes the question "Where is the universe, i.e., everything?" as of this kind. If one does include this question (which is the analogue of "Where did the universe come from?"), then the impossibility of answering it only shows something about the notion of position, and nothing about the incompleteness of our knowledge. If one excludes this question, the absence of a last stage in such a series does not show our inability to give anyone complete directions to the public library but only that the notion of completeness of such descriptions involves context criteria.

Any request for directions logically presupposes that some directions can be understood; if no directions can be understood, then the proper request is for an account of the notions of position and directions. A *complete* answer has been given when the particular object has been comprehensibly related to the directions that are understood. Similarly, then, the request for an explanation presupposes that *something* is understood, and a complete answer is one that relates the object of inquiry to the realm of understanding in some comprehensible and appropriate way. What this way is varies from subject matter to subject matter just as what makes something better than something else varies from the field of automobiles to solutions of chess problems; but the *logical function* of explanation, as of evaluation, is the same in each field. And what counts as complete will vary from context to context within a field; but the logical category of complete explanation can still be characterized in the perfectly general way just given, i.e., the logical function of "complete," as applied to "explanations," can be described. Hence the notion of the proper context for giving or requesting an explanation, which presupposes the existence of a certain level of knowledge and understanding on the part of the audience or inquirer, *automatically entails* the possibility of a complete explanation being given. And it indicates exactly what can be meant by the phrase "*the* (complete) explanation." For levels of undertanding and interest define areas of lack of understanding and interest, and the required explanation is the one which relates to these areas and not to those other areas related to the subject of the explanation but perfectly well understood or of no interest (these would be explanations which could be correct and adequate but inappropriate). It is worth mentioning that the same analogy with spatial location (or evaluation) provides a resolution of the 'problem of induction,' as a limit case of a request for a 'complete justification.'

It is also clear that calling an explanation into question is not the same as—though it includes—rejecting it as not itself explained. Type justifying involves more than showing relevance of subject matter, i.e., topical and ontological relevance; it involves showing the appropriateness of the intellectual and logical level of the content; a proposed explanation may be inappropriate because it involves the wrong kind of true statements from the right field, e.g., trivial generalizations of the kind of event to be explained, such that they fulfill the deductive model's requirements but succeed only in generalizing the puzzlement. One cannot explain why this bridge failed in this storm by appealing to a law that all bridges of this design in such sites fail in storms of this strength (there having been only two such cases, but there being independent evidence for the law, not quoted). This might have the desirable effect of making the maintenance boss feel responsible, but it surely does not explain why this bridge (or any of the other bridges of the same design) fails in such storms. It may be because of excessive transverse wind pressure, because of the waves affecting the foundations or lower members, because of resonance, etc.

So mere deduction from true general statements is again seen to be less than a sufficient condition for explanation; but what interests us here is that our grounds for *rejecting* such an explanation are not suspicions about its *truth* or its *adequacy*, which are the usual grounds for doubting an explanation, but only its failure to *explain*. Certainly it fails to explain if incorrect or inadequate, but then one feels it fails in a genuine attempt, that the slip is then between the cup and the lip; whereas irrelevance of type is a slip between the hand and the cup—the question of it being a *sound* explanation never even arises. One may react to this situation by declaring with Hempel and Oppenheim that the only *logical* criteria for an explanation are correctness and adequacy, the matter of type being psychological; or, as I think preferable, by saying that the concept of explanation is logically dependent on the concept of understanding, just

as the concept of discovery is logically dependent on the concept o
knowledge-at-a-particular-time. One cannot discover what one alread
knows, nor what one never knows; nor can one explain what everyon
or no one understands. These are tautologies of logical analysis (I hope
and hardly grounds for saying that we are confusing logic with psychology

Having distinguished the types of difficulty an explanation may en
counter, one can more easily see there is no reason for insisting that it i
complete only if it is armed against them in advance, since (i) to displa
in advance one's armor against all possible objections is impossible and
(ii) the value of such a requirement is adequately retained by requiring
that scientific explanations be such that scientifically sound defenses of
the several kinds indicated be available for them though not necessaril
embodied in them. Since there is no special reason for thinking that true
first-level role-justifying assumptions are any more necessary for the ex
planation than any others, it seems quite arbitrary to require that the
should be included in a complete explanation; and it is quite independ
ently an error to suppose they must take the form of laws.

4.3. The Elements of an Alternative Analysis.

Hempel and Oppenheim make clear on two occasions that they are
genuinely concerned with an analysis of the concept scientists and rational
men normally refer to as "explanation," not with the development of
some vaguely related and prejudicially labeled 'explication' of it; for ex
ample, they say, in rejecting a possible condition, "This does not appear
to accord with sound common usage . . ."[26] We are coming a little
closer to seeing that their proposal for explanations of events represents
an analysis of *one kind of deduction of a singular proposition*; it is neither
explanation nor the only kind of justification of explanation that they are
describing. Our alternative analysis of the *particular kind of scientific explanation* they discuss is also becoming clearer. A causal explanation of
an event of state X, in circumstances C, is *exemplified* by a set of statements S, such that (a) S asserts, at least, the existence of the phenomenon Y; (b) the existence of Y is comprehensible; (c) Y is the cause of X;
(d) it is understood that Y's can cause X's in C; (e) S is understandable;
(f) S is true; (g) S is of the correct type for the context. Without getting
carried away and suggesting these are necessary or even independent conditions (which they are not), there is perhaps some value in setting out

[26] *Op. cit.*, p. 322.

ne set of sufficient conditions, to demonstrate the importance of the
rucial concept of understanding.

.4. The Bearing of Understanding and Context on Completeness.

Who is to say whether S and Y are understood? The primary case of
xplanation is the case of explaining X *to someone;* if there were no cases
f this kind, there could be no such thing as "an explanation of X" in the
bstract, whereas the reverse is not true. For it makes no sense to talk of
n explanation which nobody understands now, or has understood, or
vill, i.e., which is not an explanation for someone. In the primary case,
he level of understanding is that of the person addressed. The notion of
'an (or the) explanation of X" in the abstract makes sense just insofar as
t makes sense to suppose a standardized context. When we talk of "the explanation of sunspots," we rely on the fact that to get to the point where
ne understands the term "sunspot," but does not understand what pro-
luces them, is to acquire a fairly definite minimum body of knowledge.
The explanation' will be the one appropriate for a person with this knowledge, and this secondary sense of explanation is quite strong enough to
survive certain types of failure of comprehension, i.e., a number of cases
where 'the explanation of X' is produced without the least understanding by those who hear it. In particular, the failure of two-year-old children
to understand the explanation of sunspots is no ground for supposing
the explanation to be unsatisfactory; it is 'the explanation,' not because
it is understandable to everyone, but because it is understandable to the
group that meets the conditions (a) and (b) just mentioned. It is important to remember how limited a use the notion of *the explanation* has;
there is nothing that naturally comes to mind as *the explanation* of the
nature of light (too general for any single account to be preferable at all
points for any substantial body of people) or of enzyme activity (of interest to several fields, in each of which an explanation can be given). When
more than one explanation is available, certain relations between them
can be inferred which we shall discuss later; but that there is more than
one is due to the fact that there is more than one way in which a phenomenon may be hard for someone to understand.

The analysis of 4.3 does not exclude giving causal laws which connect
X and Y, neither does it require it. There are other occasions, not covered
by 4.3, when the explanation of an event simply consists in giving some
causal laws. Here the role-justifying grounds may be assertions about the

occurrence of events. And there are yet other occasions when an explanation is given in physics which does not consist in the production of cause
or of causal laws. These include cases in which the explanation consist
simply in demonstrating *how* the known laws lead to the unexplained
effect; but there are other cases, including those in which the explanation
consists simply in *denying* what was believed to be a law or fact, those
in which the explanation consists in identification, and of course those
already mentioned in which a theory is explained.

Finally, it is perhaps clearer from the specimen conditions given in 4.
than in the deductive model's conditions that there may be a difference
between *doubting an explanation* and *not understanding the phenomena*
to which it refers (both being quite different from not understanding
the statements in the explanation, or the derivations involved, etc.). The
usual doubts of an explanation apply to (c) and (f); whereas, understanding is directly involved in (b) and indirectly in (d) (in the sense mentioned). This clarifies the difference between the infinite regress of requests for further explanation (which is the one often said by theists to
limit scientific explanations and comprehension of the world based on
them) and the infinite regress of requests for further justification of *this*
explanation, i.e., of the claims made in the explanation itself (the phenomenalists' path). The deductive model insists on the inclusion of something that is simply one of the first steps in the justification regress. As
far as I can see there is no good reason either for including it or, if it is
included, for refusing to insist on others. What is *supposed* to be a reason
for rejecting further steps is commonly produced, but would actually apply only if this were the comprehension regress. It is this: once we have
subsumed the event in question under a general law, we have explained
it, and any request for an explanation *of that law* is a different question
from the one with which we began. Indeed, it is said, we *can* go on and
answer this further question, but we have no obligation to do so in order
to be said to have explained the matter first raised. This move is quite
similar to the correct defense against the comprehension regress, but it
is not satisfactory as it stands, and anyway does not apply to the justification regress. It is the doubts which generate the latter which are responsible for the inclusion of a first-level role-justifying ground (assuming that
we are dealing with the explanation of one event by reference to another—some reformulating is necessary for other cases). For it is the doubt
"How can you be sure that Y is connected with X (granted that it hap-

ens)?" which is used to justify including the causal law, not the doubt
How does any of this explain X (granted that Y happens and Y's cause
X's)?" Hence the usual deductivist answer to the theist's complaint that
the deductive model provides incomplete explanations is not only too
strong, because the objection is often valid, but also leaves them open to
similar pressures to include justification of truth or type, or the various second-level grounds. Let us set out the proper answers with some
care.

The proper answer to the complaint of 'incompleteness' is twofold.
With respect to complaints which are based on failure to understand
part of what has been put forward, we reply in one way; with respect to
complaints based on doubts about the contents of the explanation, we reply in another. In terms of the distinctions introduced earlier, the former
involves type-justifying grounds, the latter truth- and particularly role-
justifying grounds. The first complaint sometimes reflects a fault of the inquirer but by no means always so: sometimes the "explanation" produced
is based on a misapprehension of the type of difficulty involved and this
may be the fault of the explainer. It then does not explain the matter to
his person at this time; if it can nonetheless be called an explanation,
this is because for someone else or for other circumstances (to which it
would be related by contextual propriety), it would explain the matter.
What cannot be dogmatically said is that we have explained the matter,
on this occasion, merely because we have gone through the motions appropriate on some other occasion. The decision will depend on what
legitimate inferences were possible from the context. For example, it
requires the most profound searching of the imagination to discover
any occasion on which the 'mere generalization' type of 'explanation'
("Bridges always do that") could qualify. There are certainly occasions
on which the error lies with the inquirer, because he has asked a question
inappropriate to his actual problem, suggesting either too much, too little,
or the wrong kind of knowledge or interest on his part. In such cases we
can rightly claim to have produced the correct explanation or an adequate
reply to his request for an explanation (though from charity we may
undertake to go ahead and deal with his real worry).

Hempel and Oppenheim's[27] first mistake, then, lies in the supposition
that by subsumption under a generalization one has automatically ex-

[27] I attribute this to the joint authors on the basis of exegetical discussion by Hempel
alone; the point is not made in their article.

plained *something*, and that queries about this 'explanation' represent
request for *further* and *different* explanation. Sometimes these queries
merely echo the original puzzlement, and it is wholly illicit to argue that
the original matter has been explained. It is as if I asked the way to the
town hall from the post office and you replied, "It's in the same relation to
the post office as town halls always are"; and upon my indicating dissati
faction you were to say, "Ah! What you *really* wanted to know (what you
now want to know) is the geographic relation of all town halls to their
post offices." This was not at all what I wanted to know, though in the
light of the facts an answer to this will provide me with the answer I seek.
You have produced an interesting regularity related to my question, but
you have not, in this case, adequately done what I asked you to do. "Ex
plaining the way" is so called just because it is not successfully accom
plished until understanding is conveyed; and there are various ways to do
this, none of them identifiable on purely syntactical grounds. It depends
on what I know; and only when there are strong contextual reasons for
supposing that I know the general relation of post offices to town halls
could you continue to sustain any kind of claim to have explained the way
to me "though it wasn't understood."

In the second place, the above debate concerns the comprehension re
gress, not the justificatory regress; and hence Hempel and Oppenheim's
defense in terms of the "That's a further question" routine, which re
quires modification before it can be accepted even in the former case, is
not an answer to the series of doubts about justification. In connection
with *those* doubts, we are accusing them of inconsistency. The inconsist
ency, once more, is between (i) insisting that laws connecting X and Y
be included in a complete explanation and (ii) denying the necessity of
including evidence for the truth of these or other assertions in it, or the
type of explanation given. Each of these is necessary to the same extent
(i.e., for the explanation to be a good explanation), yet they legislate for
the inclusion of only one of them in the explanation.

*4.5. The Relation between Deduction from True Premises and Sound
Inference.*

By requiring deduction from true premises, Hempel and Oppenheim
impose a pair of conditions which exclude their own examples and most
scientific explanations of events. The reasons for this are, I think, clear and
interesting. Deduction looks as if it is the only watertight connection be

ween the premises and the conclusion, and it appears obvious that a watertight connection must be insisted on. But the matter is much more complicated, and eventually we must abandon even the idea that they are proposing a useful *ideal* for explanation.

I want to begin with the explanation of events and only then go on to the explanation of laws. Hempel and Oppenheim never deal with the explanation of a particular event instance, but only with events of a certain kind; naturally success in this task will enable one to explain particular events of this kind, but it is worth remembering that a degree of generality is already present which is absent in the case of many scientific explanations, e.g., of the formation of the earth, the emergence of Homo sapiens, and the extinction of the dinosaurs. This degree of generality in their examples is one of the factors which lends what I take to be spurious plausibility to their explanation-prediction correlation. For an explanation of events of *kind* X is necessarily couched timelessly and hence may more reasonably be thought applicable prospectively as well as retrospectively.

Turning to the example with which they begin, and which they give in more detail than any other, we read, "A mercury thermometer is rapidly immersed in hot water; there occurs a temporary drop of the mercury column, which is then followed by a swift rise. How is this phenomenon to be explained? The increase in temperature affects at first only the glass tube of the thermometer; it expands and thus provides a larger space for the mercury inside, whose surface therefore drops. As soon as by heat conduction the rise in temperature reaches the mercury, however, the latter expands, and as its coefficient of expansion is considerably larger than that of glass, a rise of the mercury level results."[28] What can be said about this example in the terms of their analysis?

In the first place, it undoubtedly does represent a certain kind of scientific explanation. Not only is what is given, to the best of my knowledge, an explanation of the phenomenon described, but it could often perfectly well be called *the* explanation of it, since the appropriate contextual conditions for that description are usually met. Hence, an analysis of it will have some claim to be an analysis of scientific explanation. However, it is equally indubitable that it contains false statements, some of them being the alleged laws, and that the statements given do not entail the desired consequence. Their example therefore violates all three of their own

[28] *Op. cit.*, p. 320.

209

Michael Scriven

criteria (one trivially, since general statements are 'essentially' involve
even if they are not true and hence not laws). This rather extraordinar
situation was not altogether unnoticed by the authors. Referring to th
deduction requirement, they say of the explanation as they give it that "i
adequately and completely formulated," it would "entail the consequenc
that the mercury will first drop, then rise," which implies that the formu
lation given will *not* do this.

I shall not elaborate in detail the various physical errors in this an
their other examples.[29] An example or two will suffice, since I do not tak
these errors to invalidate the explanation, only the deductivist analysis o
it. There are two kinds of error in their explanation. First, they say: "Th
increase in temperature affects at first only the glass tube of the the
mometer; it expands and thus . . ." No physicist would be willing to ac
cept this as literally true. Radiation effects reach the mercury at the spee
of light and are causing it to expand while the slower conduction effect
are expanding the glass; but even these reach the mercury long before th
glass has expanded enough to produce a visible drop in the mercury col
umn. Second, they do not allow for the fact that glass itself is a physicall
unique substance, with highly anisotropic *multiple* coefficients of expan
sion, the relationship between them being dependent upon the minutia
of its chemical composition and details of the annealing process, which
vary even from batch to batch of the closely controlled thermometer glass

Surely these complaints are not serious; a little research or rewording
would take care of them, would it not? They are not serious for the ex
planation: it is correct. But they are serious for the analysis because i
requires literal truth. The only way in which literal truth could be salvage
here would be by invoking an extremely complicated inequality, involv
ing all the coefficients mentioned, the specific heats, the radiation rates
energy distribution factors, upper limits obtained from the heat-transfe
equations, and so on. And with every complication at the theoretica
level—as we try to employ laws that are exactly correct—there is a corre
sponding complication at the experimental end, since new measurements
are required in order to apply the more complex theory.

The upshot of this is not merely that Hempel and Oppenheim's exam
ple is 'incomplete' according to their own criteria of completeness; it is
that it is not an explanation at all, on their analysis. Now could the situa

[29] See pp. 255–350 of my "Explanations," D. Phil. thesis, Oxford, 1956, micro
filmed by the University of Illinois library.

tion be redeemed by turning to some textbook and using the more complex relationships referred to above? A serious research effort would be required to obtain the material, and it has certainly never been done. That is, there *is* as yet no 'complete' explanation of this simple phenomenon, on their criterion of completeness. All we have are 'explanation sketches,' which they think are characteristic of the *social* sciences, by *contrast* with physics. But in physics, just as in economics, virtually all the explanations given are of this kind, and they are usually just as 'complete' as they should be, i.e., as complete as is necessary to attain the requisite degree of reliability. Hempel and Oppenheim have an oversimple logical model: an explanation that fails to measure up to its standards may be a great deal more complete than one that does, i.e., it may identify the relevant effective variables and ignore the ineffective ones, as does their example in this case. Other processes are going on here, but nobody asked for a complete description of the thermodynamic process; they asked for the explanation of a particular effect. Giving an explanation requires *selecting* from among the variables that are involved those whose activities are unknown to the inquirer and crucial for the phenomenon; it does not involve dragging in or constructing *all* the relations of these variables to the others involved. The explanation won't be an explanation unless the variables selected are crucial. To *show* they are would thus be appropriate only if the explanation were challenged in this respect.

But what *harm* can it do to say that an explanation is complete only when its role-justifying grounds are included? Apart from the answers of the earlier sections, it must now be stressed that there are ways of supporting an explanation that do not involve anything as simple as quoting physical laws of the kind Hempel and Oppenheim have in mind (i.e., justification is not the same as deduction from true premises). In the first place, when we do quote laws, they are usually not literally true. In the second place, we often support the explanation indirectly by what I shall call elimination analysis or—to use a shorter term—*detection*. The important difference between the social sciences and the physical sciences is not that one has and one does not have complete explanations, but that one has more quantitative laws than the other. Naturally, these will often be used in giving and defending explanations. But they confer no benefit on the explanations that cannot be obtained in other ways, and in particular they do not convey the blessing of deduced truth, since they are usually only approximations.

211

Michael Scriven

4.6. The Nature of Physical Laws.

I have discussed this topic in greater detail elsewhere[30] and wish to stress here only the positive side of that treatment. The examples of physical laws with which we are all familiar are distinguished by one feature of particular interest for the traditional analyses—they are virtually all known to be in error. Nor is the error trifling, nor is an amended law available which corrects for all the error. The important feature of laws cannot be their literal truth, since this rarely exists. It is not their closeness to the truth which replaces this, since far better approximations are readily constructed. Their virtue lies in a compound out of the qualities of generality, formal simplicity, approximation to the truth, and theoretical tractability.[31]

These qualities, not truth unvarnished, are extremely important for the plausibility of the Hempel and Oppenheim model, which unfortunately excludes the laws which exemplify them, if carefully examined. Deduction from a 'mere empirical generalization' is very rarely explanatory, and it is only because laws usually involve more than this (as well as less) that they carry explanatory force. (The fact that they commonly reflect some underlying processes, albeit imprecisely, accounts for much of the inductive reliability we ascribe to them, and hence for much of our willingness to allow contrary-to-fact inferences from them.) Were it not for this fact, that what we call laws are usually thought to reflect the "inner workings" of the world, the deductive model would be singularly implausible. This fact manifests itself in another equally helpful way. Laws provide a framework for events which we use as a convenient grid for plotting phenomena that may need explanation. When we are trying to locate events with respect to what we know and understand, we often look to see whether they represent departures from patterns we know and understand, and these patterns are the laws. Their importance lies not in the *precision* with which they trace the characteristics of events or substances but in the fact that they provide a readily identifiable pattern. The event in question either conforms to a known pattern or it does not. If it does not, it (probably) needs explaining; if it does, then either it is not puzzling or the

[30] "The Key Property of Physical Laws—Inaccuracy," in *Current Issues in the Philosophy of Science*, H. Feigl and G. Maxwell, eds. (New York: Holt, Rinehart, and Winston, 1961).

[31] This does not mean I require their actual incorporation into a theory, but only considerable prospective cases of incorporation, something which requires good metatheoretical judgment for identification.

puzzle involves the origin or relation of the patterns. This visual meta-phor survives the discovery that descriptions of actual physical events and scientific laws are usually not related deductively, and it gives a different twist to the role of laws; they often serve as the *starting point* from which we survey the events, looking for the nonconformists, not only as the *rules under which we try to bring them*. Thus they often have a crucial double role in the process of explanation, but not an ultimate role; the only ultimate element in the logic of explanation is understanding itself, and that comes in many ways. Its exact relationship to the perception of regularities I shall briefly discuss later (5).

When we turn to puzzles about the patterns themselves, i.e., to the explanation of laws, we find the irrelevance of inaccuracy further demonstrated. Can we explain why bodies near the earth's surface fall according to the law $s = kt^2$? According to Hempel and Oppenheim, who use this example, this is done "by deducing it from a more comprehensive set of laws, namely Newton's laws of motion and his law of gravitation, together with some statements about particular facts namely the mass and radius of the earth."[32] There would be circumstances under which this would be a satisfactory explanation, though there are others where it would be hopeless (e.g., where the explanation is "Atmospheric density varies less than 10% over the earth's surface, and $s = 16t^2$ when measured from the top of the campanile at Pisa"). But the points to be made here are that (1) the Newtonian laws are known to be in error, and (2) even if true, they would not entail $s = kt^2$, since (a) the actual relationship varies from point to point and height to height, so no such formula is derivable, and (b) the premises quoted are inadequate for the proposed deduction anyway (the earth has no single radius, air resistance is not considered,[33] etc.).

I conclude that, where it is appropriate to invoke Newton's laws for this explanation, the point is not at all that deduction from these as true premises is possible, but that these relations are the crucial factors for this inquiry, in this context, i.e., the only *important* ones (with respect to the degree of accuracy judged appropriate from the context, and the level of knowledge of the inquirer(s)). We know this to be true, and we could

---

[32] *Op. cit.*, p. 321.

[33] Hempel and Oppenheim refer to "Galileo's Law for the free fall of bodies," not $s = kt^2$. In the *Dialogue Concerning Two New Sciences* (Third Day), it is made clear that $s = kt^2$ is intended, and that it is held to apply to motion in air. (Details are in my "Explanations," p. 346.)

even go out and do the experiments to show it is true. It does not follow from the claim that we know it and that we have to be able to show i true by deduction from true premises now in our possession.

4.7. Reduction.

These comments have obvious consequences for the analysis of reduc tion, coinciding to an interesting extent with those of Paul Feyerabend [34] in his essay in this volume. We are in agreement against Nagel, Hempel and Oppenheim about the errors of the 'principle of deducibility.' How ever, on the principle of meaning invariance, his position seems to me overstated; I would certainly agree (and I think Nagel might also) that important changes in the meaning of certain words and sentences come about with the adoption of new scientific theories. It is equally important and this is what I think Nagel is stressing in the passage Feyerabend cites (see p. 33, this volume), that important elements in the meaning of terms, given by its primary 'unreduced' relationship to observation claims, remain unchanged. This is necessary for the claim of reduction to make sense.[35]

Reduction is simply one kind of explanation, namely unilateral inter field explanation, and hence not essentially or even usually deductive. It is a requirement of adequacy for a microtheory that it be capable of "reducing" (explaining) all macrophenomena. In no stronger sense is re duction possible.

4.8. Determinism.

If we have discovered that the concept of the law has feet of clay, they are those on which the thesis of determinism stands. The best formula tion of this seems to me to be that every physical event or state is entirely governed by categorical laws: predictability in principle is neither a neces sary nor a sufficient condition. This is a difficult claim to refute, for well-

---

[34] The interest lies not least in his belief that his thesis somehow controverts those of linguistic analysts. I am not sure exactly who is in this ignorant (p. 85, fn. 102), misguided (p. 60, fn. 70) group, but if it is indeed the unhappy disciples from Oxford (as fn. 102 suggests, but see p. 60, fn. 69), it is clear that we share with him almost everything except the recognition of our similarity. (Smart, Presley, Toulmin, and Waisman would be the obvious sources from which to support this claim, along with more study of Wittgenstein and Hanson, whom he does cite; but I certainly do not wish to deny the use of a straw man to a good friend!)

[35] This point is elaborated in my "Definitions, Explanations, and Theories," in *Minnesota Studies in the Philosophy of Science*, Vol. II, H. Feigl, M. Scriven, G. Maxwell, eds. (Minneapolis: University of Minnesota Press, 1958).

known reasons; but its refutability, in the eyes of most physicists, is demonstrated by the reasoned acceptance of quantum theory as a counterexample. However, the fact that almost all quotable laws in classical physics were known to be inaccurate for almost all values of their variables has long meant that the support for determinism lay not in the possibility of immediately showing its application in any given case by deductive subsumption under known laws, but in the possession of good reasons for thinking that laws could be given or uncovered which would account for any specified feature of any phenomenon to any relevant degree of precision. Notice the essentially promissory character of this claim, identical with the promissory character of explanations on my analysis. The term 'promissory' as I use it here does not necessarily weaken the claim; a promissory note is not necessarily less valuable than cash—it is more valuable in some deflationary phases of the economy—but it is important to learn to distinguish them. One cannot support a precise determinism by producing inexact laws; the backing for it, though of good value, is usually not in the form of cash. And anything less than categorical determinism is a triviality since it is a fact of common experience that events exhibit a considerable degree of regularity, i.e., that "statistical determinism" is true.[36]

4.9. Causes.

An incidental consequence of these remarks is that genuine examples of physically sufficient conditions are almost unknown, a fact of some significance for many analyses of 'cause.' When we identify a cause, we are doing so on the basis of a contextual inference as to what type of cause is sought, with what degree of precision it must be described in order to be completely described, with what is already known about the surrounding circumstances (from which we infer what counts as an abnormal or notable circumstance), etc. Speaking loosely, we could say that a cause is a nonredundant member of a set of conditions jointly sufficient for the effect (i.e., the remaining set if not itself sufficient), the choice between the several candidates that usually meet this requirement being based on considerations of context. However, many qualifications must be made to this analysis, partly in view of the comments in this paper, and in the

[36] What would show it to be false? Presumably total randomness in the order of nature. Note that if determinism is defined in terms of predictability, the thesis of (qualitative) statistical determinism becomes a tautology (because distributions are predictable in random series).

215

Michael Scriven

end probably the best view of a cause is that it is any physical explanation which involves reference to only one state or event (or a few) other than the effect, which is independently variable.

4.10. The Creation and Application of Theories.

As a final illustration of the present thesis, I turn to the problem of interpreting theories. It is clear that one of the great virtues of a theory is its insulation from the effect of observations; but like an optimistic nature, this virtue becomes a vice in excess. Popper is fond of stressing the importance of falsifiability, i.e., pointing out the evils of excessive insulation. I am here stressing, for the moment, the importance of incorrigibility—though never *complete* incorrigibility, hosanna to the prophets.

A theory is usually based on and supported by a very wide range of phenomena: it should therefore not be possible to reject it on the basis of a single observation until the certainty of that observation, described in the usual theory-impregnated way,[37] is higher than that of the theory. So much is clear and can be handled straightforwardly. But the importance of salvaging the useful parts of a theory, when it suffers some disconfirmation, by readjusting its existential and relational claims, demands another dimension of insulation. It is this which distinguishes the usual (perhaps not every) theory from a 'mere set of empirical laws': the use of a vocabulary and grammar different from that of the (local) observation language. Of course, the primary reason for theoretical terms is the necessity for postulating the existence of the entities to which they refer if we are to explain what we see, discover what else is the case, etc.; when we say that a concern for insulation "requires" a special theoretical vocabulary, we are giving the perspective of the metatheoretician rather than the scientist.

Certainly there are no exact logical rules for inferring the right theoretical axioms from observations, but it has been held that the converse is true. The relationship of the axioms to the observations has sometimes been thought to be deductive. But it is clear from any attempt to set out the deduction, using any theories more sophisticated than the gas laws, that this is not true. I wish to make a further point. The natural substitute for deduction is "probabilistic meaning rules (correspondence rules,

---

[37] The theories which affect the way we formulate observations, which in fact partly determine what we observe, are usually but not always different theories from those which the observations are intended to test. Observations are typically dependent on the theories underlying instruments, meters, etc.

bridge laws, etc.)." This involves an equally serious error since it suggests these rules are simply laid down by the theoretical terms. On the contrary: to a large extent, though not wholly, they are empirical claims. I think the theoretical terms get their original meaning mainly from an analogy which inspires the scientific theory and is usually incorporated in its phraseology; to this the theoretician adds precision and sometimes extensions and modifications by specifying some usually quantitative properties of a relationship between the concepts. There is no 'uninterpreted theory' in theory construction. The terms have enough meaning to suggest empirical consequences for the axioms, and it is rare to find *any*, let alone *complete*, specification of interpretation rules as such. The puzzle is to find confirmation of the presence of a neutrino, gene, field, etc., when we know roughly what would count as confirmation, e.g., tracks, bulges on the chromosomes, gradients of activity, but aren't sure when or where or how these should be detectable. One can't deduce observational consequences from theoretical claims, in general; one can only infer them as probable. (The situation is no different in the formal sciences: axioms for probability itself must, if adequate, involve terms such as "random" which can only be applied probabilistically. In this sense, probability is essentially unformalizable.)

Is this kind of probable inference essentially different from the kind whereby we go from the observations to the first formulation of the theory? Remember that observations are not *absolutely* nontheoretical, but only nontheoretical with respect to the level or type of theory we are now considering, and remember that the inferences to new theories are not purely creative but essentially comparative. The question we ask ourselves is What is this phenomenon most *like*? Then it may seem more plausible to suppose that the inference to an explanatory theory from observational statements[38] and the inference from theoretical statements to observational consequences are highly comparable. I find it unenlightening to regard these inferences as dissimilar simply because of the different purposes they serve, and I cannot see any internal logical criteria for distinguishing them. Both are examples of inductive inference, using probabilistic laws or inference licenses, and I do see an important difference between this kind of inference and pure cases of deduction. A thorough

[38] Logicians have objected to the use of the term "inference" in this case, but it will be seen that I have here given reasons for rejecting the logical analyses on which their reasons are based.

Michael Scriven

analysis of the difficulties in identifying enthymemes and inferences from cluster concepts to the presence of their indicators[39] makes the supposed sharp borderline nebulous.

Thus the insulation of theoretical claims from observations consists of (a) the probability barrier erected by the multiple confirmations of the theoretical claim and (b) the doubts whether a particular observational test is really crucial for the theoretical claim, due to both (i) the empirical inferential difficulties ("Does this test really identify laevorotary molecules?") and (ii) the logical inferential difficulties due to the incompleteness ('open texture') of the theoretical claim ("Does a neutrino have to have spin energy?"). These doubts may be rephrased by saying that we can suspect an alleged observational disconfirmation in three ways: error in making the observations, error in interpreting them, and error in interpreting the theory (either mechanical slips or misconstructions).

Again we see the reduced emphasis on deduction, which nevertheless retains an important place *within the theory itself*. Its place is taken by a kind of prima-facie inference, about which I shall say a little more shortly. It is not very illuminating to think of this as simple probability inference, since it is a kind of probability not previously analyzed in any detail. As far as the deductive model of explanation goes, we find in this subsection a further difficulty in the way of providing deductively adequate premises. It is quite uncommon for a physicist to be able to give explicitly the assumptions which connect his theories with observations. He proceeds on a less formal basis, not because he is careless, but because it is usually impossible to do otherwise and highly—though not wholly—reliable to do so.

An analogy from mathematics may be useful. The definition of a limit L for a sequence may be construed as involving a dialectic. Let anyone choose a very small quantity, ϵ, as small as he chooses. If a sequence has the limit L, and only if this is so, it will then be possible, and only then (after the choice of arms, so to speak), for one to calculate a value of n such that all terms of the sequence after the nth will be nearer than ϵ to the fixed value L. It is *not* possible to give a value for n which will ensure that beyond the nth the terms will all be nearer to L than *any* small quantity ϵ, subsequently chosen. Similarly, the physicist's procedures are entirely rigorous, not because he can guarantee his inferences as beyond all

[39] See my paper, "The Logic of Criteria," *Journal of Philosophy*, 61:857–868 (1959).

doubt (which deduction does), but because he can achieve any defensible level of reliability for them that is antecedently given. He can't meet *all possible* doubts in advance, but he undertakes to meet any specific one when advanced. In a particular context it is usually clear what kind of doubts must be met, and hence appropriate to give either the primitive form of, for example, the gas laws in one's explanations or a more refined form; not because either is true, but because the one or the other is entirely adequate with respect to doubts of a certain kind. The inescapable fact is that the physicist cannot, usually, avoid the exercise of informal judgment, about both the kind of doubts present and the extent of the inaccuracy, in the laws he quotes, judgment which he cannot justify by any deduction from a known law.

4.11. Certainty.

It is very difficult to get rid of the feeling that giving up the deductive model involves giving up proper standards of rigor in favor of a kind of sloppy intuition. To counteract this it would be most effective if we could give a demonstration not only that the deductive model is not feasible, but that the alternatives are just as good. Of course, we *would* be better off if laws were simpler and more accurate, and no proof can destroy that fact. But, given the complexity of actual physical relationships, it is possible to prove that explanations which do not require deduction from exact laws may be more certain than those which do require it. Suppose that we are trying to explain the observation and A has the property X, where A is an individual or kind of individual. Perhaps we notice that Jones is not feeling well, or that a blonde sunburns more easily than a brunette, and wish to explain it. Now we discover that certain other events or properties have happened to, or apply to, A, say that A was (or is) P, Q, R . . . Moreover, there is a well-understood connection between some of these and X; perhaps P sometimes—though by no means always—causes X. We have the rudiments of a possible explanation for the phenomenon A is X:

> A is P.
> P sometimes produces X.

However, we cannot perform the deduction required by the deductive model. We cannot even deduce the conclusion that P probably produced X in A, since, for example, there may be a higher likelihood that Q did it.

Suppose now that we look carefully for other connections with X and find that Q, R . . . do not produce X, or very rarely produce X. A has, we are pretty certain, *no* history or constituent that could account for X besides P. In this case, we are normally entitled to conclude with high reliability *that P is the explanation of X in A* (eating green apples made Jones sick, a peculiarity of skin pigmentation accounts for blondes sunburning). We can infer *this*, by elimination, but (i) we usually cannot deduce it, and (ii) more to the point, we certainly cannot deduce that X has to occur to A. Here we find a divergence in the analysis, depending on whether X is an event or an atemporal property. If X is an event, it may be that P produces X only once in a hundred A's. Hence we shall never be able to predict X; rather the contrary—we can infer that it will probably not occur to A. Once we know that it has (and we know that Q, R, etc., produce X in a far fewer number of A's), we are still entitled to complete confidence in our explanation, though antecedent deduction of the effect is impossible. So far, this is simply a stronger version of the bridge kind of example. Naturally we can deduce that X occurs from the premises we use *to support* our explanation because these include the information that X has occurred. This is appropriate enough—one does not try to explain something unless one thinks it has occurred (derivative cases apart). But it annihilates the possibility of prediction; in these situations it is *logically impossible* to predict X from what is known prior to X's occurrence, i.e., infer that it will occur (in fact, one is rationally compelled to predict its nonoccurrence), even though it is possible to explain it when it does occur. (I have added this after reading Hempel's reply to my criticism earlier in this paper, which he made after he saw an early draft of the first part.)

Now, I have granted that it is possible to deduce that X occurs from the whole set of statements from which we infer the explanation of X, since this set includes the statement A is X. However, this fact cannot be taken to confirm the applicability of the deductive model of explanation, not only because the deduction is from the grounds for an explanation, rather than the explanation, but because it is an essentially trivial one in another way. Hempel and Oppenheim face a similar problem in their paper for which they admit they can offer no solution, viz., that any law could be "explained" by deducing it for the conjunction of that same law and any other true proposition. They agree that such a deduction would not be explanatory and clearly the same conclusion applies to our present case. (It is really also applicable to subsumption under a simple generalization.) Hence the kind of deducibility that does obtain is no confirmation of the view that explanation consists of deduction from true, essentially general statements (their definition of "law"). Hence the certainty that deduction confers is not certainty that we have an explanation.

Let us explore, for a moment, the question *when* deduction *is* part of an explanation, and *why*, in order to be more clear about the way in which it contributes to the reliability of an explanation. Hempel and Oppenheim's own example, on which they base their analysis, is the explanation of the law of refraction of light. Classical physics, they say, "answers in terms of the undulatory theory of light, i.e., by stating that the propagation of light is a wave phenomenon of a certain general type and that all wave phenomena of that type satisfy the law of refraction . . . Thus, the explanation of a general regularity consists in subsuming it under another, more comprehensive regularity . . ."[40] This is a strikingly bad example. There may conceivably be circumstances in which their account is explanatory, but it is a real exercise to imagine them. I prefer to take this as an example which shows that in any sense in which explanations "go beyond" descriptions, as Hempel and Oppenheim affirm, they also "go beyond" derivations.

It so happens that their example provided the goal of one of the most fascinating quests in the history of physics; and what was sought, namely "the explanation of refraction," as Edmund T. Whittaker calls it in his *History of the Theories of Aether and Electricity*,[41] was wholly unlike Hempel and Oppenheim's empty derivation. It was a demonstration that Snell's law was a consequence of the elementary nature of light, as conceived in the wave or particle theory. The "nature of light," of course, did *not* include its conformity to Snell's law; and the great Hooke and Pardies were unable to discover the explanation, which was finally given by Huygens, and in an entirely different way by Maupertuis.[42] What makes *this* derivation explanatory and denies that honor to Hempel and Oppenheim's candidate is roughly that it does *not* beg the question, that to understand its premises is by no means to understand that this consequence follows. One may understand what light is and not understand why it obeys Snell's law; the demonstration *explains* why it does (why, being what it is, it *must*). In this case we come very close to an example of explanation wherein deduction *is* the appropriate kind of inference;

[40] *Op. cit.*, pp. 320–321.
[41] Vol. 1 (London: T. Nelson, 1951), p. 24.
[42] For details, see my "Explanations," p. 272.

but we see that, even when we have it, we are by no means guaranteed an explanation as certain. It is the derivation from an already accepted premise which does *not* obviously imply the result that is crucial—not deduction from exact laws. And, of course, very few explanations actually involve a derivation: this is one of them and it is an illuminating one just because here we see a case to contrast with the usual ones where the derivation is not the focus of interest, where in fact it is not part of the explanation at all, but part of one procedure for justifying the explanation, which may or may not be required, and if required is required in response to a request for a justification, not a request for an explanation. In such cases, the main role of deduction, when it does occur, is not to make a particular inference logically necessary but to show that an inference is possible. In such cases, it is therefore not important that deduction be the kind of inference that is involved. Are there other reasons for supposing that deduction *is* crucial?

When we turn to the explanation of particular events rather than laws, we again find clear cases where deduction from laws does not guarantee beyond doubt that we have an explanation, especially in the cases where the laws are mere generalizations of the particular event to some wider class. It is also possible to construct many formal examples which commit the Hempel and Oppenheim analysis to absurd consequences; Nicholas Rescher has invented a series of these.[43] There is no escape, then, from the conclusion that the certainty obtainable via deduction is not explanatory certainty; and I hope this will support the conclusion that the deductive model is not an *ideal*, since ideals for explanations must at least have the property that if only they could be attained they would be explanations!

Turning now to the question of *alternative* routes to reliability and certainty, I illustrated the logical pattern of elimination analysis earlier in this section. This provides us with examples of a particular kind of explanation which does not require deduction or general laws of the usual kind. Its special feature is the combination of potentiality claims (P can cause X) with exhaustiveness claims (no other possible causes of X appear to be present). These are sometimes obtained from necessary-condition causal claims (the only possible causes of X are P, Q, R). Now, there are many circumstances under which these propositions can be estab-

---

[43] In a letter to me, dated September 24, 1958.

lished very firmly (or where we can be confident that we would recognize potential causes if present, even though we cannot list them), when no propositions of the kind that Hempel and Oppenheim have in mind as laws are known. (The potentiality propositions are existential, not universal, in logical form and hence excluded.)

However, there is a more general way in which nondeductive explanations may attain higher certainty than deductive ones. Suppose that we concern ourselves with straightforward laws, i.e., laws not involving the term "cause," but include statistical as well as categorical ones. P may almost always produce X, perhaps. But using this, along with the assertion that P occurs, as a premise, we cannot *deduce* the occurrence of X. (We cannot even deduce that X will probably occur, unless we add the premise that nothing else is known about the likelihood of X.) Now suppose X occurs. It is plausible to suppose that insofar as determinism is true there is some unrecognized set of conditions P′, P″, which were also present, whose presence actually fully guarantees X's occurrence, i.e., P plus P′ plus P″ is a sufficient condition for X. They may, for example, be hereditary predispositions toward a certain disease. But they may be highly inaccessible or unknown. We shall be able to identify the cause of the disease beyond anything that would constitute a reasonable doubt in many contexts, using the statistical proposition alone, because only P and its alternate possible causes Q, R, etc., concern us at all. P might be a virus, and the discovery that a virus is (almost certainly) the cause, rather than a bacterial infection, or a higher type of parasite, is exactly what we need as an explanation. Naturally, there are other contexts in which our interest would require the identification of P′, P″, etc. But the plain fact remains: the statistical law may give us exactly what we need and it may be *much more readily applicable*. For what we lose in the difference between 95 per cent and 100 per cent internally, we gain back because we can identify P with 98 per cent reliability, while the conjunction of P with P′ and P″ will often be identified with 80 per cent reliability, *even assuming* we have a law connecting them 100 per cent with X.

An analogous point bears particularly on the rare cases where we do have true categorical laws in science (conservation of mass energy, Nernst Heat Theorem, etc.). These typically involve highly abstract theoretical concepts, and then what we gain on the swings of internal certainty we often lose on the roundabouts of getting to the actual empirical consequences, because the instruments required are more complex and the chain of inference less direct. I do not deny for a moment the existence of many important cases where the most exact experimental inferences can be made from these principles, e.g., in scattering theory. The term 'mathematical physics,' with its implications of precision, is not a misnomer. I wish only to stress that only the *over-all* probability of an explanation is the matter of ultimate concern to the physicist, and this is a (monotonic increasing) function not only of the probability of inferences internal to the theory (maximized by deduction), and of the internal probability in the general statements (maximized if they are true universal statements), but also of the probability with which the presence and magnitude of the theoretical properties can be inferred from the observations. (This is the probability of the antecedent conditions in the deductive model.) And since exact laws involve more variables, or variables that are more elusive, than statistical ones, the unreliability of our determinations of their value may cost us more than we gain.

To summarize: if the world were much simpler, the deductive model would be in one respect more appropriate. If the world were still more complicated, for example, if the fundamental parameters of physics (such as c, G, e) were time and/or space dependent, then determinism could still be true,[44] but the deductive model would be even less appropriate. The ideal for explanation is explanatory reliability, and for that end it is no good enshrining criteria that (1) guarantee neither the reliability nor the explanatory nature and (2) can hardly ever be utilized in this world. The deductive model is an excellent description of *one* way of *justifying* an assertion, and that is all. It may occasionally happen that a deductive argument will be an explanation, but when this happens, it is usually because what we did not understand was the way in which the derivation could be done and so the requirement for an explanation, even here, is not *deducibility*, but inferability. To require deduction in such cases will reduce one source of error, but typically at the cost of increasing another.

5. The Alternative Analysis

What is a scientific explanation? It is a topically unified communication, the content of which imparts understanding of some scientific phenomenon. And the better it is, the more efficiently and reliably it does this, i.e., with less redundancy and a higher *over-all* probability. What

---

[44] Contrary to the view that it means "If any configuration of the universe recurred, its subsequent history would repeat" (Frank, etc.).

is understanding? Understanding is, roughly, organized knowledge, i.e., knowledge of the relations between various facts and/or laws. These relations are of many kinds—deductive, inductive, analogical, etc. (Understanding is deeper, more thorough, the greater the span of this relational knowledge.) It is for the most part a perfectly objective matter to test understanding, just as it is to test knowledge,[45] and it is absurd to identify it with a subjective feeling, as have some critics of this kind of view. So long as we give examinations to our students, we think we can test understanding objectively. (On the other hand, it is to be hoped and expected that the subjective feeling of understanding is fairly well correlated with real understanding as a result of education.)

Explanation is not "reduction to the familiar," partly because the familiar is often not understood (rainbows, memory, the appeal of music) and partly because we may understand the unfamiliar perfectly well (pure elements, the ideal gas, absolute zero). On the other hand, (i) we do understand much of what is familiar and so much explanation is reduction to the familiar. And (ii) "familiar" can be taken as synonymous with "explained," in which case this slogan comes nearer the truth, though by using the tautology route. Finally, (iii) there is the great truth implicit in this view that at a particular stage in explanations of a certain kind, there is very little more to understanding besides familiarity. We do come to accept facts and relations which we at first viewed as wholly incomprehensible; we come to accept gravity and to reject Newton's vituperative condemnation of the view that it is inherent in matter and can act through a vacuum without mediation. The stage at which we do this is the last stage in an explanation, and the explanations which lead us to this kind of last stage are those which take us back to the most fundamental features of our knowledge.

Although it is an illusion to suppose that scientific explanation is anything as simple as deductive subsumption under a true generalization, there sometimes comes a point in scientific explanation, especially in physics, cosmology, and psychophysiology, when assimilation into some kind of over-all regularity is about all we have to offer. The weakness of the deductivist's thesis, as a general account of scientific explanation, is well demonstrated by the profound unease which affects scientists when

[45] Knowledge stands in relation to understanding rather as explanations do to theories. We know the date of our birthday, but we understand the calendrical system: we know the items and the relations which, combined, make up understanding.

225

they find themselves in this situation. Why does the distribution of matter determine the curvature of space? Why is mass energy conserved? Why are protons or electrons continually appearing throughout the universe (in the steady-state cosmologies)? Why does a particular electron impinge on the screen at *this* point? Why does this neural configuration correspond to pain? Et cetera. What can we do with these questions except answer by showing *how* these variables are related and distributed? We can do no more at the moment, but (i) we know very well what it would be like to have a more comprehensive theory within which these questions could be answered; hence there is always an incentive for and a possibility of going further, though there are no good reasons for supposing there must always be something further.[46] (ii) That we finally reach this stage does not show that at earlier stages nothing more is required than reiterating the puzzling regularities. Understanding requires knowing the relations that exist; it is not confined to cases where one of these relations is subsumption under a *higher* level generalization; it may only involve subsuming (i.e., colligation), not subsumption, but it will be the less for that. (iii) There is an alternative to vertical subsumption in these areas and that is 'horizontal' assimilation, by analogy or formal principle. (iv) Even this would leave at least one fundamental principle to be explained in something more than the colligation sense, and of this all that can be said is that if the world exists at all, it is logically necessary that it have some properties and we shall be very fortunate if we can reduce them all to one. (v) Why the physical world exists, or has *this* fundamental property rather than some other, *may* have an answer in terms of some antecedent circumstances, but this only postpones the final question and that final question is probably illegitimate in just the way that the final question in the series, Where is Oxford? Where is England? . . . is illegitimate, i.e., senseless; unaskable, not unanswerable.

6. Comments on Hempel's Account of Statistical Explanation
6.1. Ideological and Logical Differences.

Hempel still stresses the syntactical criteria of explanations rather than the psychological and contextual criteria as I do. However, the relations

---

[46] There are always good reasons for trying to go further when first we come to a limiting principle of this kind. But in the end we may grant autonomy, and abandon the attempt to reduce gravity to mechanics or magnetism, ESP to radio waves, etc., and say that we understand the phenomena all the same.

which he discusses naturally include some that I take to be required for correct understanding. He here (i) stresses the importance of inferences based on generalizations and (ii) is willing to allow the derivation of approximations to the actual facts as explanatory (though he does not probe deeply into the extent of this concession). I stress the importance of several other kinds of connection, a diversifying feature of my account, but also the unifying feature of the universal dependence on understanding. Remembering that he was always clear about the possibility of statistical explanations, it is worth stressing that in the absence, even in this paper, of criteria for identifying an explanation as putatively deductive, rather than statistical, it is never possible for Hempel to reject it for not meeting the deductive model's standards. Hence, the deductive model is virtually useless in criticizing explanations (just as syllogism theory is virtually useless in criticizing arguments unless one can identify enthymemes), and everything hinges on the statistical model, especially if I am right in arguing that there is no sense in which the deductive model offers us a better ideal, apart from its inapplicability. It should also be noted that allowing approximations to truth and approximations to deduction places a very heavy emphasis on the identification of an *adequate* degree of approximation, which in my view automatically introduces contextual considerations and the need for unformalizable judgment.

A great deal depends, then, on the "statistical model" and it seems to me to face a dilemma. Insofar as it amounts to the view that statistical inferences, when used to justify an explanation, should be valid, it is true but trivial, and hardly a model of explanation. Insofar as it involves the suggestion that explanations for which the grounds include statistical generalizations, in Hempel's sense, should include these in the explanation it is unsound for the reasons given in 3.1 above. This is no mere terminological point about what collection of statements should be *called* an explanation. It is a crucial point about what explanatory claims meet scientific standards of objective reliability and utility: Hempel appears to identify as a 'statistical generalization' virtually any qualified general claim with which we might back up an explanation. But these are often not susceptible to any statistical analysis. Their only role is entirely different, as I have suggested in "Truisms as the Grounds for Historical Explanations." Either Hempel treats their informality and hence their incapacity to fit a precise statistical model as a sign of imperfection and rejects them, or he does not. If he does, he makes in my view a mistake closely analo-

227

gous to the mistake he originally made in thinking that explanations which fit the deductive models are more scientific than statistical explanations. If he does not, then the "statistical model" becomes largely a matter of good, nonformalizable judgment about the "weight of the evidence." This is not a sign that it is unscientific, but it is a sign that much of scientific argument conforms more closely to legal and historical argument than it does to mathematical inference, whether statistical or not. Hempel's recognition of this is clear in part of his discussion of causation where he agrees with a view of mine, which he quotes, that one may be able to give good reasons for a particular causal claim other than laws. This appears to commit him to accepting nonformalizable "statistical" causal explanations, and I would take this to mark the end of any "model."

It may be clarifying to indicate a number of Professor Hempel's assumptions or conclusions with which it appears to me I would disagree for reasons that are usually apparent or implicit in my paper: that half-life laws and probability statements are statistical in nature, that any counterfactual-supporting law is explanatory, that an axiomatization of "probability," using "random" as a primitive, is logically valuable, that the probability of a deductive model explanation is the same as the probability of the conjunction of statements in it (even if we include grounds), that the degree to which a statistical explanation explains is the same as the inductive support given its explanandum by its explanans, that explanations in terms of purpose are equivalent to causal explanations, that causal explanations which do not give the relevant laws and all relevant conditions are programmatic.

I shall take up in detail just one point which I think rather more novel than the above. It is connected with the stress on the nonformal elements in scientific inference discussed earlier in this section.

6.2. The Requirement of Total Evidence.

Hempel is led by examining a certain paradox of statistical explanation to propose a distinction between it and deductive model explanations, which he describes as "the ambiguity of statistical explanation." The paradox it leads to is remedied by "the requirement of total evidence," and Hempel repeatedly stresses that such a precaution is not needed as a safeguard for the deductive model. I believe I can show that it is, and should have been included in the criteria for the model. I also believe that an extension of this point will undercut his other distinction, "the nonadditivity of statistical explanations," but I shall not be able to go into this here.

Referring to "the ambiguity of statistical explanation," Hempel says, "This peculiarity has no counterpart in nomological explanation: If an object or set of objects has a character A which is invariably accompanied by C then it cannot have a character which is invariably (or even in some cases) accompanied by non-C."[47]

This looks plausible but it happens to be invalid. X may have A and hence C, and also B which is in some cases accompanied by non-C, though not in this case. This error is connected with the much more serious fact that this formalization does not adequately represent the point at issue. For X may have A, which is a universally sufficient condition for C, hence giving us a "nomological" explanation for X having C, when in fact X has C for an entirely different reason, namely that X has the property B which is *also* a sufficient condition for C. The difference between A and B as candidates for mention in the explanation is simply that X happens to exhibit the property C^1 as well as C, and whereas all B's have both C and C^1, A's only rarely have C^1.

It seems likely that Hempel was led to give the simpler formal analysis because the explanandum is, after all, that X is C. But that we are trying to explain X being C does not of course allow us to accept an explanation which cannot account for X being C^1. And there may well be (determinists would say "must be") some other property of X, here B, which will account for both C and C^1.

A simple example can be given. If we are trying to explain how a certain bridge came to be destroyed in wartime, we could appeal to the law "whenever an atom bomb is released accurately above a bridge and explodes with full force, the bridge is destroyed," plus the appropriate antecedent conditions. However, it may also be the case that whenever 1000 kilograms of dynamite are detonated on the main span of such a bridge it will be destroyed, and that the underground movement has applied just this treatment to this bridge with the attendant destruction occurring between the release and the arrival of the atomic bomb. This in no way invalidates the bomb law, but it does invalidate the bomb explanation which cannot account for other features of the event, in this case the time of the destruction. It will be noted that a prediction based on the bomb law *would* survive, which I take to be a further sign of asymmetry.

[47] P. 133, this volume.

Hempel recommends that the "ambiguity of statistical generalization" be handled by requiring that "in the application of inductive logic to a given knowledge situation, the total evidence available must be taken as basis for determining the degree of confirmation" (Carnap's formulation of "the requirement of total evidence"). Of this requirement, Carnap says it has "no analogue in deductive logic," but Hempel comments, ". . . it seems more accurate to say that the requirement is automatically met here."[48]

In its application to explanations, it seems clear that this requirement is neither absent nor automatically present. We must in fact add it to the condition, and make it more specific by requiring that an explanation be acceptable for a phenomenon only so long as no facts are known about the circumstances surrounding the occurrence of the phenomenon which the explanation cannot accommodate. The explanandum is rarely a complete description of everything that is known to occur, and one could never support the claim that it should be since this would frequently be unfulfillable. But an explanation has to pass other tests besides (and sometimes instead of) supporting an inference to the explanandum, and this is one of them.

I conclude with a comment on the over-all argument in Hempel's discussion of the differences between statistical and nomological explanation. The paradox of ambiguity has its roots in a misuse of statistical inference that is, as he points out, quite well known in the literature. Hence "the ambiguity of statistical explanations" is only apparent and not real, since one ought to have borne in mind the requirement of total evidence. Hence it cannot be an identifying property of such explanations. And, according to my arguments, prophylactic treatment is necessary in both of Hempel's realms of explanation. I think a combination of these two kinds of argument applies—though I have not shown this—to the "non-additivity of statistical explanations." And it seems quite clear that taking account of the probability of the categorical statements in nomological explanations will show that they share with statistical ones the property of being more or less likely. It seems to me that there remains only the definitional difference that the internal inference is in one case deductive and in the other inductive.

[48] See pp. 138, 139, this volume.

Explanation, Prediction, and "Imperfect" Knowledge

The Attack on the Deductive Model

Can we explain events after they happen which we could not have predicted beforehand? Do history and the other studies of man differ from the physical sciences in this respect? Do a man's character and motives satisfactorily explain his actions or do these in turn need explaining? Questions of this sort are raised, not for the first time, by certain recent criticisms of the so-called deductive model of explanation. This model holds, briefly, that to explain an individual fact we deduce it from one or more other such statements in conjunction with one or more generalizations or laws. A law, in turn, may be explained by deducing it from other laws. Prediction, upon the model, has the same logical form as explanation. In predicting something as yet unknown, one deductively infers it from particular facts and laws already known. We infer from given premises stating what is known, what so far has not been known, whether it be in the past, present, or future, or whether, as in the case of laws, time is irrelevant. It follows that if anything can be explained deductively by a set of premises after it has occurred, it could in principle have been predicted from these premises before the event. Nor does it make any difference whether the premises are statistical or deterministic. If they are deterministic, we may predict an individual event; if they are statistical, only statements about classes of events may be either explained or predicted. Virtually all those who accept the deductive model hold that it applies not only to physical but also to human phenomena, whether individual or social, whether in the past or in the present. Recently popularized among philosophers of history by a lucid exposition of Hempel's,[1] the deductive

NOTE: This paper was read, in part, at the 1960 International Congress for Logic, Methodology, and Philosophy of Science held at Stanford University.

[1] C. G. Hempel, "The Function of General Laws in History," *Journal of Philosophy*,

model traces its lineage through John Stuart Mill and David Hume back to Galileo.

In physical science, ever since physicists abandoned the anthropomor- phic-teleological "why" in favor of quantified relations among relevant variables, explanation has meant deduction. John Stuart Mill, who had the clearest anticipatory vision of what a scientific study of man and so- ciety could be, extended the notion to these areas. The extension, always bitterly resented, is now once again under attack. A certain piquancy is added to the current attack by those who now attack the deductive model for the physical sciences as well. This global nature of the attack gives the current controversy its special interest. The assailants either radically reject the model *in toto* because it is *deductive*, or, more moderately, re- ject it not because it is deductive but because of the nature of the *premises* from which the deductions are allegedly made. In either case, the critics have given what appears to be a "new look" to all the old and tired ar- guments against a science of history and of man—the arguments from uniqueness, from freedom, from mind, from complexity, from values, and the rest. Nor is that all. The extremists challenge not only explanation by deduction but the very notion of valid deduction itself. I shall first dis- cuss the more prevalent radical position of those who reject the standard model simply because it is deductive. Since there is a common core to the radical and the moderate views, it will then be possible to treat briefly of the more moderate criticism which attacks only the nature of the prem- ises. It is indeed the broader philosophical implications of this common core, cutting much deeper than the issue at hand, that I am particularly concerned to exhibit as clearly as I can. The arguments of moderate and extremist alike arise from an overemphasis on language as it is used to communicate with others, to the neglect of language as it is used to de- scribe the world. This preoccupation in turn results in a notion of "con- ceptual analysis" with philosophically untenable consequences. I shall try to show why this is so and, in particular, why the criticisms of the de- ductive model, based on this preoccupation and that notion, are without force.

The rejection, by the extremists, of deductive explanation is but part

39:35–48 (1942), reprinted in *Theories of History*, P. Gardiner, ed. (Glencoe, Ill.: Free Press, 1959). Frequently cited also is the more detailed article by C. G. Hempel and P. Oppenheim, "The Logic of Explanation," in *Readings in the Philosophy of Science*, H. Feigl and M. Brodbeck, eds. (New York: Appleton-Century-Crofts, 1953).

and parcel of a more sweeping rejection of formal logic. Three related arguments are used to support this double rejection. First, appealing to ordinary usage they find, to no one's surprise, that 'deductive' is applied to many different kinds of inference. Whenever locutions such as "because so and so, therefore," "this, so that," or "this involves that" occur, we may accordingly "properly speak of 'deductions.'" [2] 'Explain,' too, is revealed to have many different uses. The deductive model, assuming as it does the logician's notion of 'deduction' and a particular use of 'explain,' doubly affronts ordinary usage and is, accordingly, doubly "irrelevant." Second, the critics deny that anything corresponding to the logician's narrow, tautological sense of 'deduction' actually occurs in common sense or in science. Since there is no deduction, naturally there is no such thing as deductive explanation. Finally, they maintain that an appeal to so-called conceptual analysis is as conclusive as an appeal to logical truth. This third reason, as we shall see, is the most fundamental of all.

Common-Sense Facts vs. Philosophical Claims

Do we ever predict what we cannot explain or explain what, before the event, we cannot predict? We do it all the time. We predict a future event, like the outcome of next week's election, even though we cannot deduce it. Moreover, whatever the outcome, we then proceed to explain it. Of course we do. We explain a man's sudden death by his being struck by a car. The historian too explains past events that he is in no sense trying to predict. To proclaim these uncontroversial common-sense facts is one thing. To propound a philosophical thesis is quite another. At issue now is the philosophical claim that accompanies the announcement of these everyday "explainings" and "predictings." Specifically, the claim is that no laws either are or need be deductively invoked, either explicitly or implicitly, to permit the inference from some individual facts to another which they explain or predict. Moreover, an appeal to laws is held to be "wholly unnecessary" for identifying the causes of an event. The causes of the Civil War, of my car breaking down, of why a test tube blew up, of a carpet's being stained, all can be "understood" and understood with "primeval certainty" by the appropriate people, some of whom are simply endowed with a "well-developed capacity for identifying causes,"

[2] See G. Ryle, The Concept of Mind (London: Hutchinson, 1949), p. 300; S. Toulmin, The Uses of Argument (Cambridge: Cambridge University Press, 1958), p. 121.

without appealing, either implicitly or explicitly, to laws.[3] Marshaling the evidence against the deductive model, with its implied symmetry between explanation and prediction, the critic comes triumphantly upon history. We are even treated to a dithyramb on history as the "mother subject for explanations," just because it is no different from common sense.[4] Certainly, the historian now often explains what he cannot predict. However, this obvious fact does not in itself justify making history the paradigm case of "sound explanation." Nor, of course, does the mere occurrence of deductive explanation and prediction in physical science, should one grant that it does occur, *in itself* justify making physics the paradigm case, as is customary with proponents of the deductive model. A philosophical thesis is not merely a description of what does or does not happen to occur. My primary purpose is to examine the grounds for the philosophical claim that is associated with, but I believe not supported by appeal to certain indisputable facts of common sense. In the course of this examination, I shall also try to show how the deductive model can consistently account for these facts, as well as for the present and, for all we know, perhaps permanent situation in history and related areas.

The Appeal to Ordinary Usage and to "Understanding"

The assailants, probing the idiom, discover as many senses of 'explain' as there are possible answers in common speech to the question Why? In none of these senses do they discern any appeal to deduction in the logician's sense. We explain to a child how to multiply fractions; we teach him certain rules. We explain the meaning of one term, using others. We explain the structure of the ruling class of a given society—we describe a pattern. We explain the symbols of formal logic to a class—we describe what is made to stand for what. We explain the rules of chess. We explain a man's actions by pointing out his motives or his purposes. We explain the connection between the moon and the tides. Perhaps, in this case, we merely state a constant conjunction. Or we may derive it from the Newtonian law of gravitation in conjunction with certain other statements. And so on, and so on. The ordinary uses of 'explain' are manifold indeed. Only the last case involves deduction. So there is at least one case which does involve it. In spite of this case and in spite of the admitted

[3] M. Scriven, "Truisms as the Grounds for Historical Explanations," in *Theories of History*, P. Gardiner, ed. (Glencoe, Ill.: The Free Press, 1959), p. 456 and *passim*.
[4] *Ibid.*, p. 458.

multiplicity of uses, there is a drumming insistence that all of these uses of 'explain' are "essentially" the same—that there is a true meaning of 'explain' common to all of them, except, ironically, deduction. This true meaning is "*the* ordinary sense," the only one which "*really* explains."[5] Ordinary use is thus not merely described. It is appealed to as the standard that everything else must meet if it is to "count" as an explanation.

Just as individual facts, like the Civil War or a car breaking down, can be "understood" without appeal to laws, so with other things. A teacher explaining to a student how to multiply fractions, an art critic explaining why one painting is better than another, a nuclear physicist explaining the principle of complementarity, a father explaining the rules of chess to his son—each, like the mechanic explaining why a car broke down, has his own way of telling whether or not the other person understands what is being explained to him. What is the sense of 'explain' common to all these various uses—explaining facts, rules, a symbolism, how to do something, and so on? The answer, though disappointing in its triteness, reveals the difference between the study of language as a medium of communication and its study as a description of the world.

Every teacher has a way of finding out whether or not a student has understood what he has been talking about. This is "the ordinary sense," the "essential sameness" of meaning that preoccupies the critics of the deductive model. Has someone understood what we have said? Is it "intelligible" to him? To find out, we ask him certain questions, depending upon the kind of thing we are testing his understanding about.[6] If he answers appropriately, then we have successfully communicated with him. This psychological sense of understanding what is conveyed by a communication does indeed underlie most of our ordinary uses of 'explain.' The critics, understandably concerned to deny that they are engaged in factual inquiry, insist that this meaning of 'intelligibility,' 'understanding,' or 'explain' is not psychological. To be sure, one of the ordinary uses of 'psychological' is as a synonym for 'mental'—something going on in someone's mind. Another is of something "subjective"—varying from individual to individual. The critics protest that they mean nothing either mental or subjective, in this sense, and hence nothing "psychological." They mean rather a behavioristic test. We can tell that someone under-

---

[5] W. Dray, *Laws and Explanation in History* (London: Oxford University Press, 1957), p. 79; Scriven, in Gardiner, p. 449.

[6] Scriven, in Gardiner, p. 452.

stands something by what he says or does. It is thus allegedly not psycho logical because it is behavioristic, not mentalistic, and it is not subjective because we can devise a test by which anyone can tell whether any other fellow understands or not.

Only if one is frozen in the science of the nineteenth century can one believe that psychology is concerned with the "working of men's minds' rather than their behavior, verbal or otherwise. If one also happens to fee that a science of psychology is unnecessary because we already "know well enough" by "ordinary good sense"[7] why people behave as they do, then one may persuade oneself that the suggested test for "understanding" is not psychological. But this is verbal magic. How we learn, communicate by means of, and react to, language as part of our social environment are all facts about human behavior. The appeal to ordinary usage and the interest in communication leads to a notion of 'explanation' that depends not upon the way the world is, the connections among things, but upon *something in us*, upon the way we respond in the game of question-and answer that we play, not with the world, but with each other. Later on I shall have more to say about the relevance for the attack on the deductive model of this curious combination of behaviorism with a rejection of the science of psychology. For the present, it seems clear that, like the word or not, this remains a psychological sense of 'explain.' And this sense of the term does not tell us what *justifies* saying that an event happened one way rather than another, nor—apart from a quaint appeal to "judgment" and special talents—does it tell us how it happens that we are able frequently to make correct inferences from what we know to something past, present, or future that we do not yet know. In other words, it doesn't answer the philosophical questions about explanation and prediction.

Language as Communication vs. Language as Description

Among all possible means of communication, language is the richest and most subtle. By language we convey to others our desires, plans, attitudes, and, among other things, information about how to do certain things or about the way things are or seem to be. How we learn a language and, having done so, contrive to communicate successfully with each other are, to repeat, questions about human thought or behavior in

[7] Ryle, *the Concept of Mind*, pp. 52, 324, 326. An incisive critique of some of Ryle's views on psychology may be found in M. Mandelbaum's "Professor Ryle and Psychology," *Philosophical Review*, 67: 522–530 (1958).

ocial situations. They are thus to be answered by sociopsychological investigation, whether this be our everyday experience or the controlled methods of the social sciences. Speculation from, or analysis of, how we peak can at best offer cues to the factually correct answers to questions bout the so-called higher symbolic processes which are involved in linguistic behavior. However, since much of what we try to convey to each other by means of language is information about the world, language also has a descriptive function. No matter how complicated the processes by which we understand one another may be, they can only occur because anguage is also used to describe the world, including ourselves as part of it. The study of language as description is the study of terms and statements as they are used to make such assertions, no matter how we came to acquire these uses. Given this language that, indubitably, has been socially learned, philosophical questions may be asked about it. By analysis of the language itself we try to see what there is about it that, *once it has been understood*, enables it to tell us something about the world and also what it actually does tell us about the world's content or its form.

A physicist's published report of the theoretical significance of an experiment and his laboratory conversation are two quite different things. The latter is elliptical and context bound in a way his report cannot be. He may "explain" an observed phenomenon to a co-worker by simply mentioning some other fact, very likely something that he has done. The co-worker will "understand" him. The information necessary to this understanding is supplied by the total context within which the conversation occurs, both the social situation itself and the knowledge the two men share. The published report, however, will have to be considerably more circumstantial about the connection between the experiment, the observations, and what it was performed to test. Moreover, the report will use terms with constant meanings throughout. The sentences will be complete in the sense that they will be true or false, independently of the time or place at which they are read, unlike, say, "The light just went out."

Since the physicist writes for other physicists, even his report may not explicitly mention all the information that justifies certain statements, for again there is a context of shared knowledge that may be taken for granted. But the obvious differences between the explicitness of the report and laconic laboratory conversation makes clear what a complete report would be like and, under some circumstances, must be like. Only complete report, including that which is spoken as well as that which is left unspoken, presents the observational and logical grounds for an claim. Ordinary communication relies heavily on the linguistic and non linguistic context. To examine language as description, this context mus be supplied. A context-free language is artificial only in the sense that th physicist's report is "artificial" compared with his conversation. It make explicit what, in ordinary communication, is supplied by the total context By filling in the context, we know what it is that concepts are being use to refer to and what statements about the world ordinary sentences ar being used to make.

The logician uses a similarly improved language to explicate the notion of logical truth, deductive proof, logical impossibility, and the rest. Th logical notions articulate the criteria by which, in common sense and in science, we are *justified* in asserting that someone is being inconsistent o that an observation refutes a generalization. The logician's improved lan guage is useful to the philosopher precisely because and only insofar as i is a reconstruction of a large part of the language that we speak. The log cian's notions of "logical truth," "deduction," and the rest, correspond to though they do not duplicate, certain indispensable concepts in ordinar use. The logician's term 'analytic,' for instance, is not identical with th ordinary uses of 'necessary,' but it corresponds to the use in such state ments as 'It is necessary that white horses are white.' The philosophica logician thus clarifies *one* use in common speech of 'necessary' and, at th same time, shows how it differs from other uses of the term, as in 'It i necessary that all men die.'

Why Explanation Must Be Deductive

Grant for the moment that deduction is one core meaning of explana tion, at least within science. This is an anthropological fact like any othe about the linguistic behavior of a certain group of people. Why give thi use pride of place in preference to all others? Not, as some think, be cause of some scientistic prejudice in favor of physics as the model in quiry. Rather, it goes the other way. Physics is an advanced and "model science precisely because it does provide deductive explanation and pre diction of its phenomena. To show why this is so requires explication o *this* use of 'explain.' The philosopher surely is not merely an anthropolo gist describing the linguistic habits of the natives—whether they are plai men, members of the class of historians or of physical scientists—and wha different things among them "count" as explanations. Such descriptio s at best preliminary to the task of exhibiting the structure of a reasoned xplanation, showing which structure logically justifies that which is to ›e explained. In philosophy and elsewhere an appeal either to a definition ›r to a logical truth is conclusive. One does not quarrel with a statement vhich is true merely by virtue of a convention about the use of words. ;imilarly, if a statement is backed up by reference to other statements rom which it logically follows, one does not argue, as long as one accepts he truth of the premises.

The purely verbal case is too trivial and obvious to need elaborating, ›ut what makes the appeal to premises conclusive? Briefly, because the ·onditional corresponding to the premises and conclusion of the argu- nent is a tautology or logical truth. In a context-free language we can ormally characterize the notion of "tautology." Consider "It's raining or t isn't." One must eliminate vagueness—the drizzle beloved of students— ind the event must be located in time and space before we can character- ze the statement as a tautology in the strict sense. Only when these are mplicit in context is the statement even an everyday tautology. For that natter, as has been insisted, only then is the expression even a statement.)nly for *statements* can we state the criteria, which are characteristics of *he statements themselves* and not of anyone's responses to them, that nake them tautologies or logical truths. If the generalizations and indi- /idual statements of fact serving as premises are accepted as true, then, ›ecause of the tautological connection, the conclusion *must* be true. This ind this alone is the virtue of deductive explanation. Once such terms as 'must," "guarantees," and "logically implies" are clarified, then it is clear vhy deduction and deduction alone "justifies" the conclusion. At the ame time, it is also clear why any other kind of explanation of individual acts cannot possess this conclusiveness. Either the explanation is deduc- ive or else it does not justify what it is said to explain.

All this is not to "set limits" to what may constitute a sound explana- ion, but to explicate *one* sense of sound "explanation" which is the only ›ne that is conclusive for the reasons just mentioned. For the philosopher ·oncerned with language as description, the task is to show what, so to peak, in the facts themselves, or, to be accurate, *in the statements assert- ng them*, rather than in the mind or behavior of a particular person or ;roup of persons, makes one or more statements a "reason," in a precise ogical sense of reason, for one or more others. What in the statements hemselves makes one statement a prediction from others, one confirm a

239

general hypothesis, another refute it? Deductive explanation is the only answer to this question, insofar as language is used to speak, not just about how we speak, but about the world.

One particular appeal to ordinary usage to bolster the claim that scientific explanation is not deductive verges on the bizarre. Reading 'deductive' to mean 'syllogistic' in the Aristotelian sense, it is concluded that scientific explanation, since not syllogistic in form, is not deductive.[8] Explanation in science is indeed not syllogistic, at least in the most interesting cases, namely, those involving quantification, relations, or both. But I think it may be said safely that however the half-educated plain man may use the term, no scientist since Galileo has so narrowly conceived "deduction" and certainly no logician for almost a century. Yet, insofar as these critics concede a correlate to strict deduction in ordinary speech, including science, they perceive it to mean something like Barbara. Hence if not Barbara, then not deductive. Words *do* mean only what we say they mean. But one use of a term hardly justifies the assertion that nothing corresponds to another and different use of that same term.

Like 'deductive,' the word 'explain' has, as we saw, many uses. To assail the deductive model because it does not apply to all these uses is irrelevant at best, puerile punning at worst.[9] Deduction is not put forward as the model for those uses of 'explain' when we speak, say, of explaining (i.e., describing) the structure of the Egyptian ruling class, a symbolism, or how to play chess. The model is proposed for only one but one very important use of 'explain.' In particular, the occurrence of certain events like an eclipse or an earthquake, can, upon the model, sometimes in practice and always in principle be explained and predicted deductively. Practically, we can both explain and predict the occurrence of an eclipse. We can only explain the earthquake after it happens. The critics' most plausible case rests, of course, on such all-too-common "earthquake" situations. But they are not content to rest on such cases. Though conceding at one place that Newton "did achieve a large number of mathematically precise and scientifically illuminating deductions [*sic!*] from his theory," this is, to put it moderately, not the prevailing tenor of the argument.[10] What one hand gives, the other snatches away. Or perhaps it was only Newton him-

[8] Scriven, in Gardiner, p. 462; S. Toulmin, *Philosophy of Science* (London: Hutchinson, 1953), p. 25.

[9] Scriven, in Gardiner, pp. 463, 468.

[10] M. Scriven, "Explanation and Prediction in Evolutionary Theory," *Science*, 130, 477 (1959).

240

elf who could accomplish the feat of deduction. In any case, the claim is
hat not even in physics do we really have deductive explanation. Apart
rom the untenable, because false, appeal to usage, the argument has, so to
peak, two facets or takes two different forms. The first grants that laws do
lay a role in some explanations, but such explanations are allegedly still
ot really deductive. The other holds that explanation of an individual fact
; always by other such facts without the intermediary in any way of laws.
after a few preliminary comments about the structure of classical physi-
al theory, I shall discuss these two different arguments in the order men-
ioned.

"Perfect" Knowledge and Deductive Explanation

Celestial mechanics is a paradigm of what has usefully been called "per-
ect knowledge." [11] Nonatomic thermodynamics is another. In the first
heory, mass, velocity, and distance are variables that interact only with
ach other; in the second, volume, temperature, pressure, and a few other
ariables of this sort do the same. In each case, no variables that in fact
nake a difference have been omitted from the expressions of the theory.
These theories are *complete*. Nothing that happens at any other time or
lace than that being considered affects the behavior of the properties
vith which the theory is concerned, or, at worst, we know how to take
ccount of these outside influences in our predictions or computations.
The system, in other words, is *closed*. The laws of these theories are *proc-
·ss laws*, that is, the values of any one variable at any time can be com-
uted by means of the laws from the values of all the others at any other
ime. We can predict the future course of all our variables or compute
heir entire past history; we know how changes in any one variable pro-
luce changes in any other; we know to what extent by tinkering with the
ystem we can bring about desired changes, to what extent we are power-
ess.

Such knowledge may, not unreasonably, be called "perfect," for there
s clearly nothing else we could possibly want to know, as far as these
ariables are concerned. Nor is there any unreasonable sense of "perfec-
ion" implied by this notion. Every measurement has its limits of accu-
acy. Inductive generalization may be overgeneralization, as indeed we

[11] The distinction between "perfect" and "imperfect" knowledge is Bergmann's.
'or a thorough discussion, see G. Bergmann, *Philosophy of Science* (Madison: Univer-
ity of Wisconsin Press, 1957).

241

now know about the Newtonian law of gravitation. These frailties ar
common to all empirical science and do not affect the nature of the de
ductions we can make from such theories, when we assume that the law
are true and that our measurements are accurate.

From the Newtonian law of gravitation in conjunction with the pos
tions of the sun and the planets at any one time, we can predict or post
dict their position at any other time. Thus we can deductively predic
future eclipses or explain present and past ones. The laws of Newton ma
be applied, of course, to a vast number of vastly different situations, ter
restrial as well as celestial. If we know enough about the initial conditions
then we can predict what will happen. Using these laws, we can predic
how long it will take an elephant to slide down a green, grassy hill, if w
know only his mass, the angle of inclination of the hill, and the coeffi
cient of friction. Or we can predict where and when a cannon ball wil
land, if we know but a few relevant variables. In some cases, where w
fail to predict, as when a bridge collapses under a certain load, we coul
have done so, if we had taken the trouble to find out the state of the sys
tem, that is, the initial conditions, before the catastrophe. If we did no
take the trouble or if it was too complicated to do so, then, after th
event, we deductively explain it by reference to the laws and to what th
conditions must have been to make the bridge give way.

"Certainty" of Premises vs. Deductive Validity

How, then, in the light of all this can the critics nevertheless maintai
that explanations in physics are not deductive? The physicist, it is pointe
out, might be wrong either in deciding that a particular universal law ap
plies to a given set of conditions—whether or not they are within th
scope of the law—or in describing the initial conditions. The physicis
must, therefore, like the historian, use "judgment" and even "empathy"
rather than formal deduction in making his inferences.[12] Or so the argu
ment goes.

However, the choice of premises, or even the use of false ones, is no
in the strict sense a logical matter and does not affect the validity of a
deduction. The validity of the deduction itself is here confused with th
error of using, say, a first approximation like Boyle's law where only a
more general gas law would do, or in believing that the effects of frictio

[12] Scriven, in Gardiner, pp. 459, 462.

are negligible when in fact they are not. Once the scientist has used his "judgment," which is indeed a psychological matter dependent on his skill and training, then the deduction follows. The explanation *is* deductive, whether or not we are "certain" about the premises, either the law or the initial conditions. A prediction in the same circumstances is also deductive. The scientist, in fact, tests his judgment by deductively inferring what *must* be the case if what he believes to be the applicable laws and the true initial conditions really are so. Premises deductively imply a conclusion no matter how certain or unsure we may be about their truth. Logical entailment (or inconsistency) is a property of the statements themselves, not of our knowledge about them. If the prediction were not deductive, then there would be no reason why its failure should cast doubt upon the premises. In fact, it shows that at least one of them is false. In theoretical science, when a prediction fails, this generally, though not necessarily, casts doubt upon the laws. In engineering or applied science, on the other hand, the initial conditions become suspect. Far from requiring "exact truth" for its premises, all that the deductive model requires is exact *statement* of a hypothesis about their truth. The hypothesis is then tested by the "exact deduction" which, I am sorry to have to repeat, does not mean and never has meant only the syllogistic deduction with which the critics sometimes equate it.

Hempel's careful statement of the deductive model unfortunately uses an illustration that, from the point of view of the principle involved, is inadequate. Understandably, for expository purposes, trying to avoid the mathematical complications of those areas where perfect or near-perfect knowledge exists, he uses a homely, everyday illustration, the breaking of a car's radiator in freezing weather. Hempel's so-called failure here is not one of deduction, but a failure of complete knowledge. He cannot state all the initial conditions that will permit him to deduce that the radiator broke. If one appreciates that philosophical explication is explication of the principles involved, this is no failure at all. Once again, only because there is a deductive connection between the laws of physics and, among other things, the state of the weather and insufficient antifreeze in the radiator, have we any good reason to assert that the cold rather than the sparrows who daily perch upon the hood explains the breakage. The physicist does not judge "inductively . . . what the explanation is."[13] Cor-

[13] *Ibid.*, p. 457.

rectly stated, he makes an inductive judgment about the premises. An
to make this correction is not, as is claimed, to "convert" either the prem
ises or the judgment into a deduction. We want to know whether a give
statement is deductively entailed by certain others. This is a logical ques
tion. To import considerations about what the scientist happens to know
certainly or otherwise, or even, if knowing, what he happens to asser
explicitly or to "quote," is to doubly muddy the waters by first introduc
ing irrelevant psychological factors and then confusing context-bound sen
tences and complete statements.

In his zeal to confute the deductive model, Scriven makes the remark
able statement that even in those parts of physics where what I calle
"perfect knowledge" exists, namely, classical thermodynamics, "deduc
tion of the *exact* values to be explained from such laws is a matter o
chance." [14] This conjures up an image of physicists writing down conclu
sions at random. But this is not quite what happens. A quantitative sci
ence uses measurement to obtain the values of its variables. Measurement
as I mentioned before, has its limitations. Every measurement has a so
called error component associated with it. This "error" is the deviation
from a mean value that can be expected under repeated measurements o
the same variable. In other words, in repeated measurements of, say, th
length of a bar, the quantity obtained as a result of the measurement
will tend to cluster around a certain value. The result of a measuremen
is thus not a single value but a frequency distribution determined by th
nature of the measurement. This means that the premise expressing the
measured initial conditions always has a statistical component. It mean
further that the conclusion deduced from such premises must also be
about a frequency distribution, not a single value. There is no deduction
"by chance" or any other way of a single, exact value. There is instead a
strict deduction, by the probability calculus, of a so-called "chance" vari
able, that is, a frequency distribution, which is quite another thing. [15]

But we have only begun to probe the depths of confusion that these
critics display about the use of statistical premises. To probe deeper, we
must turn to the notion of "imperfect" knowledge. Moreover, the attack
on the deductive model derives its prima-facie plausibility from those

[14] *Ibid.*, p. 460.

[15] An excellent discussion of this issue and others related to the comments below on
the use of statistical premises may be found in P. E. Meehl, *Clinical versus Statistica
Prediction* (Minneapolis: University of Minnesota Press, 1954); cf. especially pp. 56ff.

areas about which we have only imperfect knowledge. It is well to be clear, therefore, about what we can and cannot do when "imperfection" prevails.

"Imperfect" Knowledge and Deductive Explanation

Perfect knowledge is the ideal, actualized only in certain branches of physical science. Elsewhere, as in biology, economics, sociology, psychology, and the social sciences generally, knowledge is conspicuously "imperfect." We do not know all the variables that affect, say, a person's resistance to disease, or his behavior under certain circumstances, or the price fluctuations of certain commodities. Our theories in these areas are not complete, nor do we know fully the conditions for closure of such systems. Since we do not know all the factors that make a difference to the variables with which we are concerned, we also have no process laws. From the values of the variables at any one time, we cannot predict their value at all other times. Yet, the social and biological sciences have developed techniques to compensate for lapses in closure and completeness. The most important of these techniques is the use of statistical concepts. Which face will turn up when a die is cast is determined by numerous causes—the center of gravity of the die, the force with which it is thrown, and so on. An attempt to calculate the results of each throw by means of the laws of mechanics is practically hopeless because of the difficulty in precisely measuring all the initial conditions. Instead, we represent, as it were this multiplicity of causes by a probability distribution for the attribute in question. The use of the statistical concept marks our ignorance of all the influencing factors, a failure in either completeness or closure or, usually, both. Similarly, the social scientist, deliberately selecting for study fewer factors than actually influence the behavior in which he is interested, shifts his goal from predicting individual events or behaviors to predicting the frequency with which this kind of behavior occurs in a large group of individuals possessing the circumscribed number of factors. This use of statistical concepts, due to lapses in closure and completeness, differs from their use for errors of measurement. Even assuming perfect observation, they would still be necessary.

Statistical knowledge is not the only kind of imperfect knowledge. We also have nonquantified imperfection. Any law, whether it be about physical objects, persons, or societies, is "imperfect" if it does not permit us to compute (predict or postdict) the state of the system, either an individual or a group, at any moment from its state at one moment. Consider the assertion that a boom in trade is always followed by slump and depression. This is imprecise with respect to time, for it does not tell us exactly when the later events will follow the earlier one, nor how long each will last. Moreover, its concepts have a fringe of vagueness that make it difficult to tell precisely when we have instances of the kinds of events mentioned. In psychology, the laws of learning that make essential use of the past history of a person in order to predict his future behavior are also imperfect. The equilibrium laws of physics, whose concepts are not vague, are nevertheless imperfect because though they tell us that under certain conditions no change in certain respects will occur, they do not tell us what will happen if these conditions are not fulfilled. To be sure, the equilibrium laws of physics are derivable from the process laws of the theory, but for those of economics no such perfect laws are available. In general, imperfect laws are indefinite with respect to time, or hedged in by qualification, or they are statistical.

The inadequacies of such "imperfect" knowledge do not affect the possibility of deduction. Not only do we sometimes know enough to deduce some of these laws, like the law of the lever or certain statistical laws of physics, from process laws, but all kinds of deductions can be made from the imperfect laws themselves, whether or not they are in turn deducible from something else. An explanation utilizing imperfect laws as premises is not the same as Hempel's "explanation sketch." The latter he describes as "a more or less vague indication of the laws and initial conditions considered as relevant, and it needs 'filling out' in order to turn into a full-fledged explanation."[16] Imperfect laws, as here defined, need not be vague or, as explicit premises from which deductions can be made, incomplete. For instance, 'All men are mortal,' though universal, is imperfect by our criterion, since from it we can neither explain nor predict a man's death at a particular time. But such nonquantitative universal generalizations, as well as many laws of the biological and social sciences or statistical laws generally, are not necessarily "vague." Nor do they necessarily need "filling out" before they can be used for significant deductions or "full-fledged" explanations.

On the other hand, it is indeed true, as I illustrate later, that we often have to make guesses as to the appropriate imperfect laws, about either

[16] Hempel in Gardiner, p. 351.

246

individuals or groups of individuals, that will permit us to explain or predict a given event. However, the deductive model by no means requires that premises be the deterministic process laws of perfect knowledge. Once this is grasped, the admitted difficulty in formulating so-called universal laws, of which the critics make so much, no longer appears insuperable. They set a demand that is not logically required by the model. After all, deductive inference was with us centuries before Newton formulated the first process laws.

"The criterion of deduction must be abandoned if the criterion of universal (non-statistical) hypotheses is abandoned." Moreover, if we only have statistical knowledge, then though we cannot predict, we can use such knowledge to explain with "certainty" the occurrence of individual events.[17] Neither of these claims, as I have just suggested, is justified. A statistical law asserts that if each member of a certain class has a certain attribute, then a certain fraction or percentage of them will have another attribute. For instance, "60 per cent of all cigarette smokers develop lung cancer." This is a generalization or universal statement, for it says of all cigarette smokers, past, present, and future, those observed and those as yet unobserved, that 60 per cent of this group will suffer from cancer of the lungs.[18] Like all statistical generalizations, the evidence for it is a finite number of cases. The statement asserts, however, that in the class of all individuals of a certain kind, a particular attribute will turn up with a specified frequency. In this respect, a statistical generalization is as "universal" as a so-called deterministic or nonstatistical law stating that each and every individual having a certain character will also have another. In both cases, the law goes far beyond the evidence. If it did not, but was just a summary of observations, it would have neither explanatory nor predictive power. The difference between them is not that one permits prediction while the other does not, but in the nature of what can be predicted or, what amounts to the same, in how they are tested.

From a deterministic law, given the initial conditions, we can predict an individual event. From a statistical law and its initial conditions (the

---

[17] Scriven, in Gardiner, pp. 457, 464; and in *Science*, 130: 479, 480.

[18] For the sake of an illustration, I have stated the hypothesis in this very alarming way, just as one might say that 60 per cent of all Norwegians are blond. Realistically, the actual lung-cancer hypothesis is a bit less terrifying, asserting a comparison between smokers and nonsmokers and that a higher percentage of the former than the latter will develop lung cancer. The relevant logic of the situation is identical in either case. See the examples that follow.

occurrence of a large group of cigarette smokers), we can predict only a so-called mass event, that is, the frequency with which an attribute will be distributed in the given class. If an unbiased coin is tossed a large number of times, then the frequency with which heads will turn up is 50 per cent. This says something about the class of *all tosses* of a coin, though it says nothing about what will happen in any particular toss. Similarly, the lung-cancer generalization says nothing about any particular cigarette smoker, though it says a good deal about the class of all cigarette smokers. From a statistical law, then, nothing can be predicted about an individual event. On the other hand, neither can we explain an individual event by reference to such a law.

It is embarrassing to rehearse these elementary matters. But the critics force such rehearsal upon one, for they argue to the contrary. Consider the nonquantified, implicitly statistical statement that Scriven calls a "hypothetical probability prediction," namely, that if a flood occurs, then animals who can swim will be more likely to survive than those who cannot.[19] Scriven notes that unless we can predict the flood, we cannot predict which animals will survive. However, he believes that, in retrospect, "we can *explain why* certain animals and plants have survived even when we could not have *predicted that* they would." In fact, given such a law, even if we could have predicted the flood, we could not have predicted anything about individual animals. We could have predicted only that more animals who can swim would survive than those who could not, that is, an implicit frequency distribution. By the same token, we cannot explain why a particular animal happened to survive. For, as Scriven says, there may be many other unknown factors besides swimming ability that contribute to survival. Our knowledge is incomplete. That, indeed, is why we can only state a "hypothetical probability prediction" or statistical law. In other words, though we can explain and, moreover, explain deductively why more fishes than chipmunks survive, we *cannot* explain why a particular fish survived. Since many do not, the "explanation" in terms of swimming alone is clearly inconclusive, far from "certain," no matter how plausible it may appear.

From statistical generalizations, we do not deduce "with probability" that a certain event will occur, rather we deduce exactly the relative frequency or "probability" with which an event will occur in a certain group.

[19] In *Science*, 130:478.

Similarly, contrary to what Scriven maintains, both statistical and deterministic laws are falsified if the prediction fails.[20] The difference is only in the falsifying event. In the statistical case, the failure, not of an individual event, but of an attribute to occur with a certain frequency in a "mass event" falsifies the law. If our generalization is not quantified and says merely that one event is "more likely" than another, then if in other large samples that event does not turn up more times, the law is falsified. If of a specified large group of cigarette smokers, satisfying certain conditions, only 50 per cent develop lung cancer before their death, then the "law" has been falsified. The use of statistical hypotheses does not, therefore, require abandoning deduction. Quite the contrary. Just as in the deterministic case, without deduction we could neither test statistical hypotheses nor, for that matter, have any rational grounds for, say, recommending a decrease in cigarette smoking or, to change the example, to innoculate our children against poliomyelitis. No doubt we would much prefer to know that each and every child receiving a certain vaccine is immune to the disease, or the exact conditions under which cancer develops in a particular person. Yet, statistical knowledge is not to be scorned though it is imperfect rather than perfect. It is far from valueless to know the factors statistically correlated to the frequency of occurrence of an event. And it is the exact deductions from such knowledge that make it valuable.

Scriven's claim that statistical premises nondeductively permit us to explain individual events rests on no firmer foundation than converting a specious plausibility into a "certainty." But he also maintains that we can explain individual events without recourse to laws at all. In particular, an appeal to laws is held to be "wholly unnecessary" for identifying the causes of an event. Let us see.

Explanation by 'Causes' vs. Explanation by Laws

"We can explain but not predict whenever we have a proposition of the form 'The only cause of X is A,' for example, 'The only cause of

[20] Just as deterministic laws cannot be conclusively *verified*, statistical laws, by their nature, cannot be conclusively *falsified*. Accepting an observed frequency in a sample as the true probability requires an induction that this frequency will persist for indefinitely large samples; this may not be correct. Lack of conclusiveness, however, does not mean that we may not have good evidence either for accepting a deterministic law or for rejecting a statistical one. Conclusiveness, again, is not required of the premises but only of the connection between premises and conclusion.

paresis is syphilis.'"[21] Given A, we cannot predict X, for only a few cases of A develop X. Only A in conjunction with certain unknown conditions is followed by X. Therefore, Scriven maintains, given a case of A, "on the evidence" we must predict that X will not occur. He is mistaken. "On the evidence" we are in no position to predict any such thing. No such prediction is *logically* justified, no matter how soothing or useful it may be, any more than we can predict of any particular cigarette smoker that he will or will not get cancer. To make his prediction, Scriven explicitly uses the premise that only a few A's develop X. Now, if we make the decision, as is customary when action is necessary, always to assume that the statistically more likely case will occur, then of course we can predict that X will not occur. But, then, having made that assumption, the prediction of the nonoccurrence of X follows *deductively* from these premises. True, we cannot predict, under the hypothesis given, that X *will* occur, but when it does we can explain it. However, as we shall see, the deductive model can account for this practical asymmetry.

The sentence "The only cause of X is A" needs considerable unpacking. Scriven, believing that "cause" is an unanalyzable concept that everyone just naturally understands, denies any need for unpacking. He can therefore maintain that such sentences present instances where we can "explain what we could not have predicted, *even if we had had* the information about the antecedent conditions." Or, as he goes on to say, "sometimes the kind of correlation we need for prediction is absent, but a causal relationship can be identified."[22] The only way Scriven persuades himself that he can explain an event that could not even in principle be predicted is by leaving "causal" statements wholly unanalyzed. Despite the confident use of the causal idiom in everyday speech, we may still significantly ask under what conditions statements like "C is a cause of E" are true or false. I shall not take the time here to exhibit the problematic nature of the notion of "cause." Nor do I believe that to most this needs exhibiting. How then must the statement be unpacked?

To say "The only cause of X is A" is *at least* to affirm the law that X never occurs without A. In other words, A is a necessary condition for X, or "Whenever we have X, then we also have A." It is also, however, to say *more*, namely, that there is a complex of conditions, of which A is always one, under which X occurs; that is, certain other factors, b, c, d,

[21] Scriven, in *Science*, 130:480.
[22] *Ibid.*

and A are sufficient for X. In other words, A is a necessary condition and also one of several jointly sufficient conditions. This indeed is a situation in which we speak of one event as the "cause" of another. As is obvious, this causal imputation is far from being independent of any laws, known or surmised. By hypothesis, we do not know the other sufficient conditions. If we knew them, then deductive explanation *and* prediction would follow directly from a statement of the necessary and sufficient conditions. However, since we do not know the sufficient conditions, how do we account for the fact that actually we would normally explain X by A? Our only justification, and in fact the only way anyone, including Scriven, does justify doing this, is by implicitly adding to our knowledge of the necessary-condition law, our "guess" about the sufficient-condition law. Knowing that both X and A have occurred, we assume the presence of the unspecifiable b, c, and d. The explanation of X then follows deductively. That is *why* we accept A as an explanation of X. In order to predict X from A, the unknown factors must also be specified and this we cannot do. The asymmetry exists in practice, but not in principle. Nor is this an "unhelpful" sense of 'in principle.' For only by exhibiting the form of the argument that would, if we knew b, c, and d, permit the prediction, can we clarify why the purported explanation really does state "why" X occurred.

This explanation implicitly assumed that certain unspecifiable events had occurred. Even more frequently, perhaps, the implicit premise is that certain easily specifiable events did *not* occur. We explain a man's death as due to his being struck by an automobile. This is not because we grasp the meaning of the term "cause,"[23] but because we know a law to the effect that if anyone is struck by a car, then he will be either killed or badly hurt. This is imperfect because, among other things, it does not tell us *which* alternative will occur. We therefore cannot predict the death from the law and the fact that he was hit by a car. But knowing it and, therefore, knowing also that the second alternative (or whatever others there may be) did not occur, the death follows deductively from the law, the initial conditions, and the denial of the alternative. The explanation of the event is conclusive because, given the explicit and *implicit* premises, it must have occurred. Clearly, if we knew the implicit premise before the event, it could be predicted. In very many cases where we know

[23] *Ibid.*

251

such "disjunctive" laws, we can in fact also eliminate all but one of the alternatives.

Explanation and Imperfect Knowledge of History and Society

Consider now the perhaps more radical situation, where an event in the past, for which there is no question of predicting, is to be explained. Every day we give such *post hoc* explanations of unpredicted events. The issue is not whether or not we do so, but whether or not the deductive model can account for these explanations. We have already seen some cases where it can. But let us examine events more properly historical and social. Take an easy case first. The unpredicted Lisbon earthquake occurred. We explain the consequent misery and wreckage by reference to it.[24] Of course we do. Furthermore, we do so deductively. From the law 'Earthquakes lead to misery and wreckage' and the *now* known initial condition, namely, the earthquake, there is no problem about explaining the misery and wreckage. The explanation of the event after the fact is clearly deductive, by means of a law, not without it. It is no miracle or accidental conjunction of events that misery accompanied the Lisbon earthquake; we expect it to do so in all earthquakes. Only because we know that earthquakes lead to destruction, floods, and so on, which in turn result in human misery, does the earthquake explain the misery. Otherwise, we might as well say that the Lisbon earthquake was just an incidental accompaniment of the misery. Without the law, the explanation is inconclusive; maybe we can sometimes have earthquakes in heavily populated areas without misery and wreckage. Only because of the deductive connection can we say it *must* have been so. Thus only the deductive connection explains the event. But we did not and cannot predict either the earthquake or, *ipso facto*, the misery. Yet, surely an effort of will is needed to doubt that, even though our recent record is not markedly more notable than at the time of Lisbon, we might someday know enough geology to predict earthquakes. Our knowledge in such matters is imperfect: we know that areas of a certain geological structure can expect earthquakes, but we cannot predict exactly when they will happen. The ex post facto explanation is nevertheless deductive. The principle of the deductive model has not been confuted. But what of historical

[24] Scriven, in Gardiner, p. 468.

events that do not have physical causes? The situation is admittedly more difficult to reconstruct, but not on that account wholly resistant to deductive explanation.

Suppose that we wish to predict the Supreme Court's decision on a school integration case that is before it. How do we evaluate two different predictions about the outcome? We consider the reasons that, if pressed, each predictor will give. If no reasons can be given or if the reasons, when made fully explicit, do not appear to be true or, if true, do not entail the predicted decision, then, with equal justification, the opposite may as well have been predicted. What kind of statements could serve to *justify* a prediction? We might well cite approximate statistical laws about certain social variables, for instance, about the equalizing effect of war and industrialization. References to our so-called national character implicitly involve other such laws about how Americans have behaved in the past and can be expected to behave in the future. Statistical laws about individual behavior, for instance, of the judges, would also be employed. Knowledge of this kind, in conjunction with what we know about the present situation, serves as the basis for "educated guesses." The best educated guess is one that *must* be true, if certain premises are accepted. We make this guess rather than another because it is logically entailed by our various bits of "imperfect" knowledge. It is conjectural because the premises are conjectural, but the connection between them and the event predicted is deductive.

After the decision has been handed down, we explain it in the same way. Of course, if by 'prediction' we mean any prophecy, or simply "a claim that at a certain time an event will occur,"[25] then we may certainly predict without being able to explain. On the other hand, if by a prediction we mean one for which *reasons* can be given, then after the event we should also be able to explain it. If we cannot, then we know that our premises must be false. Since after the event we know what conclusion we want to draw, it is all too easy to invent the appropriate reasons as premises from which the court's decision, say, follows. Only successful prediction gives us evidence for accepting the premises. "Prediction" need not be of an event in the future, only of something not as yet known.

In history, too, we use all kinds of imperfect knowledge about societies, institutions, and individuals to make inferences not about the future but

[25] Scriven, in *Science*, 130:480.

about what we do not yet know in the past. We then look for the evidence in the usual historiographical or even archeological sources. It is true that having predicted a current political event, we can wait and see what happens. However, in history, after all, we are also given plenty of well-confirmed facts. We are frequently much more concerned to explain than to establish them. So the principle is not really different. There is no such thing as "historical" explanation, only the explanation of historical events. To be sure, such explanations, as well as those proposed for contemporary events and for complex human behavior, are all in the same boat. History, whether contemporary or past, far from being the paradigm of sound explanation that some now take it to be, must make do with conspicuously imperfect knowledge. I shall make a few more remarks about explanation in history later on. But, as we saw, deductive explanation and prediction do not require that our premises be "perfect." The critics make the philosophical claim that inference from one fact to another does not require the use of laws. Accurately as this may describe our enthymematic speech habits, the claim is not logically justified by the fact that our knowledge is, in many areas, imperfect and, for all we know, likely to remain so.

Inference Tickets and Enthymemes

Once we have distinguished between the elliptical, context-bound use of language for the purposes of communication and the use of language for description, it is clear, I believe, that insofar as an individual fact justifies the inference to one or more other such facts, laws are always implicitly or explicitly invoked. Otherwise, the justification of the inference remains a mystery. Since those who now assail the deductive model deny that the premises logically entail their conclusions, we may well ask what they believe *does* justify the inference? The answer lies in the use, in one form or another, of Ryle's notion of an "inference ticket." Attending to those idiomatic expressions that signal the use of argument, like "because so and so, therefore," "this, so that," "this involves that," as well as "if, then," Ryle concludes that individual facts are deduced from premises stating other individual facts. Ryle accepts the standard interpretation of natural laws as hypothetical or conditional statements. He denies, however, that laws state facts. "Butter melts when heated" or "Tempered steel is flexible" are, on his account, not factual assertions. This, if I may say so, is an eminently "philosophical" rather than commonsensical use of 'fact.' Philosophical uses are puzzling and must be explicated. Ryle does indeed provide an explication, one which, as I shall try to show, carries with it an untenable implicit metaphysics. The difference between singular statements (not containing dispositional predicates) and hypotheticals, according to Ryle, is in the jobs they perform. The job of singular statements is to report "facts," that is, actual happenings or occurrences. Hypotheticals, on the other hand, "narrate no incidents."[26] Their job is to serve as inference tickets, licensing inferences from one or more individual facts to others. The general hypothetical "warrants" the argument "from factual statement to factual statement," but is not itself part of that argument. Judged by the logicians' account, Ryle's is clearly enthymematic. The "inference ticket" licensing the deduction is, of course, not the general hypothetical which is instead an implicit premise, but the rule of *modus ponens*.

The inference-ticket notion plays a large role in the arguments of those who now attack deductive explanation. The universal statement is said to serve as a "warrant" or "justifying ground" making "legitimate" or "guaranteeing" the step from fact to fact. Though Scriven's boundless confidence in "judgment" and "primeval certainty" leads him to deny that "we should lose faith in an explanation" if we cannot formulate justifying grounds, he concedes that sometimes universal premises are used to "justify" an explanation. Since these warrants, justifying grounds, or inference tickets are allegedly not part of the explanation itself, explanation from individual fact to individual fact is not deductive. It is indeed held that no distinction can be drawn between deductive and inductive arguments, that the logician's formally deductive arguments are really inductive or "substantial." For instance, the astronomer can predict a future eclipse from the "standard equations of stellar dynamics" and the present position and motion of the heavenly bodies involved. Yet, this prediction is said to be "substantial," not tautologically deductive, because the prediction allegedly is made from the present and past positions of the heavenly bodies. These latter—the data and "backing" for the standard equations—"do not positively entail the conclusion."[27] Of course they don't, but no one ever said they did. Patently, the evidence we have for a uni-

---

[26] Ryle, *The Concept of Mind*, pp. 120, 125.

[27] Toulmin, *Philosophy of Science*, pp. 84–85, 93ff, and *The Uses of Argument*, pp. 101, 114, 121–122, 220; Scriven, in Gardiner, pp. 446ff, 456ff; Ryle, *The Concept of Mind*, pp. 120–125.

versal statement—the standard equations—is here confused with what that statement itself actually asserts. The conclusion follows from the equations themselves and the appropriate initial conditions. Either the equations—the so-called warrant or justifying grounds—and the initial conditions jointly entail the conclusion deductively or they can neither predict nor explain it.

The suggestion to "construe" natural laws as material rules of inference is not a new one. It has been made from time to time by various philosophers as a way out of certain philosophical problems regarding universal propositions. Various difficulties with this view, apart from the fact that this is not their actual role in scientific practice, have also been pointed out.[28] The philosophical nerve of the matter, however, is plucked by Braithwaite's comment that to treat a logically contingent statement as a principle of inference is to "mix experience and the logical methods by which we think about experience in a very confusing way."[29] Why this is confusing should be, but apparently is not, obvious. One reason it is not obvious is the promiscuous use of 'logical' for quite different things, while at the same time retaining the connotations of the term in its strict sense. Statements about the past, for instance, are said to be "logically" different from statements about the future, and *therefore* one cannot be deduced from the other.[30] This is but one instance of an appeal to "logic" as if that were what it is not, namely, a logical reason in the strict sense.

Another instance is the use of "logic" to mean function or job. Both contingent hypotheticals and logical truths are used as "inference tickets"; both therefore have the same jobs or "logic." Calling two things that may have the same function in communication "logically" the same at best blurs the issue about the structural or logical, in the strict sense, difference between analytic and synthetic statements. This difference is signaled by at least some uses of 'necessary' in common speech. These, in turn, may reflect differences in the descriptive content of each kind of statement, no matter what their "job." The issue appears to be settled in the negative

[28] See E. Nagel, "A Perspective on Theoretical Physics," *Mind*, 63:403–412 (1954), reprinted in his *Logic without Metaphysics* (Glencoe, Ill.: The Free Press, 1956); and H. G. Alexander, "General Statements as Rules of Inference," in *Minnesota Studies in the Philosophy of Science*, Vol. II.

[29] R. B. Braithwaite, *Scientific Explanation* (Cambridge: Cambridge University Press, 1953), p. 86.

[30] Toulmin, *The Uses of Argument*, pp. 13, 220.

when in fact it has not even been raised. But there is an even more funda-mental reason for this confusion of experience with reasoning about it, to which the others are mere corollaries.

In ordinary communication, if a person asserting "this, so that" can produce a corresponding general hypothetical, we say that he has "the right" to conclude that from this.[31] But suppose we go on to ask what there is about the world or the way we reason about it which justifies our saying this. That is, why does the general hypothetical give us the right to infer that from this? I mentioned before that an appeal to logical truth or to definition is conclusive in a way that appeal to contingent truth is not. If we accept the logician's notion of logical truth, then we can an-swer this question. *Modus ponens* licenses the inference and that, in turn, is justified by appeal to the notion of logical truth or tautology. Ryle re-jects this answer in part at least because he rejects the question. That is, he holds that an appeal to a statement of the form 'If P then Q' is as conclusive as an appeal to a statement of the form 'If 'If P then Q' and 'P,' *then* 'Q.'' At one place, Ryle rejects the logician's reconstruction of the "p, so q" argument as enthymematic on the Tortoise and Achilles grounds that the use of a rule as a premise leads to an infinite regress.[32]

But Ryle's use of this argument is mistaken. Since they are metalin-guistic, rules are indeed of a different logical type from premises. But the justification of, say, *modus ponens* does not involve the use of its corre-sponding hypothetical as a premise, or, therefore, the use of another rule, and so on ad infinitum. Rather, the use of the rule is justified by pointing out that the hypothetical sentence corresponding to it—which is *not* a premise of the argument for which it is used—is a tautology. All that the Tortoise proved to Achilles is that some (at least one) rules are always necessary. He did not prove that these indispensable rules could not them-selves be justified. This justification is indeed a fundamental task of the philosophy of logic, as contrasted with logic itself. Ryle rejects this task and holds instead that one cannot even ask what justifies the rule because the job of the hypothetical is to *be* a rule. It "means" a rule, and the ap-peal to "meaning" is, everyone agrees, conclusive. With the appeal to "meaning," we come to the heart of the matter, the confusions about "conceptual analysis."

[31] Ryle, *The Concept of Mind*, p. 300.
[32] Cf. G. Ryle, " 'If', 'So' and 'Because'," in *Philosophical Analysis*, Max Black, ed. (Ithaca, N.Y.: Cornell University Press, 1950).

May Brodbeck

"Conceptual Analysis" and Rejection of
Hypothetico-Deductive Theories

The critics of deductive explanation quite consistently also attack the hypothetico-deductive model of scientific theories, according to which a theory is a deductively connected set of general statements, some of which, the premises or axioms, logically imply others, the theorems. These statements of a theory are hypothetical in a double sense. First, as empirical, contingent, or logically synthetic general statements, they may be falsified; second, some of them are used as premises in a deductive argument—the theory—whose purpose is to show what else must be true *if* the premises are true. It is worth mentioning that neither of these senses of "hypothetical" has anything to do with the *process* by which universal hypotheses are formulated, whether it be by induction, a hunch, or a dream. They are "generalizations" only in the sense that they are of generalized, that is, universal form. Some believe that if laws are not arrived at by a process of generalization from observed instances, then they cannot be contingent universal or general statements.[33] This is certainly consonant with the typical process-product confusion, but otherwise has less than nothing to recommend it. But the matter goes considerably deeper than this apparently merely verbal matter. For the critics deny that scientific laws are "hypothetical" in either sense mentioned. They are, of course, held to be rules and not premises in a deduction but, more revealingly, neither are they contingent or subject to refutation by observation. They are rather true by virtue of "meaning," by virtue of the way the scientist uses the terms connected by laws. A few comments about meaning are first of all in order.

Terms themselves do not mean. We mean by their use. Yet terms, including 'meaning' itself, have many different uses. We ask for the meaning of life, wanting to know why we are here, and whither, if anywhere, we are going. We ask for the meaning of an event, like a falling star, wanting to know, perhaps, what it portends. Or we ask for the meaning of another event, like an election, wanting to know what it indicates about the temper of people. We search out the meaning of a drama or a novel, wanting to know what moral it points for man and his world. Or, more mundanely, we ask for the meaning of words, like 'rabbit' or 'acceleration,' wanting to know no more than to what they refer. In the former, possibly

[33] See Toulmin, *Philosophy of Science*, p. 49; N. R. Hanson, *Patterns of Discovery* (Cambridge: Cambridge University Press, 1958), pp. 107ff.

258

more intriguing, questions about "meaning," we are interested in significance in a common-sense use of that term that is generally clear in context. In the last example, however, we are interested in the observable referent of the term. 'Rabbit' means (is used to refer to) rabbits; 'acceleration' means (is used to refer to) the rate of change of velocity with time. In asking for "meaning" in the sense of significance, we want to know with what *other* things the events or things asked about are connected. Some terms, like 'cephalic index,' have meaning in the sense of reference but no significance. We know well enough what the term is used to refer to, but we know nothing about it, that is, there are no laws connecting a person's cephalic index with any of his behavior.

Other terms have both meaning and significance. We know that rabbits eat lettuce, and we also know, among other things, that the acceleration of free fall is a constant. If we did not know the meaning of a term in the referential sense, then we could not discover its significance. We would not, in fact, know what we were talking about, either what it was that ate lettuce or what it was that is constant. A worthwhile concept has both reference and significance. In the first sense of meaning, terms themselves do not mean. We mean by their use. To put it differently, *referential* meaning is a matter of convention; it is something we give the concept. *Significance* or lawfulness, however, is not a matter of convention, but a factual matter, that is, a matter of the way things are. We can define any concept we wish to define. But we cannot endow a concept with significance. It either has it or it has not.

To blur this distinction is to blur a contribution of the mind—the concepts we use—with what is not such a contribution, but, independently of the way we speak about it, is a matter of the way the world goes. Realism and idealism are the relevant metaphysical tags for those who, respectively, insist upon and those who blur this distinction. In particular, the formula "Meaning is Use" blurs the distinction. More particularly, the doctrine that natural laws are true by virtue of the "meaning" of their constituent terms obliterates it. For the two questions, about what a thing is and about what happens to it, are held to be not two questions but one. The terms, we are told, cannot be identified apart from the laws in which they occur.[34] In other words, given a law, "If P then Q," we cannot know that we have P unless we simultaneously know that "it" has Q. This

[34] Toulmin, *Philosophy of Science*, p. 52; Hanson, *Patterns of Discovery*, pp. 61ff.

seems to make Q a definitional property of P. But definitions are tautologies, and it is denied that laws are tautologies. Yet the "meaning" of 'P' is 'Q,' and denial of the law is "conceptually untenable."[35] Again, all terms are said to be "theory loaded." Their meaning is given by the entire theory, that is, the context, in which they occur. Since the relevant context varies with each use and each user, no two people ever use a term in the same way. Lightning and thunder are said to mean something different to a youngster than to a meteorologist. A clock means something different to Galileo's apprentice than to Galileo.[36] *Accurately* stated, the meteorologist *knows more* about flashes and rumbles than does the boy, and Galileo knew more about clocks than did his apprentice. Otherwise, how would we ever know that they were all talking about the same thing? If a term means something different in every context in which it occurs, then, as has been well pointed out, an exception to or falsification of a law becomes "conceptually impossible" because the term means something different in the law and in the statement of the exception.[37] If an exception is "conceptually impossible," then the statement itself is conceptually "necessary." Instead of our speaking being determined by the way the world is, what is possible in the world is determined by the way we speak about it.

On the hypothetico-deductive account of theories, they have the form 'If 'If P then Q' and 'If Q then R,' *then* 'If P then R''; the major "if-then" expresses a logical entailment between the premises or hypotheses of the theory and the conclusion. However, if laws are held to be rules for "deducing" one individual fact from others, then what rule justifies deducing laws themselves from other laws? The critics need not answer the question for, on their view, no such deduction occurs. A theory is not a deductively connected set of statements, as indeed, if the terms differ in meaning each time they occur, it cannot be. Not the sentences, but the terms of a theory are said to be "logically linked," in some yet further Pickwickian sense of 'logical.'[38]

The idealistic monists, it will be recalled, rejected classical syllogistic logic, the only logic they knew, because they maintained that terms differ

[35] Hanson, *Patterns of Discovery*, p. 115.

[36] *Ibid.*, pp. 56–57.

[37] In the excellent critical review of Hanson's book by P. K. Feyerabend, "Patterns of Discovery," *Philosophical Review*, 69: 247–252 (1960).

[38] Toulmin, *Philosophy of Science*, p. 85.

n meaning every time they occur, so no conclusion could "really" be
drawn by the methods of that logic. Rejecting the abstract universal, in
favor of the so-called concrete universal, they argued that not sentences
but concepts were "deductively," as they said, connected by the expansion of "meanings" into ever broader and more inclusive contexts. So it
is here. What I have called the component laws are instead statements
about the meanings of the terms. The "Q" in the first premise has a different "meaning" from that in the second because it occurs in a different
context, yet they are "logically linked" since one incorporates or expands
upon the other. The meaning of a term is thus given by all the statements
in which it occurs. These statements are alternatively either rules for the
use of the term or "linguistic," "conceptual," or even "logical" truths
about it. True by "meaning," they are yet synthetic or, at least, not analytic. Though not tautologies, they are nevertheless the last court of appeal. Used as rules, they need no further justification. That is why hypothetical or conditional statements are "nonfactual" and only singular
statements are empirical statements of fact. All generalizations, once they
have been accepted and passed into usage, become true by meaning. It
then becomes "conceptually impossible" to deny an empirical law. The
bridge from 'conceptual' to 'conceivable' is as short as that from 'conceptual' to 'logical.' The psychologically inconceivable thus becomes the
"logically" contradictory, their negations, "logically" true.

Borrowing next on the conclusiveness of the appeal to logical truth *in
the strict sense*, the appeal to these nonanalytic "conceptual truths" de-
lusively appears equally conclusive. That is why one can say that the rule
"licenses" the inference, but one cannot ask what licenses the "license."
It is also the basic reason for the rejection of formal logic, whose terms in
an argument must remain constant in meaning and which distinguishes
generalized statements that are logically true from those that are factually
true. Denying that deduction in the formal sense ever occurs, naturally
there can be no deductive explanation or prediction. If the "meaning" of
a concept is always another concept, then the job of statements to describe the world, their connection with something *nonconceptual*, becomes inexplicable. Confusion of sociopsychological description of language as communication with structural analyses of what this communication asserts about the world leads, not for the first time, to a philosophy
that loses the world in a system of "meanings." This untenable conse-

261

quence is, we shall see, further supported by analysis of the more moderate criticism of the deductive model.[39]

Criticism of the Premises of the Deductive Model

The more moderate criticism of the deductive model agrees that unless an explanation logically entails, in the strict sense, what it explains, then "it will fail to explain why what it purports to explain should have happened rather than something else."[40] It denies, however, that any of the premises need be a universal law. The appearance of paradox, if strict deduction is not abandoned, dissipates when we learn that a general premise is indeed required, though this premise is not granted the status of a natural law. Nor, as I shall try to show by examining the position as presented in a carefully argued paper by Donagan,[41] is the disagreement merely a matter of words. Since Donagan's major criticism as well as certain associated doctrines derive from some of Ryle's ideas that have already been discussed, it will be possible to be rather brief. Donagan's argument is especially worth examining because, free both of the tedium of "arguments" that amount to little more than puns from usage and of obscurantist appeals to "judgment" and "intelligibility," it points up even more sharply the structural connection between the various doctrines and the virtually universal hostility of their proponents toward a science of behavior.

Donagan's argument against the deductive model allegedly depends upon Ryle's philosophical behaviorism. Our common speech is studied with expressions which, if taken in their "ordinary sense," as Moore would say, are used to refer to mental states, to such things as conscious feelings, thoughts, desires, beliefs, and sensations. But ordinary language is fickle and here deceives us for, according to Ryle, mental states are but "mythical" entities, products of the metaphysicians' fancy. Accordingly, he "construes" statements about mental states as general hypotheticals or dispositional statements about how people behave under certain circumstances. These dispositional statements license us, in the now familiar manner, when confronted with the circumstances stated in the

---

[39] For an analysis of the metaphysical uses or misuses of this notion of "conceptual truth," see G. Bergmann, "Strawson's Ontology," *Journal of Philosophy*, 57:601–622 (1960).

[40] A. Donagan, "Explanation in History," *Mind*, 66:145–164 (1957); reprinted in Gardiner, pp. 428–443.

[41] *Ibid.*

protasis, to predict that the behavior mentioned in the apodosis will occur. Or, conversely, we are licensed to explain the behavior by reference to the protasis conditions. Donagan does not adopt the inference-ticket view. The license for him, as for the logician, lies in the entailment relation, not in the general hypothetical itself. The latter therefore must occur as a premise of the argument. The use of such hypotheticals or dispositional statements as premises in deductive explanation does not depend, however, as Donagan believes, upon the truth of Ryle's philosophy of mind. Methodological behaviorism as a program in psychology and, more generally, the scientific study of man is one thing. There it is proper and essential.

Philosophical behaviorism or materialism, the denial that there are such things as mental states, is something else again and is not implied by methodological behaviorism. Drawing his philosophical conclusions about what exists from the study of language as a vehicle of communication rather than description, Ryle finds ordinary language a poor guide to ontology. Identifying how we can tell with what there is, he revamps the old verifiability theory of meaning. The meaning of mental terms in common speech becomes the behaviors that, as we ordinarily say, testify to them. I suggest that to argue from how we communicate with each other about our own and other people's states of mind to what exists is exactly analogous to the error of those verificationists who "construed" statements about the past to be "really" about the present and future, because it was by means of such observable consequences that they were verified. The ontological price for ignoring the difference between the communicative and descriptive uses of language is high indeed. Be that as it may, we are here concerned with the possibility of scientific, deductive explanation and not with the philosophy of mind. We need therefore only inquire whether the differences Ryle discerns between explanations by so-called lawlike dispositions and explanation by laws are indeed real differences. For two things are made to depend upon them. First, only explanations by means of laws are held to be "causal" explanations. Second, explanation by means of, allegedly noncausal, "lawlike" motive or dispositional statements is held to be final or conclusive, requiring no further explanation. If true, these claims support the contention that explanation in history and the other studies of man and society is fundamentally different from that in natural science. Since the proponents of the deductive model generally deny that this is a difference of principle, the issue is more than merely verbal.

"Lawlike" Hypotheticals vs. Causal Laws

'Jones is vain' is construed by Ryle as 'Whenever Jones finds a chance of securing the admiration and envy of others, he does whatever he thinks will produce this admiration and envy.'[42] The "mentalistic" terms in this hypothetical—'admire,' 'envy,' and 'thinks'—doubtless could be replaced by behavioristic, dispositional correlates. Such statements are held to be lawlike rather than laws, because, though general with respect to time, they mention individual persons or things. It is worth noting, however, that some uses of dispositional terms are not general at all. Behavioristically defined, both 'hungry' and 'irascible' are dispositional. 'Hungry,' referring to a present state of a person, would be defined by a molecular, that is, nongeneral, hypothetical. 'Jones is hungry' means, say, 'If Jones is now presented with food, then he will eat it'; or, if I may be permitted the greater clarity of a symbolism, 'If $F_1(a,t_1)$ then $F_2(a,t_1)$.' This is a nongeneralized molecular statement. "Hunger" is thus a complex property, being attributed to a specific individual at a specific time, just as "being red and hot" would be a complex property attributed to a poker. 'Irascible,' on the other hand, referring not to a present state of a person but to his disposition to behave in certain ways under certain circumstances, has the form '(t) If $F_1(a,t)$ then $F_2(a,t)$.' It is a quantified or generalized statement. All terms referring to personality or character traits, rather than to present states, would thus be generalized over one or more variables, whether they be time, objects, or circumstances.

Just as these hypotheticals about persons always have some degree of generality, so natural laws are never completely general. Galileo's laws hold only for bodies on Earth; we have laws about the expansion of the chemical mercury and Kepler's laws for the planets; the law of radioactive decay contains a constant whose value depends on the substance involved; and so on. All known laws have some kind of scope restriction which is part of the law itself. Usage certainly does not insist upon complete generality for what is to "count" as a natural law among scientists. The logically significant distinction is generalized *versus* molecular rather than "law" *versus* "lawlike." General hypotheticals about a person are of the

[42] Ryle, *The Concept of Mind*, p. 89.

ame logical form, in the strict sense, as natural laws, while both are to
e distinguished from statements of individual fact or any molecular com-
ound of them. The issue might appear to be merely verbal or at best a
lassificatory problem in a case where the boundaries are blurred, if so
uch were not made to hang on it.

Donagan acquiesces in the standard Humean account of causality, as
xpressed by Hempel's assertion that "Every 'causal explanation' is an 'ex-
lanation by scientific laws'; for in no way other than by reference to em-
irical laws can the assertion of a causal connection between certain events
e scientifically substantiated."[43] Donagan distinguishes such causal ex-
lanation from explanation by dispositions, maintaining that this distinc-
on renders historical explanation noncausal. Ryle, we recall, holds that
"This because that" or "The window broke because struck by a stone" is
self a complete, nonenthymematic explanation. "This window broke
ecause brittle" he also holds to be a complete explanation. The former
, "causal" because it mentions a "significant prior or simultaneous hap-
ening," namely, being stoned. The latter, however, is not causal, because
being brittle" is not a happening or episode. Being brittle is, we also re-
all, not even a fact, as Ryle uses that term, about the window, since it is
xpressed by the general hypothetical, "If this window is struck sharply,
hen it will break." Donagan maintains that though causal explanation is
Hempelian," that is, fits the deductive model, the dispositional one "dif-
ers from anything recognized in the Hempelian theory, which presup-
oses that the only way of deriving the statement that certain windows
roke from the statement that they were stoned is by the allegedly buried
eneral law, 'All windows break when stoned.'"

But the claim that one fits the deductive model while the other does
ot will not bear closer examination. Nor will the further claim about the
nly way to derive the statement that certain windows broke. Ryle's philo-
ophical behaviorism and his rejection of formal deduction here reinforce
ach other. Ryle, one might venture to surmise, insists that mention of
he "nonfactual" disposition is a complete explanation because of his even
reater concern to deny that there are any events or happenings corre-
ponding to mental terms which could serve as causes of motivated events.
f, for instance, "vanity" is behavioristically construed as a personality dis-
osition term, then to say "He boasted from vanity" is not to explain his

[43] Quoted by Donagan, in Gardiner, p. 430.

May Brodbeck

boasting by reference to any "cause" or "happening." [44] The "happening" Ryle is concerned to deny is of course any *mental* happening. The denial of philosophical behaviorism, while accepting methodological behaviorism and the general scientific thesis of psychophysical parallelism, does of course not at all commit one to the belief in mind-body interactionism. What matters, however, is that Ryle, given his quaint views about the subject matter of psychology, *thinks* one is thus committed, so being vain, or in love, or brittle cannot be "facts" or "happenings."

Actually, spelling out these enthymemes, it becomes clear that *both* of them, "Broke because stoned" and "Broke because brittle," mention "significant prior or simultaneous happenings." In the first case, the implicit premise or "inference ticket" is the general law "All windows break when stoned," while the happening, being stoned, is explicitly asserted. In the second case, the lawlike hypothetical defining the dispositional predicate 'brittle' is explicitly asserted, while the implicit premise is the individual statement of fact asserting an instance, namely, that the window was struck sharply, of the protasis of this hypothetical. In all logic, Ryle should permit singular statements of fact to be "inference tickets," for that is their job in such "explanations." Ryle feels no discomfort about this "episode proposition" that, he agrees, is implicitly subsumed under the general hypothetical presumably because it is only the antecedent part of the defined motive or character term and the motive itself, the entire hypothetical, is not made a "happening." Another reason why Ryle holds hypotheticals to be "nonfactual" is that when we explain by reference to a disposition rather than by mentioning a specific event, we say we are giving a reason rather than a cause. [45] I think he is right about this use of 'reason' but only because when we explain by reference to a disposition, we are, by definition of the disposition, implicitly giving the premises of a deductive argument. And giving premises that logically entail a specific event is certainly one good use of 'giving reasons' for that event.

Spelling out Ryle's enthymematic arguments reveals not only that individual happenings are involved in each case but also the inaccuracy of Donagan's contention that the only way for the Hempelian model to derive the breaking of the window from its being stoned is by the buried law 'All windows break when stoned.' It can of course also be derived from the "lawlike" disposition 'If this window is hit by a stone, then it

---

[44] Ryle, *The Concept of Mind*, p. 86.
[45] *Ibid.*, pp. 113–114, 89.

will break.' Therefore, anyone who rejects the inference-ticket notion and opts for *deductive* explanation cannot also accept "being brittle" as *complete* explanation. He has to grant that there is an implicit singular premise. To reject the deductive model, he must therefore fall back on insisting that the difference between so-called causal and noncausal general hypotheticals is more than one of degree of generality. And, indeed, Donagan does give reasons for claiming that there is an *essential* difference between dispositional statements about material objects and those about persons which serve as premises in nonphysical explanation. This alleged difference is believed to be relevant to the deterministic thesis and the possibility of a science of man. For this reason, as well as for what it reveals about "conceptual analysis," the argument is of special interest.

Motives, Meaning, and a Science of Man

The lawlike statement 'If this window is struck sharply, then it will break' is derivable by specialization from the general law about the brittleness of windows, 'If any window is struck sharply, then it will break,' which is generalized not only with respect to time but also with respect to windows. Lawlike statements about a person's motives or character are, however, allegedly not similarly "generalizable." This limitation, in turn, is believed to justify the claim that explanations in terms of motives or character are not only "complete," but also "final" or conclusive. That is, though like Kepler's laws they completely entail what they are supposed to explain, unlike Kepler's laws they do not themselves require further explaining.[46] But, first of all, this notion of "generalizability" is ambiguous. To be sure, we neither know nor expect to know any general statement "All men are vain" from which Jones's vanity could be deduced. What is true of Jones is, in this sense, not generalizable to all men. Does it follow that there is no sense in which lawlike statements about individuals are generalizable, thus turning "non-Hempelian" into "Hempelian" explanations? Donagan believes that it does follow, for the buried assumption behind the lawlike 'The Danes who sailed south to the Irish Sea were plunderers first and settlers by afterthought' would be, he asserts, 'All men were plunderers first and settlers by afterthought.' But surely this is not the correct "buried assumption." If the Danes being plunderers first and settlers by afterthought implies, as he says, the lawlike statement

[46] Donagan, in Gardiner, p. 434; Ryle, *The Concept of Mind*, pp. 89, 325.

that 'If those Danes had opportunities of sufficient plunder in a territor
they would not settle in it,' does it not also imply the same thing fo
Italian plunderers? In other words, the correct buried assumption is tha
anyone's being a plunderer first and a settler by afterthought implies '
he has opportunities of sufficient plunder in a territory, he will not settl
in it.' This is not the same as saying that everyone is a plunderer. Th
implied general statement is the definition, in whole or in part, of the di
position term 'plunderer.' From this definition, together with the info
mation that the Danes were plunderers, we could derive some of their be
havior. To be sure, the universal premise of this explanation is a verball
necessary truth, since it reflects our definition of the term 'plunderer.'

The explanation of the Danes' behavior thus logically follows from th
empirical fact that they were plunderers and the definition of that dispe
sitional term. In the course of deductive explanations, we frequently d
utilize definitions as premises. We must do so to derive theorems tha
contain defined terms from axioms that are expressed in the primitiv
terms of a scientific theory. In science, these definitions most frequentl
take the form of generalized or universal conditional statements. Such
defining sentence, like our definition of 'irascible,' may therefore serve a
the universal premise required by the deductive model. The "explanation
is then vacuous and circular in a way in which deductive explanation
utilizing *contingent* universal premises are not. Literally speaking, the
if the "Hempelian" model requires a contingent universal premise, the
this explanation is not Hempelian. But this is at best a trivial objection
The process-product equivocation in the use of the term 'generalization
which I mentioned earlier, has probably played a role in misleading Dona
gan. Since his premise, though general in form, is a definition and not
contingent generalization, he believes that it does not fit the "Hempelian
model. But the fundamental point of the deductive model is that no in
ference, either for prediction or for explanation, can be made from one o
more individual statements of fact to other such statements, without th
use of a *general* premise. In the most interesting and informative case
the general premise is a natural law, but it may also be a definition. How
ever, no one holds that deductions from definitions are the only kind w
make in the course of an explanation. In addition to the definition w
also must know some general laws containing the defined terms. In th
case at hand, if we also know some general laws about plunderers, as de
fined, laws about, say, their religion or literature, then we could derive i

good Hempelian fashion further facts about the Danes' behavior and, of course, about plunderers of any nationality. Donagan, however, is asserting not only what is obvious, that there is no law about all men being plunderers corresponding to all windows being brittle, but rather that there can be *no general statements about plunderers*, as there are about brittle windows. That is why he believes that the premises in historical explanation must be lawlike statements about *the particular people* involved.[47] In general, explanations by reference to "the character of the agent" are held to be final and conclusive. What are the reasons for this view?

It is perfectly true that in everyday life when we explain a man's actions by reference to his motives, that is all we want to know. We then "understand" why he did what he did. Frequently, that's all we care about. The historian who succeeds in ferreting out an agent's motives has supplied an explanation adequate for his purposes. As we have seen, the reconstruction of such explanations shows them to be deductive, using the dispositional definition of the motive term and individual statements of fact as premises or reasons for the behavior. Other actions are less trivially explained by using *contingent* generalized premises about how people with such motives can be expected to behave in certain circumstances. Generalizations of this sort are, of course, imperfect. Nor is their imperfection due to their limited scope, namely, being about a single individual or group of individuals. They are imperfect because, generally, they are hemmed in by qualifying phrases which render them implicitly statistical. They are also imperfect because they are limited with respect to time. A man's motives may change with his circumstances or with, as we say, "age." While a man was prime minister, his motives were such and such; they changed when he joined the opposition. These circumstances can of course be built into the protasis as part of the initial conditions or as scope restrictions on either the generalizations or the definitions. "Age" is not a true variable, but a cover or suitcase word for all of the many things that happen to a man as he grows older. The *mot* "Character is fate" enshrines the belief that by understanding a man's character we understand what he does and what happens to him. We may sometimes explain a man's motives at a given time by reference to his character, meaning by the latter certain more permanent and less restricted dispo-

[47] Donagan, in Gardiner, p. 441.

269

sitions. To know a man's character means knowing many generalization about him from which, in given circumstances, both the motives he wi act from and his other behavior can be predicted or explained.

The historian uses these kinds of imperfect laws in order to explain th decisions made and actions taken. Nor does he necessarily "presuppose any general laws about all men. He just uses the laws about his particula man. It would be odd, though, if he were to say that another man o similar character in similar circumstances would nevertheless behave ver differently. "Character," after all, refers to a certain *kind* of man, thoug possibly, but just possibly, there might be only one of that kind. Even so it is perfectly legitimate to ask why someone is the kind of man he is, wh he has the character he has, even though in everyday life and in the stud of history we might not be interested in that question. But certainly pa ents and pedagogues frequently are interested in it. We do try to "build' character. Since we try to do so, we might well want to know what form it. To find that out, we have to go beyond generalizations about particu lar persons. Character is not only a cause of behavior, it might also b caused. Though explanations in terms of the character of the agent migh satisfy us and historians, they might not quench a broader curiosity. Fo it, they would not be final and conclusive. Such persons might try to dis cover the laws of character formation. They might, in other words, try t develop a science of psychology. The fact that most of us are most of th time satisfied with explanation by reference to motives and character doe not rule out a deeper search. The quest may well be futile, but there is nc logical necessity that it be so. The critics of the deductive model, whethe of its premises or its form, express considerable repugnance and opposi tion to this quest. This opposition is buttressed by arguments that g deeper than merely practical considerations. Once again, as we shall see confusion is bred by the notion of "conceptual analysis." The fusing o reference and significance turns a matter of contingent fact, namely, tha we cannot now explain or fully explain character, into a "necessary" anc even "logical" truth.

I have been speaking of "defining" the disposition terms by means o general hypotheticals. This was, of course, a bit disingenuous. For it fol lows from the context theory of meaning I discussed before that the law like hypothetical corresponding to the statement that Jones is vain doe: not define the character trait but instead "expands its meaning." If we wish to unpack "all that is conveyed" in describing Jones as vain, we mus produce an infinite series of different hypothetical propositions."[48] The operative phrase is "all that is conveyed." In everyday life the connotations of a term include both defining and nondefining properties, as long as the latter are fairly widely and firmly believed to be exemplified by the thing in question. This is what lends to everyday speech its so-called open texture. A concept's "meaning" or "criteria of application" expands as we learn more and more about "it." The difficulties with this view I discussed before. Our concepts may be open textured, but the world is not. If language is to be descriptive, it must indicate what there is in the world, no matter how variably we talk about it. On pain of Bergsonian ineffability, a descriptive language cannot *duplicate* the world's growth and change, including that of language. It need only *account for* it. By combining defining and nondefining properties, everything that we know about a thing becomes part of its "meaning." For instance, Donagan tells us that the Danes being plunderers implies not only that they would not settle in a territory but also a "host of further law-like statements" about their literature glorifying war rather than farming, about the kind of religion they would have, and so on.[49] If all these statements do indeed "unfold the meaning" of being a plunderer, then clearly we cannot ask whether one could possibly be a plunderer or, even, a Danish plunderer, and *not* produce this kind of literature or religion. By virtue of "meaning" nothing else could have happened, just as by virtue of meaning, Jones's vanity implies he would do the various things that, as we say, "express" his vanity.

But is everything a plunderer might do part of the meaning of the term? And if not, where do we draw the line? As we come to know more and more about Danish plunderers, more and more becomes a matter of meaning or "conceptually true." Let us show by example where this leads. Within behavior science, where definitions are necessary if the scientist is to know what he is talking about, the dispositional term 'hungry' is defined so that if a person, any person, is hungry, then when, say, he is presented with food under certain circumstances, he eats it. It is then *discovered* that hungry people are irritable. This is a fact or, if you prefer, a law about them. And it could not be discovered unless the meaning of 'hunger' was independent of that of 'irritable.' On the present view, however, this irritability becomes a lawlike hypothetical which is part of the expansion of the meaning of 'hungry.' The more such hypotheticals we

---

[48] Ryle, *The Concept of Mind*, pp. 44, 86, 113.
[49] Donagan, in Gardiner, pp. 436–437.

271

have, the more we add to the meaning of the term, until everything that we say about hunger is true by "meaning." If there are an infinite number of lawlike statements expressing Jones's "kind of vanity," then where do we draw the line between what we mean by his vanity and what happens to be connected with it, like his ambition or his gregariousness? Again, arguing from language as communication rather than description, the world that language is supposed to be about is lost—for terms are tied down not to the world but to an infinite series of other terms or "meanings." The illusory conclusiveness of the inference ticket is now transferred to the lawlike premise. Since the lawlike premises about an individual's character and motives are true by meaning, no explanation of them is either necessary or possible. They are therefore "final." "Conceptual truths" need no further explanation. Like the appeal to logical truth *in the strict sense*, appeal to these conceptual truths about motives and character are "conclusive." By an a priori argument from "meaning," an empirical science of human nature is shown to be unnecessary, impossible, or both.

Concluding Remarks

No matter what other disagreements they may have among themselves, those who now attack the deductive model share certain central and basic ideas which can and do lend themselves to a philosophically argued neo obscurantism. In particular, the two ideas I have stressed, the view of language as communication rather than description and the notion of "conceptual analysis," explain but do not justify their rejection of the deductive model of explanation and of formal logic. These ideas also structurally explain their rejection of a science of man. From the dismissal of the need for a science of psychology because "we all know well enough" why people behave as they do, to a metaphysically argued case for the impossibility of a science of man,[50] the pattern is always the same. Learning a language is a social phenomenon. Once one has learned to use the language of everyday living, by participating and sharing in the social process, then one "understands" individual and social concepts. Once one understands the "meaning" of these concepts, then one *already knows* all there is to know about man and social life. To understand the world, we need therefore not look at it, but merely analyze our concepts.

[50] P. Winch, The Idea of a Social Science (London: Routledge and Kegan Paul, 1958).

The Factual Content of Theoretical Concepts

Philosophers of science have more and more united in rejecting the older positivistic judgment that all descriptive words of an ideal language are, or are explicitly definable on the basis of, terms whose referents are phenomenally "given." It is far from surprising that tendencies toward a phenomenal reductionism should be a serious philosophical pressure within many critical thinkers, for it is indeed difficult to see how the actual "content" of thought (whatever such an expression might mean) could transcend the limits of direct experience. Yet repeated failure to realize such a program increasingly dims the likelihood that scientific or everyday language can be reduced to phenomenal terms alone. To be sure, this might be interpreted as revealing merely the semantic imperfections of existent linguistic practices, but such a gambit is tantamount to abandoning the analytical scalpel for a dogmatic bludgeon, especially since a number of highly competent philosophers have seriously questioned the very possibility of a phenomenal language.

The problems of "meaning" and reductionism come into especially sharp focus in the analysis of scientific theories, for here they find expression in that conceptual framework which we use with maximal clarity. For it is in common-sense object talk, its usage refined and molded by years of pragmatic repercussions, that philosopher and layman alike carry on the business of living. And given this everyday "observation language,"

NOTE: This essay owes its existence to the vantage point erected by the philosophical tradition currently known as "logical empiricism." This movement has with increasing penetration and acuity spotlighted the epistemic and ontological problems that underlie the use of theoretical concepts, and with the assistance of the modern renaissance in formal logic, has been developing an ever more powerful conceptual frame with which to attack these problems. The basic issues involved have been set forth with particular clarity by Feigl [5] and Hempel [7], while the reader will also profit from the articles by Carnap [4] and Hempel [8] in the earlier volumes of this series. I also wish to acknowledge my indebtedness to the National Science Foundation for the postdoctoral fellowship during whose tenure this essay was written.

William W. Rozeboom

in terms of which most practical (if not philosophical) problems appear to be resolvable, we may ask in a matter-of-fact, commonsensical way: What, if anything, can be said with the theoretical terms of a science which cannot alternately be said in the observation language? Or more or less alternately: Are or are not theoretical statements "about" the same things that observation statements are "about"?

The various answers which have been proposed to questions such as these fall into two main categories. On the one hand we find *positivistic* positions, which hold theoretical terms to be either meaningless computational devices or explicitly definable by observation terms, so that statements using theoretical terms can assert nothing inexpressible in the observation language. In contrast, there are the *realistic* interpretations, which regard the designata of theoretical terms to be (in general) beyond the scope of observational reference, a view which might seem to imply that the factual commitments of a theoretical statement are incapable of expression in the observation language. Each view has its difficulties, the former in that its application to specific cases has met with repeated failure, while the latter flirts with transcendentalism. It is my opinion that, as is so frequently true of philosophical disputes, the insights of both positions are substantially sound. I shall argue that the factual commitments of a scientific theory can be expressed—in existential hypotheses, to be sure—solely in the observation language, but that theoretical terms function in a true theory as *names* of the hypothetical entities and cannot be explicitly defined in the observation language.

The remainder of this introductory section will exhibit in greater detail the problem with which we are here concerned, the presuppositions upon which it rests, and the steps to be taken in search of a solution.

The controversy over the meanings of theoretical terms would seem to be founded on the following presuppositions:

1. There exist in the world certain "particulars" (or "objects," if one wishes to accept the additional commitments of ordinary language). These are differentiated from one another by the "properties" they possess or "classes" to which they belong, while the latter, in turn, variously exemplify or belong to still higher level properties or classes, etc. Particulars, properties, properties of properties (where relations may be regarded as properties of ordered n-tuples) and any other components of reality may collectively be referred to as "entities."

274

2. Entities in combination constitute "facts."[1] Thus if a is an entity and P is a property of a, a and P are constituents of the fact that $P(a)$ — i.e., the fact that a exemplifies P.

3. It is possible for one entity to "designate," "represent," "stand for," "refer to," or be "about" another entity. Designators belong to a wider class of entities known as "symbols," certain compounds of which are also able to designate. In particular, if (a) the symbols $a_1, \ldots, a_n$ (e.g., words) are constituents of a larger symbol S (e.g., a sentence), (b) entities $a_1, \ldots, a_n$ are the constituents of a certain fact f, (c) $a_1, \ldots, a_n$ designate $a_1, \ldots, a_n$, respectively, and (d) S conforms to certain other conditions (such as exemplifying an appropriate formal structure), then S represents the fact f in a way that we shall describe by saying that S signifies the fact f.[2]

Presuppositions 1–3 assert merely that there is some sort of reality about which we can talk, speculate, and perhaps have knowledge, and that these cognitive events are possible because certain elements of our symbolic processes stand in some sort of referential relation to components of that reality. Since these beliefs manage to subsume virtually all the problems of ontology and epistemology, they can hardly be said to call for no further explication. Nonetheless, there is an important sense in which they are philosophically neutral—some such beliefs are presupposed by any

[1] The ontological status of facts has been questioned by some philosophers, especially those of an "ordinary language" turn, who, for reasons which seem to me to be either confused or obscure, are unwilling to countenance "facts" as being among what there is, and indeed, even appear unwilling to grant the term any cognitive significance whatsoever. Since the developments in Sec. II, as they now stand, depend essentially on quantification over fact variables, it should be pointed out that it is formally possible to dispense with facts by replacing them with certain uniquely correlated sets. For example, we may replace the class of facts of form $x \, \epsilon \, y$ with the class of ordered pairs of sets such that the first member of the pair is an element of the second. Still another alternative would be to replace "facts" with true statements in a hypothetical omniexpressive metalanguage. In some such fashion, the present analysis could be reworked to arrive at the same conclusions but without depending upon any assumptions about the ontological status of facts. However, the present willingness to quantify over facts is due not merely to the additional difficulties such modification would add to an already complicated story, but even more to the observation that in natural language discussion of such topics as "events," "causes," "phenomena" (in the scientific sense), etc., quantification over fact variables is spontaneous and indispensable. In other words, there are facts, and no theory of semantics can be adequate which does not examine the relationship of sentence to (extralinguistic) fact.

[2] It is tempting to indicate the semantical relation between a sentence S and a fact f signified by S by saying that S "refers" to f. However, this would be misleading, for the relationship that we wish to indicate is a cognitive one, whereas strictly speaking, "reference" is but one of the many uses to which an expression with given cognitive

William W. Rozeboom

serious intellectual undertaking. The aim of epistemology and ontology is not to establish them but to clarify and elaborate upon them. Hence we need not feel uneasy, for present purposes, in taking "entity," "fact," "designate," etc., as primitive concepts. In particular, the reader should not try to read more into the present use of 'designate' than is necessary. We here presuppose no particular *analysis* of this concept (though the outline of a behavioral theory of reference will be suggested later), but simply recognize that if it is possible for a statement to represent a fact, there must be *some* sort of relation between the constituents of the statement and the constituents of the fact. In particular, we need not assume that there is only one kind of "aboutness"—the analysis of 'x designates y' may conceivably differ in important respects according to whether x is a primitive term, a compound phrase, a sentence, or some other component of language which may in some sense be said to point out an aspect of reality extrinsic to itself.

One further background assumption will set the stage for the problem at hand. By a "language," let us mean a stock of symbols together with certain principles of usage such that when properly used, some of the symbols ("descriptive" terms) designate various entities, and certain complexes ("sentences") of symbols can be formed which then signify facts whose constituents are designated by the descriptive terms of the sentences. Then we presuppose that languages do exist and that

4. If a person has "observed" an entity e, then he can add to his language a symbol which, when used by that person, designates e.

Just what is meant by saying a person "observes" an entity is difficult to decide. Fortunately, effective use of 'observed' as a primitive concept does not depend upon clarity in its analysis, for so far as everyday language (upon which the philosopher is no less dependent for communication than anyone else) is concerned, this term is used with as much assurance and precision as any other. There is a very important intuitive sense in which we speak of certain facts, in contrast to others, as having been "observed." For science, in particular, the concept of that which is

properties may be put (see fn. 29). Actually, ordinary English usage (which does more to confuse than to clarify the nature of semantical relations) does not seem to yield a satisfactory term for the relation between a statement and that aspect of reality in virtue of which the statement is true or false. Even to say that a statement is *about* a fact, as will sometimes be done here in informal commentary, is to stretch ordinary usage a bit, for we usually (though not always) say that what a statement is "about" is the entities referred to by its constituent descriptive terms.

observed ("data") plays an especially basic role. If we take 'e is an observed entity' to mean that e is a constituent of an observed fact, then this expression should here involve us in no major difficulties.

By 'observation term,' let us mean a term which has been introduced into a language in accordance with presupposition 4. Then the controversy over the meanings of theoretical terms is basically the question of whether or not a language can contain descriptive terms (i.e., terms that designate) which are not observation terms. Or, phrased somewhat differently: Is it possible for a symbol to designate an unobserved entity? Let a fact, all of whose constituents are observed entities, be called an "observational" fact. Then a third formulation of the positivistic-realistic issue is this: Can a sentence ever signify a nonobservational fact? Or yet again, if we call a sentence all of whose descriptive terms are observation terms an "observation sentence": Can sentences be constructed which signify facts, yet which are not observation sentences? To these questions, the positivist returns an emphatic "No." He by no means necessarily holds that nothing exists which has not been observed—such a view is absurd no matter how one restricts one's ontology. He does insist, however, that only those entities which have been observed can be talked about.[3] The realist, on the other hand, just as emphatically denies that only observational facts can be signified in our language. He not merely admits the existence of unobserved entities but insists that we can and do talk about them. It is important to note that this issue cuts across the question of what can be observed. One need not hold, for example, that only sense data are observable, to be a stanch positivist—witness the operationistic movement in contemporary science.)

The difference between these contrasting views emerges with especial clarity when we try to analyze the factual content of scientific theories. It is a well-known and disquieting fact that the most powerful theories invariably contain symbols which are not logical terms, yet apparently refer to no entity which has ever been "observed" in any intuitive sense of this notion. The positivist is forced to hold either that (a) appearances are deceptive and such "theoretical" symbols do, in fact, represent concepts definable wholly in logical and observational terms, or that (b) expressions containing theoretical terms are merely computational devices which are

---

[3] This principle permeates the writings of Bertrand Russell (e.g. [16], p. 91), although reference to his theory of descriptions [15] is usually necessary to make the thesis explicit. For a more modern statement of the positivistic position, see [2].

no more semantically *about* facts than are calculating machines. The realist, on the other hand, is able to maintain that theoretical terms may designate existent but hitherto unobserved entities, and that scientific hypotheses containing these terms may simply signify certain facts which happen to be as yet nonobservational. The realistic position is a seductive one, but is incompatible with an empirical epistemology unless it is possible to show how, given an observation-based language, a person might acquire additional terms which designate hereto unobserved entities even though his scope of observation remains unchanged. One result of the present analysis is to suggest how this might come about and, correlatively, the limits of such a language enrichment.

Section I will attempt to formalize the concept of (scientific) "theory." The analysis will be idealized in that we presuppose the theory user to have at his command a fully formalizable observation language, all descriptive terms of which designate observed entities (where 'observed' is to be understood in any appropriately broad or narrow sense). It will be heuristically helpful to regard this observation language as an idealized version of the observation language we use in everyday life. In Section II, we turn to the problem of the "factual content" of a theory. We shall be able to determine this without first prejudging whether or not the theory is itself an assertion, though not without making certain general assumptions about the nature of semantical relations and the way in which theories are actually used. In Section III we shall explore the *semantical* status of theories, and conclude that under suitable circumstances, theoretical terms do, in fact, designate unobserved entities. Finally, Section IV considers briefly the implications of this analysis for several long-standing philosophical problems.

Let us conclude this introduction with some needed semantical preliminaries. While it is all very well to undertake analysis of the possible semantical properties of theories, such an effort is especially handicapped by lack of a well-understood and generally accepted theory of semantical relations upon which the analysis can draw. In particular, classical semantics, as formulated most explicitly in the work of Tarski and Carnap, does not adequately deal with the relations between cognitively meaningful sentences and extralinguistic reality (see the next paragraph). Hence the present essay labors under the double burden of developing a semantical theory even as it argues for the meaningfulness of theoretical expressions. While suggested postulates for a generalized theory of semantics will be

278

found in Sections II and III, it is highly advisable to advance a few preliminary considerations at this point.

Any comprehensive theory of semantics must come to grips with (a) the manner in which the semantical properties of a sentence derive from those of its constituent terms, (b) the extralinguistic designata of sentences—i.e., what aspects of reality sentences themselves are about, over and above the referents of their constituent terms, and (c) the truth conditions of sentences. Of these, Tarski's [19] famous schema "S is true if and only if it is the case that p," where S is a meaningful sentence whose metalinguistic translation is 'p,' concerns only the third. In Carnap's [3] more complete theory, a sentence S designates a "proposition" or "state-of-affairs" p (which might be put more idiomatically by saying "S asserts that p") when the descriptive terms in S designate the constituents of p, and S and p show similar composition. Sentence S is then said to be true if and only if there is a proposition p such that S designates p and p is the case.

Unfortunately, this formulation is still not satisfactory for present purposes. To begin with, there is the problem of the ontological status of propositions. These cannot be identified with facts, for propositions are true or false—i.e., are or are not the case—whereas facts are what determine the truth values of propositions. Neither can we identify a true proposition as a fact, for then we have no way to cope with false propositions—to say that a false proposition is a possible but not actual fact is to propose a strange ontology in which nonexistence is a category of Being. By far the most satisfactory interpretation of propositions is to regard them as the meanings, or senses, of sentences—i.e., those aspects of the linguistic process through which sentences are able to make contact with an external reality (see Section III; also [13]). But if so, it is then incorrect to say that a sentence designates a proposition; for "designation" is the relation of aboutness between linguistic and extralinguistic entities, whereas a word or sentence expresses (i.e., produces, evokes, has) a meaning in virtue of which it may designate something else.[4] Hence to analyze the semantical properties of a sentence merely in terms of "propositions" is to leave unexamined the manner in which sentences communicate with the facts that determine their truth.

Now it might be thought that statements of form "S expresses proposi-

[4] Ordinary language is in agreement (for what this is worth) that a sentence expresses, not designates, a proposition.

William W. Rozeboom

tion p," "S is true if and only if it is the case that p," or "S asserts that p," though unable to deal with the designative properties of sentences, might still suffice to determine the truth conditions and hence the factual content of meaningful expressions. Within limits which need not detain us, this is true; however, the difficulty for present purposes is how such statements are to be obtained. For expressions translatable into our metalanguage, the matter is fairly straightforward: If S is a sentence under analysis whose translation is 'p,' there is surely nothing amiss about accepting the metalinguistic statement "S is true if and only if it is the case that p," or even the stronger claim "S asserts that p." However, our major concern here is with the *problematic* semantical status of theoretical sentences, and to presuppose their translatability into the metalanguage would simply be to beg the whole issue at the outset. We shall indeed attempt to arrive eventually at a sentence schema of the form "Theory T is true if and only if it is the case that p," but for reasons which need not be explored here, it seems possible to reach such a conclusion only through analysis of the semantical relations which may obtain between sentences and those aspects of reality which determine their truth values, namely, facts, not merely between sentences and their meanings, i.e., propositions.

When we abandon propositions in favor of facts as the designata of sentences, however, a complication arises. If it is correct to say that sentence S asserts that p, and it is a fact that p, then it seems unobjectionable to conclude that what S designates, or signifies, is the fact that p. But what shall we say when S asserts that p, but it is the case that ~p? What we *cannot* say is that S signifies the fact that p, for there is no such fact. In this instance, however, S stands in an especially intimate relation to the fact that ~p, for just as S is true in virtue of p when it is the case that p, it would seem that S is false in virtue of the fact that ~p when it is not the case that p. Apparently we need to admit two kinds of semantical relations between sentences and facts; one for sentences which make true assertions and another for sentences which make false assertions. More generally, it follows from the assumptions of classical semantics that for each cognitively meaningful sentence S, there is a fact f whose constituents are designated by the descriptive terms of S and which determines the truth value of S. If S is true in virtue of such an f, we shall say that S *signifies f truly*, whereas if S is false in virtue of f, then S *signifies f falsely*. For example, under the classical assumption that a sentence 'P(a)' in which 'P' designates the property P and 'a' designates the individual a, is true if it is the case that $P(a)$ and false if it is the case that $\sim P(a)$, it would follow in the first case that 'P(a)' truly signifies the fact that $P(a)$, and in the second case that 'P(a)' falsely signifies the fact that $\sim P(a)$. Under classical theory, then, each cognitively meaningful sentence signifies, truly or falsely, exactly one fact, namely, the fact that p when the sentence asserts that p and it is the case that p, or the fact that $\sim p$ when the sentence asserts that p and it is not the case that p. There will later be occasion to question portions of the classical view. Nonetheless, the concepts of true and false signification, as roughly sketched here and defined more precisely in Section II, should allow the reader to pass without undue intuitive strain from the more familiar notion of what a sentence expresses or asserts to the needed appreciation of semantical relations between sentences and facts.

I

If we are to determine the factual content of scientific theories, it is first necessary to decide what we mean by a "theory." If we restrict "theory" to "hypothesis formulated in the observation language," we have, of course, cut ourselves off from our problem. On the other hand, "hypothesis (or statement) in a theoretical language" poses difficulties. For in what sense is a "theoretical language" entitled to be called a *language?* It is not sufficient for a string of signs to conform to certain topographical characteristics in order for it to be a "statement," for in its normal usage, this term implies that the sign complex has *meaning.* So long as the meanings of theoretical terms are in question, we are not entitled to call the expressions in which they occur "statements," "hypotheses," or other similar concepts which presuppose a certain semantical status for their subjects. Thus we must find an identifying feature of theories which does not prejudge their meaning content.

It seems quite plain, in the final analysis, that the ultimate purpose of a theory is cash-value prediction—i.e., to assist anticipation of the truth values of observation sentences,[5] given the truth values of other observation sentences. Thus whatever else a theory may be, it is *at least* a tech-

[5] For simplicity, we shall use "sentence" in the sense of "cognitively meaningful declarative sentence," or "statement." It is important to note that an "observation sentence," as defined above (p. 277), may include logical terms, hence permitting molecular and generalized sentences. Therefore, a fact signified by an observation sentence is not necessarily "observable" in the sense in which this expression is frequently used. For example, if 'R(x,y)' is a dyadic observational predicate, the observation sentence '(x)($\exists$y)R(x,y)' cannot, except in special cases, be either verified or refuted by any

281

nique by which some observation sentences are transformed into others. This confronts us with an interesting problem area which does not seem to have been previously explored: Given certain requirements such as consistency, is any *kind* of transformation which (partially) maps the domain of observation sentences into itself acceptable, prior to empirical evaluation, as a legitimate scientific theory? For example, suppose we had (a) a set of rules for generating geometric figures from one another (e.g., "x derives from y and z if x results from the superimposition of y upon z") and (b) a set of "coordinating definitions" setting up a (not necessarily exhaustive) pairing of sentences and geometric figures. Could a transformation technique based upon such a system *in principle* be construed as a theory? If not, on what grounds do we reject it? On the other hand, if a transformation such as this counts as a theory, in what sense can a theory be regarded as an *assertion*?

It would be too lengthy a digression from the main purpose of this paper to explore further the general concept of *theory as transformation* at this time. However, those theories which have actually seen application to human affairs appear to be of a special kind which I shall refer to as "normal syntactic" theories. Such a theory is identified by (1) a set of "inference rules" which are applicable to (but not only to) observation sentences, and when so applied, yield valid deductions; and (2) a set, K, of sentencelike sign complexes, or "theoretical postulates," such that application of the inference rules to the union of K and a set of observation sentences, O, yields another set of observation sentences. By "sentencelike" I mean sign complexes which are syntactically similar to observation sentences in such a way that if certain components (the "theoretical" terms) of the theoretical postulates were to possess designative meaning in the same way that descriptive terms of the observation language have designative meaning, the theoretical postulates would themselves be mean-

finite set of observations. On the other hand, the meaning of 'observable' is difficult to pin down. In what sense, for example, is the fact signified by a statement about the current number of coconuts on an uninhabited atoll observable? Presumably, because if I were there, I would be able to observe how many there are. But how does this differ in kind from saying that if I were acquainted with all pairs of objects, I would be able to observe whether or not $(x)(\exists y)R(x,y)$? To be sure, we believe it to be *physically* impossible for me to observe all pairs of objects, but it is likewise physically impossible for me to be at some spatial position other than where I am, for (presumably) I am where I am because of physical laws controlling the motions of material bodies. If the issue were germane to present purposes, I would argue that the only tenable analysis that can be given to '*f* is an observable fact' is something with roughly the force of 'There is an observation sentence which signifies *f*.'

ingful sentences. There is thus the possibility that a normal syntactic theory is not *merely* a transformation, but that perhaps its theoretical terms do, in fact, designate; in which case the theoretical postulates semantically *express* a hypothesis, and the observation sentences which are derived from the theory are the logical consequences of that hypothesis.

(Syntactic theories of a more general kind would comprise theoretical postulates not necessarily syntactically isomorphic to observation sentences, and, perhaps, inference rules not validly applicable to observation sentences. However, there is then no reason for the theory user to think that the postulates of a nonnormal syntactic theory might themselves signify facts. The present analysis will be restricted to those theories where there *is* good reason to suspect that the theory may be more than just a transformation technique—i.e., the normal syntactic case—although the more general case is certainly of philosophical interest and in fact opens some rather exciting epistemic possibilities.)

For a completely general account of normal syntactic theories, we would have to discuss a wide variety of observation languages. Fortunately, our main point of departure, Theorem 2, can be established with a minimum set of stipulations about the syntax of the observation language which, moreover, would presumably be satisfied by any satisfactory formalization of the language we in fact use in science and everyday life. As consequences of the formation rules we require:

1. Sentences are finite concatenations of certain syntactically primitive symbols, where the latter are of three kinds: (a) *logical constants*, including the truth-functional connectives and existential quantification; (b) *descriptive constants*, which designate specific entities; and (c) *variables*, each of which ranges over entities of a specific kind.

2. All primitive symbols (the logical constants may be excluded if desired, since their inclusion here is trivial) are effectively classifiable according to "formal type," so that each symbol is of exactly one formal type, and each formal type i specifies a class of entities C^i which is the range of every variable of type i and contains all entities designated by constants of type i. (We could also allow a given term to be of more than one type, but this can be reduced to the first case.)

3. Let L_o be the observation language under consideration and L_M the metalanguage in which the present analysis is being conducted—or better, let L_M be the language in which this discussion *would* be conducted were its syntax fully formalized. Then we stipulate that expressions in L_o are

283

William W. Rozeboom

translatable into L_M. That is, the descriptive constants, variables, formal types, logical constants, and concatenation arrangements of L_o correspond distinctly to descriptive constants, variables, formal types, logical constants, and concatenation arrangements of L_M in such a way that if S_o is a sentence in L_o and S_M is the sentence in L_M formed by placing those symbols of L_M which correspond to the primitive symbols in S_o in the concatenation arrangement of L_M which corresponds to that of S_o, S_M is true in L_M if and only if S_o is true in L_o. Although this stipulation concerns more than just the syntax of L_o, it economically characterizes the latter as being isomorphic to part of the syntax of a formalized English. Since we are not concerned with the physical topography of expressions in L_o as such, we may for convenience identify expressions in L_o with their translations in L_M, thus allowing us to write expressions in L_o in the standard logical notation of formalized English (e.g., taking '∃' as the existential operator in L_o, etc.). This will also allow us, in discussion of the semantical properties of L_o, to use expressions of L_o as well as *mention* them (e.g., "The sentence 'P(a)' of L_o is true only if it is the case that $P(a)$"). Granting that the observational portion of a formalized English could be made to satisfy stipulations 1, 2, and 4, stipulation 3 is then trivially satisfied by the observational basis of English as well as by any other sufficiently similar observation language.

4. If a symbol complex of the form '$P(a^i)$' is a sentence in L_o, where 'a^i' is a syntactically *primitive* descriptive constant[6] of formal type i and 'P' abbreviates a simple or complex predicate, then there is also a corresponding existentially quantified sentence in L_o of the form '$(\exists\phi^i)P(\phi^i)$,' where 'ϕ^i' is a variable of type i.[7] Further, if 'S_i' and 'S_j' are sentences in L_o, then '$\sim S_i$' and '$S_i \cdot S_j$' are also sentences in L_o. (Hence we may also assume that L_o contains '$(\phi^i)P(\phi^i)$,' '$S_i \supset S_j$,' etc.)

(Conditions 2 and 4 call for some further comment. First of all, while the languages with which we are concerned admit abstract entities, this is not a result of stipulation 4 but of presupposition 1 of the introductory

[6] No notion of a "descriptive constant" which is *not* syntactically primitive has been or will be explicitly invoked here. However, the term 'constant' is sometimes applied to certain syntactically complex expressions (e.g., compound predicates and definite descriptions) as well as to primitive terms, and the need hence arises (in view of my arguments in [13]) to make clear that stipulation 4 authorizes quantification only over primitive terms in L_o.

[7] This stipulation can be weakened without affecting the argument to follow, so long as theoretical terms are limited to formal types over which existential generalization is authorized.

284

section. I would agree with Bergmann [1], contra Quine, that the ontological commitments of a language are made by its primitive descriptive constants rather than by its variables, though to be sure these commitments are revealed by its variables.[8] In fact, I have never understood the nominalistic thesis that only particulars exist, unless "exist" here means something like "possess space-time positions," in which case the thesis is trivial. If, as other possibilities, the nominalist wishes primarily to challenge that properties can exist without being exemplified, or that every predicate has a referent, he need not commit himself to the view that no abstract entities exist. The use of predicate variables in no way necessitates that every predicate expression constructable in the language must be a substitution instance of a predicate variable, or must be assumed to designate an abstract entity (see [13]).

(With respect to stipulation 2, it should be observed that in letting the formal type of a variable specify its range, we have not stipulated that these types are necessarily the syntactical representations of purely "logical" categories. That is, we need not assume that variables are wholly a part of the logical framework of language. For example, the variable x^s might range only over the class of swans, in which case $(\exists x^s)(x^s \text{ is blue})$ would be true if and only if there is a blue swan. In particular, predicate variables need not be construed to range over the totality of properties or classes of a given Russellian type, especially if it be maintained (as I do not) that to every cognitively meaningful predicate, no matter how complex logically, there corresponds a property or class. The possibility that some variables may be nonlogical terms allows for the possibility of "theoretical" variables. However, the explicit definition of 'theoretical postulate,' below, upon which Theorems 2–5 are based, assumes that theoretical terms function syntactically in the theoretical postulates only as constants. The extent to which this attenuates the present analysis will be discussed at the end of Section II.

(Finally, a word about the status of definite descriptions, if any, in L_o: While contemporary philosophers have yet to reach general agreement

[8] Surely Quine is correct when he proposes (e.g., [10] that whether or not a term 'A' in sentence 'F(A)' can be construed to designate anything is to be tested by judging whether or not the corresponding existential generalization, $(\exists \phi)F(\phi)$, makes sense. But if such an existential generalization would make sense if we made it, we do not nullify the possibility that 'A' carries ontological commitments merely by legislating that $(\exists \phi)F(\phi)$ is not to be considered a sentence of the language. Otherwise, we could rid philosophy of ontological problems altogether simply by refusing to countenance the use of any quantifiers.

William W. Rozeboom

on the syntactic and semantic status of definite descriptions, two main alternatives would seem to be available. Either (a) descriptions are genuine *referring* expressions which play the same syntactical role as do proper names of similar formal type, or (b) sentences containing descriptions are abbreviations for more complex statements which do *not* contain descriptions (see [15]). In the latter case, descriptions are notational conveniences which are not strictly part of the formalized language—or, more precisely, can be ignored insomuch as anything which needs be said about the syntactic or semantic status of a sentence containing a description (its deductive consequences, truth conditions, etc.) is already given by the corresponding statement about the sentence for which it is an abbreviation. On the other hand, if a definite description functions referentially in a language for which stipulations 1 and 4 hold, the description must be classed by the rules of the language as a primitive term, for if a description *refers* to an entity which satisfies it, quantification over only a part of the description yields nonsense. Hence even if L_o is construed to contain definite descriptions, their syntactic and semantic properties are already covered by the rules for primitive descriptive constants in L_o. In particular, to the extent that definite descriptions are considered part of the *observation* language, they do *not* provide a way to refer to unobserved entities, for by definition all primitive descriptive terms of L_o have observed referents. In Section IV it will be argued that definite descriptions are most properly regarded as a form of theoretical term.)

On the basis of stipulations 1–4, the theoretical postulates of a normal syntactic theory may now be characterized more precisely as certain sign complexes which could be generated under the formation rules of the observation language if its primitive descriptive terms were augmented by a set of formally typed but otherwise uninterpreted constants. Let a descriptive constant of the observation language, whatever its formal type, be represented by the notation 'c_i,' where the subscript is indexical. Similarly, let an uninterpreted but typed sign be written 'τ_i.' (For notational completeness, a superscript indicating formal type should be added, but this turns out to be unnecessarily cumbersome. We will be able to omit explicit reference to type so long as it is remembered that the members of a given set of terms are not necessarily of the same formal type unless so stipulated.) Application of the formation rules of the observation language to the observational constants augmented by the uninterpreted constants, 'τ_1,' . . ., 'τ_u,' then generates syntactically well-formed, sentencelike formulas of form '$S(c_1, \ldots, c_m, \tau_1 \ldots, \tau_n)$' ($m \geqq 0$), where '$S(\quad)$' is a sentential matrix containing only logical terms (and perhaps observation-language variables, if these be non-logical) such that if the 'τ_i' were replaced by a set of observational constants, 'c^*_1,' $\ldots$, 'c^*_n,' of corresponding types, the resulting expression, '$S(c_1, \ldots, c_m, c^*_1, \ldots, c^*_n)$,' would be an observation sentence.

For simplicity, we make the further assumption that our observation language L_o is "complete" in that the theorems of the language include all sentences which are formally valid. By "theorem," we mean a sentence which can be effectively deduced by the formal (i.e., syntactical) inference rules of the language from any other sentence. 'Formally valid' may be defined as follows: By a "model" of a language L, we mean any arbitrary assignment of ranges to the variables and designata to the descriptive constants of L, subject only to the restrictions (a) that to each formal type i is correlated a nonempty class of entities, C^i, such that C^i is the range of each variable of type i and contains all designata of descriptive constants of type i; and (b) also, perhaps, additional restrictions on the ranges assigned to the various variables.[9] Then each model of L assigns a specific truth value to each sentence of L, which may be different under different models. For example, '$P^j(a^i)$' is true under a given model if and only if the designatum assigned to 'P^j' is a property of (or, if the entity assigned to 'P^j' is a class, contains) the designatum assigned to 'a^i'; and '$(\exists x^i)P^j(x^i)$' is true if and only if the designatum assigned to 'P^j' is a property of (or contains) at least one of the entities in the range assigned to 'x^i.' We then define the "formally valid" sentences of L to be those which are true in

[9] The various alternative restrictions that can be placed on the ranges assigned by a model to the variables of L generate a whole family of concepts of "formal validity," several of which appear in the technical literature of formal logic. (Just how such formal concepts relate to the philosophical notion of "logical truth" is not at all easy to decide.) Fortunately, it is here unnecessary to be explicit about these additional restrictions, for our present concern with "formal validity" is only to characterize deducibility in L_o, and Lemma 1 is unaffected by any restrictions that may be placed on the ranges assigned by a model to the variables of L. This indifference of Lemma 1 to the ranges of the variables is rather convenient, for it thus becomes unnecessary to make any commitments here as to possible *logical* restrictions on the designata of descriptive terms of various types. In particular, we need not take a stand on whether a primitive predicate necessarily designates a property, a class, or either.

Actually, the present "definition" of 'model of L' is more a heurism than literally correct. A model does not literally assign designata to the descriptive terms of a language, for these already have referents determined by their meanings. More accurately, "model" should be understood as a purely formal concept having to do with the mapping of one domain into another.

all models of L. As is customary—and also necessary if, as assumed later, formal deducibility is to reflect a necessity relation among the truth values of sentences in L—we restrict the formal inference rules of a language to some subset of those transformations which yield only valid deductions; that is, rules such that if deduction of S_j from S_i is authorized, then S_j must be true in all models of L in which S_i is true. This restriction, together with the completeness assumption, entails that the formal inference rules (some of which may be formalized as "axioms" or "axiom schemata") of observation language L_o yield as theorems exactly those sentences in L_o which are formally valid. Recent discoveries in formal logic show this to be a reasonable assumption [see 9]. The completeness assumption is merely for convenience, however, and could be arbitrarily weakened without affecting the validity of Theorems 3–5 (see footnote 11).

For comprehensiveness, let us take as the inference rules of a normal syntactic theory all formal inference rules of the observation language (however, see footnote 11). Then each normal syntactic theory is uniquely characterized by a specified *finite* set of sentencelike formulas—the "theoretical postulates"—of the form '$S(c_1, \ldots, c_m, \tau_1, \ldots, \tau_n)$' as defined above. (The set of theoretical postulates must be finite if the theory is literally to be *believable*—axiom schemata, which are occasionally interpreted as infinite sets of axioms, and which, so construed, would entail an infinite set of theoretical postulates, are most satisfactorily conceived as rules of inference.) More specifically, we shall understand a normal syntactic theory to include in its set of postulates all formally independent sentencelike formulas containing theoretical terms, including "correspondence rules," which are in force when the theory is under consideration.[10] (The set of theoretical postulates may also include sentences wholly in the observation language, indeed, must do so under the definition of 'theory' just offered if there would otherwise be sentences with theoretical terms deducible from the theory plus extraneous observation-language postulates but not deducible from the theory alone.) Obvious-

---

[10] Actually, this may be stronger than necessary. In order to see what is involved in accepting a given theoretical formula '$S(\tau_1)$' containing the theoretical term 'τ_1,' we must determine the pragmatic force of 'τ_1' under the particular circumstances involved, whether this force be characterizable as a genuine cognitive meaning or only as that of part of a transformation mechanism. But while the force of 'τ_1,' and hence that of '$S(\tau_1)$,' will be determined by the role of 'τ_1' not only in '$S(\tau_1)$' but also in other accepted formulas in which it occurs, it does not seem to follow that in order to evaluate the (cognitive or noncognitive) significance of 'τ_1' and '$S(\tau_1)$' it is necessary to con-

ly, a set of theoretical postulates has exactly the same implicative force as the single postulate formed by logical conjunction of the members of the set. Therefore, the postulates of a normal syntactic theory may be written as a single, finite, sentencelike formula, '$T(\tau_1, \ldots, \tau_n)$,' in which the sentential matrix, '$T(\quad)$,' contains only logical and observational terms, and 'τ_1,' $\ldots$, 'τ_n' are theoretical constants. It should be noted that the predicate '$(\lambda\phi_1, \ldots, \phi_n)T(\phi_1, \ldots, \phi_n)$,' henceforth written simply '$T(\phi_1, \ldots, \phi_n)$,' which is ascribed by a normal syntactic theory to its theoretical terms, is an expression constructable wholly within the observation language prior to any consideration of the theory.

It is convenient at this point to introduce the convention that if 'S' abbreviates what is construed as the principal predicate in a sentence of L_o (or, as a degenerate case, if 'S' abbreviates the sentence itself), then 'S' is the name of that sentence in the metalanguage. Thus 'S_i' is a name for '$S_i(c_1, \ldots, c_n)$.' Similarly, 'T' refers to the theory '$T(\tau_1, \ldots, \tau_n)$.' Further, if S_i and S_j are sentences, '$S_i \cdot S_j$' is the name of the conjunction of S_i and S_j, and similarly for the other connectives. Thus we will have two alternative ways of referring to theoretical expressions and sentences in L_o; the customary procedure of putting the sentence itself in quotes, and also, when brevity is in order, italic notations. (The particular notational convention here described was adopted in response to a last-minute discovery that boldface type could not be used. Unfortunately, since metalanguage transcriptions of object-language expressions have also been italicized, there is now a certain ambiguity between the use of italics to name object-language expressions on the one hand, and to *translate* them on the other. In most instances, the context makes unhesitatingly clear what interpretation is intended; however, a mild but regrettable confusion does tend to arise in some passages discussing the relevance of certain facts $T(t_1, \ldots, t_n)$ or $\sim T(t_1, \ldots, t_n)$ to theory T.)

A normal syntactic theory is a transformation in that, in general, one or more observation sentences S_j may be deduced, through use of the inference rules of the theory, jointly from T and some observation sentence S_i

sider all theoretical postulates accepted at the time. In particular, if the total set of accepted postulates can be split up into subsets for which it can be argued that the use of each subset is totally independent of the others, it would then seem that each should be regarded as a separate theory. Just what this "total independence" of usage might consist of, however, is a question which is not easily resolved. Fortunately, the present definition of "theory" to include all the accepted theoretical assumptions does not preclude the possible autonomy of certain subsets of its postulates, and as will be seen, suitable allowance for this contingency is made in the ensuing development.

William W. Rozeboom

where S_j is not deducible from S_i alone. When S_j is deducible from T alone, we may speak of S_j as a "consequence" of T. More generally, to include the case where T is not normal syntactic, we may define 'O-consequence of T' (where the 'O' is to distinguish the observational consequences of T from expressions derivable from T containing theoretical terms) by

Definition 1. C is an O-consequence of theory T (in language L) $=_{def}$ *C is an observation sentence (of L), and for any observation sentence S (of L), T transforms S into C.*[11]

While the concept of "O-consequence," so defined, is relative to a particular language L, it will henceforth be presupposed that L is the language which results when theory T is accepted by a person whose observation language is L_o. The extent to which L differs from L_o depends, of course, upon whether or not adoption of theory T effects a genuine language enrichment.

Since we have taken the inference rules for a normal syntactic theory to be sufficiently complete, a normal syntactic theory, T, transforms S_i into S_j if and only if $S_i \supset S_j$ is an O-consequence of T. Therefore, the total syntactical force of a normal syntactic theory for observation sentences is represented by the set of its O-consequences.

II

What does it mean to "believe," "accept," or "entertain" a theory T? We feel tempted to answer that it is *at least* to believe, accept, or entertain all the O-consequences of T. This will not quite do as it stands, however, for in general, T will have an infinite number of O-consequences, and it is unlikely that an infinite set of propositions can be entertained by a human mind. Hence, it is safer to say that to accept a theory is to be (at least) *committed* to its O-consequences—i.e., to be in a state such that belief that S is an O-consequence of T is sufficient cause for belief that S. More generally, we should say that to accept a theory is to be in a state such that belief that S_i and that T transforms S_i into S_j necessitates

[11] If T were here limited to normal syntactic theories, we could instead adopt: *C is an O-consequence of T* $=_{def}$ *C is an observation sentence and T formally entails C* (i.e., $T \supset C$ *is formally valid*). With similar replacement of 'is deducible from' by 'is formally entailed by' in Definition 2, Theorems 1–5 then follow without any assumptions whatsoever about the inference rules of L. (However, it must then also be argued—as indeed it may—that to accept T is to be committed to all observation sentences formally entailed by T whether they are also deducible from T or not.)

belief that S_j, but this may be reduced to commitment to the O-consequences of T, since commitment to $S_i \supset S_j$ and belief that S_i necessitate commitment to belief that S_j.

These formulations, however, raise a further problem: In what sense can commitment to the O-consequences of T be assimilated to cognitive processes? For to say that if circumstances are such-and-such, then a person will believe so-and-so, is to describe a disposition to acquire certain beliefs, a state which, unless additional conditions are also satisfied, would not ordinarily be regarded as a form of knowledge. Two possibilities arise, beyond which interpreting an accepted theory as itself a form of (possible) knowledge seems highly tenuous. (a) The theory, in itself, may be cognitively meaningful. This is not implausible in the case of a normal syntactic theory, since '$T(\tau_1, \ldots, \tau_n)$' will be an assertion if the 'τ_i' have appropriate semantical properties, but is questionable for other forms of theories, if other transformation techniques may be so designated. (b) The theory may have among its O-consequences a finite subset which entail the remainder. For convenience, let us call the conjunction of the sentences in such a subset a "prime consequence" of the theory:

Definition 2. *C is a prime consequence of theory T* $=_{def}$ *C is an O-consequence of T, and for any sentence S, if S is an O-consequence of T, S is deducible from C.*

If a theory does not itself make an assertion, the best candidate for the cognitive content of the theory would seem to be what is asserted by a prime consequence of the theory. It should be noted that it does no harm, as a figure of speech, to speak of *the* prime consequence of a theory, if it has one, for if both C_1 and C_2 are prime consequences of the theory, C_1 formally entails C_2 and conversely. That is,

Theorem 1. *All prime consequences of a theory are formally equivalent.*

We now prove a lemma, of great importance in the formal theory of quantification, which holds for any language in which '$\supset$' and '$\exists$' have the customary interpretation. Note that the lemma is *not* restricted to languages which are translatable into our present metalanguage, nor do we need a definition of 'formally valid' more precise than that provided in Section I.

Lemma 1. *If* '$A(a^i) \supset B$' *and* '$(\exists \phi^i) A(\phi^i) \supset B$' *are sentences in a language L (where* 'a^i' *and* 'ϕ^i' *are a primitive constant and a variable, respectively, of formal type i, and* 'A' *and* 'B' *abbreviate, respectively, a predicate*

291

and a sentence neither of which contain 'a$^{i'}$'), and 'A(a^i) ⊃ B' is formally valid in L, then '(∃φ^i)A(φ^i) ⊃ B' is also formally valid in L.

Proof: By definition, a sentence is formally valid in L if and only if it is true in all models of L. Now, the sentence '(∃φ^i)A(φ^i)' is true in a given model if and only if there is an entity in the range assigned to 'φ$^{i'}$' which possesses the property, or belongs to the class, assigned to 'A.' (If 'A' is a complex predicate, the property or class assigned to it is a function of the assignments made to its constituent terms.) Hence if there exists a model, M_i in which '(∃φ^i)A(φ^i)' is true, then there exists a model M_j, differing from M_i in at most the assignment to 'a^i,' in which 'A(a^i)' is true—we simply form M_j from M_i by assigning to 'a$^{i'}$' one of the entities which possess the property, or belong to the class, assigned to 'A' by M_i. (Since 'a$^{i'}$' is primitive, we are free to do this, and since 'A' does not contain 'a^i,' the latter may be reassigned without changing the assignment to the former.) Consider, now, the set Σ_k of all possible models of L given a fixed assignment of designata to the descriptive constants and ranges to the variables in 'B.' 'B' must be either true in all models in Σ_k or false in all. (1) If 'B' is true in the models in Σ_k, then '(∃φ^i)A(φ^i) ⊃ B' is obviously true in all models in Σ_k. (2) If 'B' is false in the models in Σ_k, then 'A(a^i)' must also be false in all models in Σ_k, since by hypothesis, 'A(a^i) ⊃ B' is true in all models of L. But then '(∃φ^i)A(φ^i)' must also be false in all models in Σ_k; for if '(∃φ^i)A(φ^i)' were true in a model M_i in Σ_k, then as just shown there would also be a model M_j, differing from M_i in at most the assignment to 'a$^{i'}$' and hence also in Σ_k (since reassignment of 'a$^{i'}$' does not affect the assignment to 'B'), in which 'A(a^i)' were true. Thus '(∃φ^i)A(φ^i) ⊃ B' is true in all models in Σ_k. But Σ_k is the set of all models of L given any particular (permissible) fixed assignment of designata and ranges to the terms in 'B'; since these sets jointly exhaust the models of L, '(∃φ^i)A(φ^i) ⊃ B' must be true in all models of L and is hence formally valid in L. Q.E.D.

Theorem 2. Every normal syntactic theory has a prime consequence.

Proof: Let the sentencelike formula 'T(τ_1, . . ., τ_n)' be the conjunction of the postulates of a normal syntactic theory. Then by existential generalization over the theoretical constants, we obtain the Ramsey sentence[12] '(∃φ$_1$, . . ., φ$_n$)T(φ$_1$, . . ., φ$_n$),' which may be designated by

[12] So named after F. P. Ramsey ([11], pp. 212–215, 231), who first called attention to this construction.

'R_T.' (It is to be understood, of course, that each variable 'ϕ_i' agrees in formal type with the corresponding 'τ_i.') Since R_T contains only terms in L_o, and our earlier stipulations about the formal properties of L_o ensure that R_T is deducible from T, R_T is an O-consequence of T. To see that R_T is also a prime consequence of T, we consider the formal properties of the calculus L'_o formed from the observation language L_o by adding the theoretical constants 'τ_1,' . . ., 'τ_n' to the descriptive constants in L_o. Let 'C' be a sentence in L_o which is deducible from T. Then by the definition of 'normal syntactic theory,' '$T(\tau_1, \ldots, \tau_n) \supset C$' must be formally valid in L'_o; and since 'C' and '$T(\phi_1, \ldots, \phi_n)$' contain no theoretical terms, n applications of Lemma 1 shows that '$(\exists \phi_1, \ldots, \phi_n)T(\phi_1, \ldots, \phi_n) \supset C$' must likewise be formally valid in L'_o. But the latter formula contains no theoretical terms and is thus also a formally valid sentence in L_o. Hence '$(\exists \phi_1, \ldots, \phi_n)T(\phi_1, \ldots, \phi_n) \supset C$' must be a theorem of L_o; and since 'C' is any O-consequence of T, it follows that any O-consequence of T may be deduced, in L_o, from the Ramsey sentence of T.

Corollary. *The prime consequence of a normal syntactic theory is its Ramsey sentence.*

It should be intuitively apparent that the prime consequence of a theory must stand in a special relation to its factual content. We saw above that while there may be problems in interpreting a theory as itself constituting a knowledge claim, the most obvious alternative, that the cognitive content of a theory is the (infinite) set of its O-consequences, also meets with difficulty. The horns of this burgeoning dilemma are blunted, however, by Theorem 2, which shows that all normal syntactic theories, in which class presumably fall all theories which have actually been entertained by scientists or have been of philosophical concern, possess a prime consequence. For not only is the prime consequence of a theory a straightforward (albeit existential) hypothesis in the observation language, thus posing no more philosophical difficulties than any other quantified statement in L_o, it has also exactly the same O-consequences as does the theory. Hence the conclusion that the factual content of a theory is the same as that of its prime consequence is a most seductive one. To show that this conclusion is correct as well as seductive will be the burden of the remainder of this section. However, it will also be brought out, especially in Section III, that the relation between a theory and its prime consequence is by no means a matter of simple synonymy.

William W. Rozeboom

Since the detailed discussion and proofs of the theorems which terminate this section are somewhat involved, it will be helpful to develop these theorems intuitively before turning to a more rigorous analysis. The fundamental assumption concerning the semantical status of theories, on which the remainder of this article rests, is that the theoretical terms 'τ_1,' . . ., 'τ_n' *derive* their meanings, if any, from their occurrence in the sentential function 'T(ϕ_1, . . ., ϕ_n)' which they complete to form the theory 'T(τ_1, . . ., τ_n)' (see the Thesis of Semantic Empiricism, below). Now if it is indeed true that theoretical terms are *given* their significance by the theory in which they are imbedded, it follows that so long as distinctiveness is preserved, the theoretical terms in theory T may be exchanged for any other set of theoretical terms without altering the meaning or factual commitments of the theory. Moreover, it is intuitively convincing, and indeed can be proved with a modicum of assumptions unrelated to the question of theoretical meaningfulness (see [14]), that two theories which have no theoretical terms in common are incompatible if and only if they have incompatible observational consequences. But if T_1 and T_2 have theoretical terms in common—that is, if T_1 and T_2 make use of common theoretical sign-designs (though the meanings given to the common terms may be different in the two cases)—it follows from the fundamental assumption already noted that we can replace the theoretical terms in T_2 with new theoretical terms in such a way that the resulting theory T^*_2 is equivalent to T_2 in meaning and has no theoretical terms in common with T_1. Therefore, T^*_2, and hence T_2, is incompatible with T_1 if and only if T^*_2, and hence T_2, has observational consequences which are incompatible with those of T_1. That is, two theories, having common theoretical terms (sign-designs) or not, are incompatible if and only if they have incompatible O-consequences (Theorem 4, below), and as easily shown, the same holds for a theory and an observation sentence. Now, to *deny* a theory T is to make an assertion, theoretical or observational, which is incompatible with T. But by the conclusion just reached, this denial must therefore be incompatible with the O-consequences, and hence with the Ramsey sentence, R_T, of T. But if assertion of a theory T commits one to accept R_T (since T entails R_T), while denial of T commits one to denial of R_T (since denial of T is incompatible with R_T), then T and R_T must be factually equivalent—i.e., a theory and its Ramsey sentence have the same factual content (Theorem 3, below).

If this argument appears convincing to the reader, he may turn immediately to Section III, pausing in transit only to read the Thesis of Semantic Empiricism, Semantical Principles I–IV, and Postulates 1 and 2. However, despite the conviction sustained by this intuitive demonstration (and given the Thesis of Semantic Empiricism, its force is considerable), it nonetheless glosses over several issues which, though seldom adequately explored, are basic to semantical theory. For example, it was implicitly assumed here that a denial—and no more than this—of theory T could be constructed, either as an alternative theory or as a sentence in L_o. But is it not also conceivable that although T is stronger in content than R_T, it is impossible to extract the difference between T and R_T for separate denial while retaining R_T? Again, questions need to be raised about the status of the assumption concerning the incompatibility of two theories having no theoretical terms in common, questions which penetrate to the heart of problems about extrasyntactical incompatibility (i.e., incompatibility in virtue of the *meanings* involved). Therefore, we shall now attempt to reach the conclusions derived so expeditiously in the preceding paragraph by a surer but more arduous route.

I have so far spoken of the "factual content" of a theory or statement in an offhand manner. It now becomes necessary to give this rather vague notion a more precise definition. It has already been noted that the *semantic* status of theories is problematic. However, we speak of "believing," "accepting," or "entertaining" theories in very much the same sense that we apply these terms to observation sentences. In either case, acceptance of a theory or observation sentence has pragmatic repercussions—it involves a behavioral adjustment which, broadly speaking, is pragmatically appropriate or inappropriate according to the facts of reality. In this sense, at least, both theories and sentences stand in the same kind of relations to facts. More specifically, a theory or a sentence is unqualifiedly either "correct" or "incorrect" in virtue of some fact (though perhaps not necessarily an observable fact) which determines the adaptive value of the theory or sentence. This is obviously true of a sentence, which is correct or incorrect (i.e., its acceptance is appropriate or inappropriate) in virtue of the state of reality according to which it is true or false. If it were not also true of a theory—if there were no fact, observational or otherwise, which is a sufficient condition for the correctness or incorrectness of the theory, but rather, the theory were describable only as being more or less useful—then the use of theories would be intrinsically different from that of sentences, and there would exist not even a possibility that a theory might *signify* a fact.

It is quite apparent that whatever we mean by the "factual content," "truth conditions," "factual commitments," etc., of a sentence or theory, these notions are intimately related to the conditions under which the theory or sentence is correct or incorrect. It is also apparent that in the case of a sentence, its factual content is in some sense given by what the sentence asserts. Hence if we can define 'factual content' in terms of the conditions under which a sentence or theory is correct or incorrect without presupposing that the sentence or theory has semantic properties, and can then show the definition to be in suitable agreement with what a sentence asserts, we shall also have, correspondingly, a satisfactory definition of the "factual content" of a theory.

Let us call any fact which is a sufficient condition for a sentence S (theory T) to be correct a "verifier" of S (T). More precisely, a fact f is a verifier of S (T) if and only if it may validly be reasoned, "It is a fact that f; therefore, in view of the behavior-inducing properties of S (T)—its meaning, its transformational force, etc.—acceptance of S (T) is pragmatically appropriate."[13] Similarly, any fact which is a sufficient condition for the incorrectness of S (T) may be called a "refuter" of S (T). It is important to be clear that the verifiers and refuters of S (T) do *not* include all facts which are *evidence* for or against S (T). By "evidence," we mean any fact which (correctly) influences our belief that the sentence (or theory) *has* a verifier or refuter but which is not necessarily *itself* a verifier or refuter of S (T). For example, consider the sentence Q: 'No crows are pink.' Then the fact, say, that over 100,000 crows have been observed under a

[13] Since it is not the bare sentence shape (i.e., sign-design) in language L which has truth value, but sentence-shape-cum-meaning (i.e., statement), it is not possible to tell merely from the physical topography of a sentence S in L whether or not a given fact f refutes S. It is also necessary to have information, either implicitly in the form of our own language habits if we are actually using L, or explicitly if we are evaluating L from without, about the linguistic functioning of S and perhaps other expressions in L. Consequently, any metalinguistic argument that a sentence S of language L is refuted or verified by a fact f must appeal to a set of premises about the character of L as a (meaningful) *language*, including some propositions about the conditions under which sentences in L are true or false (correct or incorrect, adaptive or maladaptive) in accordance with extralinguistic reality. Different ways of constructing these premises amount to different theories of semantics, and a theory of what such premises should consist of is a theory of *metasemantics*, a discipline in which the ground has scarcely been broken, but which will almost certainly need extensive development before the near universal confusion that seems to exist concerning the nature of semantical concepts is appreciably diminished.

wide variety of conditions and none were pink, is strong evidence that Q is correct, but does not itself verify Q, for the existence of a pink crow is still possible.

If, now, it were the case that a sentence or theory had at most one verifier or refuter, we could simply take this fact to be its factual content. Quite the contrary, however, if a sentence or theory has one verifier or refuter it also has an indefinite number of them. For example, if the fact f is a verifier of S (T), the molecular fact $f \cdot g$ (i.e., the conjunction of f and g), where g is any other fact, is also a verifier of S (T). In general, we believe that two facts f and g may in some instances be related in such a way that f is a sufficient condition for g—i.e., that because f is a fact, g is, of necessity, also a fact (more briefly: f, hence necessarily g). That is, in any speculations about what might be the case, we should feel it not only unnecessary but logically absurd to consider the possibility that f, but not g. When such a necessity relation holds between facts f and g, we say that f entails g.

Just what the nature is of this relation of entailment, we fortunately need not attempt to determine here, except, perhaps, to observe that it is reflexive, transitive, but not symmetric. It is sufficient to recognize that there seems to be some such relation. Thus we should all agree, surely (barring the natural perversity of philosophical doubt), that if S_1 truly signifies (only) fact f, S_2 truly signifies (only) fact g, and S_2 is deducible by valid formal inference rules from S_1, then f entails g. Under such circumstances, we say that f logically entails g. The extent, if any, to which "entailment" is a broader relation than "logical entailment" is still very much an open question. If S_1 and S_2 truly signify (only) facts f and g, respectively, then f entails g if $S_1 \supset S_2$ is "analytically" true; but how the concept of "analyticity" should be analyzed is an issue which currently is raging merrily. (Note that f may logically entail g even though $S_1 \supset S_2$ is not logically true; for example, S_1 and S_2 may be syntactically independent, but signify the same fact in virtue of containing synonymous terms. Under such circumstances, $S_1 \supset S_2$ is analytically true without being logically true.) Since the present analysis does not depend upon any specific interpretation of "entailment," we may, without prejudice, leave room for the possibility that "f, hence necessarily g" may hold for certain facts f and g even though f does not logically entail g.

The reason we must here recognize the relation of entailment is its intimate connection with conditions of verification and refutation. Whether or not a given fact verifies or refutes a given sentence or theory is something which can be decided only by examining the specific case. However, the notions of "verifying" or "refuting" an entity E (where E is anything —sentence, theory, or whatever—being used to effect a behavioral adjustment which is appropriate or inappropriate with respect to facts external to the behavior) are deeply embedded in the practical core of linguistic behavior, and we may easily recognize certain principles according to which these concepts are used. Thus,

1. *E cannot have both a verifier and a refuter*—i.e., the terms 'correct' and 'incorrect' are mutually exclusive in their proper application.

2. *If the use of E is such that we can properly say, "If it were the case that p, then E would be correct (incorrect)," then E has either a verifier or refuter.* That is, in view of the behavioral role embodied by E, if there are conceivable circumstances under which E would be either correct or incorrect, then, in that role, E is either correct or incorrect; for if none of the circumstances hold under which E would be correct (incorrect), then this is itself a sufficient condition for E to be incorrect (correct). The relation of entailment enters this picture in that

3. *If a fact f is a verifier (refuter) of E, then any fact which entails f is also a verifier (refuter) of E.* This may easily be seen by translating "fact f is a verifier (refuter) of E" as "since f is the case, E is necessarily correct (incorrect) in view of its behavioral properties"—i.e., "f and the facts about E's behavioral role entail that E is correct (incorrect)." Then (3) follows by the transitivity of entailment. Such sentence forms as 'f entails g,' 'f verifies (refutes) E,' 'S signifies a fact entailed by f,' and 'S_1 signifies f, S_2 signifies g, and S_2 is validly deducible from S_1' appear to be so tightly interrelated in meaning that they are undoubtedly grounded primarily on a common underlying concept. One apparently analytic consequence of this common ground is 3. Another is that

4. *If observation sentence S_2 is validly deducible from observation sentence S_1, then any verifier of S_1 is a verifier of S_2, and any refuter of S_2 is a refuter of S_1.*[14] Whenever S_2 is validly deducible from S_1, we say that S_1 "formally entails" S_2. More generally, whether S_2 is syntactically de-

---

[14] The restriction to observation sentences would seem to be necessary here, for a sentence can apparently be incorrect not only in virtue of signifying a fact falsely, but also in virtue of containing a nonlogical constant which has no referent (see [13]). Hence if S_2 contains a nonlogical constant 'a' not contained in S_1, it is possible for S_1 to be true and S_2 to be false, even though S_1 formally entails S_2, simply because 'a' has no

ducible from S_1 or not, if every verifier of S_1 is a verifier of S_2 and every refuter of S_2 is a refuter of S_1, we say simply that S_1 entails S_2, or, to emphasize that the relation is not necessarily syntactical, that S_1 "analytically" entails S_2. (Note that 'entails' may hence be used to describe either a relation between facts or a relation between sentences. Presumably, one usage is definable in terms of the other.)

We may now define the "factual content" of an entity as the set of facts which verify or refute it. We need to retain a distinction between content which verifies and content which refutes, however, or a sentence will have the same content as its negation.

Definition 3. The positive (negative) factual content of $E =_{def}$ The set of verifiers (refuters) of E.

Definition 4. The factual content of $E =_{def}$ The partially ordered set composed of, first, the members of the positive factual content of E, then the numeral '0,' and, finally, the members of the negative factual content of E.

Inclusion of '0' in the factual content of E is merely a formal device to give a simple distinction between the content of E and that of its contradiction. Since an entity cannot be both correct and incorrect, its factual content cannot contain both verifiers and refuters. Hence '0' is either the first or the last member of E's factual content. If it is both, then E has neither a verifier nor a refuter, and we may simply say that E has no factual content. If E does have factual content, then E is correct or incorrect according to whether '0' is the last or the first member of its content.

In virtue of the relation of entailment, certain members of the factual content of an entity E stand in a special relation to the remainder; namely, those facts of minimal "strength" to verify or refute E.

Definition 5. f is a positive (negative) primary content of $E =_{def}$ f is a verifier (refuter) of E, and any fact which verifies or refutes E entails f.

Since any fact which entails a verifier or refuter of E is itself a verifier or refuter E, a fact f which is a primary content of E exhaustively specifies the factual content of E, apart from the position of '0,' in that a necessary and sufficient condition for a fact g to be a member of the factual content of E is for g to entail f. Hence the factual content of E is com-

---

referent. For example, '$(x)(x = x)$' formally entails 'Pegasus = Pegasus,' even though the latter is false (see [13]). However, the definition of 'observation term' ensures that all nonlogical constants of L_o have designata, and it then follows by any acceptable theory of semantics that 4 obtains.

William W. Rozeboom

pletely described by giving a primary content of E and stating whether it verifies or refutes E. Whether or not every E which has factual content also has a primary content is an interesting question which will not be explored here. Neither are we here able to judge whether E can have more than one primary content. It does follow from Definition 5, however, that if two facts f and g are each a primary content of E, then f and g are analytically equivalent. While this still leaves open the question whether f and g can be analytically equivalent without being identical, the relation must in any case be so intimate that little harm will be done by speaking informally of *the* primary content of E, if it has one.

To what extent does the present definition of 'factual content' agree with intuitive understanding of this term as applied to sentences? By classical semantics, a cognitively meaningful sentence signifies, truly or falsely, exactly one fact, and it is this fact which intuitively is its factual content. But if f is the only fact truly signified by S, then a fact g is a sufficient condition for S to be correct if and only if g entails f; hence by the present definitions, f is the positive primary content of S. Similarly by classical semantics, the negative primary content of a false sentence is the fact falsely signified by it. But as we have seen, the primary content of an entity determines its factual content. Hence the present definition of 'factual content' agrees in logical essentials with intuitive notions insofar as the latter are well formed.

The main purpose of this paper is to clarify the "factual content," "factual commitments," etc., of a scientific theory, and to this end, the first of these expressions has been explicitly defined in terms of conditions of verification and refutation. We have yet to say, however, what kind of an answer can be given to a question about factual content. For example, suppose we are asked for the factual content of the sentence Q: 'No crows are pink.' Such queries arise frequently during scientific and philosophical pursuits in the guise "What does Q mean?" or "What are the truth conditions of Q?" It is rather unhelpful to answer, "The factual content of Q is the set of its verifiers, followed by '0,' followed by its refuters," even though by Definition 4 this is literally correct. Rather, what one wants to know is *what* are the verifiers or refuters of Q. But this cannot be answered without first determining what the facts of the case *are*. We cannot correctly say, for example, that the nonexistence of a pink crow is the primary content of Q unless it is a *fact* that no pink crows exist. It would seem that so long as we restrict "factual content" to include only that which exists and abjure appeal to a nebulous realm of "possibility," we cannot identify the factual content of a sentence or theory without at the same time determining whether it is correct or incorrect. Thus we cannot expect to give *this* kind of an answer in general explication of the factual contents of theories.

The fact remains, however, that in actual practice we do manage effectively to discuss the truth conditions of a sentence without attempting to judge its correctness. We simply resort to considerations of *meaning*, rather than factual content and say that "state-of-affairs" p is a necessary and sufficient condition for sentence S to be true—thereby indicating that *if* it is a fact that p, then p verifies S, and *if* ~p is a fact, then ~p refutes S. What is involved here is discovery in our metalanguage of a sentence 'p' which is related in meaning to sentence S in such a way that we can recognize that any verifier or refuter of 'p' must also be a verifier or refuter of S, and conversely. Hence to say (truthfully) that p is a necessary and sufficient condition for S to be correct, even though it is not known whether it is the case that p, is to use a sentence 'p,' with the same factual content as S, in the analysis of S. In this way we alleviate our uncertainties about the truth conditions of S by reducing the problem of its content to the equivalent problem for another sentence whose meaning, presumably, arouses no puzzlement. In like manner, if we are able to reason, "Theory T is verified if it is the case that p, and is refuted if it is not the case that p," then we may conclude that theory T has the same factual content as 'p.'

What are the conditions under which a fact may properly be said to verify or to refute a theory? To determine this we can only appeal to the way in which theories are actually used. It will be profitable to concentrate first on the conditions of refutation, for while the history of science is littered with abandoned theories, seldom if ever is a theory judged to be unconditionally verified. Thus while it might be difficult to determine when we would consider a theory correct, it is little trouble to discover conditions of incorrectness.

We saw earlier that to accept a theory T involves commitment to the O-consequences of T. But if one of the latter is false, then to accept T is to be led into error. That is, having a false O-consequence is a sufficient condition for T to be incorrect, a conclusion which is amply substantiated by actual practice, in which we feel compelled to revise or discard a theory whenever it leads to an erroneous conclusion that cannot be written off

William W. Rozeboom

as an "error of measurement," "approximation error," or the like. Thus the refuters of a theory T include all facts which refute an O-consequence of T.

What can we say about the relation of a fact f to theory T if T does not have an O-consequence which is refuted by f? One is tempted to answer that at least in the case where f is an *observational* fact (i.e., signifiable in L_o), f refutes T only if it also refutes an O-consequence of T. For do not the O-consequences of a theory constitute its observational force? Unfortunately, this claim prejudges the *semantic* status of the theory. For if T may itself be an assertion, as the realist insists, we must consider the possibility that T falsely signifies a fact which refutes no O-consequence of T, but which is entailed by, or is itself, an observational fact. For example, if investigation of certain known entities $t_1, \ldots, t_n$ eventually reveals it to be the case that $T(t_1, \ldots, t_n)$, the realist may wish to claim that $T(t_1, \ldots, t_n)$ is what the theory '$T(\tau_1, \ldots, \tau_n)$' signified all along even though nothing signified by an O-consequence of T entails that $T(t_1, \ldots, t_n)$; hence we may also wonder, if it had turned out that $\sim T(t_1, \ldots, t_n)$, whether this would not have refuted T.

This point is so important that I will try to clarify it a bit further. When the realist insists that a theory may itself make an assertion which goes beyond what is asserted by its O-consequences, he has raised the possibility that the theory may be false even though all its O-consequences are true—i.e., that it may be possible for a theory to signify falsely, and hence be refuted by, a fact f even though f refutes none of its O-consequences. But if theories can themselves signify facts, they are surely not limited to signifying only facts which *cannot* be signified in L_o, for then a theory which signifies a certain fact could be deprived of its ability to do so simply by enriching L_o suitably. The mechanism by which a theory is able to signify a fact would indeed be peculiar if it could work only when the observation language is sufficiently impoverished. Hence if a theory can signify a fact at all, we must also suspect that it can signify an observational fact, and the realistic interpretation of theories hence carries with it, at least prior to further analysis, the possibility that a theory may be refuted by an observational fact even when all O-consequences of the theory are true.

Now, there must be some principle according to which we can judge, in at least some cases, that an observational fact is *not* a refuter of a given

302

theory, or theories would be utterly useless—we should then never have reason to doubt that a given theory had already been refuted by known facts. Nor is it legitimate, if the realistic interpretation of theories has any intuitive plausibility at all, simply to take for granted that a theory is refuted by an observational fact only if this refutes an O-consequence of the theory. (Oddly enough, this matter—circumscription of the conditions of refutation—has been almost totally overlooked in philosophical analyses of the meanings of theories, although it would now appear that this is actually the core of the problem.) On the other hand, to claim that theories have designative meaning while not also at least implicitly adducing some principle about the *limitations* of possible theoretical reference is sheer philosophic irresponsibility. For given any nontrivial theory '$T(\tau_1, \ldots, \tau_n)$,' there will almost certainly exist some set of entities t_1, $\ldots, t_n$ such that $\sim T(t_1, \ldots, t_n)$; hence if we are given no grounds upon which to deny that T asserts that $T(t_1, \ldots, t_n)$—i.e., to deny that T falsely signifies the fact $\sim T(t_1, \ldots, t_n)$—we should never have reason to doubt that T has a refuter.

Now, it seems to me indisputable that the way in which theories are actually used with the observation language *does* impose limits on the possible meanings of theoretical expressions. The fact, for example, that theory T does *not* have an O-consequence which is refuted by observational fact f, while perhaps insufficient grounds for deciding that f does not refute T, is nonetheless *relevant* for judging the factual content of T, whether T is itself an assertion or not. Unless we yield to an unbridled transcendentalism, it is difficult to see how the referential ability, if any, of theoretical terms (which, after all, are only signs with no intrinsic meanings, and whose semantic properties must hence be acquired) could derive from anything other than their effective observational import, which, in turn, is determined by the O-consequences of the theory in which they are contained. These rather vague conclusions may be given somewhat more coherent form as a tenet which might be called

The Thesis of Semantic Empiricism: The semantic properties, if any, of theoretical expressions derive, in a potentially useful and syntactically general manner, wholly from their use with the observation language.

That is, 'τ_1,' $\ldots$, 'τ_n' and expressions which contain them have whatever ability to designate that they do have because the 'τ_i' occur in the (perhaps provisionally) accepted theory '$T(\tau_1, \ldots, \tau_n)$.'

William W. Rozeboom

The Thesis of Semantic Empiricism by no means claims that all designata of expressions containing theoretical terms are accessible to the observation language—i.e., that the 'τ_i' are definable in L_o—though neither is this possibility ruled out. It insists merely that a theoretical term has no meaning that it *brings to* the theory, so that two theories differing only in their theoretical symbols have the same factual content. As for the stipulation of generality, this is both important and intuitively inescapable: if theories are themselves able to make assertions, the mechanism by which this occurs must be a basic feature of linguistic processes which does not depend upon the theory's being of a special, restricted syntactical form.

The Thesis of Semantic Empiricism expresses the minimum restriction upon the possible meanings of theoretical expressions that can be demanded by anyone who seriously believes that knowledge is based, in some important sense, upon experience; indeed, in its loosely worded form, the Thesis scarcely seems a restriction at all. Yet properly exploited, it suffices to determine the limits of refutation, and hence the factual content, of a theory. In order to show this at all effectively, however, we need to make more explicit what until now has been left to intuitive understanding; namely, the manner in which cognitively meaningful sentences are given truth value through the semantic properties of their constituent terms. The factual content of a meaningful sentence cannot, in general, be determined wholly by its transformational force for other sentences; otherwise, any attempt to judge truth values would precipitate an infinite regress. Hence if we are to pass judgment on the factual content of a theory under the supposition that the theory might itself be meaningful, we can reach no conclusion without some explicit assumptions about the verifiers and refuters of meaningful sentences. The semantical assumptions which follow are very similar in underlying form to the Tarski-Carnap approach, except for expressing a relationship between sentences and their designata, i.e., facts, rather than between sentences and their meanings, i.e., propositions.

Definition 6. *Fact f has the form* $F(\phi_1, \ldots, \phi_n)$ =$_\text{def}$ *There exist entities* $t_1, \ldots, t_n$ *such that f is the fact that* $F(t_1, \ldots, t_n)$.

(It will be noted that the "form" of a fact, as defined here, is not unique. For example, the fact signified by a true observation sentence '$P(a_1, a_2)$' has both the form $P(\phi_1, \phi_2)$ and the form $P(\phi_1, a_2)$. Whether a fact's *logical* form—i.e., a form whose description contains no extralogical terms—is in some sense unique is a question on which there is here no need to take a stand.)

Semantic Principle I[15] *(SP I). If '$S(s_1, \ldots, s_n)$' is a sentence formed by substituting the symbols 's_1,' . . ., 's_n' for the variables 'ϕ_1,' . . ., 'ϕ_n,' respectively, in the observation predicate '$S(\phi_1, \ldots, \phi_n)$,' and '$S(s_1, \ldots, s_n)$' is "semantically proper"—i.e., its meaning, or designative potentiality, if any, is governed by the syntax of L_o—then '$S(s_1, \ldots, s_n)$' signifies a fact f only if there exist entities $t_1, \ldots, t_n$ such that 's_1,' . . ., 's_n' designate $t_1, \ldots, t_n$, respectively, and f is either the fact that $S(t_1, \ldots, t_n)$ or the fact that $\sim S(t_1, \ldots, t_n)$. If '$S(s_1, \ldots, s_n)$' signifies f, it does so truly or falsely according to whether f has the form $S(\phi_1, \ldots, \phi_n)$ or $\sim S(\phi_1, \ldots, \phi_n)$, respectively.*

That is, in more conventional (and in some respects misleading) terms, a sentence '$S(s_1, \ldots, s_n)$' can assert only that $S(t_1, \ldots, t_n)$, where $t_1, \ldots, t_n$ are entities designated by 's_1,' . . ., 's_n.' By saying that the meaning of '$S(s_1, \ldots, s_n)$' is "governed by the syntax of L_o," we rule out the possibility that the meaning of '$S(s_1, \ldots, s_n)$' has been arbitrarily assigned without regard for the meaning of its constituents. It should be noted that SP I does not stipulate that the 's_i' are observation terms. The distinction between "observational" and "theoretical" concerns possible differences in the conditions under which terms acquire their referential powers, whereas SP I deals with the more basic relation between the designative properties of a sentence and those of its constituents. In justification of SP I, one should observe that it merely formalizes part of the classic semantical belief that a sentence S makes an assertion in virtue of S's attributing a certain property (in the broad sense) to a set of entities designated by the subject terms of S, and that if these entities exemplify the property, this is the fact signified by S, whereas if these entities do not exemplify the property, this is the fact in virtue of which S is false. Of course, the classical view may be in error in various respects; in fact, SP I is stated only as a conditional, rather than as the biconditional authorized by the classical view, because further developments will indicate that the latter may be in some respects too strong. But it is incumbent upon anyone who might wish to challenge SP I to provide an alternative statement

[15] Strictly speaking, SP I is not itself an assertion, but is only a schema which generates a set of assertions. We could obtain a single assertion by stating that every sentence (in L_M) obtained by proper substitutions for '$S(\phi_1, \ldots, \phi_n)$' in SP I is true. However, this statement would have to be constructed with care, and need not detain us.

about the conditions under which sentences stand in semantical relations to extralinguistic reality.

There is, to be sure, one objection to SP I which might be raised on more or less orthodox grounds. SP I places no restrictions on the logical forms of facts—indeed, it allows us to invoke facts as needed, corresponding to the true statements in our metalanguage—and this will undoubtedly give offense to one whose tolerance for "facts" does not extend, say, to general or molecular facts. This is not the occasion, even if I were prepared to do so, to take inventory of reality's ingredients, and so I shall abstain from reciting the familiar difficulties which arise when only "atomic" facts are countenanced. Neither shall I explore the possibility that to be surprised that nature should so obligingly fit a fact to each true statement we can construct is very like being perplexed over why the world should be articulated in length units which precisely match our concepts of "inch," "centimeter," etc. Instead, I shall merely venture that if and when it becomes possible to account adequately for the factual content of true observation sentences without presupposing that each signifies a fact, it will not be difficult to accommodate the proof of Theorem 3 to whatever semantical assumptions replace the present ones.

Semantic Principle II (SP II). If a sentence S signifies a fact f truly, S is true and f is a verifier of S. If S signifies f falsely, S is false and f is a refuter of S.

This could also be put by saying that S is correct or incorrect, respectively, under the conditions stated, for when a sentence ascribes a predicate to a set of entities, it is correct or incorrect according to whether or not those entities satisfy the predicate. It should be observed that SP II does *not* say that a formula is false *only* when it signifies a fact falsely. As will be discussed further in Section IV, a descriptive term may be meaningful even when it does not designate anything. Since a sentence which contains such a term may thus be meaningful even though it does not signify any fact, we must allow for the possibility that such sentences should be called false (cf. footnote 14).

It will be noted that SP I, II partially define the conditions under which a sentence signifies a fact; namely, a semantically proper sentence '$S(s_1, \ldots, s_n)$' signifies fact f truly (falsely) only if there exist entities $t_1, \ldots, t_n$ designated by 's_1,' $\ldots$, 's_n,' respectively, such that f is the fact that $S(t_1, \ldots, t_n)$ (the fact that $\sim S(t_1, \ldots, t_n)$), and f verifies

306

(refutes) '$S(s_1, \ldots, s_n)$.' Now, while the intuitive notion of the manner in which a sentence makes semantical contact with extralinguistic reality is unfortunately vague, it would seem to consist essentially in the idea that a sentence S designates, or signifies, a fact f when S and f are related in such a way that the descriptive terms in S designate the constituents of f, and S is true, or false as the case may be, in virtue of the fact that f. Hence the conditions of signification entailed by SP I, II should be sufficient as well as necessary, and we may further assume:

Semantic Principle III (SP III). If '$S(s_1, \ldots, s_n)$' *is a semantically proper sentence formed from the observational predicate* '$S(\phi_1, \ldots, \phi_n)$' *and* '$s_1$,' $\ldots$, 's_n' *designate entities* $t_1, \ldots, t_n$, *respectively, then: (a) if it is a fact that* $S(t_1, \ldots, t_n)$ *and this verifies* '$S(s_1, \ldots, s_n)$,' *then* '$S(s_1, \ldots, s_n)$' *signifies* $S(t_1, \ldots, t_n)$ *truly; (b) if it is a fact that* $\sim S(t_1, \ldots, t_n)$ *and this refutes* '$S(s_1, \ldots, s_n)$,' *then* '$S(s_1, \ldots, s_n)$' *signifies* $\sim S(t_1, \ldots, t_n)$ *falsely.*

Since SP I–III together give necessary and sufficient conditions for a semantically proper sentence to "signify" a fact, they may, then, be construed essentially as a definition of this concept. (It would be simple to extend the definition to semantically *improper* sentences such as coded abbreviations, but this is irrelevant for present purposes.) It might seem that the clauses ". . . *and this verifies [refutes]* '$S(s_1, \ldots, s_n)$' . . ." in SP III are redundant, for by classical semantics, if 's_1,' $\ldots$, 's_n' designate $t_1, \ldots, t_n$, respectively, the fact that $S(t_1, \ldots, t_n)$ (the fact that $\sim S(t_1, \ldots, t_n)$) is a sufficient condition for '$S(s_1, \ldots, s_n)$' to be true (false) and hence verifies (refutes) the sentence. However, we shall see in Section III that the classical view may not be wholly correct in this assumption, whereas if the classical view *is* correct, the redundancy does not hurt.

Semantic Principle IV (SP IV). No sentence can be both true and false.

This merely makes explicit for sentences the point made earlier, that as we normally use the notion of "correctness," an entity cannot be both correct and incorrect. This does not, of course, preclude the possibility that a sign-design may change its truth value if the meanings of its constituent terms change. But the present analysis presupposes throughout that all linguistic entities concerned remain stable in their various behavioral roles so long as the accepted theory is not altered, and the terms "sentence," "symbol," "expression," etc., here denote not bare sign-designs (which are neither correct nor incorrect as such) but sign-designs

William W. Rozeboom

cum roles. That an entity cannot be both true and false, or both correct and incorrect, within a fixed context of usage would seem to be an analytic truth about these concepts.

Let us now examine the Thesis of Semantic Empiricism in the light of these principles. According to the Thesis, the semantic properties of a theoretical expression, E, admitted into use through adoption of a theory T are determined wholly by the use of E with respect to the observation language. That is, for every such E, its use with L_o provides a criterion by which can be determined whether or not it designates a given entity. In particular, whether or not T itself signifies a given fact must be so determinable.

Now, the use of an accepted normal syntactic theory 'T$(\tau_1, \ldots, \tau_n)$' with L_o would seem to be comprehensively (albeit schematically) described by saying that T is *formed from* the observational predicate 'T$(\phi_1, \ldots, \phi_n)$' and *used* to generate its O-consequences, where both formation of T and deductions from T *conform* to the syntax of L_o. But that T has the particular O-consequences it does have is a logical result of its formation from 'T$(\phi_1, \ldots, \phi_n)$' and use in conformity with the syntax of L_o. Hence whether or not T signifies a fact f must be determined wholly by whether or not f stands in a certain determinate relation to 'T$(\phi_1, \ldots, \phi_n)$' and the syntax of L_o. One restriction of the possible significata of T on these grounds is obvious: since the semantic properties of T derive from its use with L_o, and this use conforms to the syntax of L_o, the designative potentialities of T, if any, must be governed by the syntax of L_o. Hence by SP I, if T signifies a fact, this must either be a fact of form $T(\phi_1, \ldots, \phi_n)$ truly signified by T, or a fact of form $\sim T(\phi_1, \ldots, \phi_n)$ falsely signified by T. That is, if T itself makes an assertion, it must do so by ascribing the predicate 'T$(\phi_1, \ldots, \phi_n)$' to some set of entities $t_1, \ldots, t_n$ designated, respectively, by 'τ_1,' $\ldots$, 'τ_n,' a conclusion which should be intuitively evident even without appeal to the Thesis. Being of either of these two forms cannot be a *sufficient* condition for a fact to be signified by theory T, however, or T would simultaneously signify truly all facts of form $T(\phi_1, \ldots, \phi_n)$ and falsely all facts of the form $\sim T(\phi_1, \ldots, \phi_n)$, in flagrant violation of SP IV. Hence the predicate 'T$(\phi_1, \ldots, \phi_n)$' and the syntax of L_o must impose on the possible significata of T an additional condition which excludes this possibility. And since it is 'T$(\phi_1, \ldots, \phi_n)$' which gives the theoretical terms their particular character, contrasted to the semantic properties, if any, which would have been imparted by their normal syntactic use with a semantically different predicate of L_o, it is clear that this additional condition must stem essentially from the linguistic force of '$T(\phi_1, \ldots, \phi_n)$.' More precisely, since any one fact of the form $T(\phi_1, \ldots, \phi_n)$ or $\sim T(\phi_1, \ldots, \phi_n)$ is syntactically as legitimate a significatum of T as any other, then if T signifies the one but not the other, it must be because the one is related to '$T(\phi_1, \ldots, \phi_n)$' in virtue of the latter's linguistic properties—i.e., is "semantically" related to '$T(\phi_1, \ldots, \phi_n)$'—in a manner in which the other is not.

Now, there are a great number of possible ways in which a fact may be semantically related to the predicate '$T(\phi_1, \ldots, \phi_n)$' that might be considered as potential criteria for whether or not a given fact is signified by theory T. For example, there are an indefinite number which depend on the predicate's being of some special syntactic structure, such as the relation that obtains between a predicate of the form '$P(\phi) \cdot Q(\phi)$' and a fact of form $P(\phi)$, or the relation between a fact $P(a)$ and a complex predicate '$P(\phi)$' when the latter contains a constant which refers to a (e.g., the fact that $2 \leqq 2$, and the predicate '$2 \leqq \phi$'). It would be excessively tedious to examine these various and for the most part artificial possibilities in any detail. Fortunately, most are immediately dismissed by the demand of the Thesis that theoretical expressions derive their meanings, if any, in a syntactically *general* manner. This rules out the possibility that if theoretical expressions can designate at all, only those formed from predicates of special syntactical characteristics can do so. Hence an acceptable criterion for whether or not a fact is signified by T cannot specify any special syntactical structure for '$T(\phi_1, \ldots, \phi_n)$.'

The next point to be considered is that in (presumed) contrast to relations between facts and observation *sentences*, it does not seem possible to find any one-many semantic relations between facts and observation *predicates* which do not constrain the predicate to special characteristics. That is, if the Thesis of Semantic Empiricism is correct, there appears to be no way in which '$T(\tau_1, \ldots, \tau_n)$' can be assured a unique significatum.[16] Any criterion for what is signified by T, based only on the rela-

---

[16] An example of a potential criterion which singles out at most a unique significatum for T by violating the condition of syntactical generality is the following: A theory T signifies a fact f if and only if T is of syntactical form '$T(\tau_1, \ldots, \tau_n) \cdot (c_1 = c_1) \cdot \ldots \cdot (c_n = c_n)$' and there exist entities $t_1, \ldots, t_n$ such that 'c_1,' $\ldots$, 'c_n' designate $t_1, \ldots, t_n$, respectively, while f is either the fact that $T(t_1, \ldots, t_n) \cdot (t_1 = t_1) \cdot \ldots \cdot (t_n = t_n)$ or the fact that $\sim[T(t_1, \ldots, t_n) \cdot (t_1 = t_1) \cdot \ldots \cdot (t_n = t_n)]$. This prospective criterion guarantees T a unique significatum so long as T is of the necessary syntactical form even though there may be many facts of the form $T(\phi_1, \ldots,$

William W. Rozeboom

tion of a fact to the predicate '$T(\phi_1, \ldots, \phi_n)$' in virtue of the linguistic properties of the latter and without regard for any special form the predicate may have, will in principle be satisfiable by an indefinite number of facts. Now, we have already seen that in order for SP IV to be satisfied, our criterion must exclude the possibility that it is simultaneously satisfied both by a fact of the form $T(\phi_1, \ldots, \phi_n)$ and a fact of the form $\sim T(\phi_1, \ldots, \phi_n)$, since if this possibility were realized, T would signify the one fact truly and the other falsely. But since the criterion allows T in principle to signify more than one fact, the only apparent way to ensure against the significata of T including both facts of the form $T(\phi_1, \ldots, \phi_n)$ and facts of the form $\sim T(\phi_1, \ldots, \phi_n)$ and still have the criterion concern only the semantic relation of a fact to '$T(\phi_1, \ldots, \phi_n)$'[17] is for it to be impossible for facts of one of the two forms to satisfy the criterion. That is, it must be either that facts of the form $T(\phi_1, \ldots, \phi_n)$ or that facts of the form $\sim T(\phi_1, \ldots, \phi_n)$ are excluded as possible significata of T; and since a fact must be of one of these forms in order to be signified by T, it must be that either T can signify only facts of the form $T(\phi_1, \ldots, \phi_n)$ or only facts of the form $\sim T(\phi_1, \ldots, \phi_n)$. Now, if a fact had to be of the latter form in order to be signified by T, then, since T could signify such a fact only falsely, a theory could itself signify a fact only if the theory were false. But it would be absurd to propose seriously that despite a theory user's patent effort to make a true assertion thereby, or at the very least to make correct commitments as to what is the case, that the only way a theory can acquire cognitive meaning is through be-

$\phi_n) \cdot (c_1 = c_1) \cdot \ldots \cdot (c_n = c_n)$ or its negation; however, it would be absurd to propose this seriously as the criterion for what a theory can assert. For another illustration, consider a potential criterion which limits T to the form '$T(\tau_1, \ldots, \tau_n) \cdot (\tau_1 = c_1) \cdot \ldots \cdot (\tau_n = c_n)$' if T is to have a significatum. Again, this assures a unique significatum to any theory which meets the special syntactic requirement, but would limit semantically meaningful theories to those which are formally equivalent to an observation sentence—namely, '$T(c_1, \ldots, c_n)$'—and hence categorically denies the possibility of a realistic interpretation of theories in the more general case.

[17] One can always get around SP IV by adding a safety clause which brings in extraneous material. For example, if '$T(f)$' is a predicate describing a potential criterion for what is signified by T but which is simultaneously satisfiable by facts of both forms in question, the revised criterion '$T(f) \cdot [(\phi_1, \ldots, \phi_n)T(\phi_1, \ldots, \phi_n) \vee \sim(\exists \phi_1, \ldots, \phi_n)T(\phi_1, \ldots, \phi_n)]$' no longer violates SP IV, but neither does its satisfaction depend only on a semantic relation of the fact f to the predicate '$T(\phi_1, \ldots, \phi_n)$.' It should be clear that introducing safety clauses of this sort into potential criteria for what is expressed by a theory violates the intent of the Thesis of Semantic Empiricism when the latter stipulates that the meaning of a theory depends only on the theory's use with T, and not, in addition, upon whether or not a number of other conditions, which have nothing to do with this usage, obtain.

coming false. For this reason, the Thesis of Semantic Empiricism as formulated above rejects this possibility by including the stipulation that a theory derives its meaning, if any, in a potentially useful manner. Hence facts of the form $\sim T(\phi_1, \ldots, \phi_n)$ are excluded from what can be signified by T, and we conclude that SP I–IV and the Thesis of Semantic Empiricism imply that a theory '$T(\tau_1, \ldots, \tau_n)$' can itself signify a fact only if the fact is of the form $T(\phi_1, \ldots, \phi_n)$. That is,

Postulate 1 (P 1). If '$T(\tau_1, \ldots, \tau_n)$' is an accepted normal syntactic theory in which '$T(\phi_1, \ldots, \phi_n)$' is an observational predicate and 'τ_1,' $\ldots$, 'τ_n' are theoretical terms, then '$T(\tau_1, \ldots, \tau_n)$' signifies a fact f only if there exist entities $t_1, \ldots, t_n$ designated by 'τ_1,' $\ldots$, 'τ_n,' respectively, such that f is the fact that $T(t_1, \ldots, t_n)$.

It is an immediate consequence of SP I and P 1 that a theory can itself signify a fact only truly.[18] This conclusion might at first seem somewhat counterintuitive; and since Theorems 3–5 depend crucially upon it (more accurately, upon the slightly weaker premise that a theory can falsely signify only a fact which refutes one of its O-consequences), I shall try to show that it claims no more than lies implicit in our normal use of theories. As pointed out earlier, if a theory can signify a fact falsely, then we must expect that a theory may also be refuted by an observational fact even when all its O-consequences are true. Now, while we modify or abandon theories for a variety of reasons (excessive formal complexity, too many ad hoc accretions, probabilistic disconfirmation of an O-consequence, etc.), it is, I think, an indisputable fact about natural usage that we never regard a known fact to be an actual refuter of a theory unless the fact disproves an O-consequence of the theory. Thus we behave as though it were impossible for a theory to be refuted by an observational fact which does not refute an O-consequence of the theory, and hence as though it were impossible for a theory itself to signify a fact falsely (or, at any rate, impossible for the theory to signify falsely a fact which does not refute one of its O-consequences).

Again, suppose a person who accepts theory '$T(\tau_1, \ldots, \tau_n)$' is shown a set of entities $t_1, \ldots, t_n$ and asked whether by proposing this theory, he means to claim that $t_1, \ldots, t_n$ satisfy '$T(\phi_1, \ldots, \phi_n)$.' If T is really the totality of that person's theoretical assumptions, so that he has no

[18] This by no means implies that a theory cannot be incorrect, or even semantically false (see the comment following SP II). The conclusion is only that if a theory is incorrect, it must be so in a way other than by signifying a fact falsely—as, for example, by having an O-consequence which signifies falsely.

William W. Rozeboom

commitments to the identities of the theoretical entities other than what is made explicit in T, the fact that $T(t_1, \ldots, t_n)$ is not the case suffices for him to reject the supposition that T falsely signifies the fact that $\sim T(t_1, \ldots, t_n)$—otherwise, we should be able to make him doubt that T is correct merely by finding a set of entities (of appropriate types) which did *not* satisfy '$T(\phi_1, \ldots, \phi_n)$.' Now, the fact that in practice we *recognize* as refuters of a theory only those facts which refute one of its O-consequences does not in itself prove that the theory *has* no observational refuters beyond this class, since we are not entitled to take for granted that our use of theories gives them no factual content beyond that which we normally recognize. What natural usage does show, however, is that *if* actual practice is correct, a theory can signify a fact only truly. Hence what P 1 amounts to is an assertion that natural usage *is* correct.

Returning to the Thesis of Semantic Empiricism, our analysis so far has left us only a step from an important conclusion which, though unnecessary for Theorems 3–5, will be needed in Section III. We have seen that in order to be signified by theory T, a fact must be of form $T(\phi_1, \ldots, \phi_n)$. But it should also be apparent that the use of T with L_o is unable to differentiate further among facts of this form unless, contrary to the Thesis, one takes into consideration special characteristics of '$T(\phi_1, \ldots, \phi_n)$.' For as already seen, the predicate '$T(\phi_1, \ldots, \phi_n)$' (together with the syntax of L_o) fully characterizes the observational use of T, and it does not seem possible to find a difference in the way two facts, both of form $T(\phi_1, \ldots, \phi_n)$, could be related to '$T(\phi_1, \ldots, \phi_n)$' which does not draw upon special features of the predicate.[19] That is, there are no plausible grounds upon which a person who had accepted T could say, after extending the range of facts with which he is observationally acquainted, "Although I now know both that $T(t_1, \ldots, t_n)$ and that $T(t'_1, \ldots, t'_n)$, it was the former, rather than the latter, that I intended to signify by '$T(\tau_1, \ldots, \tau_n)$.'" Similarly, he could not legitimately claim, "Although I now know it is the case that $T(t_1, \ldots, t_n)$, this isn't what I meant by '$T(\tau_1, \ldots, \tau_n)$,'" unless he would make the same judgment about any fact of the form $T(\phi_1, \ldots, \phi_n)$. The conclusion seems inescap-

---

[19] For example, two facts $T(t_1, \ldots, t_n)$ and $T(t^*_1, \ldots, t^*_n)$ could differ in whether or not the predicate '$T(\phi_1, \ldots, \phi_n)$' contains a constant which designates one of the t_i or t^*_i. If this kind of difference could matter with respect to what is signified by a theory, then certain theories would be barred from the possibility of signification simply because they fail to exhibit certain features of formal complexity.

312

able that all facts of the form $T(\phi_1, \ldots, \phi_n)$ must equally satisfy any acceptable criterion for what is signified by T, and hence

Postulate 2. (Not to be used until Section III.) *If an accepted normal syntactic theory '$T(\tau_1, \ldots, \tau_n)$' is able to signify any fact at all, it signifies all facts of form $T(\phi_1, \ldots, \phi_n)$.*

Although the detailed arguments of the past few pages may seem a bit rarified in spots, the basic development so far is actually quite robust: If the nonobservational components of an accepted theory acquire their meanings, if any, in virtue of this usage, then the observational predicate which characterizes the theory must provide the criterion for its factual significance, and there are presently no intelligible grounds on which to suspect that such a criterion would admit as a designatum of the theory a fact which did *not* comply with the form specified therein. But now the argument becomes more delicate, for while P 1 takes an important step toward determining the refuters of a theory by excluding the possibility that the theory signifies falsely, limiting conditions yet remain to be imposed on facts which refute without being signified. Such limits will now be developed by a four-stage postulate which commands assent by drawing on our intuitive feel for the meanings of certain ill-defined but basic semantical notions. If the considerations which follow appear esoteric, they are no more so than the traditional gambits of philosophical speculation which they are designed to forestall.

Postulate 3a. If a sentencelike formula is not itself cognitively meaningful when accepted, then the only way in which its acceptance can be pragmatically inappropriate is for the formula to be inappropriate as a transformation mechanism. That is, an accepted sentencelike but cognitively meaningless formula has a refuter only if it transforms a true cognitively meaningful sentence in its accepter's language into a false one.

While we shall not here attempt a formal definition of "cognitively meaningful," we may roughly interpret it as "having pragmatic import in virtue of its semantic properties." And if a formula is not given pragmatic import by its semantic properties, if any, it is difficult to imagine what could be relevant to the appropriateness of its acceptance except its efficiency at transforming expressions which are cognitively meaningful. It must be confessed that the second sentence in P 3a is not altogether a mere clarification of the one which precedes it. For it may fairly be asked whether a transformation technique which does not lead to error when operating only upon its user's present language might not still be ultimately incorrect through having false consequences in an augmented language whose resources T's user may eventually attain through subsequent language enrichments (e.g., by addition to his stock of observation terms as a result of new experiences). The extent to which acceptance of a transformation technique involves commitments in languages not actually in use at the time of the adoption, and whether this entails that the acceptability of a cognitively meaningless transformation procedure cannot be decided on the basis of its import for the existent language alone, is an issue calling for somewhat more extensive analysis than seems profitable on this occasion. It can be shown with the help of Lemma 1, however, that so long as L_o is as powerful syntactically as stipulated here, P 3a is not disturbed by such considerations.

According to classical views on semantics, P 3a plus a simple extension of P 1 to cover theoretical consequences of T in addition to T itself are sufficient to determine the refuters of a theory. For in the classic tradition, a sentencelike formula '$S(s_1, \ldots, s_n)$' is held to be cognitively meaningful only if there exist entities $t_1, \ldots, t_n$ designated by 's_1,' $\ldots$, 's_n,' respectively, and through which '$S(s_1, \ldots, s_n)$' is then verified or refuted according to whether or not it is the case that $S(t_1, \ldots, t_n)$. However, a persistent, though less clearly conceptualized, alternative interpretation of the conditions of meaningfulness has been that a sentence which is cognitively meaningful can nonetheless fail at factual reference through lack of designata for some of its descriptive components—in fact, such a view will later be proposed in Section III, below. Hence we should also consider the possibility that a sentencelike formula may fail to signify a fact and yet escape meaninglessness. But what could refute such a formula? It seems inescapable that if the linguistic attributes of a cognitively meaningful sentence S do not establish some immediate semantical correspondence between S and an external reality which determines its correctness, the pragmatic import of S can come only through its meaning connections with other cognitive states which establish eventual reference to the conditions of verification or refutation. That is, to employ notions which have not previously been introduced here and which will appear only briefly as an intuitive justification for Postulate 3b, if S expresses a proposition which does not itself signify a fact, S has a refuter only if S commits its believer to some false proposition p which does so signify. Now if S makes commitment to p, this is accomplished through some relation of

314

syntax and meaning between S and p. But if S is an accepted normal syntactic theory T, the Thesis of Semantic Empiricism implies that the theoretical terms in '$T(\tau_1, \ldots, \tau_n)$' contribute no extrasyntactical implicative force beyond what is already contained in the predicate '$T(\phi_1, \ldots, \phi_n)$.' Hence if T makes commitment to p, so for like reason should any other sentence formed by instantiation of '$T(\phi_1, \ldots, \phi_n)$.' But then any disjunction of instantiations of '$T(\phi_1, \ldots, \phi_n)$' and hence also, as a limiting case, '$(\exists \phi_1, \ldots, \phi_n)T(\phi_1, \ldots, \phi_n)$,' should likewise make commitment to p; so any refuter of p must also refute T's prime consequence, R_T. Accordingly,

Postulate 3b. If an accepted normal syntactic theory T is cognitively meaningful but does not signify a fact, then T has a refuter only if its prime consequence has a refuter.

There is yet one further possibility to be reckoned with, namely, that some of the theoretical consequences of T—i.e., sentencelike formulas containing theoretical terms and deducible from T—are cognitively meaningful when T is accepted, even though T itself is not. To be sure, if the notion of "cognitive meaningfulness" is already vague, speculation about the circumstances under which T could have meaningful theoretical consequences while itself remaining meaningless is vagueness compounded. Nonetheless, this is one of the logical alternatives which must be accounted for by any exhaustive attempt to delimit the factual content of theories, and fortunately, by groping cautiously, we can work our way across to firm ground. To begin with, we may safely assume that a syntactically complex, well-formed expression is cognitively meaningful if and only if its formal constituents are cognitively meaningful. Hence,

Postulate 3c. If E_1 is an expression containing theoretical terms 'τ_1,' $\ldots$, 'τ_n' which is given cognitive meaning by acceptance of a normal syntactic theory T, and E_2 is a well-formed expression (of any type) containing no theoretical terms other than 'τ_1,' $\ldots$, 'τ_n,' then E_2 is also cognitively meaningful when T is accepted.

Let a meaningful consequence M of accepted theory T such that all meaningful consequences of T are also deducible from M be called a "meaning abstract" of T. More precisely,

Definition 7. M is a meaning abstract of normal syntactic theory $T =_{\text{def}}$ M is a sentencelike formula which is cognitively meaningful when T is accepted as a normal syntactic theory, and any sentencelike formula S which is cognitively meaningful when T is accepted is deducible from T if and

315

William W. Rozeboom

only if S is also deducible from M when M replaces T as the accepted (normal syntactic) theory.

Suppose, now, that T does, in fact, have a meaning abstract M. In what way, if any, is the meaning given to M through acceptance of T different from the meaning that M would receive if it were to replace T as the accepted theory? It is easy to see that the only difference between the syntactical accomplishments of T and M is the omission, when M is the accepted theory, of whatever cognitively meaningless expressions are introduced by acceptance of T. All theoretical terms, consequences, and transformation pairs which are given meaning by acceptance of T are preserved when T is replaced by M. To our admittedly primitive understanding of these matters, it seems strange to suspect that the meaning of a portion of an accepted theory would be altered by deletion of another portion which is neither cognitively meaningful nor necessary for the remainder to maintain the same syntactical interconnectedness as before. In fact, the Thesis of Semantic Empiricism can be construed to rule out this possibility in that if deleting part of an accepted theory does not change the relation between the remainder of the theory and the observation language, it should not change the meaning of the remainder, either. Hence there is ample reason to assume that

Postulate 3d. If M is a meaning abstract of normal syntactic theory T, the cognitive meaning, and hence factual content, of M when T is accepted are the same as the cognitive meaning and factual content, respectively, of M when M replaces T as the accepted (normal syntactic) theory.

Lemma 2. Every normal syntactic theory has a meaning abstract.

Proof: Let '$T(\tau_1, \ldots, \tau_n)$' be a normal syntactic theory in which m $(0 \leqq m \leqq n)$ of the theoretical terms are cognitively meaningful when T is accepted, while the remainder are not. Then by P 3c, the sentencelike formulas deducible from T which are cognitively meaningful when T is accepted are exactly those consequences of T which contain no theoretical terms other than the m meaningful ones. Call these formulas the M-class of T. Now let R^m_T be the sentencelike formula deduced from T by existential quantification over the theoretical terms which are left meaningless when T is accepted—i.e., when the meaningful terms are 'τ_1,' $\ldots$, 'τ_m,' $R^m_T =_{def} '(\exists \phi_{m+1}, \ldots, \phi_n)T(\tau_1, \ldots, \tau_m, \phi_{m+1}, \ldots, \phi_n).$' Then R^m_T belongs to T's M-class, and it is simple to show, by Lemma 1, that any formula which belongs to the M-class of T is also deducible from R^m_T

when T is replaced by $R^m{}_T$ as the accepted theory. Hence $R^m{}_T$ is a meaning abstract of T. Q.E.D.

In particular, if all of the theoretical terms in T are cognitively meaningful when T is accepted, T is its own meaning abstract; while if none are, the meaning abstract of T is its prime consequence, R_T.

While the concept of "meaning abstract" bears sufficient technical interest to warrant discussion for its own sake, its present use is solely to provide a proof for Lemma 3. Consequently, the reader who feels uncomfortable with P 3c or P 3d may instead treat Lemma 3 as a postulated generalization of P 3b to replace P 3a–d.

Lemma 3. If an accepted normal syntactic theory T does not itself signify a fact, then T has a refuter only if its prime consequence has a refuter.

Proof: If T is itself cognitively meaningful, the lemma follows immediately from P 3b. Conversely, suppose that T is not cognitively meaningful. Then by P 3a, T has a refuter only if it transforms a true cognitively meaningful sentence S_1 into another, S_2. Since T transforms S_1 into S_2 if and only if $S_1 \supset S_2$ is a consequence of T, and S_1 and S_2 are true and false, respectively, if and only if $S_1 \supset S_2$ is false, T then has a refuter only if it has a false cognitively meaningful consequence. Now, every cognitively meaningful consequence of accepted theory T is also deducible from its meaning abstract, $R^m{}_T$; so if T has a false consequence, $R^m{}_T$ must also be false when T is accepted.[20] Hence T, when accepted, has a refuter only if its meaning abstract also has a refuter. But by P 3d, $R^m{}_T$ has a refuter when T is accepted only if it has a refuter when $R^m{}_T$ replaces T as the accepted theory. But then, by P 3b, the prime consequence of $R^m{}_T$ also has a refuter; and since the prime consequence of $R^m{}_T$ is identical with the Ramsey sentence, R_T, of T, the prime consequence of T likewise has a refuter. Q.E.D.

Postulate 4. Any refuter of an O-consequence of an accepted theory T is also a refuter of T.

This simply brings forward in official form the obvious fact about theory usage noted earlier. By construing an observation sentence to be a theory with no theoretical terms, P 4 also subsumes the point noted earlier (p. 298) that an observation sentence is likewise refuted by any refuter of its O-consequences.

---

[20] This step is not quite automatic if the possibility raised in fn. 14 is taken seriously. However, the fact that $R^m{}_T$ contains every theoretical term in any meaningful consequence of T obviates the difficulty.

317

William W. Rozeboom

Postulate 5. An accepted theory has a verifier if and only if it has no refuter.

This postulate merely formalizes for theories what was observed earlier, namely, that the terms 'correct' and 'incorrect,' unless generically inapplicable, are mutually exclusive and exhaustive. P 5 asserts that the behavioral role of an accepted theory is such that it *has* factual content. To say that an accepted theory would be incorrect if such-and-such were the case makes sense only if it is also the case that the theory would be correct if no condition were to obtain under which it is incorrect. Hence any fact which, together with any needed information such as that contained in SP I–IV and P 1–5 about the behavioral role of T, authorizes the conclusion that T has no refuter, then also authorizes, by P 5, the conclusion that T has a verifier and is hence itself a verifier of T.[21]

We have now extracted the factual content of an accepted normal syntactic theory, and it only remains to put the results in polished form.

Theorem 3. An accepted normal syntactic theory has the same factual content as its prime consequence.

Proof: Let R_T be a prime consequence of accepted normal syntactic theory T. Then we have to show that the verifiers and refuters of T are identical with the verifiers and refuters, respectively, of R_T. Since R_T is an O-consequence of T, any refuter of R_T is also, by P 4, a refuter of T, and hence by P 5, any verifier of T must also be a verifier of R_T. Therefore, to complete the proof, it suffices to show that under SP I–IV and P 1–5, T can have no refuter when R_T is true. For then, given a fact f that verifies R_T, it follows by P 5 that T has a verifier. Since this reveals that f and the facts about the behavioral role of T entail that T is correct, a verifier of R_T is also a verifier of T—which also shows, by P 5, that a refuter of T must also be a refuter of R_T.

Suppose, now, that T signifies a fact. Then by P 1, T has a verifier and hence, by P 5, no refuter. On the other hand, suppose that T does not signify a fact. Then if T has a refuter, it follows by Lemma 3 that R_T must

---

[21] If the sense of this claim is not readily apparent, it would be advisable to review the discussion on p. 296, including fn. 13, of how the verifiers or refuters of an expression are in principle to be identified. Formalizing this identification procedure would have raised the logical complexity of the present arguments, already difficult enough, to a prohibitive level. Without a grasp of the unformalized procedure, however, the reader will be unable to appreciate how, in what ensues, assumption that fact f is a verifier or refuter of one linguistic entity leads to a conclusion that f is also a verifier or refuter of another.

also have a refuter, a situation incompatible with R_T's being true. Consequently, whether an accepted normal syntactic theory T itself signifies a fact or not, our premises (SP I–IV and P 1–5) about the behavioral role of theories show that T can have no refuter when its prime consequence is true. But as already pointed out, this suffices, by P 5, for a theory and its prime consequence to have the same factual content. Q.E.D.

One of the particularly important issues which demands attention by any serious methodological analysis of scientific theories is that of rivalry, conflict, or opposition among variously proposed theories. That is, what are the circumstances under which two alternative theories are incompatible? For we know from extended historical experience that many of the quarrels which arise in science and philosophy spring more from verbal misunderstandings and discoordinated interests than from genuine cognitive disagreement. When scientist A insists that his theory challenges the theory of scientist B, it would be highly useful to have means of determining in precisely what way, if at all, this is an actual clash of factual commitments and not just of personalities, especially if it is difficult to discern any testable differences between these theories. A major virtue of Theorem 3 is the illumination it brings to this question.

According to any intuitive understanding of the notion of "incompatibility," Theorem 3 implies that two theories are incompatible if and only if they have incompatible observational consequences. For if a theory is factually equivalent to its prime consequence, then two theories are incompatible if and only if their prime consequences are incompatible. To demonstrate this clearly, however, calls for a definition of 'incompatibility' as applied to theories without necessarily assuming that theories are themselves cognitively meaningful, and this is not quite so simple as it might at first appear.

What do we mean by saying that two theories are incompatible? It will be helpful first to examine the concept as it applies to observation sentences, and then seek a suitable extension to theories. A condition of "incompatibility" which immediately comes to mind is that S_1 and S_2 are incompatible when a contradiction can be deduced from them jointly. This is not a necessary condition, however, for "incompatibility" is more than just a syntactical relation, and S_1 and S_2 may be incompatible even though they formally entail no contradiction. For example, 'F(a)' and '~F(b)' are formally consistent, but are incompatible if 'a' and 'b' are synonymous. Thus a more satisfactory explication might be that two sentences are in-

319

compatible when and only when they jointly make a self-contradictory assertion—i.e., their conjunction is necessarily false and hence refuted by every fact. But the analogous claim for theories, namely, that theories T_1 and T_2 are incompatible when and only when their conjunction is refuted by every fact, will not do at all. For according to the views developed here, a theory acquires factual commitments only through being accepted, and the force of a given sentencelike formula may be expected to be influenced by whatever additional theoretical postulates have also been adopted.

Thus to ask whether theories T_1 and T_2 are incompatible is, more precisely, to ask whether T_1, when accepted as a totality of theoretical assumptions, makes factual commitments which are incompatible with those made by T_2 when the latter is accepted as an alternative to T_1. Consequently, we cannot test the incompatibility of T_1 and T_2 merely by examining the force of simultaneous acceptance of the formulas comprising T_1 and T_2, for $T_1 \cdot T_2$ is still another (alternatively acceptable) theory in which the factual content of sentencelike constituent T_1 or T_2 may not be altogether the same as its content when accepted as a theory by itself. In particular, *it must not be presupposed that two sentencelike formulas which formally entail a contradiction when accepted jointly are necessarily contradictory in their factual commitments when (alternatively) accepted singly.* For felicity of expression, we shall frequently speak, henceforth, of the factual content of a theory T without explicitly stipulating that T is normal syntactic and accepted. It should be understood, however, that what is thereby meant is not the factual content, if any, that sign-design T may actually have at the moment, but, subjunctively, the content that T would have if T were the totality of theoretical postulates accepted for normal syntactic use with L_0. This does not preclude the possibility that sign-design T might have the same factual content when part of a more inclusive accepted theory as it does when accepted by itself, but it forbids such an assumption being made without special argument (such as, for example, the one which supported P 3d).

The preceding considerations, however, have been deliberately roundabout in order to warn against seductive pitfalls. A perfectly natural way to explain "incompatibility" is to say that two sentences are incompatible if and only if they cannot both be true. To extend this notion to any pair of expressions E_1 and E_2 which have factual content (or which would have, if used in a stipulated way), we have but to substitute "have verifiers" for "be true"—i.e., E_1 and E_2 are incompatible if and only if they cannot both have verifiers in the roles allocated to them. That is, if the (potential) factual commitments of E_1 are incompatible with those of E_2, then any fact which verifies E_1 when the latter is used in its prescribed manner also refutes E_2 when the latter is used in its prescribed manner. This is not a sufficient condition for the incompatibility of E_1 and E_2, however, for if E_1 and E_2 are (or would be) both incorrect, it is vacuously true that all (potential) verifiers of E_1 (potentially) refute E_2 and all (potential) verifiers of E_2 (potentially) refute E_1. For a definition of "incompatibility" in terms of factual content, then, we need to identify something about the refuters of two incorrect expressions which reveals that they could not both be correct. The following is offered as one possibility which may or may not require modification as the ontology of "facts" becomes more clearly understood. For purposes of Theorem 4, however, any alternative definiens would do, so long as it is wholly a condition on the verifiers and refuters of the expressions involved.

Definition 8. Expressions E_1 and E_2 are incompatible (relative to usage U) $=_{def}$ E_1 (under U) has a verifier and every verifier of E_1 (under U) refutes E_2 (under U); or E_2 (under U) has a verifier and every verifier of E_2 (under U) refutes E_1 (under U); or there exists a tautologous fact which is the disjunction of a refuter of E_1 (under U) and a refuter of E_2 (under U).

By the verifiers or refuters of E_1 "under U" is meant, of course, the verifiers or refuters that E_1 would have if used in accordance with procedure U. The definition does not presuppose that U necessarily allows E_1 and E_2 to be used jointly.

To appreciate that the final clause in Definition 8 preserves the notion that if E_1 and E_2 are incompatible in their stipulated roles, they must not be, rather than merely are not, both incorrect, suppose that f is a refuter of E_1 (under U) and g is a refuter of E_2 (under U). Now, a tautological fact is such that we would say that it must be the case. Hence, if $f \vee g$ is tautologous, it must be the case that either f or g; thus either E_1 (under U) or E_2 (under U) must have a refuter.

Theorem 4. (a) Theories T_1 and T_2 are incompatible if and only if their prime consequences are incompatible. (b) Theory T and observation sentence S are incompatible if and only if S is incompatible with the prime consequence of T.

Proof: Let S_1, S_2, and S_3 be observation sentences or theories. Since application of Definition 8 depends only upon the verifiers and refuters concerned, if S_1 has the same verifiers and refuters, respectively, as S_2, then

William W. Rozeboom

S_1 and S_3 are incompatible if and only if S_2 and S_3 are also incompatible. But by Theorem 3, the verifiers and refuters of a theory (when accepted for normal syntactic usage) are identical with those of its prime consequence. Hence two theories are incompatible if and only if their prime consequences are incompatible, and similarly for a theory and an observation sentence. Q.E.D.

A theorem of great philosophical importance follows immediately from Theorem 3. If two theories have identical O-consequences, their prime consequences must be formally equivalent. Hence from Theorem 3,

Theorem 5. If two theories have identical observational consequences, they have the same factual content.

Theorems 4 and 5 dispel a number of perplexities that have traditionally been associated with the epistemic status of scientific theories. Much philosophical *Angst* and operationistic impatience have been vented over the prima-facie possibility that two conflicting theories might have no observational disagreement, or that nonequivalent or even incompatible theories might have identical observational consequences and be equally supported (or disconfirmed) by any empirical evidence. The present analysis suggests that the famous pragmatic dictum "A difference which makes no difference is no difference," should be put even more strongly as a logical contention: "There *is* no difference which makes no difference." Theorems 4 and 5 show that given the Thesis of Semantic Empiricism, it is not possible for two theories to be incompatible without being observationally incompatible as well, or to be nonequivalent without differing in their observational consequences. (Note that I am not saying that two theories with the same O-consequences are necessarily *synonymous*, but only that any verifier or refuter of one is also a verifier or refuter of the other.) The fallacy has been to presume that the same theoretical *symbol* necessarily has the same meaning in one theory as it has in another, overlooking that it is the particular theoretical usage which *gives* the symbol its meaning.

Technical Note: It will be recalled that the theorems which have been developed in this section rest upon certain assumptions in addition to those made explicit in the theorems proper. All but one of these, which were set forth in Section I, concern the character of the observation language. However, we also made one stipulation that might seem to be a gratuitous restriction on the form of a normal syntactic theory; namely,

322

that theoretical terms enter the theoretical postulates syntactically only as *constants*. It would appear that at least some of the variables of a language may be descriptive (i.e., nonlogical) terms in the sense that the classes over which they range are nonlogical categories. One might then wonder whether a theoretical concept might not find expression as the range of a variable. Syntactically, this would be accomplished by introducing variables and perhaps constants of a new, uninterpreted formal type. If it is possible to introduce theoretical terms in this way—and indeed, so long as variables need not be purely logical, it seems unreasonable to deny that they can—we might question whether the present theorems apply to such theories.

A satisfactory discussion of formal types and the ranges of variables is too lengthy to be undertaken here. There are, however, compelling reasons for believing that a language which contains nonlogical variables and is also adequate to formulate and cope with the various problems which arise from the use of these variables must be such that for any nonlogical variable 'x^i' in the language, it must contain or permit introduction of (a) a descriptive constant which refers to the class (or the defining property thereof) ranged by 'x^i,' and (b) a purely logical variable, 'x^j,' whose range includes that of 'x^i.'[22] But in such a language, for every sentence S_i containing nonlogical variables, there is an analytically equivalent sentence, S^*_i, containing descriptive constants corresponding to the ranges of the nonlogical variables in S_i and whose variables are only logical. Let such an S^*_i be called the "L(ogically)-normal form" of S_i. Then if T is a theory, containing theoretical variables as well as constants, whose factual content we wish to analyze, we simply find the L-normal form of T and proceed as before. In short, our assumption that the theoretical terms of a normal syntactic theory are constants places a restriction on the applicability of Theorems 2–5 only for an artificially weak language, just as Theorem 2 may fail for a language which does not permit existential quantification over all primitive descriptive constants.

[22] Thus, if formal type i represents a nonlogical category, an empirical problem immediately arises as to whether or not a given entity t can be designated by a constant of type i. But if the language contains the predicate 'x^i can be designated by a symbol of type i,' this is equivalent in force to 'x is a member of C,' where 'C' designates the class corresponding to formal type i. Moreover, if 'x^i' is also a nonlogical variable, we would have to determine that t belongs to the range of 'x^i' before using this predicate to inquire whether or not t belongs to the range of 'x^i'; so if the language is to be able to formulate the question about an entity's membership in the range of a nonlogical variable, it must contain a logical variable to terminate the regress.

William W. Rozeboom

In this section, we shall investigate the semantical status of theoretical terms and postulates. Before proceeding further, however, let us introduce a simplification in notation. Unless the analysis turns on the number of theoretical terms, there is no need to maintain explicit reference to *n* theoretical constants. Hence, with the exception of a few places needing careful formulation, we may restrict discussion to theories with only one theoretical term.

In preceding sections, the question repeatedly arose whether a (normal syntactic) theory is merely an instrument for generating its O-consequences, or whether T is itself cognitively meaningful. My first contention in this section is that the latter is indeed the case. For if we deny that a normal syntactic theory is, in some fashion, itself an assertion, we find ourselves committed to the view that no expression containing a descriptive term which does not designate a sense datum can be an assertion. If we take 'observed' not in the phenomenalistic sense of 'directly experienced,' but in the broader usage of science and everyday life, there is no hard and fast distinction between observational (i.e., "empirical") and theoretical concepts. The cytologist, for example, considers cells and their grosser properties as "observable," even though the observation depends upon an intervening distortion of light rays by the lens of a microscope. More generally, it has long been accepted that our access to the events in the objective world to which our observational terms are commonsensically presumed to refer is only through the medium of causal chains which bridge between these events and our nervous systems. Most of our "observational" concepts, upon philosophical scrutiny, may be seen to lose their halo of immediacy and to stand in very much the same relation to more immediately given events as theoretical concepts stand to events in the commonplace world. Now, I am by no means convinced that the phenomenally "given" is as mythological as some would have it, nor do I deny that *some* terms of ordinary language appear to designate phenomenal entities. However, I think it would be absurd to maintain that sentences in the everyday observation language, except for a proper phenomenal subset, are nothing but instruments for committing one to a set of statements wholly in a phenomenal language. This would be plausible only if we did, in fact, habitually use ordinary language for this purpose. The fact is, of course, that purely phenomenal statements, at least those *recognized as* such, play a minor if not virtually nonexistent role in linguistic practice. If only sentences known to be wholly phenomenal were able to make assertions, virtually the whole of our linguistic machinery would lack cognitive meaning. I would thus maintain that ordinary observation sentences, whether recognizably phenomenal or not, do in general have semantic properties, and that by the same token, such properties are also possessed by theoretical postulates.

However, in arguing from the negligible incidence of *recognizably* phenomenal statements in ordinary discourse to the conclusion that theoretical postulates and observation sentences in the broad sense must, in general, themselves be assertions, a possibility which must not be overlooked is that while a symbol can designate only a previously experienced sense datum, we may construct, by intricate and presumably to a large extent unwitting definitional processes, sentences which are wholly about phenomenal facts but which are not readily identified to be so. This, of course, is the standard phenomenalistic move—not to deny that everyday sentences are assertions, but to claim that upon analysis, they can be *discovered* to be wholly phenomenal. In like manner, the positivist, however he construes the "observed," can argue that while a theory is indeed an assertion, it is analyzable into a sentence of the observation language. Thus granting that theories may be assertions, we must consider whether there might not be some sentence 'S' constructable in L_o, for which it can be argued that $T(\tau) =_{def} S$—i.e., that '$T(\tau)$,' in virtue of its definition, has the same meaning as 'S.'

Let us first dispose of the possibility that the way in which '$T(\tau)$' signifies a fact is by being a peculiar, syntactically improper, notational form for some more orthodox sentence in the observation language, similar to the way, for example, that a code signal may be said to signify a fact because, while the code signal is not syntactically a sentence, it has been stipulated to abbreviate an ordinary sentence. What we are now asking is, if '$T(\tau)$' not merely carries factual content but also *signifies* a fact, whether it could do this in any way other than by 'τ' standing in a relation of reference to some entity of appropriate type. It would be *possible*, for example, to stipulate that '$T(\tau)$' is to mean that p, where 'p' is some observation sentence of syntactical form different from '$T(\tau)$.' Actually, the Thesis of Semantic Empiricism and SP I confute this; however, since we now wish to consider possible alternatives to the Thesis, we must look further into the possibility that the meaning of '$T(\tau)$' is *not* governed by the syntax of

L_o. In particular, since natural theory usage seems to show that we normally regard a theory to have the same observational force as its prime consequence, we are especially interested in the possibility that $T(\tau) =_{def} (\exists \phi) T(\phi)$.

That this is untenable, however, may be shown in at least two ways:

1. If '$T(\tau)$' is an assertion, then presumably the reason that acceptance of T commits one to the O-consequences of T is because the latter are logically entailed by '$T(\tau)$.' But if '$T(\tau)$' asserts that $(\exists \phi) T(\phi)$, then applying formal inference rules to the sentence form '$T(\tau)$' *is not logical deduction.* That logical inference is *more* than merely a set of operations within an arbitrary formal calculus is due to certain relations which obtain among the cognitive meanings of statements as a result of their logical forms, as normally mirrored by their syntactic forms. Hence it is incorrect to stipulate that a syntactically sentencelike formula S is to assert the same fact as another, syntactically different, statement, and then to claim that the syntactic consequences of S must be its *logical* consequences. To be sure, if it were true that $T(\tau) =_{def} (\exists \phi) T(\phi)$, we could *vindicate* taking a sentence in L_o, syntactically deducible from '$T(\tau)$,' to be a logical consequence of '$(\exists \phi) T(\phi)$' and hence of '$T(\tau)$,' because as it happens, the two formulas have the same O-consequences. But the existence of Ramsey sentences has been known only since 1929, and even since then has been virtually ignored. Hence, if it were the case that $T(\tau) =_{def} (\exists \phi) T(\phi)$, users of theories would heretofore have been *unjustified* in taking acceptance of T as necessary commitment to the O-consequences of T.

2. It seems inacceptable to grant semantic status to '$T(\tau)$' and withhold it from its theoretical consequences, especially since in the *de facto* use of theories, the full conjunction of theoretical postulates, '$T(\tau)$,' is seldom, if ever, actually constructed. For example, suppose that T is the conjunction of two theoretical postulates, '$S_1(\tau)$' and '$S_2(\tau)$.' If '$S_1(\tau) \cdot S_2(\tau)$' is a statement, we should certainly wish also to say that '$S_1(\tau)$' and '$S_2(\tau)$' are statements. But if $S_1(\tau) \cdot S_2(\tau) =_{def} (\exists \phi)[S_1(\phi) \cdot S_2(\phi)]$, it is not the case that $S_1(\tau) = (\exists \phi) S_1(\phi)$ and $S_2(\tau) = (\exists \phi) S_2(\phi)$ except in degenerate instances. Without entering into formal details, let me simply assert that if $T(\tau) =_{def} (\exists \phi) T(\phi)$, there appears to be no satisfactory translation for the constituent postulates in T or their theoretical consequences. We may raise analogous objections against any other sentence S which might be proposed as a definiens for '$T(\tau)$' unless S has the same syntactical structure as '$T(\tau)$.' Since in this case, S must be of the form '$T(d)$,' where 'd' is a

referring expression in L_o, setting $T(\tau) =_{\text{def}} T(d)$ is the same as setting $\tau =_{\text{def}} d$. We may therefore conclude that if '$T(\tau)$' signifies a fact, it does so because 'τ' functions referentially in a syntactically proper context.

Our next problem is to determine the extent, if any, to which theoretical terms *extend* the referential capabilities of language. In particular, we must now consider whether 'τ' might not be equivalent to some descriptive expression lying wholly within L_o. (By saying that one descriptive expression is "equivalent" to another, I mean that the one may be replaced in any sentence by the other without change in the factual content of the sentence.) For a positivist would in no way be dismayed by the fact that '$T(\tau)$' is a cognitively meaningful statement in which 'τ' functions referentially so long as he were allowed to argue that there existed an expression 'd' in L_o with the same designative force as 'τ.' He would then contend that '$T(\tau)$' is equivalent to the observation sentence '$T(d)$,' and similarly, that any other expression '$E(\tau)$' is equivalent to '$E(d)$.'

What conditions must obtain if 'τ' may be regarded as equivalent to some observational expression 'd'? (Note that we cannot, for the moment, draw upon the Thesis of Semantic Empiricism for assistance, for the positivistic contention is an *alternative* to the Thesis.) An obviously necessary condition is that '$T(\tau)$' and '$T(d)$' have the same factual content. But what, according to the positivist, *is* the factual content of a theory? If we are to proceed further, he must tell us *his* criterion for an observational fact *not* to be a refuter of '$T(\tau)$.' Only two alternatives seem available to him:

1. He might assert that the refuters of '$T(\tau)$' are exactly the refuters of '$T(d)$,' where 'd' is that observational expression to which he claims 'τ' is equivalent. But to select 'd' for this purpose, rather than some other of the many expressions of the same formal type as 'τ' available in L_o, is simply to take the alleged equivalence of 'τ' and 'd' as a *premise* from which to analyze the force of '$T(\tau)$,' which would be justified only if 'τ' had been explicitly introduced to have the same meaning as 'd.' But in this case, 'τ' is merely an ordinary defined term in the observation language, whereas we are here concerned with a class of terms which have *not* been introduced in this way. Hence this alternative is inacceptable.

2. He may hold that '$T(\tau)$' is refuted by an observational fact f only if f refutes an O-consequence of T, and hence by the corollary to Theorem 2, only if f refutes '$(\exists\phi)T(\phi)$.' But then, if 'τ' is equivalent to 'd,' '$(\exists\phi)T(\phi)$' must analytically entail '$T(d)$' (and also, of course, conversely), since any

327

William W. Rozeboom

refuter of 'T(d)'—which thus also refutes 'T(τ)'—must then also refute '($\exists\phi$)T(ϕ).' And indeed, whether the positivist wishes to adopt this position or not, it would seem that in the natural usage of theories, the fact that $T(d)$ is not the case would not be taken as disproof of 'T(τ)' unless the theory has an O-consequence which is refuted by $\sim T(d)$; otherwise, by parity of reasoning, any observational fact $\sim T(d_1)$ should disprove the theory. Hence, unless ($\exists\phi$)T(ϕ) entails that $T(d)$, the force of 'T(d)' is stronger than that of 'T(τ),' and 'τ' and 'd' cannot be equivalent. Consequently, a prerequisite of the positivist's thesis is that whenever a theory 'T(τ)' is itself a cognitively meaningful sentence, *there exists a descriptive expression 'd' in L_o such that '($\exists\phi$)T(ϕ)' and 'T(d)' are analytically equivalent.* Let us refer to the italicized condition as the *Positivistic Criterion.*

Now, there can be no denial that for some theories, the Positivistic Criterion is satisfied. For example, if '(x)[τ(x) $\supset$ P(x)] · τ(a)' is a theory in which 'P' and 'a' are observational constants, then '($\exists\phi$)[(x)[ϕ(x) $\supset$ P(x)] · ϕ(a)]' is formally equivalent to '(x)[P(x) $\supset$ P(x)] · P(a),' and a positivist could contend that 'τ' is equivalent to 'P.' However, it is obviously *not* the case that for *any* predicate 'F(ϕ),' a descriptive expression 'd' can be found such that 'F(d)' has the same force as '($\exists\phi$)F(ϕ)'—otherwise, logical quantifiers could be entirely eliminated from the language without attenuating its strength. Hence, if the positivistic thesis is to provide a *general* account of the meaning of theories, it must be the case either that (a) the only expressions, 'T(τ),' which are ever legitimately regarded as theories are those which satisfy the Positivistic Criterion, or that (b) only when the Positivistic Criterion is satisfied can 'T(τ)' itself be an assertion. But (a) is obviously false—not only do we fail to invoke the Positivistic Criterion when passing judgment upon theories, it is highly unlikely that it is met by any theory of current scientific importance. As for (b), this contention would also apply to the reduction of common-sense "observational" terms to purely phenomenal phrases, and would imply that ordinary-language observation sentences are dichotomized into those which are genuine— i.e., phenomenal—statements, and those which are merely mechanisms for passing from one phenomenal statement to another. But this is wholly implausible. Sentences which are *recognized* as being purely phenomenal play at best a minor role in actual linguistic usage, while it is just not true that of observation sentences which are not recognizably phenomenal, we differentiate in use between "genuine" statements which we think could be given a phenomenal reduction, and "formal devices" which have a second-class linguistic status. It seems to me highly gratuitous to postulate a semantic distinction which corresponds neither to a difference in use nor to any feature in our normal conceptualization of language. I can only conclude that the Positivistic Criterion for the cognitive meaningfulness of theoretical expressions is untenable, and that in general we must be prepared to find that a theoretical term, though meaningful, need not be equivalent to any phrase in the observation language.

But how, then, are we to analyze the meanings of theoretical terms? The solution lies, I believe, in regarding such meanings *not* as something brought *to* the theory by the theoretical constants, but as something acquired by the theoretical terms *in virtue of* their participation in the theory. This, of course, is simply the Thesis of Semantic Empiricism which was invoked in proof of Theorem 3, except that we are now adding that theoretical terms *do* have cognitive meaning acquired in this way. I do not see how we could possibly hold otherwise if we wish both to maintain the empiricist tradition and yet to grant extraobservational reference to theoretical terms. Hence,

Postulate 6. The semantic properties imparted to a normal syntactic theory T by its acceptance are such that T is itself able to signify a fact.

From P 1, P 2 and P 6, it follows immediately that

Theorem 6. If '$T(\tau_1, \ldots, \tau_n)$' *is an accepted normal syntactic theory in which* '$T(\phi_1, \ldots, \phi_n)$' *is an observational predicate and* 'τ_1,' $\ldots$, 'τ_n' *are theoretical terms, then: (a)* '$T(\tau_1, \ldots, \tau_n)$' *signifies a fact f if and only if there exist entities $t_1, \ldots, t_n$ such that f is the fact that $T(t_1, \ldots, t_n)$. (b) If t is an entity such that* $(\exists \phi_1, \ldots, \phi_{i-1}, \phi_{i+1}, \ldots, \phi_n) T(\phi_1, \ldots, \phi_{i-1}, t, \phi_{i+1}, \ldots, \phi_n)$, *then* '$\tau_i$' *designates t.*

That is, reverting to the simplest case, accepted theory '$T(\tau)$' signifies every fact of form $T(\phi)$, and 'τ' designates every entity which satisfies '$T(\phi)$.' It should be noticed, however, that Theorem 6 supplies sufficient but not necessary conditions for 'τ_i' to designate t. Neither do Ps 1–6 suffice to determine the factual significata of all theoretical consequences of T. We shall attempt to do something about this deficiency shortly.

The plausibility of the present interpretation of theoretical reference will be strengthened, perhaps, by the behavioral theory of designation to be sketched at the end of this section. For the present, let us consider the obvious objection which arises. I say 'the objection' because it seems to me that basically there is only one—the fact that according to the present

329

formulation, a theoretical expression may have more than one referent. For if there were at most one entity which satisfies 'T(ϕ),' we could regard 'T(τ)' as assigning a referent to 'τ' by means of description—i.e., we could assume that $\tau =_{def} (\iota\phi)T(\phi)$.[23] And while the analysis of descriptions is far from agreed upon, it is not implausible that under suitable conditions, descriptions and expressions which contain them do, in some sense, designate. Hence the present view should appear at least reasonable, so long as it is possible to develop a workable semantical theory of multiple designation.

Since the notion that a theoretical term may have more than one referent is the key idea to emerge from the present analysis, it is very important to have a clear understanding of what is being contended. The relation that obtains between a predicate and an entity which satisfies the predicate is occasionally known as 'denotation.' Thus we might say that 'human' denotes Tom, James, Elmer, etc. Under this usage, a predicate may be said to have "multiple denotata," and it is crucial that this be sharply distinguished from multiple designation. A primitive predicate designates, or refers to, an abstract entity which is exemplified (or belonged to, if the referent of a predicate is a class) by its denotata, and hence will in general have many denotata even though it has but one designatum. Since theoretical terms are syntactically primitive, they may be said to name the entities to which they refer. Then to say that a theoretical term may have multiple designata is to imply that a term may simultaneously name more than one entity, thus departing radically from classical semantics.

It is, moreover, most important to appreciate that this unorthodox suggestion which has emerged, namely, that theoretical expressions may designate without designating uniquely, is due neither to a personal perversity nor to some special, restrictive, arbitrary assumption during the earlier stages of the argument. Quite the contrary, it is an apparently inescapable joint consequence of two popular and highly plausible epistemological beliefs, namely, (a) that a theoretical sentence may genuinely signify a fact which cannot be signified by a sentence in the observation language; and (b) that the semantic properties of theoretical expressions are given to them by their use with the observation language. For even if we grant that observation terms have unique referents, there just does not

[23] Owing to limitations of the Linotype font the regular instead of the inverted iota is used.

seem to be any way for the observation language to provide a criterion which may admit an unobserved entity as a referent of a theoretical term and yet also guarantee uniqueness. It would seem, therefore, that a theory of multiple designation is an inescapable correlate to any coherent form of empirical realism,[24] where by the latter we mean epistemological theories which affirm both that knowledge about unobserved entities is possible and that this knowledge is given only through what is observed. If this be so, however, it becomes binding upon philosophically responsible empirical realists to carry through a comprehensive analysis and at least partial reformulation of basic semantical principles, for it must frankly be admitted that a theory of multiple designation is not, prima facie, wholly compatible with classical semantics.

It is a cornerstone tenet of semantics that a statement has at most one truth value—i.e., that it is not both true and false. It is further customary to hold that if 'S' designates the property P, and 's' designates entity t, then the sentence 'S(s)' is true if it is the case that $P(t)$, and false if it is the case that $\sim P(t)$. But this would constitute a fatal objection to any semantical theory which allows a term to have more than one designatum. For suppose that 's' designates both t_1 and t_2. If t_1 and t_2 are different entities, there must be some property P such that $P(t_1)$ and $\sim P(t_2)$. But if 'S' is a predicate which designates P, it would then appear that 'S(s)' must be both true and false. Applied to the present contention that theoretical terms designate all entities which satisfy the observation predicate characterizing the theory, this objection charges that it admits truth-inconsistent statements in violation of the Principle of Contradiction. To be sure, so far as theory 'T(τ)' itself and any theoretical sentences derivable from it are concerned, no ambiguities in truth value arise, for it is a condition on the designata of 'τ' that they satisfy 'T(ϕ)'; hence if 'T(τ)' formally entails 'E(τ),' the case cannot arise where 'τ' designates t and it is not the case that $E(t)$ (since by an easily proved corollary to Lemma 1, 'E(τ)' is deducible from 'T(τ)' only if every satisfier of 'T(ϕ)' also satisfies 'E(ϕ)'). However, if 'F(τ)' is a theoretical sentence not entailed by 'T(τ),' then if 'τ' has more than one designatum, say t_1 and t_2, it is entirely possible that $F(t_1)$ while not $F(t_2)$, which would seem to imply that 'F(τ)' may be both true and false.

Now, it should first of all be noted that the semantical assumption just

[24] For an informal discussion of this point through common-sense examples, see [14].

331

William W. Rozeboom

employed, namely, that if 'S' designates P and 's' designates t, then 'S(s)' is true if it is the case that $P(t)$ and false if it is the case that $\sim P(t)$, is significantly stronger than SP I–IV, in which were formalized the semantical principles on which the present analysis is based. For SP I–IV leave open the possibility that even though 'S' designates P and 's' designates t, 'S(s)' may not assert that $P(t)$—i.e., signify $P(t)$ truly or $\sim P(t)$ falsely, depending on which is the case—and hence that a sentence may fail to signify a fact even though all its descriptive terms have designata. We now see that if both multiple designation and the Principle of Contradiction (SP IV) are to be maintained, this possibility must remain. However, this does not leave matters in a very satisfactory state, for if 'S(s)' does not necessarily truly signify the fact that $P(t)$, or falsely signify the fact that $\sim P(t)$, even though 'S' designates P and 's' designates t, what then are the conditions sufficient for a fact to be a verifier or refuter of a sentence in virtue of the semantic properties of the latter?

To understand the origins of the predicament in which our analysis now finds itself, and to sympathize with its departure from classical semantics, it is necessary to remain sensitive to a truistic but not always properly appreciated prerequisite for semantical relations to obtain. This is, simply, that it is not words and sentences qua sign-designs which stand in semantical relations to entities, but words and sentences in use—i.e., symbols which have come to play a suitable role in language behavior. It is customary and quite proper for "pure" semantics to axiomatize certain properties of semantical relations abstracted from the total linguistic situation, but it must not be forgotten that when semantical relations obtain between symbols and extralinguistic entities, it is because these symbols are being used in a certain way. While it is perfectly acceptable for a semanticist to lay down sentences of the form 's designates x' as postulates for analysis without committing himself to the nature of this relation, the results of his analysis are not applicable to either de facto or idealized language practices unless the bare signs of the language are embedded in a pragmatic context by virtue of which a coordination is established between signs and their designata.

Now this point may seem trivial at first, but it ceases to do so when •one reflects that the "use," or "linguistic role," of a sign-design is more clearly described as some aspect of the psychological state of a language user o at time t with respect to that sign-design, and that it is by no means necessarily the case that the psychological state of person o at time t with

332

respect to a sentencelike sign-design S is such as to endow S (in its linguistic role for o at t) with all the semantic properties presupposed by classical semantics, even though S has an appreciable incidence in o's linguistic behavior. On the other hand, merely because the psychological state of o at t with respect to certain expressions in his language does not fully qualify these expressions for analysis by classical semantics, it would be most rash to conclude that these expressions are not in any way cognitively meaningful or do not function referentially for o at t. For example, classical semantics has no place for vague concepts; yet it would be absurd to argue that because the "borderline fuzziness" of most if not all terms in actual use reveals them to be more or less vague, ordinary language is cognitively meaningless. Moreover, it would be jeopardous to construe the discrepancies between actual languages and classical semantics as due wholly to noncognitive contaminations of a theoretically pure semantical state described by the classical postulates. It seems much more reasonable to suspect, or at the very least to entertain as a possibility, that the classical account is a limiting form of what is generally a more complex pattern of semantical relations, while the latter is just as much a pure cognitive system as its classical limit and may likewise (though more comprehensively) represent a theoretical ideal to which actual languages are but an approximation.

If one does admit the possibility that classical semantics may be only a special instance of more general semantical principles, however, then clearly we should expect that in order to analyze the cognitive function of theoretical expressions it will be necessary to develop a semantical theory adequate to the broader case. For as will shortly be examined in greater detail, concepts which qualify as "theoretical" are transient stages of a linguistic growth process and are hence incomplete in a way that the concepts presupposed by classical semantics are not. Consequently, if the present analysis of theoretical expressions is basically sound, P 1–6 provide a framework within which we may begin to explore the nonclassical dimensions of cognitive processes.

It has already been pointed out that P 1–6 do not fully delimit the designative properties of theoretical expressions. While any reasonably adequate development of a generalized theory of semantics, and discussion of its relation to the classical limit, is far beyond the present scope, let me at least offer a provisional set of hypotheses which seem to make a

William W. Rozeboom

certain amount of intuitive sense, and which, moreover, reconcile the possibility of multiple designation with the Principle of Contradiction.[25]

Definition 9. ‘$E(\tau_1, \ldots, \tau_m)$’ *is an autonomous subtheory of theory* T $=_{\text{def}}$ ‘$E(\tau_1, \ldots, \tau_m)$’ *is a sentencelike formula deducible from* T *in which* ‘$E(\phi_1, \ldots, \phi_m)$’ *is an observational predicate and* ‘τ_1,’ $\ldots$, ‘τ_m’ *are m of the n* $(0 < m \leqq n)$ *theoretical terms contained in* T; *and the cognitive meaning imparted to* ‘$E(\tau_1, \ldots, \tau_m)$’ *by (normal syntactic) acceptance of* T *is the same as the meaning acquired by* ‘$E(\tau_1, \ldots, \tau_m)$’ *when accepted as a (normal syntactic) theory by itself.*

The purpose of this definition is to facilitate handling of the possibility suggested earlier that a totality, T, of accepted theoretical postulates may contain subsets which function independently of the remainder. Whether there are, in fact, autonomous subtheories of T which are not formally equivalent to T, and what the conditions must be for a subtheory to be autonomous, we shall not here attempt to explore. There is reason to believe that a theoretical consequence E of T is an autonomous subtheory of T if every consequence of T containing one or more of the theoretical terms in E is equivalent to a sentence of form $C \cdot S_E$, in which C contains no theoretical terms in E and S_E is deducible from E alone. However, this may not be a necessary condition for autonomy.

Definition 10. E *is a unified subtheory of theory* $T =_{\text{def}}$ E *is a theoretical consequence of* T *and there is no autonomous subtheory,* E_a, *of* T *such that* E_a *is deducible from* E *and* E *is not deducible from* E_a.

That is, a unified subtheory of T cannot be resolved into components whose meanings are acquired independently of the remainder. The units of meaning acquisition when T is accepted are then those theoretical consequences of T which are autonomous and unified.

Hypothesis A. If ‘$E(\tau_1, \ldots, \tau_m)$’ *is an autonomous and unified subtheory of an accepted normal syntactic theory* T *in which* ‘τ_1,’ $\ldots$, ‘τ_m’ *are theoretical terms and* ‘$E(\phi_1, \ldots, \phi_m)$’ *is an observational predicate, then* ‘τ_i’ *designates an entity t if and only if it is the case that* $(\exists \phi_1, \ldots, \phi_{i-1}, \phi_{i+1}, \ldots, \phi_m) E(\phi_1, \ldots, \phi_{i-1}, t, \phi_{i+1}, \ldots, \phi_m)$.

Since a set of entities $t_1, \ldots, t_n$ will satisfy ‘$T(\phi_1, \ldots, \phi_n)$’ only if its subset $t_1, \ldots, t_m$ satisfies ‘$E(\phi_1, \ldots, \phi_m)$,’ an entity t will qualify as a designatum of ‘τ_i’ under Theorem 6 only if it also qualifies under

---

[25] While these hypotheses concern only the acquisition of designata by theoretical terms through their use with observation language L_o, a similar set of principles would be expected to govern the endowment of expressions in L_o with meanings derived from immediate experience in the event that L_o is not a phenomenal language (cf. fn. 32).

Hypothesis A. Hence this hypothesis extends Theorem 6 in such a way as to provide necessary as well as sufficient conditions for the designata of theoretical terms.

Hypothesis B. If 'E(τ_1, . . ., τ_m)' is a theoretical sentence in which E(ϕ_1, . . ., ϕ_m)' is an observational predicate and 'τ_1,' . . ., 'τ_m' are theoretical terms introduced by an accepted normal syntactic theory T, then 'E(τ_1, . . ., τ_m)' signifies a fact f if and only if there exist entities t_1, . . ., t_m such that 'τ_1,' . . ., 'τ_m' designate t_1, . . ., t_m, respectively, and f is the fact that $E(t_1, . . ., t_m)$.

From this and SP I it follows that a theoretical sentence can signify a fact only truly. Hypothesis B is an extension of P 1 to all theoretical sentences, deducible from T or not. Actually, it needs to be generalized to describe what a set of theoretical sentences simultaneously signify (see footnote 27), but this is a further development which may be forgone here.

Hypothesis C. If E is a theoretical sentence whose theoretical terms have been introduced by an accepted normal syntactic theory T, E is true or false according to whether or not there exists a fact signified by E.

That is, cognitive meaningfulness does not presuppose factual reference, and a sentence may be false precisely because there is nothing in external reality which conforms to the criteria built into the sentence's meaning.

It will be observed that Hypotheses A–C agree with classical semantics in the limiting case where (unified) theory 'T(τ)' is adequate to confer exactly one designatum, say t, upon 'τ' (i.e., when the situation $(\phi)[T(\phi) \equiv \phi = t]$ obtains). For then a sentence 'E(τ)' is true if it is the case that $E(t)$ and false if it is the case that $\sim E(t)$. Where they differ from classical semantics is that it is not universally the case that if 'τ' designates an entity t and $\sim E(t)$ obtains, then 'E(τ)' is false. Rather, for 'E(τ)' to be false, every designatum of 'τ' must fail to satisfy 'E(τ).' That is, the factual content of 'E(τ)' according to Hypotheses A–C is the same as that of '$(\exists \phi)[T (\phi) \cdot E(\phi)]$,' although the facts, if any, which are signified by these two sentences are by no means the same. Another prima-facie difference between classical semantics and the present generalization is rejection by the latter of the relation described earlier as "false signification." In order to deal with the semantical status of false observation sentences, we have so far implicitly assumed that when a sentence 'S(s)' of L_o is false, it is so because 'S(s)' falsely signifies a fact $\sim P(t)$ in virtue of 'S' designating P and 's' designating t. In such an interpretation, false observation sentences and

William W. Rozeboom

true observation sentences are on a par with respect to designating—both are conceived to be *about* some state of extralinguistic reality. But it may well be questioned whether such a concept of "false signification," in the sense defined by SP I–III, really can be extracted from classical semantics, which has always tended to confound the meanings of sentences with their factual significance.[26] Even without drawing upon the theoretical dimension of language, it may be argued that a sentence can be false even though—or rather, *because*—it has no designatum (see [13]). Wholly aside from the problem of theories, it may be that "falsehood" is best characterized as a derivative semantical condition wherein a sentence is false if and only if it is cognitively meaningful but fails to signify a fact. If so, then classical semantics and the present hypotheses also agree—completely, not merely in the limit—with respect to the concept of falsehood.

Because it brings out an important property of theoretical concepts, I would now like briefly to present an informal argument in favor of the contention—i.e., Hypothesis B—that it would never be correct to say that a theoretical sentence '$E(\tau)$,' not deducible from accepted theory T, falsely signifies a fact $\sim E(t)$, even though 'τ,' as introduced by theory '$T(\tau)$,' designates t. Suppose that there exists a t_1 such that $T(t_1)$ and $\sim E(t_1)$, and also a t_2 such that $T(t_2)$ and $E(t_2)$. Then by the present interpretation, 'τ,' introduced by '$T(\tau)$,' designates both t_1 and t_2 (cf. Theorem 6), and one might argue on classical grounds that if this is granted, then we should have to say that '$E(\tau)$' falsely signifies the fact that $\sim E(t_1)$ as well as truly signifying the fact that $E(t_2)$. Now, the concept of "incorrectness," of which "falsehood" is a special case, is pragmatical—an entity is "incorrect" in a certain behavioral role only if it leads, actually or potentially, to error. But a sentence can lead one into error only when it is *believed* or *accepted*, for only then does one act upon the behavioral prescriptions of the sentence. Moreover, to believe or accept '$E(\tau)$' *in addition* to accepting the theory '$T(\tau)$' is to accept the *enriched* theory '$T(\tau) \cdot E(\tau)$.' Since by hypothesis it is the case that $T(t_2) \cdot E(t_2)$, it follows that '$E(\tau)$' is then a consequence of the unambiguously true theory '$T(\tau) \cdot E(\tau)$,' and so

[26] When classical semantics analyzes the linguistic properties of the observation language through sentences of the form " '$S(s)$,' in L_o, asserts that $P(t)$," what is *primarily* being indicated is a relationship among the *meanings* of sentences in L_o and L_M; and it is necessary to be very careful in moving from this kind of an account to one analyzing the relations of expressions in L_o to their *designata*, since not all meaningful expressions have designata, even when their syntactic role is that of a descriptive term.

by Hypothesis A and SP I does not falsely signify $\sim E(t_1)$; hence there are then no grounds on which to argue that '$E(\tau)$' is incorrect. That is, when 'τ' is introduced by theory '$T(\tau)$,' so long as it is the case that $(\exists\phi)[T(\phi) \cdot E(\phi)]$, the correctness of '$E(\tau)$' is uncontaminated by the existence of a t such that 'τ' designates t and it is the case that $\sim E(t)$, for '$E(\tau)$' can lead one into error only in the course of adopting a new, improved theory '$T(\tau) \cdot E(\tau)$,' and with respect to the latter theory under the conditions stipulated, '$E(\tau)$' is in no way incorrect. But by SP II, if '$E(\tau)$' is not incorrect, it does not falsely signify a fact. Hence '$E(\tau)$' cannot signify a fact falsely so long as '$T(\phi)$' and '$E(\phi)$' are jointly satisfied. To drop the latter condition, we need but reflect that as brought out in the discussion of P 1 (see footnote 17), whether or not '$E(\tau)$' falsely signifies $\sim E(t_1)$ depends only on the relation of $\sim E(t_1)$ to the use of '$E(\tau)$,' and not, in addition, on whether or not some other entity t_2 satisfies '$E(\phi)$.' To be sure, '$E(\tau)$' is false when there is no joint satisfier of '$T(\phi)$' and '$E(\phi)$,' but only because '$E(\tau)$' then fails to signify a fact truly, not because it signifies some fact falsely.

What the preceding argument reveals—and this is its real importance— is that when a theoretical sentence '$E(\tau)$' is not entailed by the theory which gives 'τ' its meaning, then the pragmatic effectiveness and hence the truth or falsity of E is essentially given by whether or not addition of E to the postulates of the theory would yield a correctly enriched theory. Hence a theoretical sentence E not entailed by an accepted theory cannot be said to have meaning in quite the same way that the theory and its consequences have meaning; rather, the meaning of such an E is best characterized as a disposition to have the meaning it would have were the theory enriched in a certain way. This makes more palatable the rather unpleasant consequence of Hypotheses A–C that although two theoretical sentences E_1 and E_2 may each be true separately, their conjunction may be false.[27] For example, if two distinct entities t_1 and t_2 each satisfy the observational predicate '$T(\phi)$' of accepted theory '$T(\tau)$,' both '$\tau = t_1$' and '$\tau = t_2$' are separately true under Hypotheses A–C, yet '$(\tau = t_1) \cdot$

[27] It might hence be thought that Hypotheses A–C constitute a departure from classical logic, as well as from classical semantics, in that we cannot always correctly infer $S_1 \cdot S_2$ from S_1 and S_2. This difficulty is spurious, however, since Hypotheses B and C should properly be generalized to give the simultaneous significata and truth values of a set of sentences, instead of applying merely to the conjunction of the set (since one can accept a theory in the form of a set of postulates as well as in the form of their conjunction). We can then conjoin and separate sentences at will without change in their significata during a deduction from a given set of assumption formulas. On the

$(\tau = t_2)$,' which entails that $t_1 = t_2$, is false. But while this violates classical semantics, it makes a certain amount of intuitive sense upon reflection that while either 'T$(\tau) \cdot (\tau = t_1)$' or 'T$(\tau) \cdot (\tau = t_2)$' is a perfectly good (i.e., correct) enrichment of 'T(τ)' under the conditions stipulated, the stronger enrichment 'T$(\tau) \cdot (\tau = t_1) \cdot (\tau = t_2)$' is false. However, a more penetrating analysis of this situation must await another occasion.

While the preceding considerations are rather fragmentary, they nonetheless expose a particularly vital aspect of the meanings of theoretical terms. It was commented earlier that there is an important sense in which such meanings are incomplete. We now see that in order to give pragmatic effectiveness to a theoretical sentence not entailed by the theory then in force, it is necessary to augment the theory until it does entail that sentence. Given any enrichable accepted theory T, there will be sentences, containing terms whose meanings are acquired through their participation in T, whose truth or falsity cannot be judged without thereby enriching the meanings of these terms. Any enrichable theory has an inherently concomitant envelope of unresolved theoretical questions which demand that the theory be supplanted by a better, more complete theory.

In regard to such enrichments, however, there is an apparent paradox which needs resolution. Suppose that there are entities t_1 and t_2 such that $T(t_1)$ and $T(t_2)$, but that $E(t_1)$ and not $E(t_2)$. Then theory 'T(τ),' if accepted, is true, and 'τ' designates both t_1 and t_2. When it comes to enriching the theory in regard to whether or not τ satisfies 'E(ϕ),' however, it would appear that we can have it *both* ways; the theory 'T$(\tau) \cdot E(\tau)$' and the theory 'T$(\tau) \cdot \sim E(\tau)$' are both true when accepted. But this might seem paradoxical; for if we can enrich the theory in two directions, why can't we enrich it in both at once, giving us the logical inconsistency 'T$(\tau) \cdot E(\tau) \cdot \sim E(\tau)$'? Or even if we do not try both directions at once, is not 'T$(\tau) \cdot E(\tau)$' incompatible with 'T$(\tau) \cdot \sim E(\tau)$'? The answer, of course, is that 'τ' has a *different meaning* in 'T$(\tau) \cdot E(\tau)$' than it has in 'T$(\tau) \cdot \sim E(\tau)$.' By the Thesis of Semantic Empiricism, the meaning of a theory is unchanged by substitution of new theoretical terms for old; and in the present example, the enrichment 'T$(\tau) \cdot E(\tau)$' is quite compatible with the enrichment 'T$(\mu) \cdot \sim E(\mu)$.' Similarly, it *is* possible to move in both

semantical side, the significata of a given theoretical sentence will still depend, in part, on the nature of the other assumption formulas, but this is as it should be, for whether or not a given theoretical sentence is an acceptable enrichment of a theory depends in part on what other sentences are also to be added.

directions at once, only this must be done, not by assertion of a logical absurdity, but by multiplication of theoretical terms; namely, '$T(\tau) \cdot T(\mu) \cdot E(\tau) \cdot \sim E(\mu)$.' The moral, here, is that as a theory becomes explored, tested, accepted, and elaborated, not only do we find the meanings of theoretical terms enriched, we may also expect to find—and, in fact, do find—at any stage of development that a theoretical concept has suddenly fissioned into a set of concepts which share a common core of meaning but which have now become free to evolve in their own individual ways.

If there is any important distinction between theoretical and observational terms (ignoring for the present that observational terms are themselves for the most part theoretical terms whose credentials we have come to accept at face value), it must lie in the dynamic aspects of the former. It is an intrinsic part of their usage that theoretical terms are to-be-enriched terms. A theoretical concept is not merely a "promissory note" in the sense of *permitting* future elaboration, it carries theoretical problems the resolution of which *demands* meaning enrichment, if only a provisional one. At any stage in the development of a theory, the theoretical terms then in use are enveloped in a penumbra of possible extensions and multiplications. Similarly, within the harder core of meaning imparted by the theoretical postulates actually believed to be true, lie the "lines of retreat,"[28] the order in which meaning components would be relinquished as the theoretical postulates were abandoned one by one under the press of disconfirming evidence. *Theoretical terms are concepts in the act of formation.*

Suggestions for a behavioral theory of semantics. Let us conclude this section with a few words in outline of a behavioral theory of semantical relations. We saw earlier—and indeed, I do not see how it would be possible to dispute this truism—that whatever semantic properties are possessed by a set of sign-designs for a person o at time t are due to the way in which these signs are used by o at t. Since 'use' is an ambiguous and rather misleading term carrying teleological connotations of "purpose" or "intended causal effect," this is better put by saying that the semantic properties of a sign-design s for o at t depend upon the kind of behavioral effect that s has on o at t, or, more generally, the kind of effect that presentation of s *would* have on o at t. From this it is but a short step to propose that (1) the cognitive meaning of a sign-design s for a person o at time t is some

[28] I am indebted to W. Sellars for this adroit phrase.

William W. Rozeboom

aspect[29] of the behavioral effect that s has, or would have, on o at t; and that (2) the designata of s (for o at t), if any, are determined by the cognitive meaning of s (for o at t). The adjective 'behavioral' has minimal restrictive force here—in particular, it is not meant necessarily to rule out "mentalistic" interpretations of meanings, for there is no reason why behavioral and mentalistic accounts of linguistic processes may not be describing the same events (cf. [6]). Rather, it is to emphasize that meanings are to be found in the dynamics of person-symbol interactions. When s is a sentence, its "cognitive meaning" may be described as a "prescribed behavioral adjustment," since when a statement is under declarative consideration, certain behavioral tendencies or "sets" controlled by this sentence, perhaps highly removed from gross motor activities, are brought into play under a provisional status, where the degree of the latter (i.e., the degree of belief) is dependent upon factors additional to the cognitive meaning of the sentence. It should be noted that (1) does not suggest that a term must have a referent or that a sentence must signify a fact in order for the term or sentence to be meaningful.

We may elaborate this theory by two further hypotheses: (3) The cognitive meaning of a statement is compounded out of the cognitive meanings of the constituent terms in a definite way determined by the syntactical structure of the statement. This is not to imply that the meanings of constituent terms are always in some sense causally prior to the meanings of the sentences in which they occur, for we wish to interpret the meanings of theoretical terms as *derived* from the theoretical postulates. Yet, if the sense of a sentence is determined by its component terms and formal structure, as any acceptable theory of semantics must recognize,

[29] It is not true that the total "use," or behavioral force, of an expression on the occasion of a particular occurrence is relevant to its semantic properties. Thus a given sentence may be employed as a simple declaration of what is believed to be the case ("The barracks will be cleaned tonight."), as an interrogation ("The barracks will be cleaned tonight?"), or as a command ("The barracks *will* be cleaned tonight!"), to list but three broad categories of a vast number of possible uses. Yet, in each case the *cognitive* meaning of the sentence—i.e., those linguistic properties in virtue of which it is able to be *about* the external world—is the same. While we do not normally think of questions and commands as *designating* anything, if their total function did not preserve a cognitive component, they could not serve their purpose—for example, an effective command must describe the desired state of affairs which comes to exist when the command is properly executed. The total linguistic status of the occurrence of an expression would appear to require description on two dimensions: (1) the *cognitive meaning*, or *designative potential*, of the expression; and (2) the *function*—i.e., assertion, query, command, etc. in the case of sentences, reference (and other uses?) in the case of descriptive terms—for which that cognitive meaning is being employed.

340

meanings must be compoundable according to certain definite principles. (4) If a statement S signifies a fact f, it does so because the cognitive meaning of S is "appropriate" to f. Just what 'appropriate' means in this context is difficult to pin down. By saying that the meaning of S is appropriate to f, we wish to indicate that the behavioral adjustment prescribed by S somehow prepares for, or adapts to, the fact f. In those cases where it would make sense to say that a person is aware that f is the case, we might say that S signifies f (for person o at time t) when the behavioral adjustment prescribed by S is suitably similar to the behavioral adjustment that would be set off by awareness that f. More generally, if we presume there is some behavioral adjustment, which might be called the "behavioral significance" of a fact f (for o at t), which is maximally and specifically appropriate (for o at t) to f, we may then propose that a statement S (truly) signifies fact f (for o at t) when the cognitive meaning of S (for o at t) is sufficiently similar to the behavioral significance of f (for o at t).

Before the substantive details of this relation can adequately be filled in, we shall need a much more developed science of behavior than is now available. For this reason, the present theory can be no more than a crude outline. However, this is enough to make plausible the possibility that a statement may signify more than one fact. For suppose that S is a statement suitably rich in meaning that it signifies exactly one fact. Then is it not possible that by "weakening" the meaning of S—i.e., by withdrawal of a certain portion of its prescribed behavioral adjustment—S would now be "appropriate," in that way which characterizes designation, to a set of facts differing only in respect to a feature to which the weakened meaning of S no longer prescribes a differential adaptation? What I am suggesting, in other words, is that if semantic relations are grounded upon a similarity (or perhaps a more complex relation) between the behavioral prescriptions of symbols and behavioral significances of things symbolized, then designation may be a matter of degree, rather than an all-or-none affair. The more "weakly" S signifies f, the more it is possible for S also to signify other facts which are similar to f in suitable respects.

These suggestions may be sharpened by proposal of a similar analysis for the way in which a descriptive constant 's' designates an entity t. There would appear to be some sense in which an entity may be said to have "behavioral import" for an organism. This cannot be analyzed simply as the reaction produced in the organism by the entity acting as a stimulus, for organisms respond to *facts*, not stimuli as such (although the organism may respond to *the fact that* the stimulus is present—see [12]). Nonetheless, since the behavioral significance of a fact is determined by the entities it comprises—for if it were not so determined, the behavioral significances of facts sharing one or more constituents would not need to be related in the manner they in actuality are—we may regard the behavioral imports of entities as behavior elements out of which the behavioral significances of facts are compounded according to the way (i.e., logical structure) in which the entities constitute the fact. Now, we have already hypothesized that the (cognitive) meanings of statements are compounded out of the (cognitive) meanings of their constituent terms. Hence, if it is the case that when a constant 's' designates an entity *t*, the meaning of 's' is sufficiently similar to the behavioral import of *t*, it becomes clear (at least in overview) how a statement might signify a certain fact in virtue of the statement's formal structure and the meanings of its constituent terms. It is now especially easy to suggest the conditions under which multiple designata for descriptive terms, and derivatively for statements, might come about. For if the relation between the meaning of a symbol and the behavioral import of its designatum is that the former is, or closely resembles, a *part* of the latter, then the meaning of a sufficiently weak symbol might be a behavioral effect common to the behavioral imports of several entities. The process of concept formation would then consist of endowing a symbol with behavioral force—i.e., cognitive meaning—which, in turn, determines the entities, if any, to which this symbol refers. The stronger, or richer, the meaning of the symbol, the fewer the entities designated by it, while if it is possible to make the symbol sufficiently strong in meaning, it will have a unique designatum.

There are obviously many serious problems and ramifications to this theory which the present outline has not begun to explore. However, if the theory appears to have any merit at all, the purpose for which it has been suggested here has been accomplished, namely, to show it to be not unreasonable that a term might simultaneously designate more than one entity. Moreover, this sketch is further helpful in clarifying the manner in which a theory, though equivalent in force to an observation sentence, nonetheless manages to enrich the language. We have argued that although the factual content of a theory is identical to that of its Ramsey sentence, $T(\tau)$' and $(\exists\phi)T(\phi)$' do not signify the same fact; $T(\tau)$' signifies each member of a (possibly empty) set of (in the main) non-observational facts, each of which entails the observational fact, if any, signified by $(\exists\phi)T(\phi)$.'

Now, I see no reason why, if a certain complex behavioral effect can be compounded out of other effects, it might not also be compoundable in more ways than one. It is then not implausible that an organism of sufficient behavioral intricacy could take a complex effect E, compounded from behavior components acquired previously, and restructure it in such a way that some of the constituents of the restructured E are behavior elements which were not previously available. Thus it seems quite conceivable, under the present semantical theory, that a theoretical term 'τ' could be infused with just that degree of meaning which would make the behavioral force of $T(\tau)$' essentially the same as that of $(\exists\phi)T(\phi)$' when the latter already exists in the organism's behavioral repertoire. Presumably, this could be accomplished simply by using the same symbol 'τ' in the various theoretical postulates of T—it should not be necessary actually to construct the full conjunction, $T(\tau)$,' of theoretical postulates. For if each theoretical postulate contributes a meaning component to 'τ,' the combined effect should be the same as if 'τ' acquired its meaning directly from use in the conjunction, $T(\tau)$.' Further, it is important to note that while the force of $T(\tau)$' is the same as that of $(\exists\phi)T(\phi)$,' if $E(\tau)$' is entailed by $T(\tau)$' but not conversely, the meaning of $E(\tau)$' is richer than that of $(\exists\phi)E(\phi)$.' It will be recalled that one difficulty in regarding $T(\tau)$' as a peculiar way of asserting that $(\exists\phi)T(\phi)$ was that no comparable translation exists for $E(\tau)$.' But theoretical statements derived from T pose no interpretative difficulties once we realize that 'τ' functions as a name in that it has a fixed meaning (so long as the theory is not enriched or otherwise altered) which may be carried from one statement to another. The only difference between theoretical terms and observational terms, under this interpretation, is that the meanings of the former are weaker, and their referents thus possibly more numerous, than those of the latter. Hence, theoretical terms constitute a genuine enrichment of language, rather than peculiar formal devices for deriving observational predictions, and may themselves become "observational" if their meanings are given sufficient strength through an accepted, true and sufficiently forceful theory.

William W. Rozeboom

It would be highly surprising if any explication of a problem so philosophically basic as the meanings of theoretical terms did not have important implications for many other related problems as well. In closing, I would like to consider, very briefly, the import of the present analysis for certain unresolved problems of current interest.

Identification and reduction. For those who prefer a realistic interpretation of theoretical terms, it is unnecessary to conceive of theoretical entities as partaking, somehow, of a different kind of "reality" from observational entities of the same type. There is no reason why, in principle, a theoretical entity cannot become "known" in the same way that observational entities are known. In fact, the referent (or a referent) of a theoretical term may be an entity already independently accessible to the observation language. (The "phantom burglar" postulated by the police to account for a sudden upsurge in larceny may turn out to be the police chief himself.) Again, we may seek to "reduce" the theoretical terms of one theory to those of another, success in which is sometimes regarded as confirmation of the "reality" of the reduced entities. (Thus in genetics, the gene has appeared increasingly real as cytological theory has proliferated.) In either instance, we speak of finding the "identity" of the entity for which the theoretical term was at first only a "promissory note." How is such an identification to be analyzed?

In all cases where a theoretical entity is "identified," the crucial step consists in an assertion '$\tau = d$,' where 'τ' is the theoretical term whose identity is being proposed and 'd' is a designative expression which is either (1) wholly in the observation language; (2) a demonstrative (e.g., "So *that's* what τ is!"), the analysis of which case is essentially that of (1); or (3) contains other theoretical terms,'μ_1,' . . ., 'μ_n,' in which case 'τ' has been "reduced" to 'μ_1,' . . ., 'μ_n.' We need not be concerned here with the precise analysis of the identity relation (except, preferably, to assume that '$a = b$' does, in fact, make an assertion about a and b, rather than being an ellipsis for a semantical statement such as '(x)('a' designates x if and only if 'b' designates x)'). What we are now investigating is the meaning of an assertion of identity when one of the expressions involved is a theoretical term.

Suppose that 'τ' is a theoretical term whose meaning is defined by the theory '$T(\tau)$,' and that 'd' is a descriptive term in the observation lan-

344

guage. Under what circumstances would we be willing to say that d is the identity of τ—i.e., to claim that $\tau = d$? It is, of course, obvious that if it is not the case that $T(d)$, it would be most incorrect to assert '$\tau = d$,' for as was shown earlier, whatever 'τ' designates, it must be something that satisfies 'T(ϕ).' But suppose that it is the case that $T(d)$. Would we not then be justified in claiming that d is the identity of τ? It is hard to deny this claim, for what other criterion could we invoke in deciding whether or not $\tau = d$; yet the matter is not so simple as all that. First of all, we must recognize, presuming our earlier analysis to be correct, that if $T(d)$ is the case, then 'τ' designates d. But this is in itself insufficient to conclude that $\tau = d$. For suppose the fact that $T(d)$ necessitated the conclusion that $\tau = d$. Then if some other observational entity d^*, different from d, also satisfies 'T(ϕ),' we would have to conclude also that $\tau = d^*$, which by the transitivity of identity would entail, contrary to hypothesis, that $d = d^*$. The reason the fact that 'τ' designates d does not necessitate the conclusion '$\tau = d$' is that the latter adds a further restriction on the designata of 'τ' beyond that imposed by the theory 'T(τ).' Assertion that d is the identity of τ involves not only the judgment that $T(d)$, but also the decision to enrich the theory in this way.

To say that asserting '$\tau = d$' involves a decision is not to imply that the decision is a difficult one to make. For if it can be determined with high certainty that an observational entity d satisfies 'T(ϕ),' then to accept the enrichment '$\tau = d$' is to accept the theory 'T$(\tau) \cdot (\tau = d)$'—i.e., 'T(d)'— which not only is verified by the fact that $T(d)$, but also becomes supplemented by further facts known about d. That is, acceptance of the enrichment '$\tau = d$' not only changes the status of the theory from hypothesis to known fact in this case, it also increases its usefulness. Conversely, to deny the identity of τ with d is to adopt the counterenrichment 'T$(\tau) \cdot (\tau \neq d)$,' while the latter not only is unlikely to have any worthwhile consequences beyond those of 'T(d),' but also stands a reasonably good chance of being false. Hence it is an almost automatic process, and rightly so, to identify a theoretical entity with the first entity observed to satisfy the theory.

The situation is somewhat more complicated in the case of reduction, and my remarks here can be no more than fragmentary. Suppose that an accepted theory T can be written as the conjunction of two autonomous subtheories, T_1 and T_2, which contain no theoretical terms in common— i.e., that T is of the form 'T$_1(\tau) \cdot$ T$_2(\mu)$.' We may then describe T_1 and

T_2 as separate theories, say a macrotheory and a microtheory, adopted simultaneously. Suppose further that there is an expression '$d\mu$,' containing theoretical terms of T_2, such that '$T_1(d\mu)$' is entailed by '$T_2(\mu)$.' That is, suppose the microtheory T_2 implies the existence of, and supplies a descriptive expression for, an entity which exemplifies the macrotheory T_1. It follows that (a) the truth of T_1 is entailed by the truth of T_2, and (b) the entities designated by '$d\mu$' are a subset of the entities designated by 'τ.' Under such circumstances, we should be tempted to identity τ with $d\mu$—i.e., to assert '$\tau = d\mu$,' an enrichment which is equivalent simply to dropping '$T_1(\tau)$' as a separate hypothesis. (Thus as is also true in the case of observational identification, the enrichment sustained by a theory through reduction of its theoretical elements to constructs in another theory consists in assimilating the theory to a set of beliefs external to the theory, and abandoning the theory as a separate hypothesis.) And to be sure, if we are certain that '$T_2(\mu)$' is true, the reasons for identifying τ with $d\mu$ are as strong and as legitimate as identifying τ with some observational entity, d_o when it is known that $T_1(d_o)$.

But there is an important difference between observational identification and theoretical reduction. In the former case, we considered the legitimacy of asserting '$\tau = d_o$,' given knowledge that $T_1(d_o)$. In the latter case, on the other hand, we are judging the assertion of '$\tau = d\mu$' given knowledge that '$T_2(\mu)$' entails '$T_1(d\mu)$.' The difference is that while the theory '$T_1(\tau)$' is confirmed by the fact that $T_1(d_o)$, the fact that '$T_2(\mu)$' entails '$T_1(d\mu)$' does not confirm '$T_1(\tau)$'—it only shows that '$T_1(\tau)$' must be true if '$T_2(\mu)$' is true. There is thus the danger, if we accept '$\tau = d\mu$,' that '$T_2(\mu)$' is false, a contingency which, if realized, in general leaves '$d\mu$' without a designatum and hence falsifies '$T_1(d\mu)$' even though '$T_1(\tau)$' by itself may remain quite true (since $\sim(\exists\phi)T_2(\phi)$ does not entail that $\sim(\exists\phi)T_1(\phi)$ unless T_1 and T_2 are analytically equivalent). That is, to identify τ with $d\mu$ is to risk replacement of a theoretical expression which has a referent by another which does not, and thus to gamble the success of one theory upon that of another. To draw out the implications of this for the practical aspects of theory building, and to buttress the argument by citing specific examples, would require a more extensive discussion than is practical here. The conclusion which would ultimately be drawn is that although the relation between a macrotheory and a microtheory (or, for that matter, between two sets of theoretical terms on the same level) may be such as to suggest strongly that certain microstructures are the "identities" of the theoretical macroentities, it is best to remain noncommittal about the identities as long as the micro-theory sustains a reasonable doubt, or, at most, to carry the identity asser-tion as a kind of auxiliary hypothesis which may be discarded, if necessary, without otherwise necessitating any change in the macrotheory.

To summarize: To "identify" a theoretical entity is to make both a fac-tual judgment and a *decision* about the subsequent use of the theoretical term. To enrich the theory 'T(τ)' by adoption of the identity assertion '$\tau = $ d' is a legitimate and desirable move when (a) there is an entity which is designated by 'd,' and (b) an entity designated by 'd' satisfies 'T(ϕ).' To the extent there exists doubt that 'd' meets either condition, assumption that $\tau = d$ is a dubious maneuver which should never be made unless the line of retreat remains clearly visible.

Implicit definition. One of the stickier problems of analytic philosophy has been what to say about (stipulative) "implicit" definitions. Since theoretical postulates have traditionally been taken as the paradigm case of implicit definition, the present account of theoretical concepts, if ten-able, should substantially clarify this issue.

A stipulative definition is a sentence 'D(a)' (or set of sentences, the conjunction of which reduces to the first case) through which meaning is assigned to one (or more) of its constituent terms 'a.' When a stipula-tive definition is of the form '$a = $ d' (or a conjunction of sentences of the form '$a_1 = d_1$'), it is known as an "explicit" definition.[30] When 'D(a)' is of a form other than '$a = $ d,' it is known as an "implicit" definition. Since any analysis of implicit definition sufficiently broad to cover all forms of 'D(a)' other than '$a = $ d' will undoubtedly be applicable to the latter as well, it would seem more logical to regard (stipulative) explicit definition as a special case of (stipulative) implicit definition.

According to the views developed earlier, if the meaning of a term 'a' is determined (solely) by its usage in the sentence 'D(a),' then 'a' desig-nates every entity t such that $D(t)$. This account does not explain *how*

---

[30] Explicit definitions are frequently written '$a =_{def}$ d.' How '$=_{def}$' should be ana-lyzed is not easy to decide. Since the force of '$=_{def}$' does not appear to be the same as '$=$,' the subscript does not occur vacuously, yet the Identity in '$a =_{def}$ d' does not seem to differ from the Identity in '$a = $ d.' One interpretation of '$=_{def}$' is in terms of lin-guistic norms, where the sentence '$a =_{def}$ d' is regarded not as an assertion, but a *rule*, conformity to which necessitates the truth of '$a = $ d.' Still another interpretation is to regard '$a =_{def}$ d' as an ellipsis for a more complex descriptive statement relating *why* it is the case that $a = d$. We shall here treat '$a = $ d' (and more generally, 'D(a)') as the "definition" itself (cf. [17], pp. 139, 149f). Whether this is correct, or whether

'a' acquires this meaning—such an explanation lies within the province of the psychology of language. What is of philosophical relevance is that 'a' designates in this way. Thus whatever role the symbol complex '$D(a)$' may play in the acquisition by 'a' of meaning, the semantical status of '$D(a)$' is simply that of a descriptive sentence. What needs to be spelled out in greater detail, however, are the truth conditions of '$D(a)$': (1) An implicit definition is not logically true. While it is difficult to find a wholly satisfactory explication of the classical concept of "logical truth," the underlying notion is that a statement to which this term applies is true or false by virtue of its logical form. But an implicit definition '$D(a)$' is true by virtue of its logical form only if the expression formed by replacing 'a' with any descriptive constant of the same formal type is necessarily true. This obtains only when the property $D(\phi)$ is necessarily possessed by every entity of the appropriate type, in which case (as will be elaborated below) '$D(a)$' is empty of definitional force. Hence an implicit definition cannot be tautological. However, (2) an implicit definition, if true, is true ex vi terminorum. Given that 'a' designates an entity t, it is unnecessary to inquire further as to whether or not $D(t)$ is the case in order to pass judgment on the truth of '$D(a)$.' It is in the meaning of 'a' that any entity designated by 'a' satisfies '$D(a)$.'[31] On the other hand, it is not the case that '$D(a)$' has no factual content, or that '$D(a)$' is not empirically falsifiable, for (3) the empirical force of an implicit definition is contained in the defined term's success or failure at designating. While 'a' designates any entity which satisfies '$D(\phi)$,' it by no means follows that there is any such entity. Hence, '$D(a)$' is empirically true or false according to whether or not there exists an entity designated by 'a'—i.e., according to whether or not $(\exists\phi)D(\phi)$.

To summarize: A statement, '$D(a)$,' which implicitly (or, as a special case, explicitly) defines a term, 'a,' does not fit conveniently into the traditional analytic-empirical dichotomy of the truth grounds of statements. Since it is inconceivable that a (so defined) should not exemplify D, one might think that '$D(a)$' should be analytically true. On the other hand, the most important cases of implicit definition, scientific theories,

'$a =_{\text{def}} d$' (and some analogous expression in the case of implicit definition) should be treated as the "definition" proper, with '$a = d$' (or '$D(a)$') as a consequence of the definition, does not matter here so long as it is agreed that it is legitimate to analyze the force of an explicit definition in terms of a symbol 'a' being given meaning through its use in the sentence '$a = d$.'

[31] Some complications may arise here if '$D(a)$' is not unified.

reveal clearly that implicit definitions are not compatible with all possible facts, and hence must embody a factual commitment. Traditional semantical analysis presupposes that all primitive extralogical constants of cognitively meaningful statements do, in fact, designate. The present analysis suggests, on the other hand, that it is not true that all sentences which violate this presupposition are meaningless, and that *there is an important class of empirically significant statements whose truth values depend wholly upon whether or not all their primitive extralogical constants have designata.* (Actually, pursuit of this line of thought in light of Hypotheses A–C, above, leads to a radical reinterpretation of the traditional concepts of "analytic" and "synthetic," but this is far beyond the scope of the present discussion.)

Definite descriptions. This problem has been a philosophical headache for many years. The difficulty is not so much a lack of interpretations as it is a surfeit of them. While Russell's [15] famous analysis is perhaps the most widely accepted, and seems to reproduce most satisfactorily the intuitive truth conditions of statements using definite descriptions, it has its own drawbacks, while alternative interpretations find themselves parting with common sense or the Law of Excluded Middle in the case of unsatisfied descriptions.

Part of the difficulty in finding an intuitively convincing explication of descriptions probably lies in an ambiguity in common usage. It seems to me that in *de facto* language practices, descriptions are frequently used as *demonstratives.* After all, one can call attention to an object by naming some of its distinguishing features as well as by pointing at it, and when used in this way, asserting that the A is a B would have essentially the same force as saying that *this* is a B—it need not even be the case, in this instance, that there is only one A (cf. [18], p. 186), or even that the entity referred to *is* an A, so long as the context of usage is such that the sign sequence, 'the A,' momentarily designates the appropriate entity.

However, while descriptions may in fact occasionally be used as demonstratives, this is certainly not the case which has stimulated philosophical concern. What needs to be determined is what is meant by saying, 'The A is a B,' when the A is not necessarily accessible to a demonstrative. The Russellian analysis, which takes such an assertion to be equivalent to 'There is an x such that $B(x)$, and for any y, $A(y)$ if and only if $y = x$,' has one fatal drawback: Under this analysis, *descriptions do not designate.* Russell himself was quite explicit on this point. Although the assertion

349

'The A is a B' appears to be of the logical form '$\psi(x)$,' in which 'x' designates the (only) entity which possesses a property ϕ, Russell contends that the genuine logical form is '$(\exists x)(\psi[x] \cdot (y)[\phi(y) \equiv y = x])$,' and in the latter expression, there is no term, or complex of terms, which designates any entity which exemplifies ϕ. Hence under the Russellian analysis, definite descriptions are not designators, but syntactical condensations. But surely this seriously undermines the Russellian account as an acceptable analysis of statements involving definite descriptions, for it seems to me inescapable that a description is, in actual language practice, used syntactically in essentially the same manner that we would use a descriptive constant of the same type level, and that when we say, 'the A,' we intend to refer to the A. On the other hand, use of the definite article to assert that the A is a B when one could otherwise say that an A is a B, would seem to indicate that 'The A is a B' entails that there is one and only one entity which is an A; while conversely, existence of exactly one entity which is an A and which, moreover, is also a B, is certainly a sufficient condition for the truth of 'The A is a B.' Hence it would appear that 'The A is a B' and '$(\exists x)(B[x] \cdot (y)[A(y) \equiv y = x])$' have exactly the same truth conditions, and a thoroughly satisfactory explication of the former would seem to require the force of the latter, but the logical form '$\phi(x)$.'

While we need not here make any definite commitments as to what this explication might be, it is instructive to observe that in important respects, definite descriptions appear to be very similar to theoretical terms. According to the position developed earlier, if '$T(\tau)$' is a theory, then while '$T(\tau)$' has the same truth conditions as '$(\exists\phi)T(\phi)$,' 'τ' actually designates that entity (or entities), if any, in virtue of which the latter is true. Implied by this analysis is the idea that a language user does not necessarily require having had direct awareness of an entity in order to refer to it—by appropriate synthesis of meaning components available through other sources,[32] he is able to construct an expression which designates the entity. If this conclusion is correct, then it is conceivable that the phrase 'the A' may also be a designative expression of this kind. In

---

[32] I strongly suspect that the ultimate components from which all cognitive meanings are synthesized are those aroused by direct experience. This possibility must not be confused, however, with the question of whether all meaningful linguistic expressions are constructed from a phenomenal language. Contrary to frequent philosophic misconception, meanings are to be found among psychological processes even when there is no corresponding language framework to govern them.

particular, if 'the A' is regarded as a theoretical term introduced by the unified theory '$(x)[A(x) \equiv x = \text{the A}]$,' then by Hypotheses A–C, 'the A' has a referent if and only if there is exactly one entity which satisfies '$A(x)$,' while 'The A is a B' is true or false according to whether or not there is exactly one A which, moreover, is also a B. Thus the present analysis of theoretical concepts makes it plausible that a definite description could carry the force of an existential operator in the Russellian fashion and yet serve as a genuine designator. In fact, this line of reasoning also suggests an explication for that neglected waif of linguistic analysis, the indefinite description. Suppose we regard the phrase 'an A' as a theoretical term introduced by the unified theory '$A(\text{an A})$.' Then by Hypotheses A–C, 'an A' designates every satisfier of '$A(x)$'; while the sentence 'An A is a B' is true if and only if $(\exists x)[A(x) \cdot B(x)]$, yet is of the logical form '$\phi(x)$.'

The meaning criterion. One of the dominant themes of modern analytic philosophy—certainly a guiding motive of the logical empiristic movement—has been the search for the "meaning criterion," a principle by which can be determined whether or not a given expression is cognitively meaningful. For difficulties, if any, which reside in the meaningfulness of expressions constructed wholly in the observation language, the present views have little relevance. On the other hand, to the extent that the meaning problem is concerned with the meanings of nonobservational terms, the present analysis of theoretical concepts provides a simple and plausible solution. It has been here contended that the meaning of a theoretical term is not something brought with it to the context of usage, but is given to it by the (accepted) postulates which contain it. If this is correct, then it is misleading to construe the desired meaning criterion as a yes-or-no test to be applied to terms whose meaningfulness is in doubt. Rather, cognitive meaningfulness is better seen as a matter of degree, and we should ask *what* meanings the usage of such terms has conferred upon them.

Suppose that the (cognitive) meaning, if any, of a term 'a' is imparted to it by its use in a set of (perhaps provisionally) accepted sentences, the conjunction of which is '$D(a)$.' That is, '$D(a)$' is the implicit definition of 'a.' Under what circumstances would we say that 'a' is meaningless? One intuitively plausible criterion, that a term is cognitively meaningless when it has no designatum, does not seem to be acceptable. For, if the present interpretation of theoretical concepts is correct, the assertion '$D(a)$' is

351

false if and only if 'a' has no designatum—i.e., if and only if it is the case that $\sim(\exists\phi)D(\phi)$. Then if 'a' were meaningless when it has no designatum, (a) it would be possible for a sentence containing meaningless terms to be false, and (b) decisions about meaningfulness would necessitate appeal to extralinguistic facts, so that meaning judgments would be logically subsequent, rather than prior, to truth judgments. Moreover, not all apparently meaningful expressions in ordinary use have designata. For example, most philosophers would hold that definite descriptions are meaningful, even when there exists no entity which uniquely satisfies the description. Or again: we should surely not wish to say that 'square-circleness' is meaningless, even though it would seem most peculiar to say that there exists a property, Square-circleness.

It would thus appear, since a syntactically well-formed implicit definition is true when it succeeds in assigning designata to its definienda and false otherwise, that there must be some rudimentary sense in which all terms actually in use have meaning—which is really not so surprising, since assignment of a syntactic role to a sign-design must surely confer some behavioral effect upon it. On the other hand, it is by no means the case that all terms must have a pragmatically *significant* meaning. Suppose that 'a' is defined by '$a = a$.' We should scarcely feel that 'a' has thereby acquired any useful meaning, for the property of self-identity is possessed by any entity whatsoever. Similarly, in the other extreme, we should be reluctant to grant that a term defined by a logically inconsistent definition had been given any useful force. More generally, if '$D(\phi)$' is a predicate whose applicability can be determined on logical grounds alone, the assertion '$D(a)$' contributes no useful meaning to 'a.' Conversely, if the implicit definition of a term ascribes to it a logically contingent predicate, then any sentence containing this term has an empirically falsifiable existential commitment, and the term must have useful meaning. I propose, therefore, that *a term is "effectively meaningful"* (i.e., having pragmatic force) *when and only when its usage is logically consistent and imposes extralogical limits on its possible referents.* Thus, when '$D(\phi)$' is a logically consistent monadic predicate in which 'ϕ' is a purely logical variable, '$D(a)$' gives 'a' effective meaning if and only if, when '$D(a)$' is adopted, it is not the case that 'a' necessarily designates every entity in the range of 'ϕ'—i.e., if and only if '$(\phi)D(\phi)$' is not necessarily true. The reason for stipulating that 'ϕ' must be a logical variable is that if its range were a nonlogical category, it would not be logically decidable whether

'$D(\phi)$' is applicable to a given entity t even though the empirical fact that t is in the range of 'ϕ' logically entails that $D(t)$. Thus if 'x^h' is a variable ranging over humans, '$(x^h)(x^h = x^h)$' is logically true, but the implicit definition '$a^h = a^h$' gives 'a^h' effective meaning, since its designata are restricted to humans. (Actually, in a language which contains nonlogical variables, a term is endowed with empirical commitments simply by choosing it to be of a formal type which represents a nonlogical category. It is by tracing the implications of this and similar considerations that we can appreciate the necessity for terms whose formal types represent purely logical categories.)

For the general case of an n-adic implicit definition '$D(a_1, \ldots, a_n)$,' the formalized meaning criterion is more complex than in the monadic case, since some but not all of the defined terms may be given effective meaning. For example, suppose that '$D_1(a_1)$' effective-meaningfully defines 'a_1,' that '$D_2(a_2)$' is tautologous, and that '$D(a_1, a_2)$' is equivalent to '$D_1(a_1) \cdot D_2(a_2)$.' Then '$D(a_1, a_2)$' gives effective meaning to 'a_1' but not to 'a_2.' Here, as in the monadic case, 'a_2' is effectively meaningless since its designata remains unrestricted; yet '$D(a_1, a_2)$' is not necessarily true, and, if unified (cf. Definition 10), would give a designatum to 'a_2' only if it also provides one for 'a_1.' To be sure, '$D_1(a_1)$' and '$D_2(a_2)$' are undoubtedly autonomous subdefinitions (cf. Definition 9) of '$D(a_1, a_2)$' in this instance, and if so, permit '$D(a_1, a_2)$' to assign designata to 'a_2' whether '$D_1(\phi)$' is satisfied or not. However, we have not so far attempted to specify the conditions of autonomy, and if a plausible criterion of effective meaningfulness can be found without prior assumptions about autonomy, then this may also help to clarify the latter. It will be noticed in the present example that whether '$D(a_1, a_2)$' is unified or not, if there exists one pair of entities t_1, t_2 such that $D(t_1, t_2)$, then 'a_2' designates all entities in the range of its formal type, since for any entity t_i in that range, it is the case that $D(t_i, t_i)$. But this would seem more generally to be an adequate formalization of the notion that if a definition gives effective meaning to some but not all of its definienda, failure of an effectively meaningless definiendum to designate all entities of its type should result only from lack of designata for the effectively meaningful definienda. I suggest, therefore, that the following formal criterion (of which the monadic instance already treated may readily be seen to be a special case) is characteristic of the conditions under which a new descriptive term is given effective meaning through its usage with other terms.

353

William W. Rozeboom

Postulate 7. The term 'a_i' *is given effective meaning by a logically consistent implicit definition* '$D(a_1, \ldots, a_n)$,' *where none of the definienda,* 'a_1,' . . ., 'a_n,' *occur in the predicate* '$D(\phi_1, \ldots, \phi_n)$' *and* '$\phi_i$' *is a purely logical variable, if and only if it is not the case that* '$(\exists \phi_1, \ldots, \phi_n)D(\phi_1, \ldots, \phi_n)$' *entails* '$(\phi_i)(\exists \phi_1, \ldots, \phi_{i-1}, \phi_{i+1}, \ldots, \phi_n)D(\phi_1, \ldots, \phi_n)$.'

The postulate may be applied to the case where 'ϕ_i' has extralogical restrictions on its range by first putting '$D(a_1, \ldots, a_n)$' into L-normal form (see p. 323, above). The restriction of P 7 to logically consistent definitions is to allow for the possibility that if D can be decomposed into several autonomous subdefinitions, the logical inconsistency of one of these should not deny the remainder an opportunity to confer effective meaning on their definienda.

What can be said in justification of P 7 from the standpoint of intuitive conditions of meaningfulness? On the whole, these are so vague as to be of little assistance in this respect. However, I shall conclude with three observations which, I think, lend weight to the adequacy of the proposed criterion.

1. Postulate 7 is *not* merely a *syntactical* criterion. Those philosophers who have taken the search for a meaning criterion most seriously have usually attempted to characterize the conditions of meaningfulness in terms of syntactical relations among sentences containing the term in question and other sentences whose meaningfulness is not in doubt. But the *meanings* of expressions are by no means fully determined by their syntactical properties, and in particular, the implications of a sentence are not necessarily exhausted by its formal consequences. Hence a meaning criterion which draws upon only the syntactical features of language is bound to prove inadequate. In contrast, by making entailment—i.e., a relation between the factual contents of sentences (see p. 297, above) due to their meanings as well as to their syntax—a critical ingredient of the criterion, P 7 addresses itself directly to the full linguistic force of the term whose meaning is under consideration.

2. While P 7 does not draw specifically upon syntactical properties, it nonetheless satisfies a certain syntactical condition which has been thought to pose difficulties for a proposed meaning criterion, namely, the requirement that meaningfulness be invariant under syntactical equivalence transformations (see [4], p. 55). If 'a' is meaningless when defined by '$D(a)$,' and '$D(a)$' is formally equivalent to '$D^*(a)$,' then 'a' must also be meaningless when defined by '$D^*(a)$'; hence it is a condition on the

354

adequacy of a meaning criterion that it yield this result. To prove this follows from P 7, we have to show that if [a] '$D(a) \equiv D^*(a)$' is formally true and if [b] '$(\exists\phi)D(\phi) \supset (\phi)D(\phi)$' is analytically true, then '$(\exists\phi)D^*(\phi)$' also entails '$(\phi)D^*(\phi)$.' We observe first of all that since '$D(a)$' formally entails '$(\exists\phi)D(\phi)$,' it follows from [a] that '$D^*(a) \supset (\exists\phi)D(\phi)$' is formally true. Hence by Lemma 1, [c] '$(\exists\phi)D^*(\phi) \supset (\exists\phi)D(\phi)$' is also formally true. Now as easily proved from Lemma 1, a sentence of the form '$F(a)$,' in which the matrix '$F(\ )$' does not contain 'a,' is formally true if and only if '$(\phi)F(\phi)$' is formally true. Hence from [a], '$(\phi)[D(\phi) \equiv D^*(\phi)]$' and thus also [d] '$(\phi)D(\phi) \equiv (\phi)D^*(\phi)$' is formally true. Then from [b] and [d], '$(\exists\phi)D(\phi)$' entails '$(\phi)D^*(\phi)$,' and hence from [c], '$(\exists\phi)D^*(\phi)$' entails '$(\phi)D^*(\phi)$.' Q.E.D. Thus under P 7, the effective meaninglessness and, conversely, meaningfulness of an implicitly defined term is invariant over formally equivalent forms of the definition.

3. My final observation is less concerned with P 7 as such than with the inconsistency of certain widespread beliefs about the conditions of meaningfulness for terms not introduced into the language ostensively. It is widely held that in order for a theoretical term to have meaning, its defining postulates must lead to some *empirical* conclusion—i.e., that if a theory '$T(\tau)$' confers meaning upon 'τ,' '$T(\tau)$' must have an O-consequence which is not analytically true. This stipulation can be made precise, of course, only by defining 'analytic,' but we may presume that persons who subscribe to this belief would also include statements of form '$d = d$' among the analytic truths. It is universally agreed, moreover, that an explicit definition, '$a =_{\text{def}} d$,' is a perfectly legitimate way to confer meaning upon 'a.' Now, it has already been argued that there is no difference in kind between a (stipulative) explicit definition and a set of theoretical postulates—both give meaning to previously neutral symbols by using them in a context with other symbols which have already attained meaning. If this view is accepted, then any test of the meaningfulness of theoretical terms must also apply to explicitly defined terms. But the prime consequence of '$a = d$' is '$(\exists\phi)(\phi = d)$,' which is analytic if '$d = d$' is analytic. Hence, if 'analytic truth' applies to formally valid sentences containing nonlogical terms, the belief that a meaningful theory must have nonanalytic consequences is incompatible with the belief that explicitly defined terms are meaningful. The relevance of these remarks to P 7 is that the latter does *not* imply that an implicit definition must have

William W. Rozeboom

nonanalytic consequences in order for its definienda to receive effective meaning. That this is how matters *should* be may be appreciated in greater generality by realizing that so long as the predicate '$D(\phi)$' contains meaningful descriptive constants, the sentence '$(\exists\phi)D(\phi)$,' even when analytic, contains effective meaning components carried by its descriptive terms which may then be mobilized to give effective meaning to 'a' when defined by '$D(a)$.'

Thus not only does P 7 meet certain general and rather difficult conditions of adequacy, it also avoids the inconsistency in what is probably the most widely held intuitive condition on a meaning criterion. Moreover, the line of reasoning of which it is a culmination makes clear just what sort of semantic desiderata are lacking in a term which fails to meet the criterion. I submit, therefore, that even if P 7 fails to capture all the nuances that we might wish of a meaning criterion, it will serve at the very least to define a certain interesting *kind* of meaningfulness.

REFERENCES

1. Bergmann, G. "A Note on Ontology," *Philosophical Studies*, 1:89–92 (1950). Reprinted in G. Bergmann, *The Metaphysics of Logical Positivism*. New York: Longmans, Green and Co., 1954.
2. Bergmann, G. "Sense Data, Linguistic Conventions, and Existence," *Philosophy of Science*, 14:152–163 (1947). Reprinted in G. Bergmann, *The Metaphysics of Logical Positivism*.
3. Carnap, R. *Introduction to Semantics*. Cambridge, Mass.: Harvard University Press, 1942.
4. Carnap, R. "The Methodological Character of Theoretical Concepts," in *Minnesota Studies in the Philosophy of Science*, Vol. I, H. Feigl and M. Scriven, eds. Minneapolis: University of Minnesota Press, 1956.
5. Feigl, H. "Existential Hypotheses," *Philosophy of Science*, 17:35–62 (1950).
6. Feigl, H. "The 'Mental' and the 'Physical,'" in *Minnesota Studies in the Philosophy of Science*, Vol. II, H. Feigl, M. Scriven, and G. Maxwell, eds. Minneapolis: University of Minnesota Press, 1958.
7. Hempel, C. "Problems and Changes in the Empiricist Criterion of Meaning," *Revue Internationale de Philosophie*, 11:41–63 (1950). Reprinted in *Semantics and the Philosophy of Language*, L. Linsky, ed. Urbana: University of Illinois Press, 1952.
8. Hempel, C. "The Theoretician's Dilemma," in *Minnesota Studies in the Philosophy of Science*, Vol. II, H. Feigl, M. Scriven, and G. Maxwell, eds. Minneapolis: University of Minnesota Press, 1957.
9. Henkin, L. "Completeness in the Theory of Types," *Journal of Symbolic Logic*, 15:81–91 (1950).
10. Quine, W. V. "Designation and Existence," *Journal of Philosophy*, 36:701–709 (1939). Reprinted in *Readings in Philosophical Analysis*, H. Feigl and W. Sellars, eds. New York: Appleton-Century-Crofts, 1949.
11. Ramsey, F. P. *The Foundations of Mathematics, and Other Essays*. London: Paul, Trench, Trubner and Co., 1931.

12. Rozeboom, W. W. "Do Stimuli Elicit Behavior?—A Study in the Logical Foundations of Behavioristics," *Philosophy of Science*, 27:159–170 (1960).
13. Rozeboom, W. W. "Intentionality and Existence," *Mind*, 71:15–38 (1962).
14. Rozeboom, W. W. "Studies in the Empiricist Theory of Scientific Meaning. Part I—Empirical Realism and Classical Semantics: A Parting of the Ways. Part II—On the Equivalence of Scientific Theories," *Philosophy of Science*, 27:359–373 (1960).
15. Russell, B. "On Denoting," *Mind*, 14:479–493 (1905). Reprinted in *Readings in Philosophical Analysis*, H. Feigl and W. Sellars, eds. New York: Appleton-Century-Crofts, 1949.
16. Russell, B. *The Problems of Philosophy*. New York: Henry Holt, 1912.
17. Scriven, M. "Definitions, Explanations, and Theories," in *Minnesota Studies in the Philosophy of Science*, Vol. II, H. Feigl, M. Scriven, and G. Maxwell, eds. Minneapolis: University of Minnesota Press, 1958.
18. Strawson, P. F. *Introduction to Logical Theory*. London: Methuen, 1952.
19. Tarski, A. "The Semantic Conception of Truth," *Philosophical and Phenomenological Research*, 4:341–375 (1944). Reprinted in *Readings in Philosophical Analysis*, H. Feigl and W. Sellars, eds. New York: Appleton-Century-Crofts, 1949.

The Analytic and the Synthetic

The techniques employed by philosophers of physics are usually the very ones being employed by philosophers of a less specialized kind (especially empiricist philosophers) at the time. Thus Mill's philosophy of science largely reflects Hume's associationism; Reichenbach's philosophy of science reflects Viennese positivism with its conventionalism, its tendency to identify (or confuse) meaning and evidence, and its sharp dichotomy between "the empirical facts" and "the rules of the language"; and (coming up to the present time) Toulmin's philosophy of science is an attempt to give an account of what scientists do which is consonant with the linguistic philosophy of Wittgenstein. For this reason, errors in general philosophy can have a far-reaching effect on the philosophy of science. The confusion of meaning with evidence is one such error whose effects are well known: it is the contention of the present paper that overworking of the analytic-synthetic distinction is another root of what is most distorted in the writings of conventional philosophers of science.

The present paper is an attempt to give an account of the analytic-synthetic distinction both inside and outside of physical theory. It is hoped that the paper is sufficiently nontechnical to be followed by a reader whose background in science is not extensive; but it has been necessary to consider problems connected with physical science (particularly the definition of 'kinetic energy,' and the conceptual problems connected with geometry) in order to bring out the features of the analytic-synthetic distinction that seem to me to be the most important.

In addition to the danger of overworking the analytic-synthetic distinction, there is the somewhat newer danger of denying its existence altogether. Although, as I shall argue below, this is a less serious error (from the point of view of the scientist or the philosopher interested in the conceptual problems presented by physical theory) than the customary overworking of the distinction, it is, nevertheless, an error. Thus the present paper fights on two fronts: it tries to "defend" the distinction, while attacking its extensive abuse by philosophers. Fortunately, the two fronts are not too distant from each other; one reason that the analytic-synthetic distinction has seemed so difficult to defend recently is that it has become so bloated!

Replies to Quine. In the spring of 1951 Professor W. V. Quine published a paper entitled "Two Dogmas of Empiricism."[1] This paper provoked a spate of replies, but most of the replies did not match the paper which stimulated them in originality or philosophic significance. Quine denied the existence of the analytic-synthetic distinction altogether. He challenged doctrines which had been dear to the hearts of a great many philosophers and (in spite of the title of his paper) not only philosophers in the empiricist camp. The replies to Quine have played mostly on a relatively small number of stereotyped themes. The tendency has been to "refute" Quine by citing examples. Of course, the analytic-synthetic distinction rests on a certain number of classical examples. We would not have been tempted to draw it or to keep on drawing it for so long if we did not have a stock of familiar examples on which to fall back. But it is clear that the challenge raised by Quine cannot be met either by pointing to the traditional examples or by simply waving one's hand and saying how implausible it is that there should be no distinction at all when there seems to be such a clear one in at least some cases. I do not agree with Quine, as will be clear in the sequel. I am convinced that there is an analytic-synthetic distinction that we can correctly (if not very importantly) draw, and I am inclined to sympathize with those who cite the examples and who stress the implausibility, the tremendous implausibility, of Quine's thesis—the thesis that the distinction which certainly seems to exist does not in fact exist at all.

But to say that Quine is wrong is not in itself very fruitful or very interesting. The important question is How is he wrong? Faced with the battery of Quine's arguments, how can we defend the existence of any genuine analytic-synthetic distinction at all? Philosophers have the right to have intuitions and to believe things on faith; scientists often have no better warrant for many of their beliefs, at least not for a time. But if a philosopher really feels that Quine is wrong and has no statement to make other than the statement that Quine is wrong and that he feels this in his

[1] Reprinted in *From a Logical Point of View* (Cambridge, Mass.: Harvard University Press, 1953), pp. 20–46.

bones, then this is material to be included in that philosopher's autobiography; it does not belong in a technical journal under the pretense of being a reply to Quine. From this criticism I specifically exempt the article by P. F. Strawson and A. P. Grice,[2] who offer *theoretical* reasons for supposing that the analytic-synthetic distinction does in fact exist, even if they do not very satisfactorily delineate that distinction or shed much real light on its nature. Indeed, the argument used by them to the effect that *where there is agreement on the use of the expressions involved with respect to an open class, there must necessarily be some kind of distinction present*, seems to me correct and important. Perhaps this argument is the only one of any novelty to have appeared since Quine published his paper.

But important as it is to have a theoretical argument supporting the existence of the distinction in question (so that we do not have to appeal simply to "intuition" and "faith"), still the argument offered by Strawson and Grice does not go far toward clarifying the distinction, and this, after all, is Quine's challenge. In other words, we are in the position of *knowing* that there *is* an analytic-synthetic distinction but of not being able to make it very clear just what the nature of this distinction is.

Of course, in some cases it is not very important that we cannot make clear what the nature of a distinction is, but in the case of the analytic-synthetic distinction it seems that the nature of the distinction is far more imporant than the few trivial examples that are commonly cited, e.g., 'All bachelors are unmarried' (for the analytic side of the dichotomy) and 'There is a book on this table' (for the synthetic side). To repeat: philosophers who do not agree with Quine have found themselves in the last few years in this position: they *know* that there *is* an analytic-synthetic distinction but they are unable to give a satisfactory account of its nature.

It is, in the first place, no good to draw the distinction by saying that a man who rejects an analytic sentence is *said* not to understand the language or the relevant part of the language. For this is a comment on the use of the word 'understand' and, as such, not very helpful. There could be an analytic-synthetic distinction even in a language which did not use such words as 'analytic,' 'synthetic,' 'meaning,' and 'understanding.' We do not want, after all, to draw the analytic-synthetic distinction in terms

[2] In Defense of a Dogma," *Philosophical Review*, 65:141–158 (1956).

of dispositions to use the words 'analytic' and 'synthetic' themselves, nor dispositions to use related expressions, e.g., 'have the same meaning' and 'does not understand what he is saying.' What is needed is something quite different: We should be able to indicate the nature and rationale of the analytic-synthetic distinction. What happens to a statement when it is analytic? What do people do with it? Or if one wishes to talk in terms of artificial languages: What point is there to having a separate class of statements called analytic statements? Why mark these off from all the others? What do you do with the statements so marked? It is only in this sort of terms that I think we can go beyond the level of saying, "Of course there are analytic statements. I can give you examples. If someone rejects one of these, we say he doesn't understand the language, etc." The real problem is not to describe the language game we play with words like 'meaning' and 'understanding' but to answer the deeper question, "What is the point of the game?"

The analytic-synthetic distinction in philosophy. It should not be supposed that the axe I have to grind here is that Quine is wrong. That Quine is wrong I have no doubt. This is not a matter of philosophical argument: it seems to me there is as gross a distinction between 'All bachelors are unmarried' and 'There is a book on this table' as between any two things in the world, or, at any rate, between any two linguistic expressions in the world; and no matter how long I might fail in trying to clarify the distinction, I should not be persuaded that it does not exist. In fact, I do not understand what it would mean to say that a distinction between two things *that* different does not exist.

Thus I think that Quine is wrong. There are analytic statements: 'All bachelors are unmarried' is one of them. But in a deeper sense I think that Quine is right; far more right than his critics. I think that there is an analytic-synthetic distinction, but a rather trivial one. And I think that the analytic-synthetic distinction has been so radically overworked that it is less of a philosophic error, although it is an error, to maintain that there is no distinction at all than it is to employ the distinction in the way that it has been employed by some of the leading analytic philosophers of our generation. I think, in other words, that if one proceeds, as Quine does, on the assumption that there is no analytic-synthetic distinction at all, one would be right on far more philosophic issues and one will be led to far more philosophic insights than one will be if one accepts that heady concoction of ideas with which we are all too familiar: the

Hilary Putnam

idea that every statement is either analytic or synthetic; the idea that all logical truths are analytic; the idea that all analytic truth derives its necessity from "linguistic convention." I would even put the thesis to be defended here more strongly: ignore the analytic-synthetic distinction, and you will not be wrong in connection with any philosophic issues not having to do specifically with the distinction. Attempt to use it as a weapon in philosophical discussion, and you will consistently be wrong.

It is not, of course, an accident that one will consistently be wrong if one attempts to employ the analytic-synthetic distinction in philosophy. 'Bachelor' may be synonymous with 'unmarried man' but that cuts no philosophic ice. 'Chair' may be synonymous with 'movable seat for one with a back' but that bakes no philosophic bread and washes no philosophic windows. It is the belief that there are synonymies and analyticities of a deeper nature—synonymies and analyticities that cannot be discovered by the lexicographer or the linguist but only by the philosopher—that is incorrect.[3]

I don't happen to believe that there are such objects as "sense data"; so I do not find "sense-datum language" much more interesting than phlogiston language or leprechaun language. But even if sense data did exist and we granted the possibility of constructing sense-datum language, I do not think that the expression 'chair,' although it is synonymous with 'movable seat for one with a back,' is in the same way synonymous with any expression that one could in principle construct in the sense-datum language. This is an example of the type of "hidden" synonymy or "philosophic" synonymy that some philosophers have claimed to discover and that does not exist.

However, misuse of the analytic-synthetic distinction is not confined to translationists. I have seen it argued by a philosopher of a more contemporary strain that the hypothesis that the earth came into existence five minutes ago (complete with "memory traces," "causal remains," etc.) is a *logically* absurd hypothesis. The argument was that the whole use of time words presupposes the existence of the past. If we grant the meaningfulness of this hypothesis, then, it is contended, we must grant the possibility that there is no past at all (the world might have come into existence at *this* instant). Thus, we have an example of a statement which uses time words, but which, if true, destroys the possibility of their use.

[3] I do not wish to suggest that linguistic regularities, properly so called, are never of importance in philosophy, but only that analytic statements, properly so called, are not.

THE ANALYTIC AND THE SYNTHETIC

This somewhat fuzzily described situation is alleged to be tantamount to the meaninglessness or self-contradictoriness of the hypothesis I described.

Now I agree that the hypothesis in question is more than empirically false. It is empirically false, if by empirically false one means simply that it is false about the world—the world did not come into existence at this instant nor did it come into existence five minutes ago. It is not empirically false if one means by 'empirically false statement' a statement which can be confuted by *isolated* experiments. But while it is important to recognize that this is not the sort of hypothesis that can be confuted by isolated experiment, it is not, I think, happy to maintain that the existence of a past is analytic, if one's paradigm for analyticity is the 'All bachelors are unmarried' kind of statement.[4] And I think that, while few philosophers would explicitly make the kind of mistake I have described, a great many philosophers tend to make it implicitly. The idea that every truth which is not empirical in the second of the senses I mentioned must be a "rule of language" or that all necessity must be traced down to the obligation not to "violate the rules of language" is a pernicious one, and Quine is profoundly right in rejecting it; the reasons he gives are, moreover, the right reasons. What I maintain is that there are no further rules of language beyond the garden variety of rules which a lexicographer or a grammarian might discover, and which only the philosopher can discover.

This is not to say that there are not some things which are very much like "rules of language." There is after all a place for *stipulation* in cognitive inquiry, and truth by stipulation has seemed to some the very model of analyticity. There is also the question of linguistic misuse. Under certain circumstances a man is said not merely to be in error but to be making linguistic mistakes—not to know the meaning of the very words he is employing. Philosophers have thought that by looking at such situations we could reconstruct a codex which might constitute the "implicit rules" of natural language. For instance, they hold that, in many circumstances, to say of a man that he knows that p implies that he has, or had at some time, or can produce, or could produce at some time, evidence that p—and that such an implication is very much like the implication between being a bachelor and being unmarried. But, as I shall argue be-

[4] To accept the hypothesis that the world came into existence five minutes ago does not make it necessary to give up any *particular* prediction. But I deny (a) that it "makes no difference to prediction," and (b) that "it therefore (*sic!*) amounts to a change in our use of language."

363

low, there are differences which it is absolutely vital to recognize. It is not that the statements I have mentioned fall into a third category. They fall into many different categories. Over and beyond the clear-cut rules of language, on the one side, and the clear-cut descriptive statements, on the other, are just an enormous number of statements which are not happily classified as either analytic or synthetic.

The case of stipulation is one in point. One must consider the role of the stipulation and whether the truth introduced by stipulation retains its conventional character or whether it later figures in inquiry on a par with other truths, without reference to the way in which it was introduced. We have to consider the question of the arbitrariness versus systematic import of our stipulations. There is one kind of wholly arbitrary stipulation which does indeed produce analytic statements, but we should not be led to infer that, therefore, every stipulation produces analytic statements. The Einstein stipulation that the constancy of the light velocity should be used to "define" simultaneity in a reference system does not, Reichenbach to the contrary, generate an analytic truth of the same order as 'All bachelors are unmarried.' And even the case of *knowing* and *having* or *having had evidence* requires much treatment and involves special difficulties. I shall in the body of this paper try to draw some of the distinctions that I think need to be drawn. For the moment let me only say this: if one wants to have a model of language, it is far better to proceed on the idea that statements fall into three kinds—analytic, synthetic, and lots-of-other-things—than to proceed on the idea that, except for borderline fuzziness, every statement is either analytic or synthetic.

Of course many philosophers are aware that there are statements which are not happily classified as either analytic or synthetic. My point is not that there exist exceptional examples, but that there is a far larger class of such statements than is usually supposed. For example, to ask whether or not the principles of logic are analytic is to ask a bad question. Virtually all the *laws* of natural science are statements with respect to which it is not *happy* to ask the question "Analytic or synthetic? It must be one or the other, mustn't it?" And with respect to the framework principles that are often discussed by philosophers, the existence of the past or the implication that some time exists between knowing and having had evidence, it is especially a mistake to classify these statements as "rules of language" or "true because of the logic of the concepts involved" or "analytic" or "L-true" or . . . This is not to say that all these principles have the same nature or that they form a compact new class, e.g., framework principles (as if one were to take seriously the label I have been using). 'There is a past' is recognizably closer to the law of conservation of energy than 'If Jones knows that p, then he must have or have had evidence that p' (in the cases where the latter inference seems a necessary one); and 'If Jones knows that p, then he must have or have had evidence that p' is more like 'All bachelors are unmarried' than is 'There is a past.' But neither statement is of exactly the same kind as the law of conservation of energy, although that law too is a statement with respect to which it is not happy to say, "Is it analytic or synthetic?" and neither statement is of exactly the same kind as 'All bachelors are unmarried.' What these statements reveal are different degrees of something like convention, and different kinds of systematic import. In the case of 'All bachelors are unmarried,' we have the highest degree of linguistic convention and the minimum degree of systematic import. In the case of the statement 'There is a past,' we have an overwhelming amount of systematic import—so much that we can barely conceive of a conceptual system which did not include the idea of a past. That is to say, such a conceptual system differs so greatly from our present conceptual system that the idea of ever making a transition from one to the other seems fantastic.[5] In the case of knowing without ever having any reason to believe, still other considerations are involved. We have to ask what we would say if people appeared to be able to answer questions truthfully about a certain subject matter although they had never had any acquaintance with that subject matter as far as we could detect. *Knowing* is something that we do not have much of a theory about. It makes little difference at *present* whether we say that such people would be correctly described as "knowing" the answers to the various questions in the area in which they are able to act as an oracle, or whether we say that they have an "uncanny facility at guessing the correct answer"; although, in the light of a more advanced theory, it might very well make a good deal of difference what we say. The concept of the past, on the other hand, and the concept of time, are deeply integrated into our physical theory, and any tampering with

[5] For example, we *could* accept the hypothesis that the world came into existence January 1, 1957, without changing the *meaning* of any word; but to do so would have a crippling effect on many sciences, and on much of ordinary life. (Think of the *ad hoc* hypotheses that would have to be invented to account for the "creation." And consider the role played by data concerning the past in, say, astronomy—not to mention ordinary human relations!)

these concepts would involve a host of revisions if simple consistency is to be maintained. In the sequel I shall try to describe in somewhat more detail the diverse natures of the statements in that vast class with respect to which it is not happy to say "analytic or synthetic." But on the whole my story will resemble Quine's. That is to say, I believe that we have a conceptual system with centralities and priorities. I think the statements in that conceptual system—except for the *trivial* examples of analyticity, e.g., 'All bachelors are unmarried,' 'All vixens are foxes'—fall on a continuum, a multidimensional continuum. More or less stipulation enters; more or less systematic import. But any one of these principles might be given up, farfetched though it may seem, and perhaps without altering the meaning of the constituent words. Of course, if we give up a principle that is analytic in the trivial sense ('All bachelors are unmarried'), then we have clearly changed the meaning of a word. But the revision of a sufficient number of principles, no one of which is by itself analytic in quite the way in which 'All bachelors are unmarried' is analytic, may also add up to what we should describe as a change in the meaning of a word. With Quine, I should like to stress the monolithic character of our conceptual system, the idea of our conceptual system as a massive alliance of beliefs which face the tribunal of experience collectively and not independently, the idea that "when trouble strikes" revisions can, with a very few exceptions, come anywhere. I should like, with Quine, to stress the extent to which the meaning of an individual word is a function of its place in the network, and the impossibility of separating, in the actual use of a word, that part of the use which reflects the "meaning" of the word and that part of the use which reflects deeply embedded collateral information.

Linguistic conventionalism. One more point will terminate this rather interminable set of preliminary remarks. The focus of this paper is the analytic-synthetic distinction, not because I think that distinction is of itself of overwhelming importance. In fact, I think it is of overwhelming unimportance. But I believe that the issues raised by Quine go to the very center of philosophy. I think that appreciating the diverse natures of logical truths, of physically necessary truths in the natural sciences, and of what I have for the moment lumped together under the title of framework principles—that clarifying the nature of these diverse kinds of statements is the most important work that a philosopher can do. Not because philosophy is necessarily about language, but because we must become clear about the roles played in our conceptual systems by these diverse kinds of truths before we can get an adequate global view of the world, of thought, of language, or of anything. In particular, I think we might begin to appreciate the real problems in the domain of formal science once we rid ourselves of the easy answer that formal truth is in some sense "linguistic in origin"; and in any case I think that one's whole view of the world is deeply affected, if one is a philosopher, by one's view of what it is to have a view about the world. Someone who identifies conceptualization with linguistic activity and who identifies linguistic activity with response to observable situations in accordance with rules of language which are themselves no more than implicit conventions or implicit stipulations (in the ordinary unphilosophic sense of 'stipulation' and 'convention') will, it seems to me, have a deeply distorted conception of human knowledge and, indirectly, of some or all objects of human knowledge. We must not fall into the error of supposing that to master the total use of an expression is to master a repertoire of individual uses, that the individual uses are the product of something like implicit stipulation or implicit convention, and that the conventions and stipulations are arbitrary. (The notion of a nonarbitrary *convention* is of course an absurdity—conventions are used precisely to settle questions that are arbitrary.) For someone who uses language in the way that I have just described, there are observable phenomena at the macrolevel and there are conventional responses to these, and this is all of knowledge; one can, of course, say that "there are atoms" and that "science is able to tell us a great deal about atoms," but *this* turns out to be no more than making noises in response to macrostimuli *in accordance with arbitrary conventions.* I do not think that any philosopher explicitly maintains such a view of knowledge; and if he did it is clear that he would be a sort of mitigated phenomenalist. But I do think that a good many philosophers implicitly hold such a view, or fall into writing as if they held such a view, simply because they tend to think of use as a sum of individual uses and of linguistic use on the model suggested by the phrase 'rules of language.'

To sum up: I do not agree with Quine, that there is no analytic-synthetic distinction to be drawn at all. But I do believe that his emphasis on the monolithic character of our conceptual system and his negative emphasis on the *silliness* of regarding mathematics as consisting in some sense of "rules of language," represent exceedingly important theoretical insights in philosophy. I think that what we have to do now is to settle

Hilary Putnam

the relatively trivial question concerning analytic statements properly s
called ('All bachelors are unmarried'). We have to take a fresh look a
the framework principles so much discussed by philosophers, disabusin
ourselves of the idea that they are "rules of language" in any literal o
lexicographic sense; and above all, we have to take a fresh look at th
nature of logical and mathematical truths. With Quine's contribution
we have to face two choices: We can ignore it and go on talking abou
the "logic" of individual words. In that direction lies sterility and more
much more, of what we have already read. The other alternative is to fac
and explore the insight achieved by Quine, trying to reconcile the fac
that Quine is overwhelmingly right in his critique of what other philoso
phers have *done* with the analytic-synthetic distinction with the fact tha
Quine is wrong in his literal thesis, namely, that the distinction itsel
does not exist *at all*. In the latter direction lies philosophic progress. Fo
philosophic progress is nothing if it is not the discovery of new areas fo
dialectical exploration.

Analytic and Nonanalytic Statements

The "kinetic energy definition." As a step toward clarification of th
analytic-synthetic distinction, I should like to contrast a paradigm case o
analyticity—'All bachelors are unmarried'—with an example which super
ficially resembles it: the statement that kinetic energy is equal to one hal
the product of mass and velocity squared, '$e = \frac{1}{2}mv^2$.' I think that if w
can see the respect in which these two examples differ, we will have mad
important progress toward such a clarification.

Let us take the second statement first, '$e = \frac{1}{2}mv^2$'; this is the sort o
statement that before relativistic physics one might well have called
"definition of 'kinetic energy.'" Yet, its history is unusual. Certainly, be
fore Einstein, any physicist might have said, "'$e = \frac{1}{2}mv^2$'; that is just th
definition of 'kinetic energy.' There is no more to it than that. The ex
pression 'kinetic energy' is, as it were, a sort of abbreviation for the longe
expression 'one-half the mass times the velocity squared.'"

If this were true, then the expression 'kinetic energy' would, of course
be in principle dispensable. One could simply use '$\frac{1}{2}mv^2$' wherever on
had used 'kinetic energy.'

In the early years of the twentieth century, however, Albert Einstei
developed a theory, a physical theory—but of an unusual sort. It is un
usual because it contains words of a rather high degree of vagueness, a

least in terms of what we usually suppose the laws of physics to be like. All this notwithstanding, the theory is, as we all well know, a precise and useful theory.

What I have in mind is Einstein's principle that all physical laws must be Lorentz-invariant. This is a rather vague principle, since it involves the general notion of a physical law. Yet in spite of its vagueness, or perhaps because of its vagueness, scientists have found it an extremely useful leading principle. Of course, Einstein contributed more than a leading principle. He actually proceeded to find Lorentz-invariant laws of nature; and the search for a Lorentz-invariant law of gravitation, in particular, produced the general theory of relativity.[6]

But it would be a mistake to think of the special theory of relativity as the sum of the special laws that Einstein produced. The *general* principle that all physical laws are Lorentz-invariant is certainly a legitimate part of the special theory of relativity, notwithstanding the fact that it is stated in what some purists might call "the metalanguage." And it is no good to say that 'a physical law' means 'any true physical statement': for so interpreted Einstein's principle would be empty. Any equation whatsoever can be made Lorentz-invariant by writing it in terms of suitable magnitudes. The principle that the laws of nature must be Lorentz-invariant is without content unless we suppose that the magnitudes to be contained in laws of nature must be in some sense real magnitudes—e.g., electricity, gravitation, magnetism—and that the equations expressing the laws must have certain characteristics of simplicity and plausibility. In practice, Einstein's principle is quite precise, in the only sense relevant to physical inquiry, notwithstanding the fact that it contains a vague term. The point is that the vagueness of the term 'physical law' does not affect the applications which the physicist makes of the principle. In practice, the physicist has no difficulty in recognizing laws or putative laws: any "reasonable" equation proposed by a physicist in his right mind constitutes at least a putative law. Thus, the Einstein principle, although it might bother those logicians who are worried, and rightly worried, about the right distinction between a natural law and any true statement whatsoever, is one whose role in physical inquiry is clear-cut. It means simply that those equations considered by physicists as expressing possible laws of nature must, if they are to remain candidates for that role in the age of relativity,

[6] Of course, the general theory of relativity itself *replaces* the requirement of Lorentz-invariance with the requirement of covariance.

be Lorentz-invariant. Of course, the principle does not play only the purely negative role of ruling out what might otherwise be admissible scientific theories: the fact that laws of nature must be Lorentz-invariant has often been a valuable clue to fundamental new discoveries. The Einstein gravitation theory has already been mentioned; another famous example is Dirac's "hole" theory, which led to the discovery of the positron.

Returning now to our account of the history of the "energy definition": the principle just described led Einstein to change a great many physical laws. Some of the older laws, of course, survived: the Maxwell equations, for instance, turned out to be Lorentz-invariant as they stood. Some of the principles that Einstein revised would ordinarily be regarded as being of an empirical nature. The statements 'Moving clocks slow down' and 'One cannot exceed the velocity of light' are certainly statements which we should regard as synthetic. The interesting thing is that Einstein was to revise, and in an *exactly similar fashion*, principles that had traditionally been regarded as definitional in character. In particular, Einstein, as we all know, changed the definition of 'kinetic energy.' That is to say, he replaced the law '$e = \frac{1}{2}mv^2$' by a more complicated law. If we expand the Einstein definition of energy as a power series, the first two terms are '$e = mc^2 + \frac{1}{2}mv^2 + \ldots$' We might, of course, reply that classically speaking '$\frac{1}{2}mv^2$' defines not 'energy' in general (e.g., 'potential energy') but only 'kinetic energy'; we might try to say that the energy that a body has because of its rest mass (this is represented by the term 'mc^2') should not be counted as part of its kinetic energy, as Einstein does. The point is that even the magnitude in the theory of relativity that corresponds to the classical kinetic energy of a particle, that is, its total kinetic energy minus the energy due to its rest mass, is not equal to $\frac{1}{2}mv^2$ except as a first approximation. If you take the total relativistic kinetic energy of a particle and subtract the energy due to its rest mass, you will obtain not only the leading term '$\frac{1}{2}mv^2$' but also terms in 'mv^4,' etc.

It would clearly be a distortion of the situation to say that 'kinetic energy $= \frac{1}{2}mv^2$' was a definition, and that Einstein merely changed the definition. The paradigm that this account suggests is somewhat as follows: 'kinetic energy,' before Einstein, was *arbitrarily* used to stand for '$\frac{1}{2}mv^2$.' After Einstein, 'kinetic energy' was *arbitrarily* used to stand for '$m + \frac{1}{2}mv^2 + \frac{3}{8}mv^4 + \ldots$'[7] This account is, of course, incorrect.

[7] This formula assumes that the unit of time is chosen so that the speed of light $= 1$.

What is striking is this: whatever the status of the "energy definition" may have been before Einstein, in revising it, Einstein treated it as just another natural law. There was a whole set of pre-existing physical and mechanical laws which had to be tested for compatibility with the new body of theory. Some stood the test unchanged—others only with some revision. Among the equations that had to be revised (and formal considerations indicated a rather natural way of making the revision, one which was, moreover, borne out richly by experiments) was the equation $e = \frac{1}{2}mv^2$.'

The moral of all this is not difficult to find. The "energy definition" may have had a special status when it came into the body of accepted physical theory, although this is a question for the historian of science to answer. It may even, let us suppose, have originally been accepted on the basis of explicit stipulation to the effect that the phrase 'kinetic energy' was to be used in the sense of '$\frac{1}{2}mv^2$.' Indeed, there was some discussion between Newton and Leibniz on the question whether the term 'energy' should be applied to what we now do call 'energy' or what we call 'momentum.' Suppose, however, that a congress of scientists had been convened in, say, 1780 and had settled this controversy by legislating that the term 'kinetic energy' was to be used for $\frac{1}{2}mv^2$ and not for mv. Would this have made the principle '$e = \frac{1}{2}mv^2$' analytic? It would be true by stipulation, wouldn't it? It would be true by stipulation, yes, but only in a context which is defined by the fact that the only alternative principle is $e = mv$.'

Quine has suggested that the distinction between truths by stipulation and truths by experiment is one which can be drawn only at the moving frontier of science. Conventionality is not "a lingering trait" of the statements introduced as truths by stipulation. The principle '$e = \frac{1}{2}mv^2$' may have been introduced, at least in our fable, by stipulation; the Newtonian law of gravity may have been introduced on the basis of induction from the behavior of the known satellite systems and the solar system (as Newton claimed); but in subsequent developments these two famous formulas were to figure on a par. Both were used in innumerable physical experiments until they were challenged by Einstein, without ever being regarded as themselves subject to test in the particular experiment. If a physicist makes a calculation and gets an empirically wrong answer, he does not suspect that the mathematical principles used in the calculation may have been wrong (assuming that those principles are themselves

371

theorems of mathematics) nor does he suspect that the law 'f = ma' may be wrong. Similarly, he did not frequently suspect before Einstein that the law 'e = ½mv²' might be wrong or that the Newtonian gravitational law might be wrong (Newton himself did, however, suspect the latter) These statements, then, have a kind of preferred status. They can be over thrown, but not by an isolated experiment. They can be overthrown only if someone incorporates principles incompatible with those statements in a successful conceptual system.

Principles of geometry. An analogy may be drawn with the case of ge ometry. No experiments—no experiments with light rays or tape measures or with anything else—could have overthrown the laws of Euclidean ge ometry before someone had worked out *non*-Euclidean geometry. That is to say, it is inconceivable that a scientist living in the time of Hume might[8] have come to the conclusion that the laws of Euclidean geometry are false: "I do not know what geometrical laws are true, but I know the laws of Euclidean geometry are false." Principles as central to the con ceptual system of science as laws of geometry are simply not abandoned in the face of experiment *alone*. They are abandoned because a rival *theory* is available.

On the other hand, before the development of non-Euclidean geom etry by Riemann and Lobachevski, the best philosophic minds regarded the principles of geometry as virtually analytic. The human mind could not conceive their falsity. Hume would certainly not have been impressed by the claim that 'straight line' means 'path of a light ray,' and that the meeting of two light rays mutually perpendicular to a third light ray could show, if it ever occurred, that Euclidean geometry is false. It would have been self-evident to Hume that such an experimental situation, if it ever occurred, would be correctly explained by supposing that the light rays traveled in a curved path in Euclidean space, and *not* by supposing that the light rays traveled in two straight lines which were indeed mutually perpendicular to a third straight line but which nevertheless met. Hume had he employed the vocabulary of contemporary analytic philosophy might even had said that this follows from the "logic" of the words 'straight line.' It is a "criterion," to use another popular word, for lines being straight that if two of them are perpendicular to a third the two do

[8] This is not a historical remark. I mean that no scientist *ought* to have come to this conclusion at that time, no matter what experimental evidence might have been pre sented.

not meet. It may be another criterion that light travels in *approximately* straight lines; but only where this criterion does not conflict with the deeply seated meaning of the words 'straight line.' In short, the meaning of the words 'straight line' is such that light rays may sometimes be said not to travel in straight lines; but straight lines cannot be said to behave in such a way as to form a triangle the sum of whose angles is more than 180°. If he had used the jargon of another fashionable contemporary school of philosophy, Hume might have said that straight lines are "theoretical constructs." And that light ray paths constitute a "partial interpretation" of geometrical theory but one that is only admissible on condition that it does not render false any of the "meaning postulates" of the geometrical theory.

Of course Hume did not employ this jargon. But he employed what was for him an equivalent jargon: the jargon of conceiving, visualizing, mental imagery. One cannot form any image of straight lines that do not conform to the laws of Euclidean geometry. This, of course, was to be true because any image of lines not conforming to the axioms of Euclidean geometry is an image which is not *properly* called an image of *straight* lines at all. Hume did not put it that way, however. Rather he explained the alleged "impossibility of imagining" straight lines not conformant to the laws of Euclidean geometry in terms of a theory of relations between our ideas.

Was Hume wrong? Reichenbach[9] suggested that 'straight line,' properly analyzed, means 'path of a light ray'; and with this "analysis" accepted, it is clear that the principles of geometry always are and always were synthetic. They are and always were subject to experiment. Hume simply overlooked something which could *in principle*[10] have been seen

[9] Reichenbach actually claimed that there were various possible alternative "coordinative definitions" of 'straight line.' However he contended that this one (and the ones physically equivalent to it) "have the advantage of logical simplicity and require the least change in the results of science." Moreover: "The sciences have implicitly employed such a coordinative definition all the time, though not always consciously"—i.e., it renders the customary meaning of the term 'straight line.' *Space and Time* (New York: Dover Publications, 1956), p. 19.

[10] Reichenbach does not assert that the Greeks could (as a matter of psychological or historical possibility) have understood the "true" character of geometric statements prior to the invention of non-Euclidean geometry: in fact, he denies this. But there is nothing in Reichenbach's analysis in Ch. I of *Space and Time* which *logically* presupposes a knowledge of non-Euclidean geometry. Thus, if Reichenbach is right, then the Greeks could *in principle* have "realized" (a) that the question whether Euclidean geometry is correct for physical space presupposes the choice of a "coordinative defini-

Hilary Putnam

even by the ancient Greeks. I think Reichenbach is almost totally wrong.
If the paradigm for an analytic sentence is 'All bachelors are unmarried'—
and it is—then it is of course absurd to say that the principles of geometry
are analytic. Indeed, we cannot any longer say that the principles of Euclidean geometry are analytic; because analytic sentences are true, and
we no longer say that the principles of Euclidean geometry are true. But
I want to suggest that before the work of nineteenth-century mathematicians, the principles of Euclidean geometry were as *close* to analytic as
any nonanalytic statement ever gets. That is to say, they had the following status: no experiment that one could describe could possibly overthrow them, by itself.[11] Just plain experimental results, without any new
theory to integrate them, would not have been accepted as sufficient
grounds for rejecting Euclidean geometry by any rational scientist.[12] After
the development of non-Euclidean geometry, the position was rather different, as physicists soon realized: give us a rival conceptual system, and
some reason for accepting it, and we will consider abandoning the laws of
Euclidean geometry.

When I say that the laws of Euclidean geometry were, before the development of non-Euclidean geometry, as analytic as any nonanalytic
statements ever get, I mean to group them, in this respect, with many
other principles: the law 'f = ma' (force equals mass times acceleration),
the principle that the world did not come into existence five minutes ago,
the principle that one cannot know certain kinds of facts, e.g., facts about
objects at a distance from one, unless one has or has had evidence. These
principles play several different roles; but in one respect they are alike.
They share the characteristic that no isolated experiment (I cannot think
of a better phrase than 'isolated experiment' to contrast with 'rival theory')
can overthrow them. On the other hand, most of these principles can be
overthrown if there is good reason for overthrowing them, and such good
reason would have to consist in the presentation of a whole rival theory
embodying the denials of these principles, plus evidence of the success
of such a rival theory. Any principle in our knowledge can be revised for

tion," and (b) that once the customary definition has been chosen, the question is an
"empirical" one.
[11] As Mill very clearly states; see *System of Logic*, Ch. V, Secs. 4, 5, 6. As Mill fore-
saw, "There is probably no one proposition enunciated in this work for which a more
unfavorable opinion is to be expected" (than, that is, his denial of the a priori character
of geometrical propositions, notwithstanding the "inconceivability" of their negations).
[12] This is not a historical remark.

374

theoretical reasons; although many principles resist refutation by isolated experimentation. There are indeed some principles (some philosophers of science call them "low-level generalizations") which can be overthrown by isolated experiments, provided the experiments are repeated often enough and produce substantially the same results. But there are many, many principles—we might broadly classify them as "framework principles"—which have the characteristic of being so central that they are employed as auxiliaries to make predictions in an overwhelming number of experiments, without themselves being jeopardized by any possible experimental results. This is the classical role of the laws of logic; but it is equally the role of certain physical principles, e.g., 'f = ma,' and the principles we have been discussing: the laws of Euclidean geometry, and the law 'e = ½mv²,' at the time when those laws were still accepted.

I said that any principle in our knowledge can be revised for theoretical reasons. But this is not strictly correct. Any principle in our knowledge can be revised or abandoned for theoretical reasons unless it is really an analytic principle in the trivial sense in which 'All bachelors are unmarried' is an analytic principle. There are indeed analytic statements in science; and these are immune from revision, except the trivial kind of revision which arises from unintended and unexplained historical changes in the use of language. The point of the preceding discussion is that many principles which have been mistaken for analytic ones have actually a somewhat different role. There is all the difference in the world between a principle that can never be given up by a rational scientist and a principle which cannot be given up by rational scientists merely because of experiments, no matter how numerous or how consistent.

To summarize this discussion of geometry: I think that Hume was perfectly right in assigning to the principles of geometry the same status that he assigned to the principles of arithmetic. I think that in his time the principles of geometry had the same status as the principles of arithmetic. It is not that there is something—"an operational definition" of 'straight line'—which Hume failed to apprehend. The idea that, had he been aware of the "operational definition of straight line" on the one hand and of the "reduction of mathematics to logic" on the other hand, Hume would have seen that geometry is not really so much like arithmetic after all, that geometry is synthetic and arithmetic analytic, seems a crude error. The principle that light travels in straight lines is not a definition of 'straight line': as such, it is hopeless since it contains the geometrical term 'travels.'

The same objection arises if we say a "straight line is defined as the path of a light ray." In this case the definition of 'straight line' uses the topological term 'path.' The principle that light travels in a straight line is simply a law of optics, nothing more or less serious than that. What is often called "interpreting mathematical geometry" is more aptly described as testing the conjunction of geometric theory and optical theory. The implicit standpoint of Hume was that if the conjunction should lead to false predictions, then the optical theories would have to be revised; the geometric theory was analytic. The Reichenbachian criticism is that the geometry was synthetic and the optical theory was analytic. Both were wrong. We test the conjunction of geometry and optics indeed, and if we get into trouble, then we can alter either the geometry or the optics, depending on the nature of the trouble. Before Einstein, geometrical principles had exactly the same status as analytic principles, or rather, they had exactly the same status as all the principles that philosophers mistakenly cite as analytic. After Einstein, especially after the general theory of relativity, they have exactly the same status as cosmological laws: this is because general relativity establishes a complex interdependence between the cosmology and the geometry of our universe.

Thus, we should not say that 'straight line' has changed its meaning: that Hume was talking about one thing and that Einstein was talking about a different thing when the term 'straight line' was employed. Rather, we should say that Hume (and Euclid) had certain beliefs about straight lines—not just about mental images of straight lines, but about straight lines in the space in which we live and move and have our being— which were, in fact, unknown to them, false. But we can say all this, and also say that the principles of geometry had, at the time Hume was writing, the same status as the laws of mathematics.

Law-cluster concepts. At this point, a case has been developed for the view that statements expressing the laws of mathematics and geometry and our earlier example 'e $= \frac{1}{2}mv^2$' are not analytic, if by 'an analytic statement' one means a statement that a rational scientist can never give up. It remains to show that 'All bachelors are unmarried' *is* an analytic statement in that sense. This is not a trivial undertaking: for the "shocking" part of Quine's thesis is that there are no analytic statements in this sense—that all of the statements in our conceptual system have the character that I have attributed variously to the laws of logic, the laws of the older geometry at the time when they were accepted, and certain physi-

376

cal principles. But before considering this question, there are certain possible objections against the account just given which must be faced. The objections I have in mind are two. (1) It may be argued, especially in connection with logical principles, that revision of these principles merely amounts to a change in the meaning of the constituent words. Thus, logical principles are not *really* given up; one merely changes one's language. (2) It may be held that the case of the principle 'e = ½mv²' merely shows that we were able to "change our definition of 'kinetic energy,'" and *not* that a principle which was at one time definitional or stipulative could be later abandoned for reasons not substantially different from the reasons given for abandoning certain principles which philosophers would classify as synthetic.

The first objection I have discussed elsewhere.[13] The main point to be made is this: the logical words 'or,' 'and,' 'not' have a certain core meaning which is easily specifiable and which is *independent* of the principle of the excluded middle. Thus, in a certain sense the meaning does *not* change if we go over to three-valued logic or to intuitionist logic. Of course, if by saying that a change in the accepted logical principles is tantamount to a change in the meaning of the logical connective, what one has in mind is the fact that changing the accepted logical principles will affect the global use of the logical connectives, then the thesis is tautological and hardly arguable. But if the claim is that a change in the accepted logical principles would amount *merely* to redefining the logical connectives, then, in the case of intuitionist logic, this is demonstrably false. What is involved is the acceptance of a whole new network of inferences with profound systematic consequences; and it is a philosophical sin to say, even indirectly by one's choice of terminology, that this amounts to no more than stipulating new definitions for the logical connectives. A change in terminology never makes it impossible to draw inferences that could be validly drawn before; or, if it does, it is only because certain words are missing, which can easily be supplied. But the adoption of intuitionist logic as opposed to "classical" logic amounts to systematically forswearing certain classically valid inferences. Some of these inferences can be brought in again by redefinition. But others, inferences involving certain kinds of nonconstructive mathematical entities, are really forsworn in any form. To assimilate the change from one system of logic to another to the change that would be made if we were to use

[13] "Three-Valued Logic," *Philosophical Studies*, 8:73–80 (1957).

Hilary Putnam

the noise 'bachelor' to stand for 'unmarried woman' instead of 'unmarried man' is assimilating a mountain to a molehill. There is a use of the term 'meaning' according to which any change in important beliefs may be said to change the "meaning" of some of the constituent concepts. Only in this fuzzy sense may it be said that to change our accepted logical principles would be to change the "meaning" of the logical connectives. And the claim that to change our logical system would be *merely* to change the meaning of the logical connectives is just false. With respect to the second objection, there are some similar remarks to be made. Once again, to speak of Einstein's contribution as a "redefinition" of 'kinetic energy' is to assimilate what actually happened to a wholly false model.

Leibniz worried about the fact that statements containing a proper name as subject term seem never to be analytic. This seemed to be absurd, so he concluded that *all* such statements must be analytic—that is, that they must all follow from the nature of what they speak about. Mill took the different tack of denying that proper names connote; but this leaves it puzzling that they mean anything at all. Similarly, philosophers have wondered whether any statement containing the subject term 'man' is really analytic. Is it analytic that all men are rational? (We are no longer so happy with the Aristotelian idea that a necessary truth can have exceptions.) Is it analytic that all men are featherless? Aristotle thought not, thus displaying a commendable willingness to include our feathered friends, the Martians (if they exist), under the name 'man.' Suppose one makes a list of the attributes P_1, P_2 . . . that go to make up a normal man. One can raise successively the questions "Could there be a man without P_1?" "Could there be a man without P_2?" and so on. The answer in each case might be "Yes," and yet it seems absurd that the word 'man' has no meaning at all. In order to resolve this sort of difficulty, philosophers have introduced the idea of what may be called a *cluster concept*. (Wittgenstein uses instead of the metaphor of a "cluster," the metaphor of a rope with a great many strands, no one of which runs the length of the rope.) That is, we say that the meaning in such a case is given by a cluster of properties. To abandon a large number of these properties, or what is tantamount to the same thing, to radically change the extension of the term 'man,' would be felt as an arbitrary change in its meaning. On the other hand, if most of the properties in the cluster are present in any single case, then under suitable circumstances we should be inclined to say that what we had to deal with was a man.

In analogy with the notion of a cluster concept, I should like to introduce the notion of a *law-cluster concept*. Law-cluster concepts are constituted not by a bundle of properties as are the typical general names like 'man' and 'crow,' but by a cluster of laws which, as it were, determine the identity of the concept. The concept 'energy' is an excellent example of a law-cluster concept. It enters into a great many laws. It plays a great many roles, and these laws and inference roles constitute its meaning collectively, not individually. I want to suggest that most of the terms in highly developed science are law-cluster concepts, and that one should always be suspicious of the claim that a principle whose subject term is a law-cluster concept is analytic. The reason that it is difficult to have an analytic relationship between law-cluster concepts is that such a relationship would be one more law. But, in general, any one law can be abandoned without destroying the identity of the law-cluster concept involved, just as a man can be irrational from birth, or can have a growth of feathers all over his body, without ceasing to be a man.

Applying this to our example—'kinetic energy' = 'kinetic' + 'energy'— the kinetic energy of a particle is literally the energy due to its motion. The extension of the term 'kinetic energy' has not changed. If it had, the extension of the term 'energy' would have to have changed.[14] But the extension of the term 'energy' has not changed. The forms of energy and their behavior are the same as they always were, and they are what physicists talked about before and after Einstein. On the other hand, I want to suggest that the term 'energy' is not one of which it is happy to ask, What is its intension? The term 'intension' suggests the idea of a single defining character or a single defining law, and this is not the model on which concepts like energy are to be construed. In the case of a law-cluster term such as 'energy,' any one law, even a law that was felt to be definitional or stipulative in character, can be abandoned, and we feel that the identity of the concept has, in a certain respect, remained.[15] Thus, the conclusions of the present section still stand: A principle in-

[14] Kinetic energy is only one of several kinds of energy, and can be transformed into other kinds (and vice versa). Thus an adequate physical theory cannot change the meaning of the term "kinetic energy" without changing the meaning of the term "energy," without giving up the idea that "kinetic energy" is literally a kind of energy.

[15] Even the conservation law has sometimes been considered to be in doubt (in the development of quantum mechanics)! Yet it was the desire to preserve this law which led to the changes we have been discussing. In one context the law of the conservation of energy can thus serve to "identify" energy, whereas in another it can be the Hamiltonian equations of particular systems that do this.

Hilary Putnam

volving the term 'energy,' a principle which was regarded aş definitional, or as analytic, if you please, has been abandoned. And its abandonment cannot be explained away as mere "redefinition" or as change in the meaning of 'kinetic energy,' although one might say that the change in the status of the principle has *brought about* a change in the meaning of the term 'kinetic energy' in one rather fuzzy sense of 'meaning.'[16] It is important to see that the principle '$e = \frac{1}{2}mv^2$' might have been mistaken to have exactly the same nature as 'All bachelors are unmarried.' But 'All bachelors are unmarried' cannot be rejected unless we change the meaning of the word 'bachelor' and not even then unless we change it so radically as to change the *extension* of the term 'bachelor.' In the case of the terms 'energy' and 'kinetic energy,' we want to say, or at any rate *I* want to say, that the meaning has not changed enough to affect "what we are talking about"; yet a principle superficially very much like 'All bachelors are unmarried' has been abandoned. What makes the resemblance only superficial is the fact that if we are asked what the meaning of the term 'bachelor' is, we can *only* say that 'bachelor' means 'unmarried man,' whereas if we are asked for the meaning of the term 'energy,' we can do much more than give a definition. We can in fact show the way in which the use of the term 'energy' facilitates an enormous number of scientific explanations, and how it enters into an enormous bundle of laws.

The statement '$e = \frac{1}{2}mv^2$' is the sort of statement in physical theory that is currently called a "definition." That is to say, it can be taken as a definition, and many good authors did take it as a definition. Analyticity is often defined as "truth by definition," yet we have just seen that '$e = \frac{1}{2}mv^2$' is not and was not analytic, if by an analytic statement one means a statement that no one can reject without forfeiting his claim to reasonableness.

At this point one may feel tempted to agree with Quine. If even "definitions" turn out to be revisable in principle—and not in the trivial sense that arbitrary revision of our use of *noises* is always possible—then one might feel inclined to say that there is *no* statement which a rational man must hold immune from revision. I shall proceed to argue that this is wrong, but those who agree with me that this is wrong have often overlooked the fact that Quine can be wrong in his most "shocking" thesis

---

[16] The "fuzziness" is evidenced by the fact that although one can say that 'kinetic energy' has a new meaning, one cannot say that 'kinetic' has a new meaning, or that 'energy' has a new meaning, or that 'kinetic energy' is an idiom.

and still right about very important and very pervasive epistemological issues. To give a single example, I agree with Quine that in that context of argument which is defined by questions of necessity, factuality, of linguistic or nonlinguistic character, there is no significant distinction to be drawn between, say, the principle of the excluded middle and the principle that $f = ma$; and this is not to say that the law '$f = ma$' is analytic. (Of course we can imagine a physics based on $f = m^2a$, if we retain the identity of gravitational and inertial mass!) Nor is it to say that the laws of logic are "synthetic," if the paradigm for a synthetic sentence is 'There is a book on this table.' But still there are truths that it could never be rational to give up, and 'All bachelors are unmarried' is one of them. This thesis will be elaborated in the following section.

The Rationale of the Analytic-Synthetic Distinction

The problem of justification. Let us consider first the question How could one draw the analytic-synthetic distinction as a formal distinction in connection with at least some hypothetical formalized languages? If the inventor of a formalized language singles out from all his postulates and rules a certain subset (e.g., 'L-Postulates," "Meaning Postulates," and "logical axioms") and says that the designated statements, statements in the subset, are not to be given up, then these statements may be reasonably called "analytic" in that language. In the context of formal reconstruction, then, this is the first model of analyticity that comes to mind. We draw an analytic-synthetic distinction formally only in connection with formalized languages whose inventors list some statements and rules as "Meaning Postulates." That is, it is stipulated that to qualify as correctly using the language one must accept *those* statements and rules. There is nothing mysterious about this. A formal language has, after all, an inventor, and like any human being, he can give commands. Among the commands he can issue are ones to the effect that "If you want to speak my language, then do thus and so." If his commands have an escape clause, if he says, "Accept these statements unless you get into trouble, and then make such-and-such revisions," then his language is hardly one with respect to which we can draw a formal analytic-synthetic distinction. But if he says that certain statements are "to be accepted no matter what," then those statements in that language are true by stipulation, true by *his* stipulation, and that is all we mean when we say that they are "analytic" (in this model).

381

Hempel has proposed an answer to this sort of move. His answer is this: if by an analytic statement one means one which is not to be given up, then in science there are no such statements. Of course, an individual might invent a language and rule that in that language certain statements are not to be given up; but this is of no philosophic interest whatsoever, unless the language constructed by this individual can plausibly be regarded as reconstructing some feature which actually exists in ordinary unreconstructed scientific activity.

This brings us to our second question: If an artificial language in which a formalized analytic-synthetic distinction can be drawn is one in which there are rules of the form "Do not give up S under any circumstances," then what justification could there be for adopting such a language?

Certain philosophers have seen that the notion of a rule, in the sense of an *explicit* rule or explicit stipulation, is sufficiently clear to be worked with (Quine does not at all deny this), and they propose to define analytic statements as statements which are true by stipulation. Against this, there is Quine's remark that in the history of science a statement is often "true by stipulation" at one moment, but later plays a role which is in no way different from the role played by statements which enter the body of accepted truths through more direct experimental inquiry. Stipulation, Quine says, is a trait of historical events, not a "lingering trait" of the statements involved.

Philosophers who regard "true by stipulation" as explicating analyticity, and who take "true by stipulation" in its literal sense, that is to say, who mean by "stipulations" explicit stipulations, miss several points. In the first place, analytic statements in a natural language are not usually true by stipulation in anything but a metaphorical sense. "True by stipulation" is the nature of analytic statements only in the model. And even if we confine ourselves to the model and ignore the existence of natural languages, there is still the question What is the point of the model? But this is the question: Why should we hold certain truths immune from revision?

Suppose we can show that if we were to adopt an "official formalized language," it would be perfectly rational to incorporate into its construction certain conventions of the type described? Then I think we would have resolved the problem raised by Quine. Quine does not deny that some people may in fact hold some statements immune from revision; what he denies is that science does this, and his denial is not merely a descriptive denial: he doesn't think that science ought to do this. Thus the problem *really* raised by Quine is this: Once we have managed to make our own Quine's insight into the monolithic character of our conceptual system, how can we see why there should be any exceptions to this monolithic character? If science is characterized by interdependence of its principles and by the fact that "revision may strike anywhere," then why should any principles be held immune from revision? The question at the moment is not What is the nature of the analytic-synthetic distinction? but rather Why ought there to be an analytic-synthetic distinction?

Rationale. The reply that I have to offer to the question of the rationale of the analytic-synthetic distinction, and of strict synonymy within a language, is this: First of all, the answer to the question Why should we have analytic statements (or strict synonymies[17]) in our language? is, in essence, Why not? or more precisely, It can't hurt. And, second, the answer to the derivative question How do you know it can't hurt? is I use what I know. But it is obvious that both of these answers will need a little elaborating.

The first answer should, I think, be clear. There are obvious advantages to having strict synonyms in a language. Most important, there is the advantage of *brevity*. Also, there is the question of *intelligibility*. If some of the statements in a language are immune from revision and if some of the rules of a language are immune from revision, then linguistic usage with respect to the language as a whole is to a certain extent frozen. Now, whatever disadvantages this freezing may have, there is one respect in which a frozen language is very attractive. Different speakers of the same language can to a large extent understand each other better because they can predict in advance at least some of the uses of the other speaker.

Thus, I think we can see that if we are constructing a language, then there are some prima-facie advantages to having "fixed points" in that language. Hence the only real question is Why *not* have them? Quine, I believe, thinks that there is a reason why we should not have them. No matter what advantages in intelligibility and uniformity of usage might accrue, Quine is convinced that it would block the scientific enterprise to declare *any* statement immune from revision. And it may seem that I have provided Quine with more than sufficient ammunition. For in-

[17] The close connection between synonymy and analyticity is pointed out by Quine in "Two Dogmas."

stance, someone might have proposed, "Let's make the statement 'kinetic energy $= \frac{1}{2}mv^2$' analytic. It will help to stabilize scientific usage." And accepting this proposal, which might have seemed innocuous enough, would not have been very happy. On my own account, we would have been mistaken had we decided to hold the statement 'kinetic energy $= \frac{1}{2}mv^2$' immune from revision. How can we be sure that we will not be similarly mistaken if we decide to hold *any* statement immune from revision?

In terms of the conceptual machinery developed above, the reason that we can safely decide to hold 'All bachelors are unmarried' immune from revision, while we could not have safely decided to hold 'kinetic energy $= \frac{1}{2}mv^2$' immune from revision, is that 'energy' is a law-cluster term, and 'bachelor' is not. This is not to say that there are no laws underlying our use of the term 'bachelor'; there are laws underlying our use of any words whatsoever. But it is to say that there are no exceptionless laws of the form 'All bachelors are . . .' except 'All bachelors are unmarried,' 'All bachelors are male,' and consequences thereof. Thus, preserving the interchangeability of 'bachelor' and 'unmarried man' in all extensional contexts can never conflict with our desire to retain some other natural law of the form 'all bachelors are . . .'

This cannot happen because bachelors are a kind of synthetic "class." They are not a "natural kind" in Mill's sense. They are rather grouped together by ignoring all aspects except a single legal one. One is simply not going to find any laws, except complex statistical laws depending on sociological conditions, about such a class. Thus, it cannot "hurt" if we decide always to preserve the law 'All bachelors are unmarried.' And that it cannot hurt is all the justification we need; the positive advantages are obvious.

As remarked, there may be *statistical* laws, dependent on sociological conditions, concerning bachelors. But these cannot be incompatible with 'All bachelors are unmarried men.' For the truth of a statistical law, unlike that of a deterministic law, is not affected by slight modifications in the extension of a concept. The law '99 per cent of all A's are B's,' if true, remains true if we change the extension of the concept A by including a few more objects or excluding a few objects. Thus, making *slight* changes in the extension of the term 'bachelor' would not affect any statistical law about bachelors; but by exactly the same token, neither would *refusing* to make such changes. And if the statistical law held true only provided we were willing to make a large change in the extension or putative extension of the term 'bachelor,' then we would certainly reject the statistical law.

Let us consider one objection. I have maintained that there are no exceptionless laws containing the term 'bachelor.' But this statement is surely a guess on my part. Let us suppose that my "guess" is wrong, and that there are exceptionless laws about bachelors. Let us suppose for instance that all bachelors share a special kind of neurosis universal among bachelors and unique to bachelors. Not to be too farfetched, let us call it "sexual frustration." Then the statement 'All bachelors suffer from sexual frustration, and only bachelors suffer from sexual frustration' would express a genuine law. This law could still not provide us with a *criterion* for distinguishing bachelors from nonbachelors, unless we were good at detecting this particular species of neurosis. It is alleged that some primitive peoples can in fact do this by smell; but let us make a somewhat more plausible assumption, in terms of contemporary mores. Let us suppose that we all mastered some form of super psychoanalysis; and let us suppose that we all became so "insightful" that we should be able to tell in a moment's conversation whether someone suffered from the neurosis of "sexual frustration" or not. Then this law would indeed constitute a criterion for bachelorhood, and a far more convenient criterion than the usual one. For one cannot employ the usual criterion without asking a man a somewhat personal question concerning his legal status; whereas, in our hypothetical situation, one would be able to determine by a quick examination of the man's conversation whether he was a bachelor or not, no matter what one conversed about. Under such circumstances, possession of the neurosis might well become the dominant criterion governing the use of the word. Then what should we say, if it turned out that a few people had the neurosis without being bachelors? Our previous stipulation that 'bachelor' is to be synonymous with 'unmarried man' might well appear inconvenient!

The point of this fable is as follows: Even if we grant that 'bachelor' is not *now* a law-cluster term, how can we be *sure* that it will never become such a term? This leads to my second answer, and to a further remark, "I use what I know." It is logically possible that all bachelors should have a certain neurosis and that nobody else should have it; it is even possible that we should be able to detect this neurosis at sight. But, of course, there is no such neurosis. This I *know* in the way that I know most negative propositions. It is not that I have a criterion for as yet undiscovered neuroses, but simply that I have no good reason to suppose that there might be such a neurosis. And in many cases of this kind, *lack* of any good reason for supposing existence is itself the very best reason for supposing nonexistence.

In short, I regard my "guess" that there are no exceptionless laws about bachelors as more than a guess. I think that in a reasonable sense we may say that this is something that we *know*. I shall not press this point. But *bachelor* is not now a law-cluster concept; I think we can say that, although it is *logically* possible that it might become a law-cluster concept, in fact it will not.

Let us summarize the position at this point: I have suggested that the statement 'All bachelors are unmarried' is a statement which we might render true by stipulation, in a hypothetical formalized language. I have argued that this stipulation is convenient, both because it provides us with one more "fixed point" to help stabilize the use of our hypothetical language, and because it provides us with an expression which can be used instead of the somewhat cumbersome expression 'male adult human being who has never in his life been married'; and I have argued that we need not be afraid to accept these advantages, and to make these stipulations, because it can do no harm. It can do no harm because *bachelor* is not a law-cluster concept. Also it is not independently "defined" by standard examples, which might only contingently be unmarried men. I have admitted that my knowledge (or "state of pretty-sureness") that 'bachelor' will not become a law-cluster term is based upon what we might call, in a very broad sense, empirical argumentation. That *there are no exceptionless laws containing the term 'bachelor'* is empirical in the sense of being a fact about the world; although it is not empirical in the sense of being subject to confrontation with isolated experiments. More precisely: it occupies the anomalous position of being falsifiable by isolated experiments (since isolated experiments could verify an empirical generalization which *would* constitute a "law about all bachelors"); but it could not be verified by isolated experiments. One cannot examine a random sample of *laws*, and verify that they are all not-about-bachelors. But the statement is empirical, at least in the first sense, and it is "synthetic" to the extent that it is revisable in principle. So my position is this: a "synthetic" statement, a statement which could be revised in principle, may serve as a warrant for the decision that another statement should not be revised, no matter what. One may safely hold certain statements immune from revision; but *this* statement is itself subject to certain risks.

But there is no real paradox here at all. To say that an intention is to do something permanently is not the same as saying that the intention is permanent. To marry a woman is to legally declare an intention to remain wedded to her for life; although the bride and groom know perfectly well that there exists such an institution as divorce, and that they may avail themselves of it. The existence of divorce does not change the fact that the legal and declared intention of the persons getting married is to be wedded for life. And this is the further remark that I wish to make in connection with my second answer. It is perfectly rational to make stipulations to the effect that certain statements are never to be given up, and those stipulations remain stipulations to that effect, notwithstanding the fact that under certain circumstances the stipulations *themselves* might be given up.

All of this may sound like a bit of sophistry, if one forgets that we are still in the context of formalized languages. Thus, if one has in mind "implicit stipulations" and natural language, one might feel tempted to say: "What is the difference between having a stipulation to the effect that every statement can be revised, and having a stipulation to the effect that certain statements are never to be revised, if the latter stipulations are themselves always subject to revision?" But in connection with formalized languages, there is all the difference in the world. The rule "Let every statement be subject to revision" is not sufficiently precise to be a formal rule. It would have to be supplemented by further rules determining what revisions to make, and in what order. And there is all the difference in the world between making a decision in accordance with a pre-established plan, and making the decision by "getting together" and doing whatever seems most cogent in the light of the circumstances at the moment and the standards or codes we see accepted at the moment. The first case would arise in connection with a language in which Quine's ideas concerning priorities and centralities had been formalized—a language in which any statement may be given up and in which there are rules telling one which statements to give up first and under what circumstances. Such a language could in principle be constructed. But compare the case of a scientist who is in difficulties, and who resolves his difficulties by using a predetermined rule, with the following case: we imagine that we have a formalized language in which 'All bachelors are unmarried' is a "meaning

387

postulate." We further imagine, as in our "fable," that all bachelors suffer from a neurosis and that only bachelors suffer from that particular neurosis. Also we suppose that the neurosis is detectable at sight and that it is used as the dominant criterion. Then it is discovered that one person or a very few people have the neurosis although they are married. The question might then arise as to which would be more convenient: to preserve 'All bachelors are unmarried' or to get together and modify the rules of the language. Contrast the procedure which would be employed if the latter alternative were the one adopted, with the procedure of settling the question in accordance with a predetermined plan. There would be, let us say, a convention at which some would argue that it is better to preserve the rules that were agreed upon for the language, and to give up the psychological law that had been thought to hold without exception; there might be others who would argue that the new use of the term 'bachelor' was so standard that it would be simpler to grace the new use with the hallmark of legality and to change the rules of the language. In short, the question would be settled by informal argument.

Thus, at the level of formalized languages, there *is* a difference, and a rather radical difference between these different systems: a formal language which can be described as having rules to the effect that every statement may be revised, and a formal language having rules to the effect that certain statements are never to be revised—notwithstanding the fact that, even if one employs a formal language of the second kind, one retains the option of later altering or abandoning it. And even if one uses a system of the first kind, a "holistic" system of the sort Quine seems to envisage, there is still the possibility that one might find it desirable to revise the rules determining the nature and order of revisions, when they are to be made—the centralities and priorities of this system. And the same difference mirrors itself in the difference between those questions which one settles in accordance with the antecedently established rules and those questions which one settles by informal argument when they arise.

In short: if we think in terms of people using formalized languages, then we have to distinguish between the things that are done inside the language in accordance with whatever rules and regulations may have been previously decided upon and published, and the informal argumentation and discussion that takes place outside of the language, and which perhaps leads to a decision, in its turn to be duly formalized, to alter the language. This distinction is not the same as the analytic-synthetic distinction, but it is deeply relevant to it. If we use the model of people employing formalized languages, then we have to imagine those people as deciding upon and declaring certain rules. And it is perfectly rational in human life to make a rule that something is always to be done; and the rule is no less a rule that something is always to be done on account of the fact that the rule itself may someday be abandoned.

There are a host of examples: for instance, it is a rule of etiquette that one is not to address a person to whom one has never been introduced by his first name (with a few exceptions). The rule may someday be changed. But that does not change the fact that the present rule is to the effect that this is to be done under *all* circumstances. In the same way, a rational man may perfectly well adopt a rule that certain statements are never to be given up: he does not forfeit his right to be called reasonable on account of what he does, and he can give plenty of good reasons in support of his action.

The Analytic-Synthetic Distinction in Natural Language

The formal language model. The foregoing discussion is characterized by an air of fictionality. But this does not obliterate its relevance to Quine's difficulties. Quine does not deny that there may be some statements which some individuals will never give up. His real contention is that there are no statements which *science* holds immune from revision. And this is not a descriptive judgment; judgments by philosophers containing the word 'science' almost never are. What Quine really means is that he cannot see why science ought to hold any statements immune from revision. And this is the sort of difficulty that one may well resolve by telling an appropriate fable.

Still we are left with the problem of drawing an analytic-synthetic distinction in natural language; and this is a difficult problem. Part of the answer is clear. We commonly use formalized objects to serve as models for unformalized objects. We talk about a game whose rules have never been written down in terms of a model of a game whose rules have been agreed upon and codified, and we talk about natural languages in terms of models of formal languages; and, if a formal language means a "language whose rules are written down," then we have been doing this for a long time, and not just since the invention of symbolic logic. The concept of a rule of language is commonly used by linguists in describing even

389

Hilary Putnam

the unwritten languages of primitive peoples, just as the concept of a rule of social behavior is used by anthropologists. Such reference is sometimes heavily disguised by current jargon, but is nevertheless present. For instance, if a linguist says: "The pluralizing morphophoneme —s has the zero allophone after the morpheme *sheep*," what he is saying is that it is a rule of English that the plural of 'sheep' is 'sheep' and not 'sheeps.' And his way of saying this is not so cumbersome either: he would not really write the sentence I just quoted, but would embed the information it contains in an extremely compact morphophoneme table.

Thus I think that we may say that the concept *rule of language*, as applied to natural language, is an "almost full-grown" theoretical concept. Linguists, sent out to describe a jungle language, describe the language on the model of a formal language. The elements of the *model* are the expressions and rules of a formal language, that is, a language whose rules are explicitly written down. The corresponding elements in the real world are the expressions of a natural language and certain of the dispositions of the users of that language. The model is not only a useful descriptive device, but has genuinely explanatory power. The distinction, at present very loosely specified, between a rule of language and a *mere* habit of the speakers of the language is an essential one. Speakers of English (except very small speakers of English) rarely use the word 'sheeps.' Speakers of English rarely use the word 'otiose.' But someone who uses the word 'sheeps' is said to be speaking incorrectly; whereas someone who uses the word 'otiose' is only using a rare word. That we behave differently in the two cases is explained, and it is a genuine explanation, by saying that it is a rule of English that one is to use 'sheep' as the plural of 'sheep,' and it is not a rule of English that one is not to use the word 'otiose'; it is just that most people do not know *what* the rule for using the word 'otiose' is at all, and hence do not employ it.

But all this will not suffice. True, we have a model of natural language according to which a natural language has "rules," and a model with some explanatory and predictive value, but what we badly need to know are the respects in which the model is exact, and the respects in which the model is misleading. For example, in many circumstances it is extremely convenient to talk about electron currents on the model of water flowing through a pipe; but physical scientists know very well in which respects this model holds exactly and in which respects it is extremely misleading. The same can hardly be said in the case just described—the case wherein

390

we employ a formal language as a model for a natural language. The difficulty I have in mind is not the difficulty of determining what the rules of a natural language are. The art of describing a natural language in terms of this kind of model is one that is relatively well developed; and linguists are aware that the correspondence between this kind of model and a given natural language is not unique: there are alternative "equally valid descriptions." The dispositions of speakers of a natural language are not rules of a formal language, the latter are only used to represent them in a certain technique of representation; and the difficulty lies in being sure that other elements of the model, e.g., the sharp analytic-synthetic distinction, correspond to anything at all in reality.

To give only one example: I argued above, and it was a central part of the argument, that there is a clear-cut difference between solving a problem by relying on a pre-established rule, and solving it by methods construed on the spot. But one might wonder whether the distinction is so sharp if the pre-established rule is only an *implicit* rule to begin with. It is clear that there is a difference between stipulations allowing for revisions and stipulations prohibiting revisions, but themselves always subject to informal revision. But is it so clear that there is such a distinction if the stipulations are themselves informal and "implicit"? In view of this difficulty, and other related difficulties, it seems to me that we must look at natural language directly, and try to draw the analytic-synthetic distinction without relying on the formal language model, if we are to be sure that it exists at all.

The nature of the distinction in natural language. The statements which satisfy the criteria presented below are a *fundamental subset* of the totality of analytic statements in the natural language. They are the so-called "analytic definitions," e.g., 'Someone is a bachelor if and only if he is an unmarried man.' Other statements may be classified as "analytic," although they do not satisfy the criteria, because they are consequences of statements which do satisfy the criteria. The older philosophers recognized a related though different distinction by referring to "intuitive" and "demonstrative" truths. The distinction had a point: there is a difference, even in our formal model, between those statements whose truth follows from *direct* stipulation and statements whose truth follows from the fact that they are *consequences* of statements true by direct stipulation. The latter statements involve not only arbitrary stipulation but also logic.

391

Nevertheless, the term 'intuitive' has bad connotations. And because of these bad connotations, philosophers have been led not to reformulate the distinction between intuitive and demonstrative truths but to abandon it. So today the fashion is to lump together the analytic statements which would traditionally have been classified as intuitive with all their consequences, and to use the word 'analytic' for the whole class. *The criteria to be presented do not, however, apply equally well to the whole class, or even to all the "intuitive" analytic truths, but to a fundamental subset. This fundamental subset is, roughly speaking, the set of analytic definitions*; or less roughly, it is the set of analytic definitions which are also "intuitive" and not "demonstrative."

In short, I shall present criteria which are intended to show what is unique or different about certain analytic statements. Such criteria do not constitute a definition but one might obtain a definition, of a rough and ready sort, from them: an analytic statement is a statement which satisfies the criteria to be presented, or a consequence of such statements, or a statement which comes pretty close to satisfying the criteria, or a consequence of such statements. The last clause in this "definition" is designed to allow for the fact that there are some "borderline" cases of analyticity, e.g., 'Red is a color.' However, it is not a very important point that the analytic-synthetic distinction *is* afflicted with "borderline fuzziness." The trouble with the analytic-synthetic distinction *construed as a dichotomy* is far more radical than mere "borderline fuzziness." Yet, there are borderline cases; and the reason for their existence is that the analytic-synthetic distinction is tied to a certain model of natural language and correspondence between the model and the natural language is not unique. To say that it is not unique is not, however, to say that it is arbitrary. Some statements in natural language really are analytic; others may be construed as analytic; still others really are synthetic; others may be construed as synthetic; still other statements belong to still other categories or may be construed as belonging to still other categories.

The following are the criteria in question:

> (1) The statement has the form: "Something (Someone) is an A if and only if it (he, she) is a B," where A is a single word.[18]

[18] The requirement that A be a single word reflects the principle that the meaning of a whole utterance is a function of the meanings of the individual words and grammatical forms that make it up. This requirement should actually be more complicated

    (2) The statement holds without exception, and provides us with a *criterion* for something's being the sort of thing to which the term A applies.

    (3) The criterion is the only one that is generally accepted and employed in connection with the term.

    (4) The term A is not a "law-cluster" word.

Criteria (1) by itself is surely insufficient to separate analytic definitions from natural laws in all cases. Thus let us examine criteria (2), (3), and (4). A statement of the form "Something is an A if and only if it is a B" provides a criterion for something's being a thing to which the term A applies if people can and do determine whether or not something is an A by *first* finding out whether or not it is a B. For instance, the only generally accepted method for determining whether or not someone is a bachelor, other than putting the question itself, is to find out whether or not the person is married and whether or not he is an adult male. There are of course independent tests for both marital status (consult suitable records) and masculinity.

One objection must be faced at the outset: it might be argued that these criteria are circular in a vicious way, since knowing that the two statements, (a) "Someone is a bachelor if and only if he is an unmarried man," and (b) "Someone is a bachelor if and only if he is an unwed man," provide the same criterion for the application of the term "bachelor" is the same thing as knowing that "unmarried" and "unwed" are *synonyms*. For the present purposes, however, identity of criteria can be construed behavioristically: criteria (say, X and Y) correspond to the same way of ascertaining that a term A applies if subjects who are instructed to use criterion X do the same thing[19] as subjects who are instructed to use criterion Y. Thus, if I were instructed to ascertain whether or not Jones is unmarried, I would probably go up to Jones and ask "Are you married?"—

to take care of words which consist of more than one morpheme and of idioms, but these complications will not be considered here. We can now give another reason why 'Kinetic energy $= \frac{1}{2}mv^2$' was never an analytic statement: its truth did not follow from the meanings of the words 'kinetic' and 'energy.' On the other hand, it would be absurd to maintain that, during its tenure of office, it was an "empirical statement" in the usual sense (subject to experimental test, etc.).

[19] The use of the expression "do the same things" here will undoubtedly raise questions in the minds of certain readers. It should be noted that what is meant is not total identity of behavior (whatever that might be) but the absence of relevant and statistically significant regularities running through the behavior of the one group of subjects and not of the other. Separation of "relevant" from "irrelevant" regularities does not seem difficult in practice, however difficult it might be to "mechanize" our "institutions" in these matters.

and answer "No" to the original question if Jones' answer was "Yes," and vice versa. On any such occasion, I could truthfully say that I "would have done the same thing" if I had been instructed to ascertain whether Jones was "unwed" instead of whether Jones was "unmarried." Thus, in my idiolect,[20] "being an unmarried man" and "being an unwed man" are not two criteria for someone's being a bachelor, but one.

But let us consider a somewhat different type of objection. On what basis are we to rule out the statement "Someone is a bachelor if and only if he is either an unmarried man or a unicorn" as nonanalytic?[21] Here three grounds are relevant: (a) the statement is a linguistically "odd"[22] one, and is not clearly true; (b) the statement would not be generally accepted; (c) people do not ascertain that someone is a bachelor by first finding out that he is either an unmarried man or a unicorn. To take these in turn: (a) The English "or" and "if and only if" are not synonymous with the truth functions "∨" and "≡" of formal logic. Thus it is not even clear that the quoted statement is an intelligible English statement, let alone true. (b) Even if we grant truth, it would not be generally accepted. Many persons would reject it, and others, who might not actually reject it, might decline to accept it (e.g., they might query its intelligibility or express puzzlement). (c) People (other than formal logicians) would certainly deny that they ascertain that someone is a bachelor by first finding out that he is either unmarried or a unicorn. In fine, the quoted statement does not provide a criterion for someone's being a bachelor, in the sense in which 'criterion' is being used here; and it is not a generally accepted criterion for someone's being a bachelor.

Since a good deal of the present discussion depends upon the way in which the word 'criterion' is being used, I should like to emphasize two points. Although sufficient conditions, necessary conditions, etc., are sometimes called "criteria" (e.g., the above "criteria" for analyticity), the

---

[20] An "idiolect" is the speech of a single speaker.

[21] The difficulty here is that the class of bachelors = the sum of the class of bachelors and the class of unicorns (the latter being the null class). What has to be shown is that the so-called "intensional" difference between the two terms 'unmarried man' and 'unmarried man or unicorn' is reflected by our criteria, at least in connection with the definition of 'bachelor.'

[22] The quoted sentence is even ungrammatical, using the term in the sense of Noam Chomsky's Syntactic Structures (The Hague: Mouton and Co., 1957); for its transformational history involves the ungrammatical sentence "Someone is a unicorn." To change the example: "Someone is a bachelor if and only if he is either an unmarried man or eleven feet tall" is grammatical, but pretty clearly false, given the counterfactual force of the ordinary "if and only if."

sense of 'criterion' in which an analytic definition provides a criterion for something's being the sort of thing to which a term applies is a very strong one: (a) the "criteria" I am speaking of are necessary and sufficient conditions for something's being an A; and (b) by means of them people can and do determine that something is an A. For instance, there are various things that we might call indications of bachelorhood: being young, high spirited, living alone. Using these, one can often tell that someone is a bachelor without falling back on the criterion; but the only criterion (satisfying (a) and (b)) by means of which one can determine that someone is a bachelor is the one which is provided by the analytic definition.

Returning now to our main concern, what is the relevance of the four criteria for analyticity? Someone imbued with the view that an analytic statement is simply one which is true by the rules of the language, i.e., one who insists on stating the distinction in terms of a model, instead of discussing the relevance of the model to that vast disorderly mass of human behavior that makes up a natural language, may be wholly dissatisfied with what has been said. I can imagine someone objecting: "What you are saying is that the difference between an analytic principle and a natural law consists in the accidental fact that no laws happen to be known containing the subject term of the analytic principle." That is almost what I am saying. But the emphasis is wrong; and in any case the thing is not so implausible once one has grasped the rationale of analyticity.

In the first place it is not just that there do not happen to be any known principles concerning bachelors other than the principle that someone is a bachelor if and only if he is an unmarried man: it is reasonable to suppose that there do not exist any exceptionless (as opposed to statistical) scientific laws to be discovered about bachelors.[23] And even if there were an exceptionless law about bachelors, it is extremely unlikely that it

---

[23] It has occurred to me that someone might argue that "all bachelors have mass" is an example of an exceptionless "law about bachelors." Even if this were granted, the objection is not serious. In the first place, in deciding whether or not a word is a "law-cluster" word, what we have to consider are not all the laws (including the unknown ones) containing the word, but only those statements which are accepted as laws and which contain the word. It does not even matter if some of these are false: if a word appears in a large number of statements (of sufficient importance, interconnectedness, and systematic import) which are accepted as laws, then in the language of that time it is a "law-cluster" word. And second, if a statement would be accepted as true, but is regarded as so unimportant that it is not stated as a law in a single scientific paper or text, then it can certainly be disregarded in determining whether or not a word is a "law-cluster" word.

would have the form "Someone is a bachelor if and only if . . ."—i.e. that it would provide a *criterion* for someone's being a bachelor.

But still we have to face the questions (1) Why is the exceptionless principle that provides the criterion governing a *one-criterion* concept analytic? (2) What happens if, contrary to our well-founded beliefs and expectations, a large number of exceptionless laws of high systematic import containing the subject term are someday discovered? The second question has already been discussed. If 'bachelor' ever becomes a "law-cluster" word, then we shall simply have to admit that the linguistic character of the word has changed. The word 'atom' is an example of a word which was once a "one-criterion" word and which has become a "law-cluster" word (so that the sentence 'Atoms are indivisible,' which was once used to make an analytic statement, would today express a false proposition).

But to consider the first question: Why is a statement which satisfies the criteria analytic? Well, in the first place, *such a statement is certainly not a synthetic statement in the usual sense*; it cannot be confuted by isolated experiments, or, what amounts to the same thing, it cannot be verified by "induction" in the sense of induction by simple enumeration. To verify or confute a statement of the form 'Something is an A if and only if it is a B' in this way requires that we have *independent* criteria for being an A and for being a B. Moreover, since the subject concept is not a law-cluster concept, the statement has little or no systematic import. In short, there could hardly be *theoretical* grounds for accepting or rejecting it. It is for these reasons that such statements might plausibly be regarded as constituting the arbitrary fixed points in our natural language.

There they are, the analytic statements: unverifiable in any practical sense, unrefutable in any practical sense, yet we do seem to have them. This must always seem a mystery to one who does not realize the significance of the fact that in any rational way of life there must be certain arbitrary elements. They are "true by virtue of the rules of the language"; they are "true by stipulation"; they are "true by implicit convention." Yet all these expressions are after all nothing but metaphors: true statements, but couched in metaphor nonetheless. What is the reality behind the metaphor? The reality is that they are true because they are accepted as true, and because this acceptance is quite arbitrary in the sense that the acceptance of the statements has no systematic consequences beyond those lescribed in the previous section, e.g., that of allowing us to use pairs of expression interchangeably.

Finally, the question as to whether it is *rational* to accept as true statements satisfying the four criteria is easily answered in the affirmative. This is the question as to whether all these statements may reasonably be taken as true in a "sensible" rational reconstruction of our actual language. To discuss this point in detail would involve repeating the argument of the preceding section, since this is just the problem which was treated in that section.

Does the fact that everyone accepts a statement make it rational to go on believing it? The answer is that it *does*, if it can be *shown* that it would be reasonable to render the statement immune from revision by stipulation, *if* we were to formalize our language.

In short, analytic statements are statements which we all accept and for which we do not give reasons. This is what we mean when we say that they are true by "implicit convention." The problem is then to distinguish them from other statements that we accept, and do not give reasons for, in particular from the statements that we *un*reasonably accept. To resolve this difficulty, we have to point out some of the crucial distinguishing features of analytic statements (e.g., the fact that the subject concept is not a law-cluster concept), and we have to connect these features with what, in the preceding section, was called the "rationale" of the analytic-synthetic distinction. Having done this, we can see that the acceptance of analytic statements is *rational*, even though there are no reasons (in the sense of "evidence") in connection with them.

The Necessary and the Contingent

To begin my defense of the analytic-synthetic dichotomy I shall look askance at another dichotomy, the distinction between so-called natural languages, on the one hand, and constructed or "artificial" ones,[1] on the other. The "natural languages," for many philosophers, seem to be those to which explicit *rules* and *stipulations* are not applicable. Now of course one could (stipulatively!) define 'natural language' in such a manner; but, unfortunately, such languages exist only in the pipe dreams of philosophers. For all actual languages contain the apparatus for "explaining"—for specifying—meanings of their terms; and, I contend, statements of a form such as '"—" means . . .' should be regarded as prescriptive or stipulative in force; it is a variant of the naturalistic fallacy to regard them as descriptive assertions or, even, as functioning chiefly to convey information about how anyone, as a matter of fact, does (as opposed to should or is to) use language.[2] And the important point is that this apparatus is put to frequent use, not only by philosophers *and* by scientists, but by the ordinary man using "ordinary language." Furthermore, it either is or ought to be used in all contexts which are of any philosophic interest. I for one sub-

NOTE: For a more detailed exposition of some of the points adumbrated herein, see G. Maxwell, "Meaning Postulates in Scientific Theories," in *Current Issues in the Philosophy of Science*, H. Feigl and G. Maxwell, eds. (New York: Holt, Rinehart, and Winston, 1961), and G. Maxwell and H. Feigl, "Why Ordinary Language Needs Reforming," *Journal of Philosophy*, 58:488 (1961).

Professor Putnam has told me that his stimulating article in this volume resulted from his disagreement with some views of mine contained in an informal research memorandum. Since I have not been entirely convinced by Professor Putnam's essay, it seems not inappropriate to express my largely unregenerate views in this space.

[1] In both instances, I am concerned with *languages* and not, for example, with uninterpreted calculi.

[2] I cannot but believe that this fallacy is at the root of much of what is contained in the famous "Two Dogmas" paper. Quine conceives of statements involving *meaning* talk as statements which are ostensibly confirmable or disconfirmable by data concerning actual linguistic practices. He sees the difficulties involved in such a view and becomes inclined to throw out meanings (and, thus, analyticity) altogether.

scribe to the commonplace that many disputes and problems which are purportedly about substantive matters are actually due to conceptual (linguistic) difficulties. And many of them can be resolved only by stipulations which "tighten up" or modify certain concepts and which "create" certain other, new ones. Thus, many, many contexts of use of "natural languages" are, and many others ought to be, shot through and through with something like what is sometimes called 'rational reconstruction.' In order to avoid the somewhat narrow connotations of this term, I have used the expression 'rational reformation' to refer either to the process or to the product of issuing rules or stipulations concerning meaning and use, whether this be full-fledged formalization, explication, or merely the issuance of one simple rule or stipulation.

Next, I take any sentence of the form 'Sentence S is analytic (in the broad sense) in this context' to be a signal to the effect that, within the rational reformation to be used in the context in question, the meanings of the terms of S are to be taken as such that S is unconditionally assertable. Note that according to this, strictly speaking, it is proper to ascribe analyticity to a sentence only with reference to its being within a rational reformation. However, we might take 'S is analytic (full stop!)' to mean the same (roughly) as 'Within any reasonable reformation, S would be unconditionally assertable.'

In the sense of 'analytic' which I am recommending, all analytic sentences are totally devoid of "factual content." To use Max Black's helpful terminology, analytic sentences are (object language) surrogates for linguistic rules.[3] Of course, a sentence which is analytic in one context may be synthetic in another; in this case, it is always correct to say that some of the terms of the sentence (and, a fortiori, such sentences themselves) change their meaning from context to context.

But although analytic sentences do not even convey, in any straightforward manner, factual information (not even information about linguistic practice), facts are involved here in the following manner. When we give reasons for adopting rules or stipulations which render certain sentences analytic we usually appeal to facts. Thus we may point to the (linguistic) fact that most people have a certain set of habits involving the word 'bachelor,' and argue that if we adopt a stipulation (along with certain others) which renders 'All bachelors are unmarried' analytic, then

[3] M. Black, "Necessary Statements and Rules," *Philosophical Review*, 67:313–341 (1958).

Grover Maxwell

the resulting meaning of 'bachelor' will be such that its use in our rational
reformation will correspond quite closely to its "ordinary use." Or we
might appeal to the (nonlinguistic) fact that no causal chain can be
propagated with a velocity greater than the velocity of light in *vacuo* as
one of the reasons for adopting a stipulation which renders 'The one-
way speed of light is independent of its direction' analytic (and factually
empty). (See Professor Grünbaum's essay in this volume.)

Among the simple kinds of rules we shall consider is the *explicit definition,* which, for the sake of convenience, we shall take to have the form
exemplified as follows:

'Bachelor' is to mean the same as 'unmarried, adult, male human.

Explicit definitions are of relatively minor interest here, partly because of
their relatively transparent role and partly because they are, in principle,
entirely dispensable.

Next, there are what might be called "partial explicit definitions." Just
as in a rational reformation which embodies the explicit definition given
above, the sentence, for example, 'All bachelors are unmarried' becomes
analytic, a reformation could adopt a stipulation which would render 'All
dogs are mammalian' analytic. (Needless to say, any reasonable reformation of ordinary English would adopt both of these particular stipulations.) In the latter case the relevant stipulation might be (roughly),
"Nothing is to be taken as an element of the extension of 'dog' unless it
is mammalian and . . ." (I assume that no satisfactory ("complete")
explicit definition can be given for 'dog.') This stipulation gives some but
not all of the "defining characteristics" of dogs.

Finally, the most interesting group of analytic sentences consists of
those whose corresponding stipulations are neither explicit definitions nor
partial explicit definitions. Examples of such sentences are 'Nothing can
be red and green in the same respect at the same time' and 'The one-way
speed of light is independent of its direction.'

My "analytic sentences" are, of course, quite similar to Carnap's "meaning postulates" and, indeed, I am indebted to him for some of the terminology I shall henceforth adopt.[4]

Now there are certain rules which, it would seem, must either be explicitly adopted or conformed to in actual practice in the use of any conceptual system whatever, i.e., the basic principles of logic. For example,

[4] R. Carnap, "Meaning Postulates," *Philosophical Studies,* 3:65–73 (1953).

t follows from any reasonable meaning which can be given to 'language' and 'conceptual system' that any "system" which does not conform, say, to the principle of noncontradiction is not a language or a conceptual system. Let us, after Carnap, call sentences which are rendered analytic by these rules *L-true*. Roughly speaking, L-true sentences will be those true sentences which contain only logical terms or which contain descriptive terms only vacuously. Let us call all other analytic sentences, *A-true*.[5] Thus 'analytic' in the broad sense includes both A-true and L-true sentences.

Let us turn to a brief explication of the notion of implicit definition as it will be used in this paper. I shall say that a descriptive (i.e., nonlogical) term is implicitly defined (always relative to a given reformation, of course) by the set of all L-independent A-true sentences which contain the term nonvacuously, provided the term is one for which no ("complete") explicit definition is given.[6] In the sense thus adopted, *implicit definition never affords a complete specification of meaning*. Terms so defined either remain open-textured or enjoy whatever additional meaning they may possess by virtue of "ostension."

But have I been playing fast and loose with dangerous terms such as 'meaning' and 'factually empty'? I am sorry, but I believe that, in spite of frequent manhandling, they are perfectly good terms. It is true that no helpful explicit definitions can be given for them, but the same is true for any indispensable term. I have been trying to clarify somewhat the sense in which I use them by giving a part of their implicit definition in the rational reformation of the metalanguage which I am, at present, adumbrating.

Are all accepted lawlike sentences, particularly theoretical postulates and/or "correspondence rules," analytic in a broad sense (W. Sellars)? Or are all (A. Pap—at least, all theoretical postulates) or some (M. Scriven) of them of such a nature that questions concerning their analyticity can-

[5] R. Carnap, "Beobachtungsprache und theoretische Sprache," *Dialectica*, 12:236–248 (1957).

[6] We could have omitted the provisional clause, in which case explicit definition would, in a sense, be a special case of implicit definition. More precisely, the explicitly defined term would be implicitly defined by the object-language surrogate of the explicit definition. For example, 'bachelor' would be implicitly defined by the (one) A-true sentence, 'Something is a bachelor if and only if it is an unmarried, male, adult human.' In fact, if the form 'A $=_{df}$ B' is taken as being in the object language, it seems felicitous to think of it as an abbreviation of the corresponding biconditional, but with the '$=_{df}$' also serving as a signal that the biconditional is A-true.

not be settled and, indeed, cannot be properly asked? Or is the analytic-synthetic distinction itself a "dogmatic," "untenable dualism" (W. V. Quine, M. White) or, at best, a distinction of "overwhelming unimportance" (H. Putnam)? I do not believe that it is necessary to give an affirmative answer to any of these questions. However, it is easy to give extremely plausible arguments for some, or all, of these views, particularly as far as they concern theoretical terms. Since the more interesting of these terms are not explicitly defined and since they cannot enjoy their meaning, in any straightforward way, by virtue of "ostension," it seems that we must say that they are implicitly defined by postulates and "correspondence rules."[7] However, according to the position herein advocated, it is only a proper subset of the postulates-plus-"C-rules" which play a part in this implicit definition; and the members of this subset are all A-true and, thus, factually empty.

But, I shall be asked, is it not the case that in most theories, the postulates are all more or less on a par with each other? Which are the A-true postulates and which are the "factual ones"? The answer is, of course, that it depends upon the context—upon the rational reformation which is used or presupposed. A postulate which is A-true in one context may be contingent in another. This entails that the meanings of some of the terms involved change from context to context. Is such "conceptual instability" of theoretical terms undesirable? I do not think so; at any rate, it is unavoidable. Consider the often-used example concerning the concept of force. Arguing, even, from actual use, we may say that in some contexts, the most reasonable reformation would take 'f = ma' as A-true, in this case, 'force' would mean the same as 'mass times acceleration.' (This is, to be sure, a very simple example, for if 'f = ma' defines 'force,' it defines it explicitly rather than implicitly.) In other contexts, the most reasonable reformation would take 'f = ma' as contingent, and 'force' would "obtain" its (different) meaning from other principles, e.g., Hookes law, restricted to certain appropriately narrow limits.[8]

This approach, it seems to me, is adequate for solving not only many of the problems about the meaning of theoretical terms but, also, some

[7] The (more or less established) use of the word 'rule' here is unfortunate. In almost any reformation, some so-called correspondence rules will be factual sentences and others will be A-true. The same is true, of course, of the "postulates."

[8] See H. Feigl, "Some Major Issues and Developments in the Philosophy of Science of Logical Empiricism," in *Minnesota Studies in the Philosophy of Science*, Vol. I, H. Feigl and M. Scriven, eds. (Minneapolis: University of Minnesota Press, 1956).

of those involving observation terms. Consider the so-called cluster concepts, whether they be observational or theoretical (cf. H. Putnam's helpful notion of "law-cluster concepts" in his essay in this volume). Is 'All lemons are sour' analytic, synthetic, neither, or both?[9] Such a question should be interpreted as a request for a rational reformation—for a stipulation. In one context it might be convenient to adopt one which would render the sentence analytic; in another our reformation might be such that it would be synthetic.

Since any lawlike sentence is a possible candidate for an A-true sentence in *some* rational reformation, does it make sense to talk of the "actual, over-all meaning" of a term, a meaning which is sort of a hybrid of the various meanings it would have in the various possible reasonable reformations involving the term? And, if so, does it follow that all lawlike statements are endowed with a kind of necessity? Interpreted loosely and somewhat metaphorically, the view implicit in these questions contains an important insight. After all, the "meaning which comes to mind" when such terms are mentioned often *is* such a hybrid. But if the "systematic ambiguity"[10] of such terms is overlooked, *it is this very "hybrid meaning" which gives rise to conceptual muddles*, and, at times, outright inconsistencies. And is it really a hardship to eschew such "actual meanings"? After all, any attempt to deal with any problem must occur in a specific context and can utilize exactly one rational reformation at a time; and in a specific reformation, the term(s) in question will have one and only one meaning.

But the question persists How does one actually go about constructing a rational reformation of a specific theory? The actual course to be taken will vary so greatly, depending upon what the specific problem is, that only a few very general considerations can be mentioned here. For any reformation, it will be necessary to presuppose—perhaps to stipulate—that a certain set of sentences, or sentences having one of a number of specified forms, are synthetic (e.g., certain singular, observation sentences of a certain form, involving, perhaps, a specified list of predicates—or kinds of

pp. 3–37, for similar ideas to which I am heavily indebted. See also A. Pap, *The A Priori in Physical Theory* (New York: Kings Crown Press, 1946). For an excellent, detailed discussion of the various ways in which 'f = ma' functions in various contexts, see N. R. Hanson, *Patterns of Discovery* (Cambridge: Cambridge University Press, 1958).

[9] Pap, *loc. cit.*; Scriven, *loc. cit.*; Putnam, *loc. cit.*
[10] Feigl, *loc. cit.*

predicates—and kind words). There will be a rule to the effect that the selection of sentences which are to be taken as A-true must be such that no conjunction of A-true sentences L-implies a synthetic sentence. From this point on, as far as I can see, one must proceed to a large extent by trial and error. As Carnap has pointed out there is (certainly!) no decision procedure for A-truth. One will try to select a set of sentences from the postulates and correspondence rules which, if taken as A-true, will "fix" to a satisfactory extent the meanings of the relevant expressions. As a starting place, there will usually be a few sentences which would, clearly, be taken as A-true in any reasonable reformation, e.g., 'All electrons are electrically charged.' Usually, also, there will be a few which can be selected as contingent without appreciable hesitation. As regards the latter, however, any lawlike sentence is a possible candidate for A-truth. Several alternative reformations may be considered, and, perhaps, one of these may be selected as being the most satisfactory for the purpose at hand. Among the (not necessarily independent) *general* reasons which may be given for selecting a particular reformation are general clarity, simplicity (both "inductive" and conceptual), heuristic and didactic felicity, and, *ceteris paribus*, reasonably close correspondence to actual usage of the relevant terms.

The "necessary," then, includes only the L-true and the A-true statements. These are true by virtue of meaning alone and are completely devoid of any factual content. The "contingent" includes all other descriptive sentences. The ambiguity and vagueness inherent in actual linguistic practice, even in scientific practice, is what gives rise to the difficulties in segregating sentences into these two categories. Why this is so and what can be done about it is what I have tried to show in this brief paper.

Geometry, Chronometry, and Empiricism

1. Introduction

The moment the mathematical discovery of the non-Euclidean geometries had deprived Euclideanism of its claim to *uniqueness*, the triumph of an empiricist account of physical geometry and chronometry seemed assured. Observational findings were presumed capable, at least in principle, of establishing the unique truth of a particular kind of metric geometry. Ironically, however, it soon became clear that the very mathematical discoveries which had heralded the demise of the classical rationalist and Kantian conceptions of geochronometry were a double-edged sword. Critics were quick to marshal these mathematical results against the renascent geometric empiricists who felt emboldened by their victory over the thesis that Euclidean geometry is certifiable a priori as the true description of physical space. The challenge came from several distinct versions of *conventionalism* whose espousal has issued in a proliferous and continuing philosophical debate.

In an endeavor to resolve the issues posed by the several variants of conventionalism, the present essay aims to answer the following question: In what sense and to what extent can the ascription of a particular metric geometry to physical space and the chronometry ingredient in physical theory be held to have an *empirical* warrant? To carry out this inquiry, we must ascertain whether and how empirical findings function restrictively so as to determine a *unique* geochronometry as the true account of the structure of physical space-time.

NOTE: I am indebted to Dr. Samuel Gulden of the department of mathematics, Lehigh University, for very helpful discussions. I have also benefited from conversations with Professor Albert Wilansky of that department, with Professor E. Newman of the University of Pittsburgh, and with Professor Grover Maxwell and other fellow participants in the 1958 and 1959 conferences of the Minnesota Center for the Philosophy of Science. Grateful acknowledgment is made to the National Science Foundation of the United States for the support of research.

It will turn out that the status of the metrics of space and time is profoundly illuminated by (i) the distinction between *factual* and conventional ingredients of space-time theories and (ii) a precise awareness of the *warrant* for deeming *some* of the credentials of these theories to be conventional. Thus, we shall see that whatever the merits of the repudiation of the analytic-synthetic dichotomy and of the antithesis between theoretical and observation terms, it is grievously incorrect and obfuscating to deny as well the distinction between factual and conventional ingredients of sophisticated space-time theories in physics.

Among the writers who have held the empirical status of geochronometry to depend on whether the rigidity of rods and the isochronism of clocks are conventional, we find Riemann [75, pp. 274, 286], Clifford [14, pp. 49–50], Poincaré [61, 62, 63, 64, 65], Russell [81], Whitehead [97, 99, 101], Einstein [21, p. 161; 23, Section 1, pp. 38–40; 26, pp. 676–678], Carnap [8], and Reichenbach [72]. Their assessment of the epistemological status of *congruence* as pivotal has been rejected by Eddington [20, pp. 9–10]. On his view, the thesis that congruence (for line segments or time intervals) is conventional is true only in the trivial sense that "the meaning of every word in the language is conventional" [20, p. 9]: instead of being an insight into the status of spatial or temporal equality, the conventionality of congruence is a semantical platitude expressing our freedom to decree the referents of the *word* 'congruent,' a freedom which we can exercise in regard to any linguistic symbols whatever which have not already been preempted semantically. Thus, we are told that though the conventionality of congruence is merely an unenlightening triviality holding for the language of any field of inquiry whatever, it has been misleadingly inflated into a philosophical doctrine about the relation of spatiotemporal equality purporting to codify fundamental features endemic to the materials of geochronometry. Eddington's conclusion that only the use of the *word* 'congruent' but *not* the ascription of the congruence *relation* can be held to be a matter of convention has also been defended by a cognate argument which invokes the theory of models of uninterpreted formal calculi as follows: (i) physical geometry is a spatially interpreted abstract calculus, and this interpretation of a formal system was effected by semantical rules which are all equally conventional and among which the definition of the relation term 'congruent' (for line segments) does *not* occupy an epistemologically distinguished position, since we are just as free to give a *non*customary interpretation of the abstract sign 'point' as of the sign 'congruent'; (ii) this model theoretic conception makes it apparent that there can be no basis at all for an epistemological distinction *within* the system of *physical* geochronometry between factual statements, on the one hand, and supposedly conventional assertions of rigidity and isochronism on the other; (iii) the factual credentials of physical geometry or chronometry can no more be impugned by adducing the alleged conventionality of rigidity and isochronism than one could gainsay the factuality of genetics by incorrectly affirming the conventionality of the relation of uniting which obtains between two gametes when forming a zygote.

When defending the alternative metrizability of space and time and the resulting possibility of giving either a Euclidean or a non-Euclidean description of the same spatial facts, Poincaré had construed the conventionality of congruence as an epistemological discovery about the status of the relation of spatial or temporal equality. The proponent of the foregoing model theoretic argument therefore indicts Poincaré's defense of the feasibility of *choosing* the metric geometry as amiss, misleading, and unnecessary, deeming this choice to be automatically assured by the theory of models. And, by the same token, this critic maintains that there is just as little reason for our having posed the principal question of this essay as there would be for instituting a philosophical inquiry as to the sense in which genetics *as such* can be held to have an empirical warrant. Whereas the aforementioned group of critics has charged the conventionality of congruence with being only *trivially true*, such thinkers as Russell and Whitehead have strongly opposed that doctrine because they deemed it to be *importantly false*.

These strictures call for critical scrutiny, and their rebuttal is included among the polemical objectives of this essay. Before turning to their refutation, which will be presented in Section 5 below, I shall (a) set forth the rationale of the principal concern of this essay; (b) articulate the meaning of the contention that rigidity and isochronism are conventional; (c) assess the respective merits of the several justifications which have been given for that thesis by its advocates; and (d) develop its import for (i) the epistemological status of geochronometry and of explanatory principles in dynamics, and (ii) alternative formulations of physical theory. Sections 6 and 7 will then come to grips with the articulation of the answer to the principal inquiry of this essay.

2. The Criteria of Rigidity and Isochronism: The Epistemological Status of Spatial and Temporal Congruence

(i) The Clash between Newton's and Riemann's Conceptions of Congruence and the Role of Conventions in Geochronometry.

The metrical comparisons of separate spatial and temporal intervals required for geochronometry involve *rigid* rods and *isochronous* clocks. Is this involvement of a transported congruence standard to which separate intervals can be referred a matter of the mere *ascertainment* of an otherwise intrinsic equality or inequality obtaining among these intervals? Or is reference to the congruence standard essential to the very *existence* of these relations? More specifically, we must ask the following questions:

1. What is the warrant for the claim that a solid rod remains *rigid* under transport in a spatial region *free* from inhomogeneous thermal, elastic, electromagnetic, and other "deforming" or "perturbational" influences? The geometrically pejorative characterization of thermal and other inhomogeneities in space as "deforming" or "perurbational" is due to the fact that they issue in a dependence of the coincidence behavior of transported solid rods on the latter's *chemical composition*, and *mutatis mutandis* in a like dependence of the rates of clocks.

2. What are the grounds for asserting that a clock which is *not perturbed* in the sense just specified is *isochronous*?

This pair of questions and their far-reaching philosophical ramifications will occupy us in Sections 2–5. It will first be in Section 7 that we shall deal with the further issues posed by the logic of making *corrections* to compensate for deformations and rate variations exhibited by rods and clocks respectively when employed geochronometrically under *perturbing* conditions.

In the *Principia*, Newton states his thesis of the *intrinsicality of the metric* in "container" space and the corresponding contention for absolute time as follows:

. . . the common people conceive those quantities [i.e., time, space, place, and motion] under no other notions but from the relation they bear to sensible objects. And thence arise certain prejudices, for the removing of which it will be convenient to distinguish them into absolute and relative, true and apparent, mathematical and common [54, p. 6]. . . . because the parts of space cannot be seen, or distinguished from one another by our senses, therefore in their stead we use sensible measures of them. For

from the positions and distances of things from any body considered as immovable, we define all places; and then with respect to such places, we estimate all motions, considering bodies as transferred from some of those places into others. And so, instead of absolute places and motions, we use relative ones; and that without any inconvenience in common affairs; but in philosophical disquisitions, we ought to abstract from our senses, and consider things themselves, distinct from what are only sensible measures of them. For it may be that there is no body really at rest, to which the places and motions of others may be referred [54, p. 8]. . . . those . . . defile the purity of mathematical and philosophical truths, who confound real quantities with their relations and sensible measures [54, p. 11].
. . .

I. Absolute, true, and mathematical time, of itself, and from its own nature, flows equably[1] without relation to anything external and by another name is called duration: relative, apparent, and common time, is some sensible and external (whether accurate or unequable) measure of duration by the means of motion, which is commonly used instead of true time; such as an hour, a day, a month, a year.

II. Absolute space, in its own nature, without relation to anything external, remains always similar and immovable. Relative space is some movable dimension or measure of the absolute spaces; which our senses determine by its position to bodies; and which is commonly taken for immovable space; such is the dimension of a subterraneous, an aerial, or celestial space, determined by its position in respect of the earth. Absolute and relative space are the same in figure and magnitude; but they do not remain always numerically the same. For if the earth, for instance, moves, a space of our air, which relatively and in respect of the earth remains always the same, will at one time be one part of the absolute space into which the air passes; at another time it will be another part of the same, and so, absolutely understood, it will be continually changed [54, p. 6].
. . . Absolute time, in astronomy, is distinguished from relative, by the equation or correction of the apparent time. For the natural days are truly unequal, though they are commonly considered as equal, and used for a measure of time; astronomers correct this inequality that they may measure the celestial motions by a more accurate time. It may be, that there is no such thing as an equable motion, whereby time may be accurately measured. All motions may be accelerated and retarded, but the flowing of absolute time is not liable to any change. The duration or perseverance of the existence of things remain the same, whether the motions are swift or slow, or none at all: and therefore this duration ought to be distin-

---

[1] It is Newton's conception of the attributes of "equable" (i.e., congruent) time intervals which will be subjected to critical examination and found untenable in this essay. But I likewise reject Newton's view that the concept of "flow" has relevance to the time of physics, as distinct from the time of psychology: see [32, Sec. 4].

guished from what are only sensible measures thereof; and from which we deduce it, by means of the astronomical equation [54, pp. 7–8].

Newton's fundamental contentions here are that (a) the *identity* of points in the physical container space in which bodies are located and of the instants of receptacle time at which physical events occur is autonomous and *not* derivative: physical things and events do *not* first define, by their own identity, the points and instants which constitute their loci or the loci of other things and events, and (b) receptacle space and time each has its own *intrinsic metric*, which exists quite independently of the existence of material rods and clocks in the universe, devices whose function is *at best* the purely epistemic one of enabling us to ascertain the intrinsic metrical relations in the receptacle space and time contingently containing them. Thus, for example, even when clocks, unlike the rotating earth, run "equably" or uniformly, these periodic devices merely *record* but do *not define* the temporal metric. And what Newton is therefore *rejecting* here is a *relational* theory of space and time which asserts that (a) bodies and events first *define* points and instants by conferring their identity upon them, thus enabling them to serve as the loci of other bodies and events, and (b) instead of having an intrinsic metric, physical space and time are metrically amorphous pending explicit or tacit appeal to the bodies which are first to define their respective metrics. To be sure, Newton would *also* reject quite emphatically any identification or isomorphism of absolute space and time, on the one hand, with the *psychological* space and time of conscious awareness whose respective metrics are given by unaided ocular congruence and by psychological estimates of duration, on the other. But one overlooks the essential point here, if one is led to suppose with F. S. C. Northrop (see, for example, [56], pp. 76–77) that the relative, apparent, and common space and time which Newton contrasts with absolute, true, and mathematical space and time are the private visual space and subjective psychological time of immediate sensory experience. For Newton makes it unambiguously clear, as shown by the quoted passages, that his *relative* space and time are indeed that *public* space and time which is defined by the system of *relations* between *material* bodies and events, and *not* the egocentrically private space and time of phenomenal experience. The "sensible" measures discussed by Newton as constitutive of "relative" space and time are those furnished by the public bodies of the physicist, *not* by the unaided ocular congruence of one's eyes or by one's mood-dependent psychological estimates of duration. This interpretation of Newton is fully attested by the following specific assertions of his:

i. "Absolute and relative space are the same in figure and magnitude," a declaration which is incompatible with Northrop's interpretation of *relative* space as "the immediately sensed spatial extension of, and relation between, sensed data (which is a purely private space, varying with the degree of one's astigmatism or the clearness of one's vision)" [56, p. 76].

ii. As examples of merely "*relative*" times, Newton cites any "sensible and external (whether accurate or unequable [nonuniform]) measure of duration" such as "an hour, a day, a month, a year" [54, p. 6]. And he adds that the apparent time commonly used as a measure of time is based on natural days which are "truly unequal," true equality being allegedly achievable by astronomical corrections compensating for the nonuniformity of the earth's rotational motion caused by tidal friction, etc.[2] But Northrop erroneously takes Newton's relative time to be the "immediately sensed time" which "varies from person to person, and even for a single person passes very quickly under certain circumstances and drags under others" and asserts incorrectly that Newton identified with absolute time the public time "upon which the ordinary time of social usage is based."

iii. Newton illustrates *relative* motion by reference to the kinematic relation between a body on a moving ship, the ship, and the earth, these relations being defined in the customary manner of physics *without* phenomenal space or time.

Northrop is entirely right in going on to say that Einstein's conceptual innovations in the theory of relativity cannot be construed, as they have been in certain untutored quarters, as the abandonment of the distinction between physically public and privately or egocentrically sensed space and time. But Northrop's misinterpretation of the Newtonian conception of "relative" space and time prevents him from pointing out that Einstein's philosophical thesis can indeed be characterized epigrammatically as the enthronement of the very relational conception of the space-time framework which Newton sought to proscribe by his use of the terms "*relative*," "apparent," and "common" as philosophically disparaging epithets!

Long before the theory of relativity was propounded, a relational con-

[2] The logical status of the criterion of uniformity implicitly invoked here will be discussed at length in Sec. 4, part (i) below.

ception of the metric of space and time diametrically opposite to Newton's was enunciated by Riemann in the following words:

Definite parts of a manifold, which are distinguished from one another by a mark or boundary are called quanta. Their quantitative comparison is effected by means of counting in the case of discrete magnitudes and by measurement in the case of continuous ones.[3] Measurement consists in bringing the magnitudes to be compared into coincidence; for measurement, one therefore needs a means which can be applied (transported) as a standard of magnitude. If it is lacking, then two magnitudes can be compared only if one is a [proper] part of the other and then only according to more or less, not with respect to how much. . . . in the case of a discrete manifold, the principle [criterion] of the metric relations is already implicit in [intrinsic to] the concept of this manifold, whereas in the case of a continuous manifold, it must be brought in from elsewhere [extrinsically]. Thus, either the reality underlying space must form a discrete manifold or the reason for the metric relations must be sought extrinsically in binding forces which act on the manifold. [75, pp. 274, 286.]

Although we shall see in part (iii) of Section 2 that Riemann was mistaken in supposing that the first part of this statement will bear critical scrutiny as a characterization of continuous manifolds *in general*, he does render here a fundamental feature of the continua of *physical space* and *time*, which are manifolds whose elements, taken singly, all have zero magnitude. And this basic feature of the spatiotemporal continua will presently be seen to invalidate decisively the Newtonian claim of the intrinsicality of the metric in empty space and time. When now proceeding to state the upshot of Riemann's declaration for the spatiotemporal congruence issue before us, we shall *not* need to be concerned with either of the following two facets of his thesis: (1) the inadequacies arising from Riemann's treatment of discrete and continuous types of order as *jointly exhaustive* and (2) the prophetic character of his suggestion that the "reason for the metric relations must be sought extrinsically in the binding forces which act on the manifold" as a precursor of Einstein's original quest to implement Mach's Principle in the general theory of relativity [25; 36, pp. 526–527].

Construing Riemann's statement as applying not only to lengths but also, *mutatis mutandis*, to areas and to volumes of higher dimensions, he gives the following *sufficient* condition for the intrinsic definability and

---

[3] Riemann apparently does not consider sets which are neither discrete nor continuous, but we shall consider the significance of that omission below.

nondefinability of a metric without claiming it to be necessary as well: in the case of a discretely ordered set, the "distance" between two elements can be defined *intrinsically* in a rather natural way by the cardinality of the smallest number of intervening elements.[4] On the other hand, upon confronting the extended continuous manifolds of physical space and time, we see that neither the cardinality of intervals nor any of their other topological properties provide a basis for an *intrinsically* defined metric. The first part of this conclusion was tellingly emphasized by Cantor's proof of the equicardinality of all positive intervals independently of their length. Thus, there is no *intrinsic* attribute of the space between the end points of a line-segment AB, or any relation between these two points themselves, in virtue of which the interval AB could be said to contain the same amount of space as the space between the termini of another interval CD not coinciding with AB. Corresponding remarks apply to the time continuum. Accordingly, the continuity we postulate for physical space and time furnishes a *sufficient* condition for their *intrinsic metrical amorphousness*.[5]

This intrinsic metric amorphousness is made further evident by reference to the axioms for spatial congruence [99, pp. 42–50]. These axioms preempt "congruent" (for intervals) to be a *spatial equality predicate* by

[4] The *basis* for the discrete ordering is not here at issue: it can be conventional, as in the case of the letters of the alphabet, or it may arise from special properties and relations characterizing the objects possessing the specified order.

[5] Clearly, this does *not* preclude the existence of sufficient conditions *other than* continuity for the intrinsic metrical amorphousness of sets. But one *cannot* invoke densely ordered, *denumerable* sets of points (instants) in an endeavor to show that discontinuous sets of such elements may likewise lack an intrinsic metric: even without measure theory, ordinary analytic geometry allows the deduction that the length of a *denumerably* infinite point set is intrinsically zero. This result is evident from the fact that since each point (more accurately, each unit point set or degenerate subinterval) has length zero, we obtain zero as the *intrinsic* length of the densely ordered denumerable point set upon summing, in accord with the usual limit definition, the sequence of zero lengths obtainable by denumeration (cf. Grünbaum [33, pp. 297–298]). More generally, the measure of a denumerable point set is always zero (cf. Hobson [41, p. 166]) unless one succeeds in developing a very restrictive intuitionistic measure theory of some sort.

These considerations show incidentally that space intervals cannot be held to be merely denumerable aggregates. Hence in the context of our post-Cantorean meaning of "continuous," it is actually not as damaging to Riemann's statement as it might seem prima facie that he neglected the denumerable dense sets by incorrectly treating the discrete and continuous types of order as *jointly exhaustive*. Moreover, since the distinction between denumerable and super-denumerable dense sets was almost certainly unknown to Riemann, it is likely that by "continuous" he merely intended the property which we now call "dense." Evidence of such an earlier usage of "continuous" is found as late as 1914 (cf. Russell [80, p. 138]).

assuring the reflexivity, symmetry, and transitivity of the congruence relation in the class of spatial intervals. But although having thus preempted the use of "congruent," the congruence axioms still allow an *infinitude* of *mutually exclusive* congruence classes of intervals, where it is to be understood that any *particular* congruence class is a *class of classes* of congruent intervals whose lengths are specified by a *particular* distance function $ds^2 = g_{ik}dx^i dx^k$. And we just saw that there are no intrinsic metric attributes of intervals which could be invoked to single out *one* of these congruence classes as unique.

How then can we speak of the assumedly continuous physical space as having a metric or *mutatis mutandis* suppose that the physical time continuum has a unique metric? The answer can be none other than the following:[6] Only the choice of a particular *extrinsic* congruence standard can determine a unique congruence class, the *rigidity* or self-congruence of that standard under transport being *decreed by convention*, and similarly for the periodic devices which are held to be *isochronous* (uniform) clocks. Thus the role of the spatial or temporal congruence standard *cannot* be construed with Newton or Russell ([81]; cf. also Section 5, part (i) below) to be the mere ascertainment of an otherwise intrinsic equality obtaining between the intervals belonging to the congruence class defined by it. Unless one of two segments is a subset of the other, the congruence of two segments is a matter of convention, stipulation, or definition and not a factual matter concerning which empirical findings could show one to have been mistaken. And hence there can be no question at all of an *empirically* or factually determinate metric geometry or chronometry until *after* a physical stipulation of congruence.[7]

[6] The conclusion which is about to be stated will appear unfounded to those who follow A. N. Whitehead in rejecting the "bifurcation of nature," which is assumed in its premises. But in Sec. 5 below, the reader will find a detailed rebuttal of the Whiteheadian contention that *perceptual* space and time do have an intrinsic metric and that once the allegedly illegitimate distinction between physical and perceptual space (or time) has been jettisoned, an intrinsic metric can hence be meaningfully imputed to physical space and time. Cf. also A. Grünbaum, "Whitehead's Philosophy of Science," *Philosophical Review*, 71:218–229 (1962).

[7] A. d'Abro [16, p. 27] has offered an *unsound* illustration of the thesis that the metric in a continuum is conventional: he considers a stream of sounds of varying pitch and points out that a congruence criterion based on the successive auditory octaves of a given musical note would be at variance with the congruence defined by equal differences between the associated frequencies of vibration, since the frequency differences between successive octaves are *not* equal. But instead of constituting an example of the alternative metrizability of the *same* mathematically continuous manifold of elements, d'Abro's illustration involves the metrizations of *two different* mani-

In the case of geometry, the specification of the intervals which are stipulated to be congruent is given by the distance function

$$ds = \sqrt{g_{ik}dx^i dx^k},$$

congruent intervals being those which are assigned equal lengths ds by this function. Whether the intervals defined by the coincidence behavior of the transported rod are those to which the distance function assigns *equal* lengths ds or not will depend on our selection of the functions g_{ik}. Thus, if the components of the metric tensor g_{ik} are suitably chosen in any given coordinate system, then the transported rod will have been stipulated to be congruent to itself everywhere independently of its position and orientation. On the other hand, by an appropriately different choice of the functions g_{ik}, the length ds of the transported rod will be made to *vary* with position or orientation instead of being constant. Once congruence has been defined via the distance function ds, the geodesics (straight lines) [8] associated with that choice of congruence are determined, since the family of geodesics is defined by the variational requirement $\delta \int ds = 0$, which takes the form of a differential equation whose solution is the equation of the family of geodesics. The geometry characterizing the relations of the geodesics in question is likewise determined by the distance function ds, because the Gaussian curvature K of every surface element at any point in space is fixed by the functions g_{ik} ingredient in the distance function ds.

There are therefore alternative metrizations of the same factual coincidence relations sustained by a transported rod, and *some* of these alternative definitions of congruence will give rise to different metric geometries than others. Accordingly, via an appropriate definition of congruence we are free to choose as the description of a *given* body of spatial facts any

---

folds only *one* of which is continuous in the mathematical sense. For the auditory contents sustaining the relation of being octaves of one another are elements of a merely sensory "continuum." Moreover, we shall see in part (iii) of this section that while holding for the mathematical continua of physical space and time, whose elements (points and instants) are respectively alike both qualitatively and in magnitude, the thesis of the conventionality of the metric cannot be upheld for *all* kinds of mathematical continua, Riemann and d'Abro to the contrary notwithstanding.

[8] The geodesics are called "straight lines" when discussing their relations in the context of synthetic geometry. But this identification must *not* be taken to entail that *every* geodesic connection of two points is a line of shortest distance between them. For once we abandon the restriction to Euclidean geometry, being a geodesic connection is only a necessary and not also a sufficient condition for being the shortest distance [86, pp. 140–143].

metric geometry compatible with the existing topology. Moreover, we shall find later on (Section 3, part (iii)) that there are infinitely many incompatible definitions of congruence which will implement the choice of any one metric geometry, be it the Euclidean one or one of the non-Euclidean geometries.

An illustration will serve to give concrete meaning to this general formulation of the conventionality of spatial congruence. Consider a physical surface such as part or all of an infinite blackboard and suppose it to be equipped with a network of Cartesian coordinates. The customary metrization of such a surface is based on the congruence defined by the coincidence behavior of transported rods: line segments whose termini have coordinate differences dx and dy are assigned a length ds given by

$$ds = \sqrt{dx^2 + dy^2},$$

and the geometry associated with this metrization of the surface is, of course, Euclidean. But we are also at liberty to employ a different metrization in part or all of this space. Thus, for example, we could equally legitimately metrize the portion *above* the x-axis by means of the new metric

$$ds = \sqrt{\frac{dx^2 + dy^2}{y^2}} \, .$$

This alternative metrization is incompatible with the customary one: for example, it makes the lengths ds = dx/y of horizontal segments whose termini have the *same* coordinate differences dx depend on *where* they are along the y-axis. Consequently, the new metric would commit us to regard a segment for which dx = 2 at y = 2 as congruent to a segment for which dx = 1 at y = 1, although the customary metrization would regard the length ratio of these segments to be 2:1. But, of course, the new metric does *not* say that a transported solid rod will coincide successively with the intervals belonging to the congruence class defined by that metric; instead it allows for this noncoincidence by making the length of the rod a suitably nonconstant function of its position. And this noncustomary congruence definition, which was suggested by Poincaré, confers a *hyperbolic* geometry on the half plane y > 0 of the customarily Euclidean plane: the associated geodesics in the half plane are a family of hyperbolically related lines whose infinitude is assured by the behavior of the new metric as y → 0 and whose *Euclidean* status is that of semicircular arcs.[9] It is now clear that the hyperbolic metrization of the semi-blackboard possesses not only mathematical but also epistemological credentials as good as those of the Euclidean one.

It might be objected that although *not* objectionable *epistemologically*, there is a pedantic artificiality and even perverse *complexity* in all congruence definitions which do not assign equal lengths ds to the intervals defined by the coincidence behavior of a solid rod. The grounds of this objection would be that (a) there are no convenient and familiar natural objects whose coincidence behavior under transport furnishes a physical realization of the bizarre, noncustomary congruences, and (b) after correcting for the chemically dependent distortional idiosyncrasies of various

[9] The reader can convince himself that the new metrization issues in a hyperbolic geometry by noting that now $g_{11} = 1/y^2$, $g_{12} = g_{21} = 0$, and $g_{22} = 1/y^2$ and then using these components of the metric tensor to obtain a *negative* value of the Gaussian curvature K via Gauss's formula. (For a statement of this formula, see, for example, F. Klein [44, p. 281].)

To determine what particular curves in the semiblackboard are the geodesics of our hyperbolic metric

$$ds = \frac{\sqrt{dx^2 + dy^2}}{y},$$

one must substitute this ds and carry out the variation in the equation $\delta \int ds = 0$, which is the defining condition for the family of geodesics. The desired geodesics of our new metric must therefore be given by the equation

$$\delta \int \frac{\sqrt{1 + \left(\frac{dy}{dx}\right)^2}}{y}\, dx = 0.$$

It is shown in the calculus of variations (cf., for example [48, pp. 193–195]) that this variational equation requires the following differential equation—known as Euler's equation—to be satisfied, if we put

$$I \equiv \frac{\sqrt{1 + \left(\frac{dy}{dx}\right)^2}}{y} \; : \quad \frac{\partial I}{\partial y} - \frac{d}{dx}\frac{\partial I}{\partial \left(\frac{dy}{dx}\right)} = 0.$$

Upon substituting our value of I, we obtain the differential equation of the family of geodesics:

$$\frac{d^2 y}{dx^2} + \frac{1}{y}\left[1 + \left(\frac{dy}{dx}\right)^2\right] = 0.$$

The solution of this equation is of the form

$$(x - k)^2 + y^2 = R^2,$$

where k and R are constants of integration, and thus represents—*Euclideanly speaking*—a family of circles centered on and perpendicular to the x-axis, the upper semi-circles being the geodesics of Poincaré's remetrized half plane above the x-axis.

kinds of solids in inhomogeneous thermal, electric, and other fields, all transported solid bodies furnish the same physical intervals, and thus they realize only one of the infinitude of incompatible mathematical congruences. *Mutatis mutandis*, the same objection might be raised to any definition of temporal congruence which does not accord with the cycles of standard material clocks. The reply to this criticism is twofold:

1. The prima-facie plausibility of the demand for *simplicity* in the choice of the congruence definition gives way to second thoughts the moment it is realized that the desideratum of simplicity requires consideration not only of the congruence definition but also of the latter's bearing on the form of the associated system of geometry and physics. And our discussions in Sections 4 and 6 will show that a bizarre definition of congruence may well have to be countenanced as the price for the attainment of the over-all simplicity of the total theory. Specifically, we anticipate Section 4, part (ii) by just mentioning here that although Einstein merely alludes to the possibility of a noncustomary definition of *spatial* congruence in the general theory of relativity without actually availing himself of it [21, p. 161], he does indeed utilize in that theory what our putative objector deems a highly artificial definition of *temporal* congruence, since it is *not* given by the cycles of standard material clocks.

2. It is particularly instructive to note that the cosmology of E. A. Milne [51, p. 22] postulates the actual existence in nature of two metrically different kinds of clocks whose respective periods constitute physical realizations of *incompatible* mathematical congruences. Specifically, Milne's assumptions lead to the result that there is a nonlinear relation

$$\tau = t_0 \log\left(\frac{t}{t_0}\right) + t_0$$

between the time τ defined by periodic astronomical processes and the time t defined by atomic ones, t_0 being an appropriately chosen arbitrary constant. The nonlinearity of the relation between these two kinds of time is of paramount importance here, because it assures that two intervals which are congruent in *one* of these two time scales will be *incongruent* in the other. Clearly, it would be utterly gratuitous to regard one of these two congruences as bizarre, since each of them is presumed to have a physical realization. And the choice between these scales is incontestably conventional, for it is made quite clear in Milne's theory that their associated different metric descriptions of the world are factually equivalent and hence equally true.

What would be the verdict of the Newtonian proponent of the intrinsicality of the metric on the examples of alternative metrizability which we gave both for space (Poincaré's hyperbolic metrization of the half plane) and also for time (general theory of relativity and Milne's cosmology)? He would first note correctly that once it is understood that the term "congruent," as applied to intervals, is to denote a reflexive, symmetrical, and transitive relation in this class of geometrical configurations, then the use of this term is restricted to designating a spatial equality relation. But then the Newtonian would proceed to claim *unjustifiably* that the spatial equality obtaining between congruent line segments of physical space (or between regions of surfaces and of 3-space respectively) consists in their each containing *the same intrinsic amount of space*. And having introduced this false premise, he would feel entitled to contend that (1) it is *never* legitimate to choose arbitrarily what specific intervals are to be regarded as congruent, and (2) as a corollary of this lack of choice, there is no room for selecting the lines which are to be regarded as straight and hence no choice among alternative geometric descriptions of actual physical space, since the geodesic requirement $\delta \int ds = 0$ which must be satisfied by the straight lines is subject to the restriction that only the members of the unique class of *intrinsically equal* line segments may ever be assigned the same length ds. By the same token, the Newtonian asserts that only "truly" (intrinsically) equal time intervals may be regarded as congruent and therefore holds that there exists only *one* legitimate metrization of the time continuum, a conclusion which he then attempts to buttress further by adducing certain *causal* considerations from Newtonian dynamics which will be refuted in Section 4, part (i) below.

It is of the utmost importance to realize clearly that the thesis of the conventionality of congruence is, in the first instance, a claim concerning *structural properties of physical space and time*; only the semantical *corollary* of that thesis concerns the *language* of the geochronometric description of the physical world. Having failed to appreciate this fact, some philosophers were led to give a shallow caricature of the debate between the Newtonian, who affirms the factuality of congruence on the strength of the alleged intrinsicality of the metric, and his Riemannian conventionalistic critic. According to the burlesqued version of this controversy, the Riemannian is offering no more than a semantical truism, and the

Adolf Grünbaum

Newtonian assertion of metric absolutism can be dismissed as an evident absurdity on purely semantical grounds, since it is the denial of that mere truism. More specifically, the detractors suppose that their trivialization of the congruence issue can be vindicated by pointing out that we are, of course, free to decree the referents of the *unpreempted word* "congruent," because such freedom can be exercised with respect to *any* as yet semantically uncommitted term or string of symbols whatever. And, in this way, they misconstrue the conventionality of congruence as merely an inflated special case of a semantical banality holding for any and all linguistic signs or symbols, a banality which we shall call "trivial semantical conventionalism" or, in abbreviated form, "TSC."

No one, of course, will wish to deny that qua *uncommitted signs*, the terms "spatially congruent" and "temporally congruent" are fully on a par in regard to the *trivial* conventionality of the semantical rules governing their use with all linguistic symbols whatever. And thus a sensible person would hardly wish to contest that the unenlightening affirmation of the conventionality of the use of the *unpreempted word* "congruent" is indeed a subthesis of TSC. But it is a serious obfuscation to identify the Riemann-Poincaré doctrine that the *ascription* of the congruence or equality *relation* to space or time intervals is conventional with the platitude that the use of the *unpreempted word* "congruent" is conventional. And it is therefore totally incorrect to conclude that the Riemann-Poincaré tenet is merely a gratuitously emphasized special case of TSC. For what these mathematicians are advocating is *not* a doctrine about the semantical freedom we have in the use of the uncommitted sign "congruent." Instead, they are putting forward the initially nonsemantical claim that the continua of physical space and time each lack an intrinsic metric. And the metric amorphousness of these continua then serves to *explain* that even *after* the word "congruent" has been preempted semantically as a spatial or temporal *equality* predicate by the axioms of congruence, congruence remains ambiguous in the sense that these axioms still allow an infinitude of mutually exclusive congruence classes of intervals. Accordingly, fundamentally *nonsemantical* considerations are used to show that only a conventional choice of one of these congruence classes can provide a unique standard of length equality. In short, the conventionality of congruence is a claim *not* about the noise "congruent" but about the character of the conditions relevant to the obtaining of the relation denoted by the term "congruent." For alternative metrizability is not a matter of the freedom to use the semantically uncommitted noise "congruent" as we please; instead, it is a matter of the nonuniqueness of a relation term already preempted as the physico-spatial (or temporal) equality predicate. And this nonuniqueness arises from the lack of an intrinsic metric in the continuous manifolds of physical space and time.

The epistemological status of the Riemann-Poincaré conventionality of congruence is fully analogous to that of Einstein's conventionality of simultaneity. And if the reasoning used by critics of the former in an endeavor to establish the banality of the former were actually sound, then, as we shall now show, it would follow by a precisely similar argument that Einstein's enunciation of the conventionality of simultaneity [23, Section 1] was no more than a turgid statement of the platitude that the uncommitted word "simultaneous" (or "gleichzeitig") may be used as we please. In fact, in view of the complete epistemological affinity of the conventionality of congruence with the conventionality of simultaneity, which we are about to exhibit, it will be useful subsequently to combine these two theses under the name "geochronometric conventionalism" or, in abbreviated form, "GC."

We saw in the case of spatial and temporal congruence that congruence is conventional in a sense other than that prior to being preempted semantically, the sign "congruent" can be used to denote anything we please. Mutatis mutandis, we now wish to show that precisely the same holds for the conventionality of metrical simultaneity. Once we have furnished this demonstration as well, we shall have established that neither of the component claims of conventionality in our compound GC thesis is a subthesis of TSC.

We proceed in Einstein's manner in the special theory of relativity and first decree that the noise "topologically simultaneous" denote the relation of not being connectible by a physical causal (signal) chain, a relation which may obtain between two physical events. We now ask: Is this definition unique in the sense of assuring that one and only one event at a point Q will be topologically simultaneous with a given event occurring at a point P elsewhere in space? The answer to this question depends on facts of nature, namely on the range of the causal chains existing in the physical world. Thus, once the above definition is given, its uniqueness is not a matter subject to regulation by semantical convention. If now we

421

Adolf Grünbaum

assume with Einstein as a fact of nature that light *in vacuo* is the fastest causal chain, then this postulate entails the nonuniqueness of the definition of "topological simultaneity" given above and thereby also prevents *topological* simultaneity from being a transitive relation. On the other hand, if the facts of the physical world had the structure assumed by Newton, this nonuniqueness would *not* arise. Accordingly, the structure of the facts of the world postulated by relativity prevents the above definition of "topological simultaneity" from also serving, as it stands, as a metrical synchronization rule for clocks at the spatially separated points P and Q. Upon coupling this result with the relativistic assumption that transported clocks do *not* define an absolute metrical simultaneity, we see that the facts of the world leave the equality relation of metrical simultaneity *indeterminate*, for they do not confer upon topological simultaneity the uniqueness which it would require to serve as the basis of metrical simultaneity as well. Therefore, the assertion of that indeterminateness and of the corollary that metrical simultaneity is made determinate by convention is in no way tantamount to the purely semantical assertion that the mere uncommitted noise "metrically simultaneous" must be given a physical interpretation before it can denote and that this interpretation is trivially a matter of convention.

Far from being a claim that a mere linguistic noise is still awaiting an assignment of semantical meaning, the assertion of the factual indeterminateness of metrical simultaneity concerns *facts of nature* which find expression in the residual nonuniqueness of the definition of "topological simultaneity" once the latter has already been given. And it is thus impossible to construe this residual nonuniqueness as being attributable to taciturnity or tight-lippedness on Einstein's part in telling us what he means by the noise "simultaneous." Here, then, we are confronted with a kind of logical gap needing to be filled by definition which is precisely analogous to the case of congruence, where the continuity of space and time issued in the residual nonuniqueness of the congruence axioms.

When I say that metrical simultaneity is not wholly factual but contains a conventional ingredient, what am I asserting? I am claiming none other than that the residual nonuniqueness or logical gap *cannot* be removed by an appeal to facts but only by a *conventional* choice of a *unique* pair of events at P and at Q as metrically simultaneous from within the class of pairs that are topologically simultaneous. And when I assert that

422

it was a great philosophical (as well as physical) achievement for Einstein to have discovered the conventionality of metrical simultaneity, I am crediting Einstein *not* with the triviality of having decreed semantically the meaning of the *noise* "metrically simultaneous" (or "gleichzeitig") but with the recognition that, contrary to earlier belief, the facts of nature are such as to deny the required kind of semantical univocity to the already preempted term "simultaneous" ("gleichzeitig"). In short, Einstein's insight that metrical simultaneity is conventional is a contribution to the theory of time rather than to semantics, because *it concerns the character of the conditions relevant to the obtaining of the relation denoted by the term "metrically simultaneous."*

The *conventionality* of metrical simultaneity has just been formulated without any reference whatever to the relative motion of different Galilean frames and does *not* depend upon there being a relativity or nonconcordance of simultaneity as between different Galilean frames. On the contrary, it is the conventionality of simultaneity which provides the logical framework within which the *relativity* of simultaneity can first be understood: if *each* Galilean observer adopts the particular metrical synchronization rule adopted by Einstein in Section 1 of his fundamental paper [23]—a rule which corresponds to the value $\epsilon = \frac{1}{2}$ in the Reichenbach notation [72, p. 127]—then the relative motion of Galilean frames issues in their choosing as metrically simultaneous *different pairs of events* from within the class of topologically simultaneous events at P and Q, a result embodied in the familiar Minkowski diagram.

In discussing the definition of simultaneity [23, Section 1], Einstein italicized the words *"by definition"* in saying that the *equality* of the to and fro velocities of light between two points A and B is a matter of definition. Thus, he is asserting that metrical simultaneity is a matter of definition or convention. Do the detractors really expect anyone to believe that Einstein put these words in italics to convey to the public that the *noise* "simultaneous" can be used as we please? Presumably they would recoil from this conclusion. But how else could they solve the problem of making Einstein's avowedly conventionalist conception of metrical simultaneity compatible with their semantical trivialization of GC? H. Putnam, one of the advocates of the view that the conventionality of congruence is a subthesis of TSC, has sought to meet this difficulty along the following lines: in the case of the congruence of intervals, one would never run into trouble upon using the customary definition;[10] but in the case of simultaneity, actual contradictions would be encountered upon using the customary classical definition of metrical simultaneity, which is based on the transport of clocks and is vitiated by the dependence of the clock rates (readings) on the transport velocity. But Putnam's retort will not do. For the appeal to Einstein's recognition of the inconsistency of the classical definition of metrical simultaneity accounts only for his abandonment of the latter but does *not* illuminate—as does the thesis of the conventionality of simultaneity—the logical status of the *particular set* of definitions which Einstein put in its place. Thus, the Putnamian retort does *not* recognize that the logical status of Einstein's synchronization rules is not at all adequately rendered by saying that whereas the classical definition of metrical simultaneity was inconsistent, Einstein's rules have the virtue of consistency. For what needs to be elucidated is *the nature of the logical step* leading to Einstein's particular synchronization scheme within the wider framework of the *alternative consistent sets of rules for* metrical simultaneity any one of which is allowed by the nonuniqueness of topological simultaneity. Precisely this elucidation is given, as we have seen, by the thesis of the conventionality of metrical simultaneity.

We see therefore that the *philosophically illuminating* conventionality of an affirmation of the congruence of two intervals or of the metrical simultaneity of two physical events does *not* inhere in the arbitrariness of what linguistic *sentence* is used to express the proposition that a relation of equality obtains among intervals, or that a relation of metrical simultaneity obtains between two physical events. Instead, the important conventionality lies in the fact that even *after* we have specified what respective linguistic sentences will express these propositions, a convention is ingredient in each of the propositions expressed, i.e., in the very *obtaining* of a congruence relation among intervals or of a metrical simultaneity relation among events.

These considerations enable us to articulate the misunderstanding of the conventionality of congruence to which the proponent of its *model theoretic* trivialization (cf. Section 1) fell prey. It will be recalled that this critic argued somewhat as follows: "The theory of models of uninterpreted formal calculi shows that there can be no basis at all for an episte-

[10] We shall see in our discussion of time measurement on the rotating disk of the general theory of relativity in Sec. 4, part (ii), that there is one sense of "trouble" for which Putnam's statement would *not* hold.

mological distinction *within* the system of physical geometry (or chronometry) between factual statements, on the one hand, and supposedly conventional statements of rigidity (or isochronism), on the other. For we are just as free to give a *non*customary spatial interpretation of, say, the abstract sign "point" in the formal geometrical calculus as of the sign "congruent," and hence the physical interpretation of the relation term "congruent" (for line segments) cannot occupy an epistemologically distinguished position among the semantical rules effecting the interpretation of the formal system, all of which are on a par in regard to conventionality." But this objection overlooks that (a) the obtaining of the spatial congruence relation provides scope for the role of convention because, independently of the particular formal geometrical calculus which is being interpreted, the term "congruent" functions as a spatial *equality predicate* in its *non*customary spatial interpretations no less than in its customary ones; (b) consequently, suitable alternative spatial interpretations of the term "congruent" and correlatively of "straight line" ("geodesic") show that it is always a live option (subject to the restrictions imposed by the existing topology) to give *either* a Euclidean *or* a *non*-Euclidean description of the same body of physico-geometrical facts; and (c) by contrast, the possibility of alternative spatial interpretations of such *other* primitives of rival geometrical calculi as "point" does *not* generally issue in this option. Our concern is to note that, even disregarding inductive imprecision, the empirical facts themselves do *not* uniquely dictate the truth of either Euclidean geometry or of one of its non-Euclidean rivals in virtue of the lack of an intrinsic metric. Hence in this context the different spatial interpretations of the term "congruent" (and hence of "straight line") in the respective geometrical calculi play a philosophically different role than do the interpretations of such other primitives of these calculi as "point," since the latter generally have the *same* spatial meaning in both the Euclidean and non-Euclidean descriptions. The preeminent status occupied by the interpretation of "congruent" in this context becomes apparent once we cease to look at physical geometry as a spatially interpreted system of abstract *synthetic* geometry and regard it instead as an interpreted system of abstract *differential* geometry of the Gauss-Riemann type: by choosing a particular distance function

$$ds = \sqrt{g_{ik}dx^{i}dx^{k}}$$

for the line element, we specify not only what segments are congruent and what lines are straights (geodesics) but the entire geometry, since the metric tensor g_{ik} fully determines the Gaussian curvature K. To be sure, if one were discussing *not* the alternative between a Euclidean and non-Euclidean description of the same *spatial* facts but rather the set of *all* models (including *non*-spatial ones) of a *given* calculus, say the Euclidean one, then indeed the physical interpretation of "congruent" and of "straight line" would not merit any more attention than that of other primitives like "point." [11]

We have argued that the continuity postulated for physical space and time issues in the metric amorphousness of these manifolds and thus makes for the conventionality of congruence, much as the conventionality of nonlocal metrical simultaneity in special relativity is a consequence of the postulational *fact* that light is the fastest causal chain *in vacuo* and that clocks do not define an absolute metrical simultaneity under transport. But it might be objected that unlike the latter Einstein postulate, the postulational ascription of continuity (in the mathematical sense) to physical space and time instead of some discontinuous structure cannot be regarded, even in principle, as a *factual* assertion in the sense of being either true or false. For surely, this objection continues, there can be no *empirical* grounds for accepting a geometry postulating continuous intervals in preference to one which postulates discontinuous intervals consisting of, say, only the algebraic or only the rational points. The rejection of the latter kind of denumerable geometry in favor of a nondenumerable one affirming continuity therefore has no kind of factual warrant but is based solely on considerations of *arithmetic convenience* within the analytic part of geometry. And hence the *topology* is no less infested with features springing from conventional choice than is the metric. Hence the exponent of this criticism concludes that it is not only a misleading emphasis but outright incorrect for us to persevere in the course taken by Carnap [8] and Reichenbach [72] and to discern conventional elements in geometry only in the *metrization* of the topological substratum while deeming the latter to be *factual*.

This plea for a conventionalist conception of continuity is not con-

---

[11] We have been speaking of certain *uninterpreted* formal calculi as systems of synthetic or differential *geometry*. It must be understood, however, that *prior to* being given a *spatial* interpretation, these abstract deductive systems no more qualify as *geometries*, strictly speaking, than as systems of genetics or of anything else; they are called "geometries," it would seem, only "because the name seems good, on emotional and traditional grounds, to a sufficient number of competent people" [90, p. 17].

vincing, however. Admittedly, the justification for regarding continuity as a broadly *inductive framework principle* of physical geometry cannot be found in the direct verdicts of measuring rods, which could hardly disclose the super-denumerability of the points on the line. And prima facie there is a certain measure of plausibility in the contention that the postulation of a super-denumerable infinity of irrational points in addition to a denumerable set of rational ones is dictated solely by the desire for such arithmetical convenience as having closure under operations like taking the square root. But, even disregarding the Zenonian difficulties which may vitiate denumerable geometries logically (cf. footnote 5 above), these considerations lose much of their force, it would seem, as soon as one aplies the *acid test* of a convention to the conventionalist conception of continuity in physical geometry: the feasibility of one or more *alternate* formulations dispensing with the particular alleged convention and yet permitting the successful rendition of the same total body of experiential findings, such as in the case of the choice of a particular system of *units* of measurement. Upon applying this test, what do we find? Attempts to dispense with the continuum of classical mathematics (geometry and analysis) by providing adequate substitutes for the mathematics used by the total body of advanced modern physical theory have been programmatic rather than successful. And this *not* for want of effort on the part of their advocates. Thus, for example, the neointuitionistic endeavors to base mathematics on more restrictive foundations involve mutilations of mathematical physics whose range A. A. Fraenkel has characterized as follows: "intuitionistic restriction of the concept of continuum and of its handling in analysis and geometry, though carried out in quite different ways by various intuitionistic schools, always goes as far as to exclude vital parts of those two domains. (This is not altered by Brouwer's peculiar way of admitting the continuum *per se* as a 'medium of free growth.')" [12] The impressive difficulties encountered in these endeavors to provide a viable denumerable substitute for such topological components of the geometry as continuity thus insinuate the following suspicion: the empirical facts codified in terms of the classical mathematical apparatus in our most sophisticated and best-confirmed physical theories support continuity in a *broadly inductive* sense as a *framework principle* to the exclusion of

---

[12] A. A. Fraenkel and Y. Bar-Hillel [29, p. 200]. Chapter IV of this work gives an admirably comprehensive and lucid survey of the respects in which neointuitionist restrictions involve truncations of the system of classical mathematics.

Adolf Grünbaum

the prima-facie rivals of the continuum.[13] Pending the elaboration of a successful alternative to the continuum, therefore, the charge that in a geometry the topological component of continuity is no less conventional than the metric itself seems to be unfounded.

(ii) Physical Congruence, Testability, and Operationism.

We have grounded the conventionality of spatial and temporal congruence on the continuity of the manifolds of space and time. And, in thus arguing that "true," absolute, or intrinsic rigidity and isochronism are nonexisting properties in these respective continua, we did not adduce any phenomenalist or homocentric-operationist criterion of factual meaning. For we did not say that the actual and potential failure of human testing operations to disclose "true rigidity" constitutes either its nonexistence or its meaninglessness. It will be well, therefore, to compare our Riemannian espousal of the conventionality of rigidity and isochronism with the reasoning of those who arrive at this conception of congruence by arguing from nontestability or from an operationist view of scientific concepts. Thus W. K. Clifford writes: "we have defined length or distance by means of a measure which can be carried about without changing its length. But how then is this property of the measure to be tested? . . . Is it possible . . . that lengths do really change by mere moving about, without our knowing it? Whoever likes to meditate seriously upon this question will find that it is wholly devoid of meaning" [14, pp. 49, 50].

We saw that within our Riemannian framework of ideas, length is relational rather than absolute in a twofold sense: (i) length obviously depends numerically on the units used and is thus arbitrary to within a constant factor, and (ii) in virtue of the lack of an intrinsic metric, sameness or change of the length possessed by a body in different places and at

[13] It would be an error to believe that this conclusion requires serious qualification in the light of recent suggestions of space (or time) quantization. For as H. Weyl has noted [95, p. 43]: "so far it [i.e., the atomistic theory of space] has always remained mere speculation and has never achieved sufficient contact with reality. How should one understand the metric relations in space on the basis of this idea? If a square is built up of miniature tiles, then there are as many tiles along the diagonal as there are along the side; thus the diagonal should be equal in length to the side." And Einstein has remarked (see The Meaning of Relativity (Princeton: Princeton University Press, 1955), pp. 165–166) that "From the quantum phenomena it appears to follow with certainty that a finite system of finite energy can be completely described by a finite set of numbers (quantum numbers). This does not seem to be in accordance with a continuum theory, and must lead to an attempt to find a purely algebraic theory for the description of reality. But nobody knows how to obtain the basis of such a theory."

428

different times consists in the sameness or change respectively of the ratio (relation) of that body to the conventional standard of congruence. Whether or not this ratio changes is quite independent of any human discovery of it: the number of times which a given body B will contain a certain unit rod is a property of B that is not first conferred on B by human operations. And thus the relational character of length derives, in the first instance, not from how we human beings discover lengths but from the failure of the continuum of physical space to possess an intrinsic metric, a failure obtaining quite independently of our measuring activities. In fact, it is this relational character of length which prescribes and regulates the kinds of human operations appropriate to its discovery. Since, to begin with, there exists no property of true rigidity to be disclosed by any human test, no test could possibly reveal its presence. Accordingly, the unascertainability of true rigidity by us humans is a consequence of its nonexistence in physical space and evidence for that nonexistence but not constitutive of it.

On the basis of this nonhomocentric relational conception of length, the utter vacuousness of the following assertion is evident at once: overnight everything has expanded (i.e., increased its length) but such that all length ratios remained unaltered. That such an alleged "expansion" will elude any and all human test is then obviously explained by its not having obtained: the increase in the ratios between all bodies and the congruence standard which would have constituted the expansion avowedly did not materialize.

We see that the relational theory of length and hence the particular assertion of the vacuousness of a universal nocturnal expansion do not depend on a grounding of the meaning of the metrical concepts of length and duration on human testability or manipulations of rods and clocks in the manner of Bridgman's homocentric operationism.[14] Moreover, there is a further sense in which the Riemannian recognition of the need for a specification of the congruence criterion does not entail an operationist conception of congruence and length: as we noted preliminarily at the beginning of Section 2 and will see in detail in Section 7, the definition of "congruence" on the basis of the coincidence behavior common to all kinds of transported solid rods provides a rule of correspondence (coordi-

[14] For arguments supporting the conclusion that homocentric operationism is similarly dispensable and, in fact, unsuccessful in giving an account of the conceptual innovations of the special theory of relativity, see [35].

429

Adolf Grünbaum

native definition) *through the mediation of hypotheses and laws that are collateral* to the abstract geometry receiving a physical interpretation. For the physical laws used to compute the corrections for thermal and other substance-specific deformations of solid rods made of different kinds of materials *enter integrally* into the statement of the physical meaning of "congruent." Thus, in the case of "length" no less than in most other cases, operational definitions (in any distinctive sense of the term "operational") are a quite idealized and limiting species of correspondence rules. Further illustrations of this fact are given by Reichenbach, who cites the definitions of the unit of length on the basis of the wave length of cadmium light and also in terms of a certain fraction of the circumference of the earth and writes: "Which distance serves as a unit for actual measurements can ultimately be given only by reference to some actual distance. . . . We say with regard to the measuring rod . . . that only 'ultimately' the reference is to be conceived in this form, because we know that by means of the interposition of conceptual relations the reference may be rather remote" [72, p. 128]. An even stronger repudiation of the operationist account of the definition of "length" because of its failure to allow for the role of auxiliary theory is presented by K. R. Popper, who says: "As to the doctrine of operationalism—which demands that scientific terms, such as length . . . should be defined in terms of the appropriate experimental procedure—it can be shown quite easily that all so-called operational definitions will be circular. . . . the circularity of the operational definition of length . . . may be seen from the following facts: (a) the 'operational' definition of *length* involves *temperature* corrections, and (b) the (usual) operational definition of *temperature* involves measurements of *length*" [66, p. 440 and p. 440n].[15]

(iii) The Inadequacies of the Nongeometrical Portion of Riemann's Theory of Manifolds.

In pointing out earlier in Section 2 that the status of *spatial* and *temporal* congruence is decisively illuminated by Riemann's theory of continuous manifolds, I stated that this theory will *not* bear critical scrutiny as a characterization of continuous manifolds in general. To justify and clarify this indictment, we shall now see that continuity cannot be held with Riemann to furnish a sufficient condition for the intrinsic metric

[15] In Sec. 7, we shall see how the circularity besetting the operationist conception of length is circumvented within our framework when making allowance for thermal and other deformations in the statement of the definition of congruence.

amorphousness of any manifold *independently of the character of its elements*. For, as Russell saw correctly [79, Sections 63 and 64], there are continuous manifolds, such as that of colors (in the physicist's sense of spectral frequencies) in which the individual elements differ qualitatively from one another and have inherent magnitude, thus allowing for metrical comparison of *the elements themselves*. By contrast, in the continuous manifolds of *space* and of *time*, neither points nor instants have any inherent magnitude allowing an individual metrical comparison between them, since all points are alike, and similarly for instants. Hence in these manifolds metrical comparisons can be effected only among the *intervals* between the elements, *not* among the homogeneous elements themselves. And the continuity of *these* manifolds then assures the nonintrinsicality of the metric for their intervals.

To exhibit further the bearing of the character of the elements of a continuous manifold on the feasibility of an intrinsic metric in it, I shall contrast the status of the metric in space and time, on the one hand, with its status in (a) the continuum of real numbers, arranged according to magnitude and (b) the quasi-continuum of masses, mass being assumed to be a property of bodies in the Newtonian sense clarified by Mach's definition.[16]

The assignment of real numbers to points of physical space in the manner of the introduction of generalized curvilinear coordinates effects only a *coordinatization* but *not* a *metrization* of the manifold of physical space. No informative metrical comparison among individual points could be made by comparing the magnitudes of their real number coordinate names. On the other hand, within the continuous manifold formed by the real numbers themselves, when arranged according to magnitude, every real number is singly distinguished from and metrically comparable to every other by its inherent magnitude. And the measurement of mass can be seen to constitute a counterexample to Riemann's metrical philosophy from the following considerations.

In the Machian definition of Newtonian (gravitational and inertial) mass, the *mass ratio* of a particle B to a standard particle A is given by the magnitude ratio of the acceleration of A due to B to the acceleration of B due to A. Once the space-time metric and thereby the accelerations are fixed in the customary way, this ratio for any particular body B is *independent*, among other things, of how far apart B and A may be during

---

[16] For a concise account of that definition, see [58, pp. 56–58].

their interaction. Accordingly, any affirmations of the mass equality (mass "congruence") or inequality of two bodies will hold independently of the extent of their spatial separation. Now, the set of medium-sized bodies form a *quasi*-continuum with respect to the dyadic relations "being more massive than," and "having the same mass," i.e., they form an array which is a continuum *except* for the fact that *several* bodies can occupy the same place in the array by sustaining the mass-"congruence" relation to each other. Without having such a relation of mass-equality *ab initio*, the set of bodies do not even form a quasi-continuum. We complete the metrization of this quasi-continuum by choosing a unit of mass (e.g., 1 gram) and by availing ourselves of the numerical mass ratios obtained by experiment. There is no question here of the lack of an intrinsic metric in the sense of a choice of making the mass *difference* between a given pair of bodies equal to that of another pair or not. In the resulting continuum of real mass numbers, the elements themselves have inherent magnitude and can hence be compared individually, thus defining an intrinsic metric. Unlike the point elements of space, the elements of the set of bodies are *not* all alike mass-wise, and hence the metrization of the quasi-continuum which they form with respect to the relations of being more massive than and having the same mass can take the form of directly comparing the individual elements of that quasi-continuum rather than only intervals between them.

If one did wish for a spatial (or temporal) analogue of the metrization of masses, one should take as the set to be metrized *not* the continuum of points (or instants) but the quasi-continuum of all spatial (or temporal) *intervals*. To have used such intervals as the *elements* of the set to be metrized, we must have had a prior criterion of spatial congruence and of "being longer than" in order to arrange the intervals in a quasi-continuum which can then be metrized by the assignment of length numbers. This metrization would be the space or time analogue of the metrization of masses.

3. An Appraisal of R. Carnap's and H. Reichenbach's Philosophy of Geometry

(i) *The Status of Reichenbach's 'Universal Forces,' and His 'Relativity of Geometry.'*

In *Der Raum* [8, p. 33], Carnap begins his discussion of *physical* space by inquiring whether and how a line in this space can be identified as straight. Arguing from *testability* and not, as we did in Section 2, from the continuity of that manifold, he answers this inquiry as follows: "It is impossible in principle to ascertain this, if one restricts oneself to the unambiguous deliverances of experience and does not introduce freely-chosen conventions in regard to objects of experience" [8, p. 33]. And he then points out that the most important convention relevant to whether certain physical lines are to be regarded as straights is the specification of the metric ("Mass-setzung"), which is conventional because it could "never be either confirmed or refuted by experience." Its statement takes the following form: "A particular body and two fixed points on it are chosen, and it is then agreed what length is to be assigned to the interval between these points under various conditions (of temperature, position, orientation, pressure, electrical charge, etc.). An example of the choice of a metric is the stipulation that the two marks on the Paris standard meter bar define an interval of 100. $f(T; \phi, \lambda, h; \ldots)$ cm; ... a unit must also be chosen, but that is not our concern here which is with the choice of the body itself and with the function $f(T, \ldots)$" [8, pp. 33–34].

Once a particular function f has been chosen, the coincidence behavior of the selected transported body permits the determination of the metric tensor g_{ik} appropriate to that choice, thereby yielding a *congruence* class of intervals and the associated geometry. Accordingly, Carnap's thesis is that the question as to the geometry of physical space is indeed an *empirical* one but subject to an important proviso: it becomes empirical only *after* a physical definition of congruence for line segments has been given *conventionally* by stipulating (to within a constant factor depending on the choice of unit) what length is to be assigned to a transported solid rod in different positions of space.

Like Carnap, Reichenbach invokes testability [72, p. 16] to defend this qualified empiricist conception of geometry and speaks of "the relativity of geometry" [72, Section 8] to emphasize the dependence of the geometry on the definition of congruence. Carnap had lucidly conveyed the conventionality of congruence by reference to our freedom to choose the function f in the metric. But Reichenbach couches this conception in *metaphorical* terms by speaking of "universal forces" [72, Sections 3, 6, 8] whose metrical "effects" on measuring rods are then said to be a matter of convention as follows: the customary definition of congruence in which the rod is held to be of equal length everywhere (*after* allowance for substance-specific thermal effects and the like) corresponds to equating the universal forces to zero; on the other hand, a noncustomary definition of congruence, according to which the length of the rod varies with position or orientation (after allowance for thermal effects, etc.), corresponds to assuming an appropriately specified nonvanishing universal force whose mathematical characterization will be explained below. Reichenbach did not anticipate that this metaphorical encumbrance of his formulation would mislead some people into making the ill-conceived charge that noncustomary definitions of congruence are based on ad hoc invocations of universal forces. Inasmuch as this charge has been leveled against the conventionality of congruence, it is essential that we now divest Reichenbach's statement of its misleading potentialities.

Reichenbach [72, Section 3] invites consideration of a large hemisphere made of glass which merges into a huge glass plane, as shown in cross section by the surface G in the accompanying diagram, which consists of a plane with a hump. Using solid rods, human beings on this surface would readily determine it to be a Euclidean plane with a central hemispherical hump. He then supposes an opaque plane E to be located below the sur-

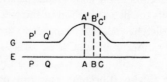

face G as shown in the diagram. Vertical light rays incident upon G will cast shadows of all objects on that glass surface onto E. As measured by actual solid rods, G-people will find A'B' and B'C' to be equal, while their projections AB and BC on the Euclidean plane E would be unequal. Reichenbach now wishes to prepare the reader for the recognition of the conventionality of congruence by having him deal with the following kind of question. Might it not be the case that (1) the inequality of AB and BC is only apparent, these intervals and other projections like them in the region R of E under the hemisphere being really equal, so that the true geometry of the surface E is spherical in R and Euclidean only outside it; (2) the equality of A'B' and B'C' is only apparent, the true geometry of surface G being plane Euclidean throughout, since in the apparently hemispherical region R' of G real equality obtains among those intervals which are the upward vertical projections of E-intervals in R that are equal in the customary sense of our daily life; and (3) on each of the two surfaces, transported measuring rods respectively fail to coincide with really equal intervals in R and R' respectively, because they do not remain truly congruent to themselves under transport, being deformed under the influence of undetectable forces which are "universal" in the sense that (a) they affect all materials alike, and (b) they penetrate all insulating walls?

On the basis of the conceptions presented in Section 2 above, which involve no kind of reference to universal forces, one can fulfill Reichenbach's desire to utilize this question as a basis for expounding the conventionality of congruence by presenting the following considerations. The legitimacy of making a distinction between the real (true) and the apparent geometry of a surface turns on the existence of an intrinsic metric. If there were an intrinsic metric, there would be a basis for making the distinction between real (true) and apparent equality of a rod under transport, and thereby between the true and the apparent geometry. But inasmuch as there is not, the question as to whether a given surface is really a Euclidean plane with a hemispherical hump or only apparently so must be replaced by the following question: on a particular convention of congruence as specified by a choice of one of Carnap's functions f, does the coincidence behavior of the transported rod on the surface in question yield the geometry under discussion or not? Thus the question as to the geometry of a surface is inherently ambiguous without the introduction of a congruence definition. And in view of the conventionality of spatial congruence, we are entitled to metrize G and E either in the customary way or in other ways so as to describe E as a Euclidean plane with a hemispherical hump R in the center and G as a Euclidean plane throughout. To assure the correctness of the latter noncustomary descriptions, we need only decree the congruence of those respective intervals which our questioner called "really equal" as opposed to apparently equal in parts (1) and (2) of his question respectively. Accordingly, without the presupposition of an intrinsic metric there can be no question of an absolute or "real" deformation of all kinds of measuring rods alike under the action of universal forces, and, mutatis mutandis, the same statement applies to clocks. Since a rod undergoes no kind of objective physical change in the supposed "presence" of universal forces, that "presence" signifies no more than that we assign a different length to it in different positions or orientations by convention. Hence, just as the conversion of the length of a table from meters to feet does not involve the action of a force on the table as the "cause" of the change, so also reference to universal forces as "causes" of "changes" in the transported rod can have no literal but only metaphorical significance. Moreover, mention of universal

Adolf Grünbaum

forces is an entirely dispensable *façon de parler* in this context as is evident from the fact that the rule assigning lengths to the transported rod which vary with its position and orientation can be given by specifying Carnap's function f.

Reichenbach, on the other hand, chooses to formulate the conventionality of congruence by first distinguishing between what he calls "differential" and "universal" forces and then using "universal forces" metaphorically in his statement of the epistemological status of the metric. By "differential" forces [72, Section 3] he means thermal and other influences which we called "perturbational" (cf. Section 2, part (i)) and whose presence issues in the dependence of the coincidence behavior of transported rods on the latter's chemical composition. Since we conceive of physical geometry as the system of metric relations which are *independent* of chemical composition, we correct for the substance-specific deformations induced by differential forces [1; 72, p. 26; 76; 77, pp. 327–329]. Reichenbach defines "universal forces" as having the twin properties of affecting all materials in the same way and as all-permeating because there are no walls capable of providing insulation against them. There is precedent for a literal rather than metaphorical use of universal forces to give a congruence definition: in order to provide a *physical realization* of a noncustomary congruence definition which would metrize the interior of a sphere of radius R so as to be a model of an infinite 3-dimensional hyperbolic space, Poincaré [63, pp. 75–77] postulates that (a) each concentric sphere of radius r < R is held at a constant absolute temperature $T \propto R^2 - r^2$, while the optical index of refraction is inversely proportional to $R^2 - r^2$, and (b) *contrary to actual fact*, all kinds of bodies within the sphere have the *same* coefficient of thermal expansion. It is essential to see that the expansions and contractions of these bodies under displacement have a *literal* meaning in this context, because they are *relative* to the actual displacement behavior of our normally Euclidean kinds of bodies and are linked to thermal sources.[17]

But Reichenbach's *metaphorical* use of universal forces for giving the congruence definition and exhibiting the dependence of the geometry on that definition takes the following form: "Given a geometry G' to which

[17] A precisely analogous *literal* use of universal forces is also made by Reichenbach [72, pp. 11–12] to convey pictorially a physical realization of a congruence definition which would confer a spherical geometry on the region R of the surface E discussed above.

436

the measuring instruments conform [*after* allowance for the effects of thermal and other "differential" influences], we can imagine a universal force F which affects the instruments in such a way that the actual geometry is an arbitrary geometry G, while the observed deviation from G is due to a universal deformation of the measuring instruments" [72, p. 33]. And he goes on to say that if g_{ik}' (i = 1,2,3; k = 1,2,3) are the empirically obtained metrical coefficients of the geometry G' and g_{ik} those of G, then the force tensor F is given mathematically by the tensor equation

$$g_{ik}' + F_{ik} = g_{ik},$$

where the g_{ik}', which had yielded the observed geometry G', are furnished experimentally by the measuring rods[18] and where the F_{ik} are the "correction factors"[19] $g_{ik} - g_{ik}'$ which are added correctionally to the g_{ik}' so that the g_{ik} are obtained.[20] But since Reichenbach emphasizes that it is a matter of *convention* whether we equate F_{ik} to zero or not [72, pp. 16, 27–28, 33], this formulation is merely a metaphorical way of asserting that the following is a matter of convention: whether congruence is said to obtain among intervals having equal lengths ds given by the metric $ds^2 = g_{ik}'dx^idx^k$—which entails G' as the geometric description of the observed coincidence relations—or among intervals having equal lengths ds given by the metric $ds^2 = g_{ik}dx^idx^k$ which yields a different geometry G.[21] Clearly then, to equate the universal forces to zero is merely to choose the metric based on the tensor g_{ik}' which was obtained from measurements in which the rod was called congruent to itself everywhere. In other

[18] For details on this experimental procedure, see, for example, [72, Secs. 39 and 40].

[19] The quotation marks are also in Reichenbach's text.

[20] We shall see in Sec. 3, part (iii), that Reichenbach was *mistaken* in asserting [72, pp. 33–34] that for a given surface or 3-space a particular metric geometry *determines* (1) a *unique* definition of congruence and, once a unit has been chosen, (2) a *unique* set of functions g_{ik} as the representations of the metric tensor in any particular coordinate system.

It will turn out that there are infinitely many *incompatible* congruence definitions and as many correspondingly different metric tensors which impart the *same* geometry to physical space. Hence, while a given metric tensor yields a unique geometry, a geometry G does *not* determine a metric tensor *uniquely* to within a constant factor depending on the choice of unit. And thus it is incorrect for Reichenbach to speak here of the components g_{ik} of a particular metric tensor as "those of G" [72, p. 33n] and to suppose that a unique F is specified by the requirement that a certain geometry G prevail in place of the observed geometry G'.

[21] It is to be clearly understood that the g_{ik} yield a congruence relation *incompatible* with the one furnished by the g_{ik}', because in any *given* coordinate system they are *different* functions of the given coordinates and *not proportional* to one another. (A difference consisting in a mere proportionality could *not* involve a difference in the congru-

Adolf Grünbaum

words, to stipulate $F_{ik} = 0$ is to choose the customary congruence standard based on the rigid body. On the other hand, apart from one exception to be stated presently, to stipulate that the components F_{ik} may not all be zero is to adopt a noncustomary metric given by a tensor g_{ik} corresponding to a specified variation of the length of the rod with position and orientation.

That there is one exception, which, incidentally, Reichenbach does not discuss, can be seen as follows: A given congruence determines the metric tensor up to a constant factor depending on the choice of unit and conversely. Hence two metric tensors will correspond to different congruences if and only if they differ other than by being proportional to one another. Thus, if g_{ik} and g_{ik}' are proportional by a factor different from 1, these two tensors furnish metrics differing only in the choice of unit and hence yield the same congruence. Yet in the case of such proportionality of g_{ik} and g_{ik}', the F_{ik} cannot all be zero. For example, if we consider the line element

$$ds^2 = a^2 d\phi^2 + a^2 \sin^2\phi d\theta^2$$

on the surface of a sphere of radius $a = 1$ meter $= 100$ cm, the mere change of units from meters to centimeters would change the metric $ds^2 = d\phi^2 + \sin^2\phi d\theta^2$ into $ds^2 = 10000 d\phi^2 + 10000 \sin^2\phi d\theta^2$. And if these metrics are identified with g_{ik}' and g_{ik} respectively, we obtain

$$F_{11} = g_{11} - g_{11}' = 10000 - 1 = 9999$$
$$F_{12} = F_{21} = g_{12} - g_{12}' = 0$$
$$F_{22} = g_{22} - g_{22}' = 10000\sin^2\phi - \sin^2\phi = 9999\sin^2\phi.$$

ence classes but only in the unit of length used.) The incompatibility of the congruences furnished by the two sets of g's is a necessary though not a sufficient condition (see the preceding footnote) for the nonidentity of the associated geometries G and G'.

The difference between the two metric tensors corresponding to incompatible congruences must not be confounded with a mere difference in the representations in different coordinate systems of the one metric tensor corresponding to a single congruence criterion (for a given choice of a unit of length): the former is illustrated by the incompatible metrizations $ds^2 = dx^2 + dy^2$ and

$$ds^2 = \frac{dx^2 + dy^2}{y^2}$$

in which the g's are different functions of the same rectangular coordinates, whereas the latter is illustrated by using first rectangular and then polar coordinates to express the same metric as follows: $ds^2 = dx^2 + dy^2$ and $ds^2 = d\rho^2 + \rho^2 d\theta^2$. In the latter case, we are not dealing with different metrizations of the space but only with different coordinatizations (parametrizations) of it, one or more of the g's being different functions of the coordinates in one coordinate system from what they are in another but so chosen as to yield an invariant ds.

It is now apparent that the F_{ik}, being given by the difference between the g_{ik} and the g_{ik}', will not all vanish and that also these two metric tensors will yield the same congruence, if and only if these tensors are proportional by a factor different from 1. Therefore, a necessary and sufficient condition for obtaining incompatible congruences is that both there be at least one nonvanishing component F_{ik} and that the metric tensors g_{ik} and g_{ik}' not be proportional to one another. The exception to our statement that a noncustomary congruence definition is assured by the failure of at least one component of the F_{ik} to vanish is therefore given by the case of the proportionality of the metric tensors.[22]

Although Reichenbach's metaphorical use of "universal forces" issued in misleading and wholly unnecessary complexities which will be pointed out presently, he himself was in no way victimized by them. Writing in 1951 concerning the invocation of universal forces in this context, he declared: "The assumption of such forces means merely a change in the coordinative definition of congruence" [73, p. 133]. It is therefore quite puzzling that, in 1956, Carnap, who had lucidly expounded the same ideas nonmetaphorically as we saw, singled out Reichenbach's characterization and recommendation of the customary congruence definition in terms of equating the universal forces to zero as a praiseworthy part of Reichenbach's outstanding work The Philosophy of Space and Time [72]. In his Preface to the latter work, Carnap says: "Of the many fruitful ideas which Reichenbach contributed . . . I will mention only one, which seems to me of great interest for the methodology of physics but which has so far not found the attention it deserves. This is the principle of the elimination of universal forces. . . . Reichenbach proposes to accept as a general methodological principle that we choose that form of a theory among physically equivalent forms (or, in other words, that definition of 'rigid body' or 'measuring standard') with respect to which all universal forces disappear" [9, p. vii].

The misleading potentialities of including metaphorical "universal forces" in the statement of the congruence definition manifest them-

[22] There is a most simple illustration of the fact that if the metric tensors are not proportional while as few as one of the components F_{ik} is nonvanishing, the congruence associated with the g_{ik} will be incompatible with that of the g_{ik}' and will hence be noncustomary. If we consider the metrics $ds^2 = dx^2 + dy^2$ and $ds^2 = 2dx^2 + dy^2$, then comparison of an interval for which $dx = 1$ and $dy = 0$ with one for which $dx = 0$ and $dy = 1$ will yield congruence on the first of these metrics but not on the second.

selves in three ways as follows: (1) The formulation of noncustomary congruence definitions in terms of deformations by universal forces has inspired the erroneous charge that such congruences are *ad hoc* because they involve the *ad hoc* postulation of nonvanishing universal forces. (2) In Reichenbach's statement of the congruence definition to be employed to explore the spatial geometry in a *gravitational* field, universal forces enter in both a *literal* and a *metaphorical* sense. The conflation of these two senses issues in a seemingly contradictory formulation of the customary congruence definition. (3) Since the variability of the curvature of a space would manifest itself in alterations in the coincidence behavior of all kinds of solid bodies under displacement, Reichenbach speaks of bodies displaced in such a space as being subject to universal forces "destroying coincidences" [72, p. 27]. In a manner analogous to the gravitational case, the conflation of this literal sense with the metaphorical one renders the definition of rigidity for this context paradoxical. We shall now briefly discuss these three sources of confusion in turn.

1. If a congruence definition itself had factual content, so that alternative congruences would differ in factual content, then it would be significant to say of a congruence definition that it is *ad hoc* in the sense of being an *evidentially unwarranted* claim concerning facts. But inasmuch as ascriptions of spatial congruence to noncoinciding intervals are not factual but conventional, neither the customary nor any of the noncustomary definitions of congruence can possibly be *ad hoc*. Hence the abandonment of the former in favor of the latter kind of definition can be no more *ad hoc* than the regraduation of a centigrade thermometer into a Fahrenheit one or than the change from Cartesian to polar coordinates. By formulating noncustomary congruence definitions in terms of the metaphor of universal forces, Reichenbach made it possible for people to misconstrue his metaphorical sense as literal. And once this error had been committed, its victims tacitly regarded the customary congruence definition as factually true and felt justified in dismissing other congruences as *ad hoc* on the grounds that they involved the *ad hoc* assumption of (literally conceived) universal forces. Thus we find that Ernest Nagel, for example, overlooks that the invocation of universal forces to preserve Euclidean geometry can be no more *ad hoc* than a change in the *units* of length (or of temperature) whereby one assures a particular numerical value for the length of a given object in a particular situation. For after granting that, if necessary, Euclidean geometry can be retained by an appeal to universal

forces, Nagel writes: "Nevertheless, universal forces have the curious feature that their presence can be recognized only on the basis of geometrical considerations. The assumption of such forces thus has the appearance of an ad hoc hypothesis, adopted solely for the sake of salvaging Euclid." [55, p. 264.]

2. Regarding the geometry in a gravitational field, Reichenbach says the following: "We have learned . . . about the difference between universal and differential forces. These concepts have a bearing upon this problem because we find that gravitation is a universal force. It does indeed affect all bodies in the same manner. This is the physical significance of the equality of gravitational and inertial mass" [72, p. 256]. It is entirely correct, of course, that a uniform gravitational field (which has not been transformed away in a given space-time coordinate system) is a universal force in the literal sense with respect to a large class of effects such as the free fall of bodies. But there are other effects, such as the bending of elastic beams, with respect to which gravity is clearly a differential force in Reichenbach's sense: a wooden book shelf will sag more under gravity than a steel one. And this shows, incidentally, that Reichenbach's classification of forces into universal and differential is not mutually exclusive. Of course, just as in the case of any other force having differential effects on measuring rods, allowance is made for differential effects of gravitational origin in laying down the congruence definition.

The issue is therefore twofold: (1) Does the fact that gravitation is a universal force in the literal sense indicated above have a bearing on the spatial geometry, and (2) in the presence of a gravitational field is the logic of the spatial congruence definition any different in regard to the role of metaphorical universal forces from what it is in the absence of a gravitational field? Within the particular context of the general theory of relativity—hereafter denoted by "GTR"—there is indeed a literal sense in which the gravitational field of the sun, for example, is causally relevant geometrically as a universal force. And the literal sense in which the coincidence behavior of the transported customary rigid body is objectively different in the vicinity of the sun, for example, from what it is in the absence of a gravitational field can be expressed in two ways as follows: (i) relatively to the congruence defined by the customary rigid body, the spatial geometry in the gravitational field is not Euclidean— contrary to pre-GTR physics—but is Euclidean in the absence of a gravitational field; (ii) the geometry in the gravitational field is Euclidean if

441

and only if the customary congruence definition is supplanted by one in which the length of the rod varies *suitably* with its position or orientation,[23] whereas it is Euclidean relatively to the customary congruence definition for a vanishing gravitational field.

It will be noted, however, that formulation (i) makes no mention at all of any deformation of the rod by universal forces as that body is transported from place to place in a given gravitational field. Nor need there be any metaphorical reference to universal forces in the statement of the customary congruence definition ingredient in formulation (i). For that statement can be given as follows: in the presence no less than in the absence of a gravitational field, congruence is conventional, and hence we are free to adopt the customary congruence in a gravitational field as a basis for determining the spatial geometry. By encumbering his statement of a congruence definition with metaphorical use of "universal forces," Reichenbach enables the unwary to infer incorrectly that a rod subject to the universal force of gravitation in the specified *literal* sense *cannot* consistently be regarded as *free* from deforming universal forces in the *metaphorical* sense and hence cannot serve as the congruence standard. This conflation of the literal and metaphorical senses of "universal force" thus issues in the mistaken belief that in the GTR the customary spatial congruence definition cannot be adopted consistently for the gravitational field. And those who were led to this misconception by Reichenbach's metaphor will therefore deem self-contradictory the following consistent assertion by him: "We do not speak of a change produced by the gravitational field in the measuring instruments, but regard the measuring instruments as 'free from deforming forces' in spite of the gravitational effects" [72, p. 256]. Moreover, those victimized by the metaphorical part of Reichenbach's language will be driven to reject as inconsistent Einstein's characterization of the geometry in the gravitational field in the GTR as given in formulation (i) above. And they will insist erroneously that formulations (i) and (ii) are *not* equally acceptable alternatives on the grounds that formulation (ii) is uniquely correct.

The confounding of the literal and metaphorical senses of "universal force" by reference to gravity is present, for example, in Ernest Nagel's treatment of universal forces with resulting potentialities of confusion. Thus, he incorrectly cites the force of gravitation in its role of being *liter-*

---

[23] For the gravitational field of the sun, the function specifying a noncustomary congruence definition issuing in a Euclidean geometry is given in [8, p. 58].

ally a "universal force" as a *species* of what are only metaphorically "universal forces." Specifically, in speaking of the assumption of universal forces whose "presence can be recognized only on the basis of geometrical considerations" because they are assumed "solely for the sake of salvaging Euclid"—i.e., "universal forces" in the *metaphorical* sense—he says: "'Universal force' is not to be counted as a 'meaningless' phrase, for it is evident that a procedure is indicated for ascertaining whether such forces are present or not. Indeed, gravitation in the Newtonian theory of mechanics is just such a universal force; it acts alike on all bodies and cannot be screened" [55, p. 264, n. 19]. But this misleading formulation suggests the incorrect conclusion that a rod subject to a gravitational field cannot be held to be "free" from "universal forces" in performing its metrical function.

3. In a manner analogous to the gravitational case just discussed, we can assert the following: since congruence is conventional, we are at liberty to use the customary definition of it without regard to whether the geometry obtained by measurements expressed in terms of that definition is one of *variable* curvature or not. And thus we see that without the intrusion of a metaphorical use of "universal forces," the statement of the congruence definition need not take any cognizance of whether the resulting geometry will be one of constant curvature or not. A geometry of constant curvature or so-called congruence geometry is characterized by the fact that the so-called axiom of free mobility holds in it: for example, on the surface of a sphere, a triangle having certain angles and sides in a given place can be moved about without any change in their magnitudes relatively to the customary standards of congruence for intervals and angles. On the other hand, on the surface of an egg, the failure of the axiom of free mobility to hold can be easily seen from the following indicator of the variability of the curvature of that 2-space: a circle and its diameter made of any kind of wire are so constructed that one end P of the diameter is attached to the circle while the other end S is free *not* to coincide with the opposite point Q on the circle though coinciding with it in a given initial position on the surface of the egg. Since the ratio of the diameter and the circumference of a circle *varies* in a space of variable curvature such as that of the egg surface, S will no longer coincide with Q if the circular wire and its attachment PS are moved about on the egg so as to preserve contact with the egg surface everywhere. The indicator

thus exhibits an objective destruction of the coincidence of S and Q which is wholly independent of the indicator's chemical composition (under uniform conditions of temperature, etc.). One may therefore speak *literally* here, as Reichenbach does [72, Section 6] of universal forces acting on the indicator by destroying coincidences. And since the customary congruence definition is entirely permissible as a basis for geometries of variable curvature, there is, of course, no inconsistency in giving that congruence definition by equating universal forces to zero in the *metaphorical* sense, even though the destruction of coincidences attests to the quite *literal* presence of causally efficacious universal forces. But Reichenbach invokes universal forces *literally* without any warning of an impending metaphorical reference to them in the congruence definition. And the reader is therefore both startled and puzzled by the seeming paradox in Reichenbach's declaration [72, p. 27] that "Forces *destroying coincidences* must also be set equal to zero, if they satisfy the properties of the universal forces mentioned on p. 13; only then is the problem of geometry uniquely determined." Again the risk of confusion can be eliminated by dispensing with the metaphor in the congruence definition.

While regarding Reichenbach's *The Philosophy of Space and Time* as still the most penetrating single book on its subject, the analysis we have given compels us to dissent from Nagel's judgment that, in that book, Reichenbach employs "The distinction between 'universal' and 'differential' forces . . . with great clarifying effect" [55, p. 264, n. 18].

Having divested Reichenbach's statements about universal forces of their misleading potentialities, we shall hereafter safely be able to discuss other issues raised by statements of his which are couched in terms of universal forces.

The first of these is posed by the following assertion by him: "We obtain a statement about physical reality only if in addition to the geometry G of the space its universal field of force F is specified. Only the combination

$$G + F$$

is a testable statement" [72, p. 33]. In order to be able to appraise this assertion, consider a surface on which some set of generalized curvilinear (or "Gaussian") coordinates has been introduced and onto which a metric $ds^2 = g_{ik} \, dx^i dx^k$ is then put quite arbitrarily by a capricious choice of a set of functions g_{ik} of the given coordinates.[24] Assume now that the latter specification of the geometry G is *not* coupled with any information regarding F. Is it then correct to say that since this metrization provides no information at all about the coincidence behavior of a rod under transport on the surface, it conveys no factual information whatever about the surface or physical reality? That such an inference is mistaken can be seen from the following: depending upon whether the Gaussian curvature K associated with the stipulated g_{ik} is positive (spherical geometry), zero (Euclidean geometry), or negative (hyperbolic geometry), it is an objective fact about the surface that through a point outside a given geodesic of the chosen metric, there will be respectively 0, 1, or infinitely many other such geodesics which will not *intersect* the given geodesic. Whether or not certain lines on a surface intersect is, however, merely a *topological* fact concerning it. And hence we can say that although an arbitrary metrization of a space without a specification of F is not altogether devoid of factual content pertaining to that space, such a metrization can yield no objective facts concerning the space not already included in the latter's topology.

We can therefore conclude the following: if the description of a space (surface) is to contain empirical information concerning the coincidence behavior of transported rods in that space and if a metric $ds^2 = g_{ik} \, dx^i dx^k$ (and thereby a geometry G) is chosen whose congruences do *not* accord with those defined by the application of the transported rod, then indeed Reichenbach's assertion holds. Specifically, the chosen metric tensor g_{ik} and its associated geometry G must then be coupled with a specification of the different metric tensor g_{ik}' that would have been found experimentally, if the rod had actually been chosen as the congruence standard. But Reichenbach's provision of that specification via the universal force F is quite unnecessarily roundabout. For F is defined by $F_{ik} = g_{ik} - g_{ik}'$ and cannot be known without already knowing *both* metric tensors. Thus, there is a loss of clarity in pretending that the metric tensor g_{ik}', which

[24] The coordinatization of a space, whose purpose it is merely to number points so as to convey their topological neighborhood relations of betweenness, does not, as such, presuppose (entail) a metric. However, the statement of a *rule* assuring that different people will independently effect the *same* coordinatization of a given space may require reference to the use of a rod. But even in the case of coordinates such as ·rectangular (Cartesian) coordinates, whose assignment is carried out by the use of a rigid rod, it is quite possible to ignore the manner in which the coordinatization was effected and to regard the coordinates purely topologically, so that a metric very different from $ds^2 = dx^2 + dy^2$ can then still be introduced quite consistently.

codifies the empirical information concerning the rod, is first obtained from the identity

$$g_{ik}' = g_{ik} - F_{ik}.$$

(ii) Reichenbach's Theory of Equivalent Descriptions.

A metric geometry in conjunction with a specified congruence definition given by a statement about the F_{ik} describes the coincidence behavior of a transported rod. Since *the same facts* of coincidence can be described in a linguistically alternative way by a different geometry when coupled with a suitably different congruence definition, Reichenbach speaks of different geometric descriptions which have the same factual content as "logically equivalent" [69, pp. 374–375] or simply as "equivalent" [73, pp. 133ff]. More generally, since not only spatial and temporal congruence but also metrical simultaneity are conventional in the sense of Section 2, part (i) above, geochronometries based on different definitions of congruence and/or simultaneity are equivalent. Among the equivalent descriptions of space, Reichenbach calls the one employing the customary definition of congruence the "normal system" and the particular geometry appropriate to it the "natural geometry" [73, p. 134]. The choice of a particular member from *within* a class of equivalent descriptions is a matter of convention and hence decided on the basis of convenience. But, as Reichenbach points out correctly, the decision as to which class among nonequivalent classes of equivalent descriptions is true is not at all conventional but is a matter of *empirical fact*. Thus, a Euclidean description and a certain nonEuclidean one *cannot* both be the *natural* geometry of a given space simultaneously: if a normal system is employed, they cannot both be true but will be the respective "normal" *representatives* of two *non*equivalent classes of equivalent descriptions. On the other hand, all of the members of a particular class of equivalent descriptions obviously have the same truth value.

Reichenbach's characterization of the empirical status of various geochronometries in terms of equivalence classes of descriptions is seen to be correct in the light of our preceding analysis, and it is summarized here because of its usefulness in the further pursuit of our inquiry.[25]

[25] Our endorsement of Reichenbach's theory of equivalent descriptions in the context of geochronometry should *not* be construed as an espousal of his application of it to the theory of the states of macro-objects like trees during times when no human being observes them [70, pp. 17–20]. Since a critical discussion of Reichenbach's account of unobserved macro-objects would carry us too far afield in this essay, suffice it

(iii) An Error in the Carnap-Reichenbach Account of the Definition of Congruence: The Non-Uniqueness of Any Definition of Congruence Furnished by Stipulating a Particular Metric Geometry.

Upon undertaking the mathematical implementation of a given choice of a metric geometry in the context of the topology of the space we must ask: does the topology in conjunction with the desired metric geometry determine a *unique* definition of congruence and a *unique* metric tensor, the latter's uniqueness being understood as uniqueness to within a constant factor depending on the choice of unit but *not* as unique representability of the metric tensor (cf. footnote 21) by a particular set of functions g_{ik}? As we shall see, the answer to this question has an immediate bearing on the validity of a number of important assertions made by Carnap and Reichenbach in their writings on the philosophy of geometry. Preparatory to appraising these claims of theirs, we shall show that given the topology, the desired geometry does *not* uniquely specify a metric tensor and thus, in a particular coordinate system, *fails* to determine a unique set of functions g_{ik} to within a constant factor depending on the choice of unit. Our demonstration will take the form of showing that besides the customary definition of congruence, which assigns the same length to the measuring rod everywhere and thereby confers a Euclidean geometry on an ordinary table top, there are infinitely many *other* definitions of congruence having the following property: they likewise yield a Euclidean geometry for that surface but are incompatible with the customary one by making the length of a rod depend on its orientation and/or position.

Thus, consider our horizontal table top equipped with a network of Cartesian coordinates x and y, but now metrize this surface by means of the *non*-standard metric

$$ds^2 = \sec^2\theta \, dx^2 + dy^2,$$

where $\sec^2\theta$ is a constant greater than 1. Unlike the standard metric, this metric assigns to an interval whose coordinates differ by dx *not* the length dx but the *greater* length $\sec\theta$ dx. Although this metric thereby makes the length of a given rod dependent on its orientation, we shall show that the *infinitely* many different *non*-standard congruences generated by values

to say the following here. In my judgment, Reichenbach's application of the theory of equivalent descriptions to unobserved objects lacks clarity on precisely those points on which its relevance turns. But as I interpret the doctrine, I regard it as fundamentally incorrect because its ontology is that of Berkeley's *esse est percipi*.

447

Adolf Grünbaum

of secθ which exceed 1 each impart a Euclidean geometry to the table top no less than does the standard congruence given by

$$ds^2 = dx^2 + dy^2.$$

Accordingly, our demonstration will show that the requirement of Euclideanism does *not* uniquely determine a congruence class of intervals but allows an *infinitude* of incompatible congruences. We shall therefore have established that there are infinitely many ways in which a measuring rod could squirm under transport on the table top as compared to its familiar *de facto* behavior while still yielding a Euclidean geometry for that surface.

To carry out the required demonstration, we first note the preliminary fact that the geometry yielded by a particular metrization is clearly independent of the particular coordinates in which that metrization is expressed. And hence if we expressed the standard metric

$$ds^2 = dx^2 + dy^2$$

in terms of the primed coordinates x′ and y′ given by the transformations

$$x = x' \sec\theta$$
$$y = y',$$

obtaining

$$ds^2 = \sec^2\theta \, dx'^2 + dy'^2,$$

we would obtain a Euclidean geometry as before, since the latter equation would merely express the original standard metric in terms of the primed coordinates. Thus, when the same *invariant* ds of the standard metric is expressed in terms of both primed and unprimed coordinates, the metric coefficients g_{ik}' given by $\sec^2\theta$, 0 and 1 yield a Euclidean geometry no less than do the unprimed coefficients 1, 0 and 1.

This elementary ancillary conclusion now enables us to see that the following non-standard metrization (or remetrization) of the surface in terms of the *original*, unprimed rectangular coordinates must likewise give rise to a Euclidean geometry:

$$ds^2 = \sec^2\theta \, dx^2 + dy^2.$$

For the value of the Gaussian curvature and hence the prevailing geometry depends *not* on the particular coordinates (primed or unprimed) to which the metric coefficients g_{ik} pertain but only on the *functional form* of the g_{ik} [42, p. 281], which is the same here as in the case of the g_{ik}' above.

448

More generally, therefore, the geometry resulting from the *standard* metrization is *also* furnished by the following kind of *non*-standard metrization of a space of points, expressed in terms of the same (unprimed) coordinates as the standard one: the *non*-standard metrization has unprimed metric coefficients g_{ik} which have the same *functional form* (to within an arbitrary constant arising from the choice of unit of length) as those primed coefficients g_{ik}' which are obtained by expressing the *standard* metric in some set or other of *primed* coordinates via a suitable coordinate transformation. In view of the large variety of allowable coordinate transformations, it follows at once that the class of *non*-standard metrizations yielding a Euclidean geometry for a table top is far wider than the already infinite class given by

$$ds^2 = \sec^2\theta\, dx^2 + dy^2, \text{ where } \sec^2\theta > 1.$$

Thus, for example, there is *identity* of *functional form* between the *standard* metric in *polar* coordinates, which is given by

$$ds^2 = d\rho^2 + \rho^2 d\theta^2,$$

and the *non*-standard metric in *Cartesian* coordinates given by

$$ds^2 = dx^2 + x^2 dy^2,$$

since x plays the same role *formally* as ρ, and similarly for y and θ. Consequently, the latter *non*-standard metric issues in a *Euclidean* geometry just as the former standard one does.

It is clear that the multiplicity of metrizations which we have proven for Euclidean geometry obtains as well for each of the non-Euclidean geometries. The failure of a geometry of two or more dimensions to determine a congruence definition uniquely does *not*, however, have a counterpart in the *one*-dimensional time continuum: the demand that Newton's laws hold in their customary metrical form determines a *unique* definition of temporal congruence. And hence it is feasible to rely on the law of translational or rotational inertia to define a time metric or "uniform time."

On the basis of this result, we can now show that a number of claims made by Reichenbach and Carnap respectively are false.

1. In 1951, Reichenbach wrote: "If we change the coordinative definition of congruence, a different geometry will result. This fact is called the *relativity of geometry*" [73, p. 132]. That this statement is false is evident from the fact that if, in our example of the table top, we change our con-

gruence definition from $ds^2 = dx^2 + dy^2$ to any one of the infinitely many definitions incompatible with it that are given by $ds^2 = sec^2\theta dx^2 + dy^2$, precisely the same Euclidean geometry results. Thus, contrary to Reichenbach, the introduction of a nonvanishing universal force corresponding to an alternative congruence does *not* guarantee a change in the geometry. Instead, the correct formulation of the relativity of geometry is that in the form of the ds function the congruence definition uniquely determines the geometry, though *not* conversely, and that any one of the congruence definitions issuing in a geometry G′ can always be replaced by infinitely many *suitably* different congruences yielding a specified different geometry G. In view of the unique fixation of the geometry by the congruence definition in the context of the facts of coincidence, the repudiation of a given geometry in favor of a different one does indeed require a change in the definition of congruence. And the new congruence definition which is expected to furnish the new required geometry can do so in one of the following two ways: (i) by determining a system of geodesics *different* from the one yielded by the original congruence definition, or (ii) if the geodesics determined by the new congruence definition are the *same* as those associated with the original definition, then the angle congruences must be different, i.e., the new congruence definition will have to require a different congruence class of *angles*. (For the specification of the magnitudes assigned to angles by the components g_{ik} of the metric tensor, see [27, pp. 37–38].)

That (ii) constitutes a genuine possibility for obtaining a different geometry will be evident from the following example in which two *incompatible* definitions of congruence

$ds_1{}^2 = g_{ik}dx^i dx^k$, and
$ds_2{}^2 = g_{ik}{}'dx^i dx^k$,

yield the *same* system of geodesics via the equations $\delta \int ds_1 = 0$ and $\delta \int ds_2 = 0$ and yet determine *different* geometries (Gaussian curvatures) because they require incompatible congruence classes of *angles* appropriate to these respective geometries. A horizontal surface which is a Euclidean plane on the *customary* metrization can alternatively be metrized to have the geometry of a hemisphere by projection from the *center* of a sphere through the lower half of the sphere whose south pole is resting on that plane. Upon calling congruent on the horizontal surface segments and angles which are the projections of equal segments and angles respectively on the lower hemisphere, the great circle arcs of the hemisphere map into the Euclidean straight lines of the plane such that every straight of the Euclidean description is *also* a straight (geodesic) of the new hemispherical geometry conferred on the horizontal surface. But the *angles* which are regarded as congruent on the horizontal surface in the new metrization are *not* congruent in the original metrization yielding a Euclidean description.

It must be pointed out, however, that if a change in the congruence definition *preserves* the geodesics, then its issuance in a *different* congruence class of *angles* is only a necessary and *not* a sufficient condition for imparting to the surface a metric geometry *different* from the one yielded by the original congruence definition. This fact becomes evident by reference to our earlier case of the table top's being a model of Euclidean geometry *both* on the customary metric $ds^2 = dx^2 + dy^2$ and on the different metric $ds'^2 = \sec^2\theta dx^2 + dy^2$: the geodesics as well as the geometries furnished by these incompatible metrics are the same, but the angles which are congruent in the new metric are generally *not* congruent in the original one. That these two metrics issue in *incompatible* congruence classes of *angles* though in the *same* geometry can be seen as follows: a Euclidean triangle which is equilateral on the new metric ds' will *not* be equilateral on the customary one ds, and hence the three angles of such a triangle will all be congruent to each other in the former metric but not in the latter.

It is clear now that an *arbitrary* change in the congruence definition for either line segments or angles or both cannot as such guarantee a different geometry.

2. Reichenbach explicitly asserts incorrectly that the geometry uniquely determines a congruence definition appropriate to it. Says he: "There is nothing wrong with a coordinative definition established on the requirement that a certain kind of geometry is to result from the measurements. . . . A coordinative definition can also be introduced by the prescription what the result of the measurements is to be. 'The comparison of length is to be performed in such a way that Euclidean geometry will be the result'—this stipulation is a possible form of a coordinative definition" [72, pp. 33–34]. And in reply to Hugo Dingler's contention [18, p. 50] that the rigid body is uniquely specified by the geometry and only by the latter, Reichenbach mistakenly agrees [68, p. 52; 74, p. 35] that the geom-

etry is sufficient to define congruence and contests only Dingler's further claim that it is necessary.[26]

Carnap [8, p. 54] discusses the dependencies obtaining between (a) the metric geometry, which he symbolizes by "R" in this German publication; (b) the topology of the space and the facts concerning the coincidences of the rod in it, symbolized by "T" for "Tatbestand"; and (c) the metric M ("Mass-setzung"), which entails a congruence definition and is given by the function f (and by the choice of a unit), as will be recalled from the beginning of our Section 3.[27] And he concludes that the functional relations between R, M, and T are such "that if two of them are given, the third specification is thereby uniquely given as well" [8, p. 54]. Accordingly, he writes:

$$R = \phi_1(M,T)$$
$$M = \phi_2(R,T)$$
$$T = \phi_3(M,R).$$

While the first of these dependencies does hold, our example of imparting a Euclidean geometry to a table top by each of two incompatible congruence definitions shows that not only the second but also the third of Carnap's dependencies fails to hold. For the mere specification of M to the effect that the rod will be called congruent to itself everywhere and of R as Euclidean does not tell us whether the coincidence behavior T of the rod on the table top will be such as to coincide successively with those intervals that are equal according to the formula $ds^2 = dx^2 + dy^2$ or with the *different* intervals that are equal on the basis of one of the metrizations $ds^2 = \sec^2\theta \, dx^2 + dy^2$ (where $\sec^2\theta \neq 1$). In other words, the stated specifications of M and R do *not* tell us whether the rod behaves on the table top as we know it to behave in actuality or whether it squirms in any one of infinitely many different ways as compared to its actual behavior.

3. As a corollary of our proof of the nonuniqueness of the congruence definition, we can show that the following statement by Reichenbach is

[26] In Sec. 7, we shall assess the merits of Reichenbach's denial that the geometry is necessary, which he rests on the grounds that rigidity can be defined by the elimination of differential forces.

[27] Although both Carnap's metric M and the distance function

$$ds = \sqrt{g_{ik}dx^i dx^k}$$

can provide a congruence definition, they cannot be deduced from one another without information concerning the coincidence behavior of the rod in the space under consideration.

false: "If we say: actually a geometry G applies but we measure a geometry G', we define at the same time a force F which causes the difference between G and G'" [72, p. 27]. Using our previous notation, we note first that instead of determining a metric tensor g_{ik} uniquely (up to an arbitrary constant), the geometry G determines an infinite class a of such tensors differing other than by being proportional to one another. But since $F_{ik} = g_{ik} - g_{ik}'$ (where the g_{ik}' are furnished by the rod prior to its being regarded as "deformed" by any universal forces), the failure of G to determine a tensor g_{ik} uniquely (up to an arbitrary constant) issues in there being as many *different* universal forces F_{ik} as there are different tensors g_{ik} in the class a determined by G. We see, therefore, that contrary to Reichenbach, there are infinitely many different ways in which the measuring rod can be held to be "deformed" while furnishing the same geometry G.

4. Some Chronometric Ramifications of the Conventionality of Congruence

(i) *Newtonian mechanics.*

On the conception of time congruence as conventional, the preference for the customary definition of isochronism—a preference not felt by Einstein in the general theory of relativity, as we shall see in part (ii)— can derive only from considerations of convenience and elegance so long as the resulting form of the theory is *not* prescribed. Hence, the thesis that isochronism is conventional *precludes* a difference in factual import (content) or in *explanatory power* between two descriptions one of which employs the customary isochronism while the other is a "translation" (transcription) of it into a language employing a time congruence incompatible with the customary one.

As a test case for this thesis of *explanatory parity*, the following kind of counterargument has been suggested in outline. On the Riemannian analysis, congruence must be regarded as conventional in the time continuum of Newtonian dynamics no less than in the theory of relativity. We shall therefore wish to compare in regard to explanatory capability the two forms of Newtonian dynamics corresponding to two different time congruences as follows.

The first of these congruences is defined by the requirement that Newton's laws hold, as modified by the addition of very small corrective terms expressing the so-called relativistic motion of the perihelia. This time congruence will be called "Newtonian," and the time variable whose values represent Newtonian time after a particular unit has been chosen will be denoted by "t." The second time congruence is defined by the rotational motion of the earth. It does not matter for our purpose whether we couple the latter congruence with a unit given by the mean solar second, which is the 1/86400 part of the mean interval between two consecutive meridian passages of the fictitious mean sun, or with a different unit given by the sidereal day, which is the interval between successive meridian passages of a star. What matters is that both the mean solar second and the sidereal day are based on the periodicities of the earth's rotational motion. Assume now that one or another of these units has been chosen, and let T be the time variable associated with that metrization, which we shall call "diurnal time." The important point is that the time variables t and T are *non*linearly related and are associated with *incompatible* definitions of isochronism, because the speed of rotation of the earth *varies relatively to the Newtonian time metric* in several distinct ways [13, pp. 264–267]. Of these, the best known is the relative slowing down of the earth's rotation by the tidal friction between the water in the shallow seas of the earth and the land under it. Upon calculating the positions of the moon, for example, via the usual theory of celestial mechanics, which is based on the Newtonian time metric, the observed positions of the moon in the sky would be found to be *ahead* of the calculated ones *if* we were to identify the time defined by the earth's rotation with the Newtonian time of celestial mechanics. And the same is true of the positions of the planets of the solar system and of the moons of Jupiter in amounts all corresponding to a slowing down on the part of the earth.

Now consider the following argument for the lack of explanatory parity between the two forms of the dynamical theory respectively associated with the t and T scales of time: "*Dynamical* facts will discriminate in favor of the t scale as opposed to the T scale. It is granted that it is *kinematically* equivalent to say

> (a) the earth's rotational motion has slowed down relatively to the "clocks" constituted by various revolving planets and satellites of the solar system,

or

> (b) the revolving celestial bodies speed up their periodic motions relatively to the earth's uniform rotation.

454

But these two statements are *not* on a par explanatorily in the context of the *dynamical* theory of the motions in the solar system. For whereas the slowing down of the earth's rotation in formulation (a) *can* be understood as the dynamical effect of nearby masses (the tidal waters and their friction), no similar dynamical *cause* can be supplied for the accelerations in formulation (b). And the latter fact shows that a theory incorporating formulation (a) has greater explanatory power or factual import than a theory containing (b)." In precisely this vein, D'Abro, though stressing on the one hand that apart from convenience and simplicity there is nothing to choose between different metrics [16, p. 53], on the other hand adduces the provision of *causal* understanding by the t-scale as an argument in its favor and thus seems to construe such differences of simplicity as involving *nonequivalent* descriptions:

If in mechanics and astronomy we had selected at random some arbitrary definition of time, if we had defined as congruent the intervals separating the rising and setting of the sun at all seasons of the year, say for the latitude of New York, our understanding of mechanical phenomena would have been beset with grave difficulties. As measured by these new temporal standards, free bodies would no longer move with constant speeds, but would be subjected to periodic accelerations *for which it would appear impossible to ascribe any definite cause*, and so on. As a result, the law of inertia would have to be abandoned, and with it the entire doctrine of classical mechanics, together with Newton's law. Thus a change in our understanding of congruence would entail far-reaching consequences.

Again, in the case of the vibrating atom, had some arbitrary definition of time been accepted, we should have had to assume that the same atom presented the most capricious frequencies. *Once more it would have been difficult to ascribe satisfactory causes to these* seemingly haphazard fluctuations in frequency; and a simple understanding of the most fundamental optical phenomena would have been well-nigh impossible [16, p. 78, my italics].

To examine this argument, let us set the two formulations of dynamics corresponding to the t and T scales respectively before us mathematically in order to have a clearer statement of the issue.

The differences between the two kinds of temporal congruence with which we are concerned arise from the fact that the functional relationship

$$T = f(t)$$

relating the two time scales is *nonlinear*, so that time intervals which are congruent on the one scale are generally incongruent on the other. It is clear that this function is monotone increasing, and thus we know that permanently

$$\frac{dT}{dt} \neq 0.$$

Moreover, in view of the nonlinearity of $T = f(t)$, we know that dT/dt is *not* constant. Since the function f has an inverse, it will be possible to translate any set of laws formulated on the basis of either of the two time scales into the corresponding other scale. In order to see what form the customary Newtonian force law assumes in diurnal time, we must express the acceleration ingredient in that law in terms of diurnal time. But in order to derive the transformation law for the accelerations, we first treat the velocities. By the chain rule for differentiation, we have, using 'r' to denote the position vector

$$(1) \quad \frac{d\mathbf{r}}{dt} = \frac{d\mathbf{r}}{dT} \frac{dT}{dt}.$$

Suppose a body is *at rest* in the coordinate system in which **r** is measured, *when Newtonian time is employed*, then this body will also be held to be *at rest diurnally*: since we saw that the second term on the right-hand side of equation (1) cannot be zero, the left-hand side of (1) will vanish if and only if the first term on the right-hand side of (1) is zero. Though rest in a given frame in the t-scale will correspond to rest in that frame in the T-scale as well, equation (1) shows that the *constancy* of the non-vanishing Newtonian velocity $d\mathbf{r}/dt$ will *not* correspond to a constant diurnal velocity $d\mathbf{r}/dT$, since the derivative dT/dt changes with both Newtonian and diurnal time. Now, differentiation of equation (1) with respect to the Newtonian time t yields

$$(2) \quad \frac{d^2\mathbf{r}}{dt^2} = \frac{d\mathbf{r}}{dT} \frac{d^2T}{dt^2} + \frac{dT}{dt} \frac{d}{dt}\left(\frac{d\mathbf{r}}{dT}\right).$$

But, applying the chain-rule to the second factor in the second term on the right-hand side of (2), we obtain

$$(2a) \quad \frac{d}{dt}\left(\frac{d\mathbf{r}}{dT}\right) = \frac{d^2\mathbf{r}}{dT^2} \frac{dT}{dt}.$$

Hence (2) becomes

$$(3) \quad \frac{d^2\mathbf{r}}{dt^2} = \frac{d\mathbf{r}}{dT} \frac{d^2T}{dt^2} + \frac{d^2\mathbf{r}}{dT^2}\left(\frac{dT}{dt}\right)^2.$$

456

Solving for the diurnal acceleration, and using equation (1) as well as the abbreviations

$$f'(t) \equiv \frac{dT}{dt} \text{ and } f''(t) \equiv \frac{d^2T}{dt^2},$$

we find

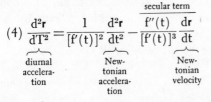

$$(4) \quad \frac{d^2\mathbf{r}}{dT^2} = \underbrace{\frac{1}{[f'(t)]^2} \frac{d^2\mathbf{r}}{dt^2}}_{\substack{\text{New-}\\\text{tonian}\\\text{accelera-}\\\text{tion}}} - \overbrace{\underbrace{\frac{f''(t)}{[f'(t)]^3} \frac{d\mathbf{r}}{dt}}_{\substack{\text{New-}\\\text{tonian}\\\text{velocity}}}}^{\text{secular term}}$$

with the left-hand side labeled *diurnal acceleration*.

Several ancillary points should be noted briefly in regard to equation (4) before seeing what light it throws on the form assumed by causal explanation within the framework of a diurnal description. When the Newtonian force on a body is not zero because the body is accelerating under the influence of masses, the diurnal acceleration will generally also not be zero, save in the unusual case when

$$(5) \quad \frac{d^2\mathbf{r}}{dt^2} = \frac{f''(t)}{f'(t)} \frac{d\mathbf{r}}{dt}.$$

Thus the causal influence of masses, which gives rise to the Newtonian accelerations in the usual description, is seen in (4) to make a definite contribution to the diurnal acceleration as well. But the new feature of the diurnal description of the facts lies in the possession of a *secular acceleration* by all bodies not at rest, even when no masses are inducing Newtonian accelerations, so that the first term on the right-hand side of (4) vanishes. And this secular acceleration is numerically not the same for all bodies but depends on their velocities $d\mathbf{r}/dt$ in the given reference frame and thus also on the reference frame.

The character and existence of this secular acceleration calls for several kinds of comment.

Its dependence on the velocity and on the reference frame should neither occasion surprise nor be regarded as a difficulty of any sort. As to the velocity dependence of the secular acceleration, consider a simple numerical example which removes any surprise: if instead of calling two successive hours on Big Ben equal, we remetrized time so as to assign the measure ½ hour to the second of these intervals, then all bodies having uniform speeds in the usual time metric will double their speeds on the new scale after the first interval, and the *numerical* increase or *acceleration* in the speeds of initially faster bodies will be greater than that in the speeds of the initially slower bodies. Now as for the dependence of the secular acceleration on the reference frame, in the context of the physical facts asserted by the Newtonian theory *apart from* its metrical philosophy it is a mere prejudice to require that, to be admissible, a formulation of that theory must agree with the customary one in making the acceleration of a body at any given time be the same in all Galilean reference frames ("Galilean relativity"). For not a single bona fide physical fact of the Newtonian world is overlooked or contradicted by a kinematics not featuring this Galilean relativity. It is instructive to be aware in this connection that even in the *customary* rendition of the *kinematics of special relativity*, a constant acceleration in a frame S' would *not* generally correspond to a constant acceleration in a frame S, because the component accelerations in S depend not only on the accelerations in S' but also on the component velocities in that system which would be changing with the time.

But what are we to say, apart from the dependence on the velocity and reference system, about the very presence of this "dynamically unexplained" or causally baffling *secular acceleration*? To deal with this question, we first observe merely for comparison that in the *customary* formulation of Newtonian mechanics, constant *speeds* (as distinct from constant *velocities*) fall into *two* classes with respect to being attributable to the dynamical action of perturbing masses: constant *rectilinear* speeds are affirmed to prevail in the absence of any mass influences, while constant *curvilinear* (e.g., circular) speeds are related to the (centripetally) accelerating actions of masses. Now in regard to the presence of a secular acceleration in the diurnal description, it is fundamental to see the following: Whereas on the version of Newtonian mechanics employing the *customary* metrizations (of time and space) *all* accelerations whatsoever in Galilean frames are of *dynamical* origin by being attributable to the action of specific masses, *this feature of Newton's theory is made possible not only by the facts but also by the particular time metrization chosen to codify them.* As equation (4) shows upon equating d^2r/dt^2 to zero, the dynamical character of *all* accelerations is not vouchsafed by any causal facts of the world with which every theory would have to come to terms. For the diurnal description encompasses the objective behavior of bodies (point events and coincidences) as a function of the presence or absence of other bodies no less than does the Newtonian one, thereby achieving full explanatory parity with the latter in all logical (as distinct from pragmatic!) respects.

Hence the provision of a dynamical basis for *all* accelerations should *not* be regarded as an *inflexible epistemological requirement* in the elaboration of a theory explaining mechanical phenomena. *Disregarding* the pragmatically decisive consideration of convenience, there can therefore be no valid explanatory objection to the diurnal description, in which accelerations fall into *two* classes by being the superpositions, in the sense of equation (4), of a dynamically grounded *and* a *kinematically* grounded term. And, most important, since there is no *slowing down* of the earth's rotation on the diurnal metric, there can be no question in that description of specifying a *cause* for such a nonexistent deceleration; instead, a frictional cause is now specified for the earth's diurnally *uniform* rotation *and* for the liberation of heat accompanying this kind of uniform motion. For in the T-scale description it is *uniform* rotation which requires a dynamical cause constituted by masses interacting (frictionally) with the uniformly rotating body, and it is now a law of nature or entailed by such a law that all diurnally uniform rotations issue in the dissipation of heat. Of course, the mathematical representation of the frictional interaction will not have the customary Newtonian form: to obtain the diurnal account of the frictional dynamics of the tides, one would need to apply transformations of the kind given in our equation (4) to the quantities appearing in the relevant Newtonian equations for this case, which can be found in [43, Ch. 8] and in [88].

But, it will be asked, what of the Newtonian conservation principles, if the T scale of time is adopted? It is readily demonstrable by reference to the simple case of the motion of a free particle that while the Newtonian kinetic energy will be constant in this case, its formal diurnal homologue (as opposed to its diurnal equivalent!) will *not* be constant. Let us denote the constant Newtonian velocity of the free particle by "v_t," the subscript "t" serving to represent the use of the t scale, and let "v_T" denote the diurnal velocity corresponding to v_t. Since we know from equation (1) above that $v_t = v_T \, dT/dt$, where v_t is constant but dT/dt is *not*, we see that the diurnal *homologue* $\frac{1}{2}mv_T^2$ of the Newtonian kinetic energy cannot be constant in this case, although the diurnal equivalent $\frac{1}{2}m(v_T \, dT/dt)^2$ of the constant Newtonian kinetic energy $\frac{1}{2}mv_t^2$ is necessarily constant. Just as in the case of the Newtonian equations of motion themselves, so also in the case of the Newtonian conservation principle of mechanical energy, the diurnal equivalent or transcription explains all the facts explained by the Newtonian original. Hence our critic can derive no support at all from the fact that the formal diurnal homologues of Newtonian conservation principles generally do not hold. And we see, incidentally, that the time invariance of a physical quantity and hence the appropriateness of singling it out from among others as a form of "energy," etc., will depend not only on the facts but also on the time metrization used to render them. It obviously will not do, therefore, to charge the diurnal description with inconsistency via the *petitio* of *grafting* onto it the requirement that it incorporate the homologues of Newtonian conservation principles which are incompatible with it: a case in point is the charge that the diurnal description violates the conservation of energy because in its metric the frictional generation of heat in the tidal case is *not* compensated by any reduction in the speed of the earth's rotation! Whether the diurnal time metrization permits the deduction of conservation principles of a *relatively simple type* involving diurnally based quantities is a rather involved mathematical question whose solution is *not* required to establish our thesis that, apart from pragmatic considerations, the diurnal description enjoys explanatory parity with the Newtonian one.

We have been disregarding pragmatic considerations in assessing the explanatory capabilities of two descriptions associated with different time metrizations as to parity. But it would be an error to infer that in pointing to the equivalence of such descriptions in regard to factual content, we are committed to the view that there is no criterion for choosing between them and hence no reason for preferring any one of them to the others. Factual adequacy (truth) is, of course, the cardinal *necessary* condition for the acceptance of a scientific theory, but it is hardly a *sufficient* condition for accepting any one particular formulation of it which satisfies this necessary condition. As well say that a person pointing out that equivalent descriptions can be given in the decimal (metric) and English system of units cannot give telling reasons for preferring the former! Indeed, after first commenting on the *factual basis* of the existence of the Newtonian time congruence, we shall see that there are weighty *pragmatic* reasons for preferring that metrization of the time continuum. And these reasons will turn out to be entirely consonant with our twin contention

460

that alternative metrizability allows linguistically different, equivalent descriptions and that geochronometric conventionalism is not a subthesis of trivial semantical conventionalism.

The *factual basis* of the Newtonian time metrization will be appreciated by reference to the following two considerations: (i) As we shall prove presently, it is a highly fortunate empirical fact, and not an a priori truth, that there *exists* a time metrization *at all* in which *all* accelerations with respect to inertial systems are of dynamic origin, as claimed by the Newtonian theory, and (ii) it is a further empirical fact that the time metrization having this remarkable property (i.e., "ephemeris time") is furnished physically by the earth's annual revolution around the sun (not by its diurnal rotation) albeit not in any observationally simple way, since due account must be taken computationally of the irregularities produced by the gravitational influences of the other planets.[28] That the existence of a time metrization in which all accelerations with respect to inertial systems are of *dynamical* origin cannot be guaranteed a priori is demonstrable as follows.

Suppose that, contrary to actual fact, it were the case that a free body did accelerate when its motion is described in the metric of ephemeris time t, it thus being assumed that there are accelerations in the customary time metric which are not dynamical in origin. More particularly, let us now posit that, contrary to actual fact, a free particle were to execute one-dimensional simple harmonic motion of the form

$$r = \cos\omega t,$$

where r is the distance from the origin. In that hypothetical eventuality, the acceleration of a free particle in the t scale would have the time-dependent value

$$\frac{d^2r}{dt^2} = -\omega^2\cos\omega t.$$

And our problem is to determine whether there would then exist some other time metrization $T = f(t)$ possessing the Newtonian property that our free particle has a zero acceleration. We shall now find that the answer is definitely negative: under the hypothetical empirical conditions which we have posited, there would indeed be no admissible, single-valued

[28] For details on the so-called ephemeris time given by this metric, see [10, 11, 12, 13].

Adolf Grünbaum

time metrization T at all in which all accelerations with respect to inertial systems would be of dynamical origin.

For let us now regard "T" in equation (4) of this subsection as the time variable associated with the sought-after metrization $T = f(t)$ in which the acceleration d^2r/dT^2 of our free particle would be zero. We recall that equation (5) of this subsection was obtained from equation (4) by equating the T-scale acceleration d^2r/dT^2 to zero. Hence *if* our sought-after metrization exists at all, it would have to be the solution $T = f(t)$ of the scalar form of equation (5) as applied to our one-dimensional motion. That equation is

$$(6)\ \frac{d^2r}{dt^2} = \frac{f''(t)}{f'(t)} \frac{dr}{dt} .$$

Putting $v \equiv dr/dt$ and noting that

$$\frac{d}{dt}\log f'(t) = \frac{f''(t)}{f'(t)}, \text{ and } \frac{d}{dt}\log v = \frac{1}{v}\frac{dv}{dt} ,$$

equation (6) becomes

$$\frac{d}{dt}\log v = \frac{d}{dt}\log f'(t).$$

Integrating, and using log c as the constant of integration, we obtain

$$\log v = \log c\, f'(t),$$

or

$$v = c\, f'(t),$$

which is

$$\frac{dr}{dt} = c\, \frac{dT}{dt}.$$

Integration yields

$$(7)\ r = cT + d,$$

where d is a constant of integration. But, by our earlier hypothesis, $r = \cos\omega t$. Hence (7) becomes

$$(8)\ T = \frac{1}{c}\cos\omega t - \frac{d}{c}.$$

It is evident that the solution $T = f(t)$ given by equation (8) is *not* a one-to-one function: the *same* time T in the sought-after metrization would correspond to all those *different* times on the t scale at which the

462

oscillating particle would return to the *same place* $r = \cos\omega t$ in the course of its periodic motion. And by thus violating the basic topological requirement that the function $T = f(t)$ be one-to-one, the T scale which does have the sought-after Newtonian property under our *hypothetical* empirical conditions is physically a quite inadmissible and hence unavailable metrization.

It follows that there is no a priori assurance of the existence of at least one time metrization possessing the Newtonian property that the acceleration of a free particle in inertial systems is zero. So much for the *factual basis* of the Newtonian time metrization.

Now, inasmuch as the employment of the time metrization based on the earth's annual revolution issues in Newton's relatively *simple* laws, there are powerful reasons of *mathematical tractability* and *convenience* for greatly preferring the time metrization in which *all* accelerations are of dynamical origin. In fact, the various refinements which astronomers have introduced in their physical standards for temporal congruence have been dictated by the demand for a definition of temporal congruence (or of a so-called invariable time standard) for which Newton's laws will hold in the solar system, including the relatively simple conservation laws interconnecting diverse kinds of phenomena (mechanical, thermal, etc.). And thus, as Feigl and Maxwell have aptly put it, one of the important criteria of descriptive simplicity which greatly restrict the range of "reasonable" conventions is seen to be the *scope* which a convention will allow for mathematically tractable laws.

ii) The General Theory of Relativity.

In the special theory of relativity, only the customary time metrization is employed in the following sense: At any given point A in a Galilean frame, the length of a time interval between two events at the point A is given by the *difference* between the time coordinates of the two events as furnished by the readings of a standard clock at A whose periods are defined to be congruent. This is, of course, the precise analogue of the customary definition of spatial congruence which calls the rod congruent to itself everywhere, when at relative rest, after allowance for substance-specific perturbations. On the other hand, as we shall now see, there are contexts in which the *general* theory of relativity (GTR) utilizes a criterion of *temporal* congruence which is an analogue of a *noncustomary* kind of spatial congruence in the following sense: the length of a time

Adolf Grünbaum

interval separating two events at a clock depends not only on the differ
ence between the time coordinates which the clock assigns to these event
but also on the *spatial* location of the clock (though not on the time it
self at which the interval begins or ends).

A case in point from the GTR involves a *rotating disk* to which we
apply those principles which GTR takes over from the special theory o
relativity. Let a set of standard material clocks be distributed at variou
points on such a disk. The infinitesimal application of the special rela
tivity clock retardation then tells us the following: a clock at the center C
of the disk will maintain the rate of a contiguous clock located in ar
inertial system I with respect to which the disk has angular velocity ω
but the same does *not* hold for clocks located at other points A of the
disk which are at positive distances r from O. Such A clocks have variou
linear velocities ωr relatively to I in virtue of their common angular ve
locity ω. Accordingly, all A clocks (whatever their chemical constitution
will have readings lagging behind the corresponding readings of the re
spective I-system clocks adjacent to them by a *factor* of

$$\sqrt{1 - \frac{r^2\omega^2}{c^2}},$$

where c is the velocity of light. What would be the consequences of using
the *customary* time metrization everywhere on the rotating disk and let
ting the duration (length) of a time interval elapsing at a given point A
be given by the *difference* between the time coordinates of the termin
of that interval as furnished by the readings of the standard clock at A?
The adoption of the customary time metric would saddle us with a *most*
complicated description of the propagation of light in the rotating system
having the following undesirable features: (i) time would enter the de
scription of nature explicitly in the sense that the one-way velocity of
light would depend on the time, since the lagging rate of the clock at A
issues in a temporal change in the magnitude of the one-way transit time
of a light ray for journeys between O and A, and (ii) the number of light
waves emitted at A during a unit of time on the A clock is *greater* than
the number of waves arriving at the center O in one unit of time on the
O clock [53, pp. 225–226]. To avoid the undesirably complicated laws en
tailed by the use of the simple customary definition of time congruence,
the GTR jettisoned the latter. In its stead, it adopted the following more
complicated, noncustomary congruence definition for the sake of the

implicity of the resulting laws: at any point A on the disk the length
duration) of a time interval is given *not* by the *difference* between the
clock coordinates of its termini but by the *product* of this increment
and the rate factor

$$\frac{1}{\sqrt{1 - \dfrac{r^2\omega^2}{c^2}}}$$

which depends on the spatial coordinate r of the point A. This rate factor
erves to assign a *greater* duration to time intervals than would be ob-
ained from the customary procedure of letting the length of time be
given by the increment in the clock readings. In view of the *dependence*
of the metric *on the spatial position r*, via the rate factor entering into it,
we are confronted here with a *noncustomary* time metrization fully as
onsonant with the temporal order of the events at A as is the customary
metric.

A similarly nonstandard time metric is used by Einstein in his GTR
paper of 1911 [24, Section 3] in treating the effect of gravitation on the
propagation of light. Analysis shows that the very same complexities in
he description of light propagation which are encountered on the rotat-
ng disk arise here as well, if the standard time metric is used. These complexities are eliminated here in quite analogous fashion by the use of a
noncustomary time metric. Thus, if we are concerned with light emitted
on the sun and reaching the earth and if "−Φ" represents the negative
lifference in gravitational potential between the sun and the earth, then
we proceed as follows: prior to being brought from the earth to the sun,
a clock is set to have a rate *faster* than that of an adjoining terrestrial clock
by a factor of

$$\frac{1}{1 - \dfrac{\Phi}{c^2}}$$

(to a first approximation), where

$$\frac{\Phi}{c^2} < 1.$$

(iii) *The Cosmology of E. A. Milne.*

E. A. Milne, whose two logarithmically related t and τ scales of time
were mentioned in Section 2, part (i), has attempted to erect the usual

465

space-time structure of special relativity on the basis of a light signal kinematics of particle observers purportedly dispensing with the use of rigid solids and isochronous material clocks [51, 52]. In his *Modern Cosmology and the Christian Idea of God* [52, Ch. III], Milne begins his discussion of time and space by *incorrectly* charging Einstein with failure to realize that the concept of a rigid body as a body whose rest length is invariant under transport contains a conventional ingredient just as much as does the concept of metrical simultaneity at a distance.[29] Milne then proposes to improve upon a rigid body criterion of spatial congruence by proceeding in the manner of radar ranging and using instead the round trip times required by light to traverse the corresponding closed paths these times *not* being measured by material clocks but, in outline, as follows.[30] Each particle is equipped with a device for ordering the genidentical events belonging to it temporally in a linear Cantorean continuum Such a device is called a "clock," and the single observer at the particle using such a local clock is called a "particle observer." If now A and B are two particle observers and light signals are sent from one to the other, then the time of arrival $\bar{t}'$ at B can be expressed as a function $\bar{t}' = f(t)$ of the time t of emission at A, and likewise the time of arrival t' at A is a function $t' = F(\bar{t})$ of the time $\bar{t}$ of emission at B. Particle observers equipped with clocks as defined are said to be "*equivalent*," if the so-called signal functions f and F are the same, and the clocks of equivalent particle observers are said to be *congruent*. It can be shown that if A and B are not equivalent, then B's clock can be regraduated by a transformation of the form $\bar{t}' = \psi(\bar{t})$ so as to render them equivalent [52, pp. 39–41]. The congruence of the clocks at A and B does not, of course, assure their synchronism. Milne now uses Einstein's definition of simultaneity [52, p. 42]: the time t_2 assigned by A to the arrival of a light ray at B which is emitted at time t_1 at A and returns to A at time t_3 after instantaneous reflection at B is defined to be

$$t_2 = \frac{1}{2}(t_1 + t_3).$$

And he defines the distance r_2 of B, *by A's clock, upon the arrival of the light from A at B* to be given by the relation

$$r_2 = \frac{1}{2}c(t_3 - t_1),$$

[29] That Einstein was abundantly aware of this point is evident from his definition of the "practically rigid body" in [22, p. 9].

[30] For more detailed summaries of Milne's light-signal kinematics see [92, pp. 309–310; 59, pp. 78–85].

where c is an *arbitrarily* chosen constant [52, p. 42]. Since

$$\frac{r_2}{t_2 - t_1} = \frac{r_2}{t_3 - t_2} = c,$$

the constant c represents the velocity of the light signal in terms of the conventions adopted by A for measuring distance and time at a remote point B. Milne gives the following statement of his epistemological objections to Einstein's use of rigid rods and of his claim that his light-signal kinematics provides a philosophically satisfactory alternative to it:

. . . the concept of the transport of a rigid body or rigid length measure is itself an indefinable concept. In terms of one given standard metre, we cannot say what we mean by asking that a given 'rigid' length measure shall remain 'unaltered in length' when we move it from one place to another; for we have no standard of length at the new place. Again, we should have to specify standards of 'rest' everywhere, for it is not clear without consideration that the 'length' will be the same, even at the same place, for different velocities. The fact is that to say of a body or measuring-rod that it is 'rigid' is no definition whatever; it specifies no 'operational' procedure for testing whether a given length-measure after transport or after change of velocity is the same as it was before [52, p. 35]. . . .

It is part of the debt we owe to Einstein to recognize that only 'operational' definitions are of any significance in science . . . Einstein carried out his own procedure completely when he analysed the previously undefined concept of simultaneity, replacing it by tests using the measurements which have actually to be employed to recognize whether two distant events are or are not simultaneous. But he abandoned his own procedure when he retained the indefinable concept of the length of a 'rigid' body, i.e. a length unaltered under transport. The two indefinable concepts of the transportable rigid body and of simultaneity are on exactly the same footing; they are fog-centres, inhibiting further vision, until analysed and shown to be equivalent to conventions [52, p. 35]. . . .

It will be one of our major tasks to elucidate the type of graduation employed for graduating our ordinary clocks; that is to say, to inquire what is meant by, and if possible to isolate what is usually understood by, 'uniform time'. In other words, we wish to inquire which of the arbitrarily many ways in which the markings of our abstract clock may be graduated can be identified with the 'uniform time' of physics [52, p. 37]. . . .

The question now arises: is it possible to arrange that the mode of graduation of observer B's clock corresponds to the mode of graduation of A's clock in such a way that a meaning can be attached to saying that B's clock is a copy of A's clock? If so, we shall say that B's clock has been made congruent with A's [52, p. 39]. . . .

It will have been noticed that we have succeeded in making B's clock a copy of A's without bringing B into permanent coincidence with A. We have made a copy of an arbitrary clock *at a distance*. This is something we cannot do with metre-scales or other length-measures. The problem of copying a clock is in principle simpler than the problem of copying a unit of length. We shall see in due course that with the construction of a copy of a clock at a distance we have solved the problem of comparing lengths [52, p. 41]. . . .

The important point is that epoch and distance (which we shall call coordinates) are purely conventional constructs, and have meaning only in relation to a particular form of clock graduation . . . But it is to be pointed out that when the mode of clock graduation reduces to that of ordinary clocks in physical laboratories, our coordinate conventions provide measures of epoch and distance which coincide with those based on the standard metre [52, pp. 42–43]. . . .

The reason why it is more fundamental to use clocks alone rather than both clocks and scales or than scales alone is that the concept of the clock is more elementary than the concept of the scale. The concept of the clock is connected with the concept of 'two times at the same place', whilst the concept of the scale is connected with the concept of 'two places at the same time'. But the concept of 'two places at the same time' involves a convention of simultaneity, namely, simultaneous events at the two places, but the concept of 'two times at the same place' involves no convention; it involves only the existence of an ego [52, p. 46]. . . .

Length is just as much a conventional matter as an epoch at a distance. Thus the metre-scale is not such a fundamental instrument as the clock. In the first place its length for any observer, as measured by the radar method, depends on the clock used by the observer; in the second place, different observers assign different lengths to it even if their clocks are congruent, owing to the fact that the test of simultaneity is a conventional one. The clock, on the other hand, once graduated, gives epochs at itself which are independent of convention.

Once we have set up a clock, arbitrarily graduated, distances for the observer using this clock become definite. If a rod, moved from one position of rest relative to this observer to another position of rest relative to the same observer, possesses in the two positions the same length, as measured by this observer using his own clock, as graduated, then the rod is said to have undergone a rigid-body-displacement by this clock. In this way we see that once we have fixed on a clock, a rigid-body-displacement becomes definable. But until we have provided a clock, there is no way of saying what we mean by a rigid body under displacement [52, pp. 47–48].

Now, if Milne is to make good his criticism of Einstein by erecting the space-time structure of special relativity on alternative epistemological foundations, he must provide us with inertial systems by means of the resources of his light-signal kinematics as well as with the measures of length and time on which the kinematics of special relativity is predicated. This means that he must be able to characterize inertial systems *within* the confines of his epistemological program as some kind of dense assemblage of equivalent particle observers filling space such that each particle observer is at rest relative to and synchronous with every other. We have already seen that he was wholly in error in charging Einstein with lack of awareness of the conventionality of spatial congruence as defined by the rigid rod. But that, much more fundamentally, he is mistaken in believing to have erected the kinematics of special relativity on an epistemologically more satisfactory base than Einstein did will now be made clear by reference to the following result pointed out by L. L. Whyte [102]: Using only light signals and temporal succession without either a solid rigid rod or an isochronous material clock, it is *not* possible to construct ordinary measures of length and time. For "a physicist using only light signals cannot discriminate inertial systems from these subjected to arbitrary 4-D similarity transformations.[31] The system of 'resting' mass points which can be so identified may be arbitrarily expanding and/or contracting relatively to a rod, and these superfluous transformations can only be eliminated by using a rod or a clock" [102, p. 161].

The significance of the result stated by Whyte is twofold: (i) If Milne dispenses with material clocks and bases his chronometry only on the congruences yielded by his light-signal clocks, then he cannot obtain inertial systems without a rigid rod in the following sense. The rigid rod is *not* needed for the definition of *spatial congruence* within the system but *is* required to assure that the distance between one particular pair of points connected by it *at one time* t_0 is the same as *at some later time* t_1. In other words, the rod is rigid *at a given place* by remaining congruent to itself (by convention) as time goes on. And in this way the rod assures the time constancy of the distance between the two given points connected by it. This reliance on the rigid rod thus involves the use of the definition of simultaneity. Hence, if Milne were right in charging that the use of a rigid rod is beset by philosophic difficulties, then he indeed would be incurring these liabilities no less than Einstein does. On the other hand, suppose that (ii) Milne does use a material clock to define the time metric at a space point and thereby to particularize his clock graduations

[31] For a brief account of similarity transformations, and a further articulation of Whyte's point here, cf. [72, pp. 172–173].

Adolf Grünbaum

to the kind required for the elimination of the unwanted reference systems described by L. L. Whyte. This procedure is a far cry from his purely topological clock which "involves only the existence of an *ego*" [52, p. 46] in contradistinction to the rigid scale's involvement of a definition of simultaneity. And, in that case, his measurement of the equality of space intervals by means of the equality of the corresponding round-trip time intervals involves the following conventions: (a) the tacit use of a definition of simultaneity of noncoinciding events. For although a *round*-trip time on a given clock does *not*, of course, itself require such a simultaneity criterion, the measurement of a spatial distance in an inertial system by means of this time does: the distance yielded by the round-trip time on a clock at A is the distance r_2 between A and B *at the time* t_2 on the A clock *when the light pulse from A arrives at B* on its round-trip ABA, (b) successive equal differences in the readings of a given local clock are stipulated to be measures of equal time intervals and thereby of equal space intervals, and (c) equal differences on separated clocks of identical constitution are decreed to be measures of equal time intervals and thereby of equal space intervals.

To what extent then, if any, does Milne have a case against Einstein? It would appear from our analysis that the only justifiable criticism is not at all epistemological but concerns an innocuous point of axiomatic economy: once you grant Milne a material clock, he does not require the rigid rod at all, whereas Einstein utilizes the spatial congruence definition based on the rigid rod *in addition* to all of the conventions needed by Milne. Thus, Milne's kinematics, as supplemented by the use of a material clock, is constructed on a slightly narrower base of conventions than is Einstein's.[32]

It will be recalled that if measurements of spatial and temporal extension are to be made by means of solid rods and material clocks, allowance must be made computationally for thermal and other perturbations of these bodies so that they can define rigidity and isochronism. Calling attention to this fact and believing Milne's light-signal kinematics to be essentially successful, L. Page [59, pp. 78–79] deemed Milne's construction more adequate than Einstein's, writing: "the original formulation of the relativity theory was based on undefined concepts of space and time

[32] The preceding critique of Milne supplants my earlier brief critique [36, pp. 531–533] in which Milne's arguments were *misinterpreted* as indicative of lack of appreciation on his part of the conventionality of *temporal* congruence.

470

intervals which could not be identified unambiguously with actual observations. Recently Milne has shown how to supply the desired criterion [of rigidity and isochronism] by erecting the space-time structure on the foundations of a constant light-signal velocity." It is apparent in the light of our appraisal of Milne's kinematics that Page's claim is vitiated by Milne's need for a rigid rod or material clock as specified.

It should be noted, however, as Professor A. G. Walker has pointed out to me, that if Milne's construction is interpreted as applying *not* to special relativity kinematics but to his cosmological world model, then our criticisms are no longer pertinent. In terms of his logarithmically related τ and t scales of time, it turns out that upon measuring distances by the specified chronometric convention, the galaxies are at relative rest in τ-scale kinematics and in uniform relative motion in the t scale. Each of these time scales is unique up to a trivial change of units, and their associated descriptions of the cosmological world are equivalent in Reichenbach's sense. In this *cosmological* context, the problem of eliminating the superfluous transformations mentioned by Whyte therefore does not arise.

5. Critique of Some Major Objections to the Conventionality of Spatio-Temporal Congruence

(i) The Russell-Poincaré Controversy.

During the years 1897–1900, B. Russell and H. Poincaré had a controversy which was initiated by Poincaré's review [62] of Russell's *Foundations of Geometry* of 1897, and pursued in Russell's reply [81] and Poincaré's rejoinder [65]. Russell criticized Poincaré's conventionalist conception of congruence and invoked the existence of an intrinsic metric as follows:[33]

It seems to be believed that since measurement [i.e., comparison by means of the congruence standard] is necessary to discover equality or inequality, these cannot exist without measurement. Now the proper conclusion is exactly the opposite. Whatever one can discover by means of an operation must exist independently of that operation: America existed before Christopher Columbus, and two quantities of the same kind must be equal or unequal before being measured. Any method of measurement [i.e., any congruence definition] is good or bad according as it yields a result which is true or false. Mr. Poincaré, on the other hand, holds that

[33] This argument is implicitly endorsed by Helmholtz [91, p. 15].

measurement creates equality and inequality [i.e., that there is no intrinsic metric]. It follows [then] . . . that there is nothing left to measure and that equality and inequality are terms devoid of meaning [81, pp. 687–688].

We have argued that the Newtonian position espoused by Russell is untenable. But our critique of the model-theoretic trivialization of the conventionality of congruence shows that we must reject as inadequate the following kind of criticism of Russell's position, which he would have regarded as a *petitio principii*: "Russell's claim is an absurdity, because it is the denial of the truism that we are at liberty to give whatever physical interpretations we like to such abstract signs as 'congruent line segments' and 'straight line' and then to inquire whether the system of objects and relations thus arbitrarily named is a model of one or another abstract geometric axiom system. Hence, these linguistic considerations suffice to show that there can be no question, as Russell would have it, whether two non-coinciding segments are truly equal or not and whether measurement is being carried out with a standard yielding results that are true in that sense. Accordingly, awareness of the model-theoretic conception of geometry would have shown Russell that alternative metrizability of spatial and temporal continua should never have been either startling or a matter for dispute. And, by the same token, Poincaré could have spared himself a polemic against Russell in which he spoke misleadingly of the conventionality of congruence as a philosophical doctrine pertaining to the structure of space."

Since this model-theoretic argument fails to come to grips with Russell's root assumption of an intrinsic metric, he would have dismissed it as a *petitio* by raising exactly the same objections that the Newtonian would adduce (cf. Section 2, part (i)) against the alternative metrizability of space and time. And Russell might have gone on to point out that the model theoretician cannot evade the spatial equality issue by (i) noting that there are axiomatizations of each of the various geometries dispensing with the abstract relation term "congruent" (for line segments), and (ii) claiming that there can then be no problem as to what physical interpretations of that relation term are permissible. For a metric geometry makes metrical comparisons of equality and inequality, however covertly or circuitously these may be rendered by its language. It is quite immaterial, therefore, whether the relation of spatial equality between line segments is designated by the term "congruent" or by some other term or terms. Thus, for example, Tarski's axioms for elementary Euclidean geometry [87] do not employ the term "congruent" for this purpose, using instead a quaternary predicate denoting the equidistance relation between four points. Also, in Sophus Lie's group-theoretical treatment of metric geometries, the congruences are specified by groups of point transformations [3, pp. 153–154]. But just as Russell invoked his conception of an intrinsic metric to restrict the permissible spatial interpretations of "congruent line segments," so also he would have maintained that it is never arbitrary what quartets of physical points may be regarded as the denotata of Tarski's quaternary equidistance predicate. And he would have imposed corresponding restrictions on Lie's transformations, since the displacements defined by these groups of transformations have the logical character of spatial congruences. These considerations show that it will not suffice in this context simply to take the model-theoretic conception of geometry for granted and thereby to dismiss the Russell-Helmholtz claim peremptorily in favor of alternative metrizability. Rather what is needed is a refutation of the Russell-Helmholtz root assumption of an intrinsic metric: to exhibit the untenability of that assumption as we have endeavored to do in Section 2 is to provide the justification of the model-theoretic affirmation that a given set of physico-spatial facts may be held to be as much a realization of a Euclidean calculus as of a non-Euclidean one yielding the same topology.

The refutation presented in Section 2 requires supplementation, however, to invalidate A. N. Whitehead's perceptualistic version of Russell's argument. We therefore now turn to an examination of Whitehead's philosophy of congruence.

(ii) A. N. Whitehead's Unsuccessful Attempt to Ground an Intrinsic Metric of Physical Space and Time on the Deliverances of Sense.

Commenting on the Russell-Poincaré controversy [97, pp. 121–124], Whitehead maintains the following: (i) Poincaré's argument on behalf of alternative metrizability is unanswerable only if the philosophy of physical geometry and chronometry is part of an epistemological framework resting on an illegitimate bifurcation of nature, (ii) consonant with the rejection of bifurcation, we must ground our metric account of the space and time of nature not on the relations between material bodies and events as fundamental entities but on the more ultimate metric deliverances of sense perception, and (iii) perceptual time and space exhibit an intrinsic metric. Specifically, Whitehead proposes to point out "the factor in nature which issues in the preeminence of one congruence relation over the indefinite herd of other such relations" [97, p. 124] and writes:

The reason for this result is that nature is no longer confined within space at an instant. Space and time are now interconnected; and this peculiar factor of time which is so immediately distinguished among the deliverances of our sense-awareness, relates itself to one particular congruence relation in space [97, p. 124]. . . . Congruence depends on motion, and thereby is generated the connexion between spatial congruence and temporal congruence [97, p. 126].

Whitehead's argument is thus seen to turn on his ability to show that *temporal* congruence cannot be regarded as conventional in physics. He believes to have justified this crucial claim by the following reasoning in which he refers to the conventionalist conception as "the prevalent view" and to his opposing thesis as "the new theory":

The new theory provides a definition of the congruence of periods of time. The prevalent view provides no such definition. Its position is that if we take such time-measurements so that certain familiar velocities which seem to us to be uniform are uniform, then the laws of motion are true. Now in the first place no change could appear either as uniform or nonuniform without involving a definite determination of the congruence for time-periods. So in appealing to familiar phenomena it allows that there is some factor in nature which we can intellectually construct as a congruence theory. It does not however say anything about it except that the laws of motion are then true. Suppose that with some expositors we cut out the reference to familiar velocities such as the rate of rotation of the earth. We are then driven to admit that there is no meaning in temporal congruence except that certain assumptions make the laws of motion true. Such a statement is historically false. King Alfred the Great was ignorant of the laws of motion, but knew very well what he meant by the measurement of time, and achieved his purpose by means of burning candles. Also no one in past ages justified the use of sand in hour glasses by saying that some centuries later interesting laws of motion would be discovered which would give a meaning to the statement that the sand was emptied from the bulbs in equal times. Uniformity in change is directly perceived, and it follows that mankind perceives in nature factors from which a theory of temporal congruence can be formed. The prevalent theory entirely fails to produce such factors [97, p. 137]. . . . On the orthodox theory the position of the equations of motion is most ambiguous. The space to which they refer is completely undetermined and so is the measurement of the lapse of time. Science is simply setting out on a fishing expedition to see whether it cannot find some procedure which it can call the measurement of space and some procedure which it can call the measurement of time, and something which it can call a system of forces, and something which it can call masses, so that these formulae may be satisfied. The only reason—on this theory—why anyone should want to satisfy these formulae is a sentimental regard for Galileo, Newton, Euler and Lagrange. The theory, so far from founding science on a sound observational basis, forces everything to conform to a mere mathematical preference for certain simple formulae.

I do not for a moment believe that this is a true account of the real status of the Laws of Motion. These equations want some slight adjustment for the new formulae of relativity. But with these adjustments, imperceptible in ordinary use, the laws deal with fundamental physical quantities which we know very well and wish to correlate.

The measurement of time was known to all civilised nations long before the laws were thought of. It is this time as thus measured that the laws are concerned with. Also they deal with the space of our daily life. When we approach to an accuracy of measurement beyond that of observation, adjustment is allowable. But within the limits of observation we know what we mean when we speak of measurements of space and measurements of time and uniformity of change. It is for science to give an intellectual account of what is so evident in sense-awareness. It is to me thoroughly incredible that the ultimate fact beyond which there is no deeper explanation is that mankind has really been swayed by an unconscious desire to satisfy the mathematical formulae which we call the Laws of Motion, formulae completely unknown till the seventeenth century of our epoch [97, pp. 139–140].

After commenting that purely mathematically, an infinitude of incompatible spatial congruence classes of intervals satisfy the congruence axioms, Whitehead says:

This breakdown of the uniqueness of congruence for space . . . is to be contrasted with the fact that mankind does in truth agree on a congruence system for space and a congruence system for time which are founded on the direct evidence of its senses. We ask, why this pathetic trust in the yard measure and the clock? The truth is that we have observed something which the classical theory does not explain.

It is important to understand exactly where the difficulty lies. It is often wrongly conceived as depending on the inexactness of all measurements in regard to very small quantities. According to our methods of observation we may be correct to a hundredth, or a thousandth, or a millionth of an inch. But there is always a margin left over within which we cannot measure. However, this character of inexactness is *not* the difficulty in question.

Let us suppose that our measurements can be ideally exact; it will be still the case that if one man uses one qualifying [i.e., congruence] class γ and the other man uses another qualifying [i.e., congruence] class δ, and if they both admit the standard yard kept in the exchequer chambers to be their unit of measurement, they will disagree as to what other distances [at other] places should be judged to be equal to that standard distance in the exchequer chambers. Nor need their disagreement be of a negligible character [99, pp. 49–50]. . . .

When we say that two stretches match in respect to length, what do we mean? Furthermore we have got to include time. When two lapses of time match in respect to duration, what do we mean? We have seen that measurement presupposes matching, so it is of no use to hope to explain matching by measurement [99, pp. 50–51]. . . .

Our physical space therefore must already have a structure and the matching must refer to some qualifying class of quantities inherent in this structure [99, p. 51].

. . . there will be a class of qualities γ one and only one of which attaches to any stretch on a straight line or on a point, such that matching in respect to this quality is what we mean by congruence.

The thesis that I have been maintaining is that measurement presupposes a perception of matching in quality. Accordingly in examining the meaning of any particular kind of measurement we have to ask, What is the quality that matches? [99, p. 57].

. . . a yard measure is merely a device for making evident the spatial congruence of the [extended] events in which it is implicated [99, p. 58].

Let us now examine the several strands in Whitehead's argument in turn. We shall begin by inquiring whether his historical observation that the human race possessed a time metric prior to the enunciation of Newton's laws during the seventeenth century can serve to invalidate Poincaré's contentions [64] that (1) time congruence in physics is conventional, (2) the definition of temporal congruence used in refined physical theory is given by Newton's laws, and (3) we have no direct intuition of the temporal congruence of nonadjacent time intervals, the belief in the existence of such an intuition resting on an illusion.

To see how unavailing Whitehead's historical argument is, consider first the spatial analogue of his reasoning. We saw in Section 3, part (iii) that although the demand that Newton's laws be true does uniquely define temporal congruence in the one-dimensional time continuum, it is *not* the case that the requirement of the applicability of Euclidean geometry to a table top similarly yields a unique definition of spatial congruence for that two-dimensional space. For the sake of constructing a *spatial* analogue to Whitehead's historical argument, however, let us assume that, *contrary to fact*, it were the case that the requirement of the Euclidean-ism of the table top did uniquely determine the *customary* definition of perfect rigidity. And now suppose that a philosopher were to say that the latter definition of spatial congruence, like all others, is conventional. What then would be the force of the following kind of Whiteheadian assertion: "Well before Hilbert rigorized Euclidean geometry and even much before Euclid less perfectly codified the geometrical relations be-tween the bodies in our environment, men used not only their own limbs but also diverse kinds of solid bodies to certify spatial equality"? Ignoring now refinements required to allow for substance-specific distortions, it is clear that, under the assumed hypothetical conditions, we would be con-fronted with *logically independent* definitions of spatial equality[34] issu-ing in the *same* congruence class of intervals. The concordance of these definitions would indeed be an impressive empirical fact, but it could not possibly refute the claim that the one congruence defined alike by all of them is conventional.

Precisely analogous considerations serve to invalidate Whitehead's his-torical argument regarding time congruence, if we discount Milne's hy-pothesis of the incompatibility of the congruences defined by "atomic" and "astronomical" clocks (cf. Section 4, part (iii)) and consider the agreement obtained after allowance for substance-specific idiosyncrasies between the congruences defined by a class of devices located in vanish-ing or stationary gravitational fields. A candle always made of the same material, of the same size, and having a wick of the same material and size burns very nearly the same number of inches each hour. Hence as early as during the reign of King Alfred (872–900), burning candles were used as rough time keepers by placing notches or marks at such a distance apart that a certain number of spaces would burn each hour [50, pp. 53–54]. Ignoring the relatively small variations of the rate of flow of water with the height of the water column in a vessel, the water clock or clep-sydra served the ancient Chinese, Byzantines, Greeks, and Romans [67], as did the sand clock, keeping very roughly the same time as burning candles. Again, an essentially frictionless pendulum oscillating with con-

---

[34] As noted in fn. 26, we shall show in Sec. 7 in what sense the criterion of rigidity based on the solid body is logically independent of Euclidean geometry when cog-nizance *is taken* of substance-specific distortions.

477

stant amplitude at a point of given latitude on the earth defines the same time metric as do "natural clocks," i.e., quasi-closed periodic systems [5]. And, ignoring various refinements, similarly for the rotation of the earth, the oscillations of crystals, the successive round trips of light over a fixed distance in an inertial system, and the time based on the natural periods of vibrating atoms or "atomic clocks" [93; 42; 2; 46].

Thus, unless Milne is right, we find a striking concordance between the time congruence defined by Newton's laws and the temporal equality furnished by several kinds of definitions logically independent of that Newtonian one. This agreement obtains as a matter of *empirical fact* (cf. Section 4, part (i)) for which the GTR has sought to provide an explanation through its conception of the metrical field, just as it has endeavored to account for the corresponding concordance in the coincidence behavior of various kinds of solid rods [16, pp. 78–79]. No one, of course, would wish to deny that of all the definitions of temporal congruence which yield the same time metric as the Newtonian laws, some were used by man well before these laws could be invoked to provide such a definition. Moreover, Whitehead might well have pointed out that it was only because it was possible to measure time in one or another of these pre-Newtonian ways that the discovery and statement of Newton's laws became possible. But what is the bearing of these genetic considerations and of the (presumed) fact that the same time congruence is furnished alike by each of the aforementioned logically independent definitions on the issue before us? It seems quite clear that they cannot serve as grounds for impugning the thesis that the equality obtaining among the time intervals belonging to the one congruence class in question is conventional in the Riemann-Poincaré sense articulated in this essay: this particular equality is no less conventional in virtue of being defined by a plethora of physical processes in addition to Newton's laws than if it were defined merely by one of these processes alone or by Newton's laws alone.

Can this conclusion be invalidated by adducing such agreement as does obtain under appropriate conditions between the metric of psychological time and the physical criterion of time congruence under discussion? We shall now see that the answer is decidedly in the negative.

Prior attention to the source of such concordance as does exist between the psychological and physical time metrics will serve our endeavor to determine whether the metric deliverances of psychological time furnish any support for Whitehead's espousal of an intrinsic metric of physical time.[35]

It is well known that in the presence of strong emotional factors such as anxiety, exhilaration, and boredom, the psychological time metric exhibits great variability as compared to the Newtonian one of physics. But there is much evidence that when such factors are not present, physiological processes which are geared to the periodicities defining physical time congruence impress a metric upon man's psychological time and issue in rhythmic behavior on the part of a vast variety of animals. There are two main theories at present as to the source of the concordance between the metrics of physical and psychobiological time. The older of these maintains that men and animals are equipped with an internal "biological clock" *not* dependent for its successful operation on the conscious or unconscious reception of sensory cues from outside the organism. Instead the success of the biological clock is held to depend only on the occurrence of metabolic processes whose rate is *steady in the metric of physical clock time* [30, 39, 40]. As applied to humans, this hypothesis was supported by experiments of the following kind. People were asked to tap on an electric switch at a rate which they judged to be a fixed number of times per second. It was found over a relatively small range of body temperatures that the temperature coefficient of counting was much the same as the one characteristic of chemical reactions: a two or threefold increase in rate for a $10°$ C rise in temperature [17]. The defenders of the conception that the biological clock is purely internal further adduce observations of the behavior of bees: both outdoors on the surface of the earth and at the bottom of a mine, bees learned to visit at the correct time each day a table on which a dish of syrup was placed daily for a short time at a fixed hour. Having been found to be hungry for sugar all day long, neither the assumption that the bees experience periodic hunger, nor the ap-

---

[35] It will be noted that Whitehead does *not* rest his claim of the intrinsicality of the temporal metric on his thesis of the *atomicity* of becoming. We therefore need not deal here with the following of his contentions: (i) becoming or the transiency of "now" is a feature of the time of physics, the bifurcation of nature being philosophically illegitimate, and (ii) there is no continuity of becoming but only becoming of continuity [101, p. 53]. But the reader is referred to F. S. C. Northrop's rebuttal to Whitehead's attack on bifurcation [57], to my demonstration of the irrelevance of becoming to physical (as distinct from psychological) time [32, Sec. 4], to my critique [37] of Whitehead's use of the "Dichotomy" paradox of Zeno of Elea to prove that time intervals are only potential and not actual continua, and to my essay "Whitehead's Philosophy of Science," *Philosophical Review*, April 1962, for a defense of bifurcation.

pearance of the sun nor yet the periodicities of the cosmic ray intensity can explain the bees' success in time keeping. But dosing them with substances like thyroid extract and quinine, which affect the rate of chemical reactions in the body, was found to interfere with their ability to appear at the correct time.

More recently, however, doubt has been cast on the adequacy of the hypothesis of the purely internal clock. A series of experiments with fiddler crabs and other cold-blooded animals [6, 7] showed that these organisms hold rather precisely to a 24-hour coloration cycle (lightening-darkening rhythm) regardless of whether the temperature at which they are kept is 26 degrees, 16 degrees, or 6 degrees centigrade, although at temperatures near freezing, the color clock changes. It was therefore argued that if the rhythmic timing mechanism were indeed a biochemical one wholly inside the organism, then one would expect the rhythm to speed up with increasing temperature and to slow down with decreasing temperature. And the exponents of this interpretation maintain that since the period of the fiddler crab's rhythm remained 24 hours through a wide range of temperature, the animals must possess a means of measuring time which is independent of temperature. This, they contend, is "a phenomenon quite inexplicable by any currently known mechanism of physiology, or, in view of the long period-lengths, even of chemical reaction kinetics" [7, p. 159]. The extraordinary further immunity of certain rhythms of animals and plants to many powerful drugs and poisons which are known to slow down living processes greatly is cited as additional evidence to show that organisms have daily, lunar, and annual rhythms impressed upon them by external physical agencies, thus having access to outside information concerning the corresponding physical periodicities. The authors of this theory admit, however, that the daily and lunar-tidal rhythms of the animals studied do not depend upon any now known kind of external cues of the associated astronomical and geophysical cycles [7, pp. 153, 166]. And it is postulated [7, p. 168] that these physical cues are being received because living things are able to respond to additional kinds of stimuli at energy levels so low as to have been previously held to be utterly irrelevant to animal behavior. The assumption of such sensitivity of animals is thought to hold out hope for an explanation of animal navigation.

We have dwelled on the two current rival theories regarding the source of the ability of man (and of animals) to make successful estimates of duration introspectively in order to show that, on either theory, the metric of psychological time is tied causally to those physical cycles which serve to define time congruence in physics. Hence when we make the judgment that two intervals of physical time which are equal in the metric of standard clocks also appear congruent in the psychometry of mere sense awareness, this justifies only the following innocuous conclusion in regard to physical time: the two intervals in question are congruent by the physical criterion which had furnished the psychometric standard of temporal equality both genetically and epistemologically. How then can the metric deliverances of psychological time possibly show that the time of physics possesses an intrinsic metric, if, as we saw, no such conclusion was demonstrable on the basis of the cycles of physical clocks?

As for spatial congruence, what are we to say of Whitehead's argument [100, p. 56] that just as it is an objective datum of experience that two phenomenal color patches have the same color, i.e., are "color-congruent," so also we see that a given rod has the same length in different positions, thus making the latter congruence as objective a relation as the former? As Whitehead puts it: "It is at once evident that all these tests [of congruence by means of steel yard measures, etc., are] dependent on a direct intuition of permanence" [101, p. 501]. He would argue, for example, that in the accompanying diagram the horizontal segment AC could not be stipulated to be congruent to the vertical segment AB. For the deliverances of our visual intuition unequivocally show AC to be shorter than AB and AB to be congruent to AD, a fact also attested by the finding that a solid rod coinciding with AB to begin with and then rotated into the horizontal position would extend over AC and coincide with AD.

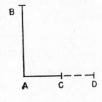

On this my first comment is to ask: What is the significance for the status of the metric of *physical* as distinct from *visual* space of these observational deliverances? And I answer that their significance is entirely consonant with the conventionalist view of *physical* congruence expounded above. The criterion for ocular congruence in our visual field was presumably furnished both genetically and epistemologically by ocular adaptation to the behavior of transported solids. For when pressed as to what it is about two congruent-*looking* intervals that enables them to sustain the relation of spatial equality, our answer will inevitably have to be this: the fact of their capacity to coincide successively with a transported solid rod. Hence when we make the judgment that two intervals of physical space with which transported solid rods coincide in succession also *look* congruent even when compared frontally purely by inspection, what this proves in regard to *physical space* is only that these intervals are congruent on the basis of the criterion of congruence which had furnished the basis for the ocular congruence to begin with, a criterion given by solid rods. But the visual deliverance of congruence does *not* constitute an ocular test of the "true" rigidity of solids under transport in the sense of establishing the factuality of the congruence defined by this class of bodies. Thus, it is a *fact* that in the diagram AD extends over (includes) AC, thus being longer. And it will be recalled that Riemann's views on the status of measurement in a spatial continuum require that every definition of "congruent" be consistent with this kind of inclusional fact. How then can visual data possibly interdict our calling AC congruent to AB and then allowing for the *de facto* coincidence of the rotated rod with AB and AD by assigning to the rod in the horizontal position a length which is suitably *greater* than the one assigned to it in the vertical orientation?

It will be recalled (cf. Section 5, part (i)) that Russell had unsuccessfully sought to counter Poincaré's position by answering the question "What is it that is measured?" on the basis of the affirmation of an intrinsic metric. Whitehead believes himself to have supplied the missing link in Russell's answer by having adduced the deliverances of *visual* space and of psychological time. It remains for us to consider briefly the further reasons put forward by Whitehead in his endeavor to show the following: transported rods and the successive periods of clocks can be respectively held to be *truly* unaltered or congruent to themselves, thereby rendering testimony of an intrinsic metric, because an intuitively apprehended matching relation obtains between the visual and psychotemporal counterparts of the respective space and time intervals in question.

Invoking visual congruence, Whitehead claims [97, p. 121] that an immediate perceptual judgment tells us that whereas an elastic thread does *not* remain unaltered under transport, a yard measure does. And from this he draws three conclusions [97, p. 121]: (i) "immediate judgments of congruence are presupposed in measurement," (ii) "the process of measurement is merely a procedure to extend the recognition of congruence to cases where these immediate judgments are not available," and

(iii) "we cannot define congruence by measurement."[36] The valid core of assertions (i) and (iii) is that measurement presupposes (requires) a congruence criterion in terms of which its results are formulated. But this does not, of course, suffice to show that the congruence thus presupposed is nonconventional. Neither can the latter conclusion be established by Whitehead's contention that we apprehend a *matching* relation among those intervals which are congruent according to the customary standards of rigidity (or isochronism). For the matching among intervals which are congruent relatively to a rigid rod is only with respect to such metrically nonintrinsic properties as the coincidence of each of them with that transported rod, or as the round trip times required by light to traverse them in both directions in an inertial system. To this, Whitehead retorts with the declaration "There is a modern doctrine that 'congruence' *means* the possibility of coincidence. . . . although 'coincidence' is used as a *test* of congruence, it is not the *meaning* of congruence" [101, p. 501]. The issue raised by Whitehead here is, of course, one of *uniqueness* and *intrinsicness of equality* among intervals. We are therefore *not* concerned with the *separate* point that no one physical criterion can exhaustively specify "*the* meaning" of the open cluster concept of congruence as applied to any particular congruence class of intervals. Accordingly, we ignore here refinements that would allow for the open cluster character of physical congruence. And we point out that if there were a meaning in the ascription of an intrinsic metric to space, then it would be quite correct to regard coincidence as only a test of congruence in Whitehead's sense of ascertaining the existence of intrinsic equality. For in that case one would be able to speak of two separated intervals as matching spatially in the sense of containing the same intrinsic amount of space. But it is the existence of an intrinsic metric which is first at issue. And the position taken by Whitehead on that issue cannot be justified *without begging the question* by simply *asserting* that coincidence is only the test but not the meaning of congruence. Hence, his argument from visual congruence having failed, as we saw, Whitehead has not succeeded in refuting the conception that (1) the matching lies wholly in the objective coincidences of each of two (or more) intervals with the same transported rod and (2) the self-congruence of the rod under transport is a matter of convention.

It is significant, however, that there are passages in Whitehead where

[36] Cf. also [100, pp. 54–55].

he comes close to the admission that the preeminent role of certain classes of physical objects as our standards of rigidity and isochronism is *not* tantamount to their making evident the *intrinsic* equality of certain spatial and temporal intervals. Thus speaking of the space-time continuum, he says: "This extensive continuum is one relational complex in which all potential objectifications find their niche. It underlies the whole world, past, present, and future. Considered in its full generality, apart from the additional conditions proper only to the cosmic epoch of electrons, protons, molecules, and star-systems, the properties of this continuum are very few and do not include the relationships of metrical geometry" [101, p. 103]. And he goes on to note that there are competing systems of measurement giving rise to alternative families of straight lines and correspondingly alternative systems of metrical geometry of which no one system is more fundamental than any other [101, p. 149]. It is in our present cosmic epoch of electrons, protons, molecules, and star systems that "more special defining characteristics obtain" and that "the ambiguity as to the relative importance of competing definitions of congruence" is resolved in favor of "one congruence definition" [101, p. 149]. Thus Whitehead maintains that among competing congruence definitions, "That definition which enters importantly into the internal constitutions of the dominating . . . entities is the important definition for the cosmic epoch in question" [101, p. 506]. This important concession thus very much narrows the gap between Whitehead's view and the Riemann-Poincaré conception defended in this essay, viz., that once a congruence definition has been given conventionally by means of the customary rigid body (or otherwise), then, assuming the usual physical interpretation of the remainder of the geometrical vocabulary, the question as to which metric geometry is true of physical space is one of objective physical fact. That the gap between the two views is narrowed by Whitehead's concession here becomes clear upon reading the following statement by him in the light of that concession. Speaking of Sophus Lie's treatment of congruence classes and their associated metric geometries in terms of groups of transformations between points (cf. part (i) of Section 5), Whitehead cites Poincaré and says:

The above results, in respect to congruence and metrical geometry, considered in relation to existent space, have led to the doctrine that it is intrinsically unmeaning to ask which system of metrical geometry is true of the physical world. Any one of these systems can be applied, and in an indefinite number of ways. The only question before us is one of convenience in respect to simplicity of statement of the physical laws. This point of view seems to neglect the consideration that science is to be relevant to the definite perceiving minds of men; and that (neglecting the ambiguity introduced by the invariable slight inexactness of observation which is not relevant to this special doctrine) we have, in fact, presented to our senses a definite set of transformations forming a congruence-group, resulting in a set of measure relations which are in no respect arbitrary. Accordingly our scientific laws are to be stated relevantly to that particular congruence-group. Thus the investigation of the type (elliptic, hyperbolic or parabolic) of this special congruence-group is a perfectly definite problem, to be decided by experiment [98, p. 265].

(iii) A. S. Eddington's Specious Trivialization of the Riemann-Poincaré Conception of Congruence and the Elaboration of Eddington's Thesis by H. Putnam and P. K. Feyerabend.

Though Whitehead's argument that observed matching relations attest the existence of an intrinsic metric is faulty, as we saw, that argument can no more be dismissed on the basis of the model-theoretic trivialization of the congruence issue than Russell's argument against Poincaré (cf. Section 5, part (i)). In fact, one should have supposed that those who maintain with Eddington that GC is a subthesis of TSC (cf. Section 2, part (i)) would have suspected that their critique had missed the point. For what should have given them pause is that Russell, Whitehead, and, for that matter, Poincaré were clearly aware of the place of geometry in the theory of models of abstract calculi and yet carried on their philosophical polemic regarding the status of congruence. According to the triviality thesis, the stake in their controversy was no more than the pathetic one that Russell and Whitehead were advocating the customary linguistic usage of the term "congruent" (for line segments) while Poincaré was maintaining that we need not be bound by the customary usage but are at liberty to introduce bizarre ones as well. Thus, commenting on Poincaré's statement that we can always avail ourselves of alternative metrizability to give a Euclidean interpretation of any results of stellar parallax measurements ([63, p. 81]; cf. Section 6 below), Eddington writes:

Poincaré's brilliant exposition is a great help in understanding the problem now confronting us. He brings out the interdependence between geometrical laws and physical laws, which we have to bear in mind continu-

485

ally.[37] We can add on to one set of laws that which we subtract from the other set. I admit that space is conventional—for that matter, the meaning of every word in the language is conventional. Moreover, we have actually arrived at the parting of the ways imagined by Poincaré, though the crucial experiment is not precisely the one he mentions. But I deliberately adopt the alternative, which, he takes for granted, everyone would consider less advantageous. I call the space thus chosen *physical space*, and its geometry *natural geometry*, thus admitting that other conventional meanings of space and geometry are possible. If it were only a question of the meaning of space—a rather vague term—these other possibilities might have some advantages. But the meaning assigned to length and distance has to go along with the meaning assigned to space. Now these are quantities which the physicist has been accustomed to measure with great accuracy; and they enter fundamentally into the whole of our experimental knowledge of the world. . . . Are we to be robbed of the terms in which we are accustomed to describe that knowledge? [20, pp. 9–10.]

We see that Eddington objects to Poincaré's willingness to guarantee the retention of Euclidean geometry by resorting to an alternative metrization: in the context of general relativity, the retention of Euclideanism would indeed require a congruence definition different from the customary one. Regarding the possibility of a remetrizational retention of Euclidean geometry as merely illustrative of being able to avail oneself of the conventionality of all language, Eddington would rule out such a procedure on the grounds that the customary definition of spatial congruence which would be supplanted by it retains its usefulness.

Earlier in this essay (cf. Section 2, part (i)), we presented a critique of Eddington's claim that GC is a subthesis of TSC by giving an analysis of the sense in which ascriptions of congruence (and of simultaneity) are conventional.

In the present section, we shall examine the following corollary of Eddington's contention as elaborated by H. Putnam and P. K. Feyerabend[38]: GC must be a subthesis of TSC because GC has bona fide analogues in every branch of human inquiry, such that GC cannot be construed as an insight into the structure of space or time. As Eddington puts it: "The law of Boyle states that the pressure of a gas is proportional to its density. It is found by experiment that this law is only approximately

---

[37] This interdependence will be analyzed in Sec. 6 below.

[38] Professor Feyerabend no longer endorses Eddington's position. I cite his name in this context merely to acknowledge his further articulation of the Eddington-Putnam thesis.

true. A certain mathematical simplicity would be gained by conventionally redefining pressure in such a way that Boyle's law would be rigorously
obeyed. But it would be high-handed to appropriate the word pressure in
this way, unless it had been ascertained that the physicist had no further
use for it in its original meaning."

P. K. Feyerabend has noted that what Eddington seems to have in
mind here is the following: instead of revising Boyle's law

$$pv = RT$$

in favor of van der Waals' law

$$\left(p + \frac{a}{v^2}\right)(v - b) = RT,$$

we could preserve the statement of Boyle's law by merely redefining "pressure"—now to be symbolized by "P" in its new usage—putting

$$P =_{\text{Def}} \left(p + \frac{a}{v^2}\right)\left(1 - \frac{b}{v}\right).$$

In the same vein, H. Putnam maintains that instead of using phenomenalist (naive realist) color words as we do customarily in English, we could
adopt a new usage for such words—to be called the "Spenglish" usage—
as follows: we take a white piece of chalk, for example, which is moved
about in a room, and we lay down the rule that depending (in some specified way) upon the part of the visual field which its appearance occupies,
its color will be called "green," "blue," "yellow," etc., rather than "white"
under constant conditions of illumination.

It is a fact, of course, that whereas actual scientific practice in the GTR,
for example, countenances and uses remetrizational procedures based on
noncustomary congruence definitions,[39] scientific practice does not contain any examples of Putnam's "Spenglish" space-dependent (or time-
dependent) use of phenomenalist (naive realist) color predicates to denote the color of a given object in various places (or at various times)
under like conditions of illumination. According to Eddington and Put-

[39] The proponents of ordinary language usage in science, to whom the "ordinary
man" seems to be the measure of all things, may wish to rule out noncustomary congruence definitions as linguistically illegitimate. But they would do well to remember
that it is no more incumbent upon the scientist (or philosopher of science) to use the
customary scientific definition of congruence in every geochronometric description
than it is obligatory for, say, the student of mechanics to be bound by the familiar
common sense meaning of "work," which contradicts the mechanical meaning as
given by the space integral of the force.

nam, the existence of noncustomary usages of "congruent" in the face of there being no such usages of color predicates is no more than a fact about the linguistic behavior of the members of our linguistic community. We saw that the use of linguistic alternatives in the specifically geochronometric contexts reflects fundamental *structural properties* of the facts to which these alternative descriptions pertain. And we must now show that the alleged Eddington-Putnam analogues of GC are *pseudo-analogues.*

The essential point in assessing the cogency of the purported analogues is the following: do the domains from which they are drawn (e.g., phenomenalist or naive realist color properties, and pressure phenomena) exhibit *structural* counterparts to (a) those factual properties of the world postulated by relativity which entail the nonuniqueness of topological simultaneity, and (b) the postulated topological properties of physical space and time which make for the nonuniqueness of the congruence axioms by assuring the nonexistence of an intrinsic metric in these manifolds? Or are the examples cited by Eddington and Putnam analogues of the conventionality of metrical simultaneity or of congruence *only* in the impoverished, trivial sense that they feature linguistically alternative equivalent descriptions *while lacking* the following decisive property of the geochronometric cases: the alternative metrizations are the linguistic renditions or reverberations, as it were, of the *structural properties* assuring the aforementioned two kinds of nonuniqueness enunciated by GC? If the examples given are analogues only in the superficial, impoverished sense—as indeed I shall show them to be—then what have Eddington and Putnam accomplished by their examples? In that case they have merely provided unnecessary illustrations of the correctness of TSC *without* proving their examples to be on a par with the geochronometric ones. In short, their examples will then have served in no way to make good their claim that GC is a subthesis of TSC.

We shall find presently that their examples fail because (a) the domains to which they pertain do *not* exhibit structural counterparts to those features of the world which make the definitions of topological simultaneity and the axioms for spatial or temporal congruence *nonunique,* and (b) Putnam's example in "Spenglish" is indeed an illustration *only* of the trivial conventionality of all language: *no* structural property of the domain of phenomenal color (e.g., in the appearances of chalk) is rendered by the feasibility of the Spenglish description.

To state my objections to the Eddington-Putnam thesis, I call attention to the following two sentences:

(A) Person X does not have a gall bladder.

(B) The platinum-iridium bar in the custody of the Bureau of Weights and Measures in Paris (Sèvres) is 1 meter long everywhere rather than some other number of meters (after allowance for "differential forces").

I maintain that there is a *fundamental difference* between the senses in which each of these statements can possibly be held to be conventional, and I shall refer to these respective senses as "A-conventional" and "B-conventional": in the case of statement (A), what is conventional is only the use of the given *sentence* to render the *proposition* of X's not having a gall bladder, *not* the factual *proposition* expressed by the sentence. This A-conventionality is of the trivial weak kind affirmed by TSC. On the other hand, (B) is conventional not merely in the *trivial* sense that the English sentence used could have been replaced by one in French or some other language but in the much *stronger* and *deeper* sense that it is *not* a factual proposition that the Paris bar has everywhere a length *unity* in the meter scale even *after* we have specified what sentence or string of noises will express this proposition. In brief, in (A), semantic conventions are *used*, whereas, in (B), a semantic convention is *mentioned*. Now I maintain that the alleged analogues of Eddington and Putnam illustrate conventionality only in the sense of A-conventionality and therefore cannot score against my contention that geochronometric conventionality is *nontrivial* by having the character of B-conventionality. Specifically, I assert that statements about phenomenalist colors are empirical statements pure and simple in the sense of being *only* A-conventional and *not* B-conventional, while an important class of statements of geochronometry possesses a different, deeper conventionality of their own by being B-conventional. What is it that is conventional in the case of the color of a given piece of chalk, which appears white in various parts of the visual field? I answer: only our customary decision to use the *same* word to refer to the various qualitatively same white chalk appearances in different parts of the visual field. But it is *not* conventional whether the various chalk appearances do have the same phenomenal color property (to within the precision allowed by vagueness) and thus are "color congruent" to one another or not! Only the *color words* are conventional, *not* the *obtaining* of specified color properties and of color congruence.

489

And the *obtaining* of color congruence is *non*conventional quite inde
pendently of whether the various occurrences of a particular shade of colo
are denoted by the same color word or not.

In other words, there is *no* convention in whether two objects or two
appearances of the same object under like optical conditions have the
same phenomenal color property of whiteness (apart from vagueness) bu
only in whether the *noise* "white" is applied to both of these objects o
appearances, to one of them and not to the other (as in Putnam's chall
example) or to neither. And the alternative color descriptions *do not ren
der any structural facts of the color domain* and are therefore purely trivial
Though failing in this decisive way, Putnam's chalk color case is falsely
given the *semblance* of being a bona fide analogue to the spatial congru
ence case by the device of laying down a rule which makes the use of
color names *space dependent*: the rule is that *different noises* (color
names) will be used to refer to the same *de facto* color property occurring
in different portions of visual space. But this stratagem cannot over
come the fact that while the assertion of the possibility of assigning a
space-dependent length to a transported rod reflects linguistically the
objective nonexistence of an intrinsic metric, the space-dependent use
of color names does *not* reflect a corresponding property of the domain
of phenomenal colors in visual space. In short, the phenomenalist color
of an appearance is an intrinsic, objective property of that appearance (to
within the precision allowed by vagueness), and color congruence is an
objective relation. But the length of a body and the congruence of noncoinciding intervals are *not* similarly *non*conventional. And we saw in
our critique of Whitehead that this conclusion cannot be invalidated by
the fact that two noncoinciding intervals can *look* spatially congruent no
less than two color patches can appear color congruent.

Next consider Eddington's example of the preservation of the language
of Boyle's law to render the *new facts* affirmed by van der Waals' law by
the device of giving a new meaning to the word "pressure" as explained
earlier. The customary concept of pressure has geochronometric ingredients (force, area), and any alterations made in the geochronometric congruence definitions will, of course, issue in changes as to what pressures
will be held to be equal. But the conventionality of the geochronometric
ingredients is *not* of course at issue, and we ask: Of what *structural feature* of the domain of pressure phenomena does the possibility of Eddington's above linguistic transcription render testimony? The answer clearly

s of none. Unlike GC, the thesis of the "conventionality of pressure," if put forward on the basis of Eddington's example, concerns only A-conventionality and is thus merely a special case of TSC. We observe, incidentally, that two pressures which are equal on the customary definition will also be equal (congruent) on the suggested redefinition of that term: apart from the distinctly geochronometric ingredients not here at issue, the domain of pressure phenomena does not present us with any structural property as the counterpart of the lack of an intrinsic metric of space which would be reflected by the alternative definitions of "pressure."

The absurdity of likening the conventionality of spatial or temporal congruence to the conventionality of the choice between the two above meanings of "pressure" or between English and Spenglish color discourse becomes patent upon considering the expression given to the conventionality of congruence by the Klein-Lie group-theoretical treatment of congruences and metric geometries. For their investigations likewise serve to show, as we shall now indicate, how far removed from being a semantically uncommitted noise the term "congruent" is while still failing to single out a unique congruence class of intervals, and how badly amiss it is for Eddington and Putnam to maintain that this nonuniqueness is merely a special example of the semantical nonuniqueness of all uncommitted noises.

Felix Klein's Erlangen Program (1872) of treating geometries from the point of view of groups of spatial transformations was rooted in the following two observations: (1) the properties in virtue of which spatial congruence has the logical status of an equality relation depend upon the fact that displacements are given by a group of transformations, and (2) the congruence of two figures consists in their being intertransformable into one another by means of a certain transformation of points. Continuing Klein's reasoning, Sophus Lie then showed that, in the context of this group-theoretical characterization of metric geometry, the conventionality of congruence issues in the following results: (i) the set of all the continuous groups in space having the property of displacements in a bounded region fall into three types which respectively characterize the geometries of Euclid, Lobachevski-Bolyai, and Riemann [3, p. 153], and (ii) for each of these metrical geometries, there is not one but an infinitude of different congruence classes [99, p. 49] (cf. Section 3, part (iii), where this latter result is obtained without group-theoretical devices). On the Eddington-Putnam thesis, Lie's profound and justly celebrated re-

491

Adolf Grünbaum

sults no less than the relativity of simultaneity and the conventionalit
of temporal congruence must be consigned absurdly to the limbo of trivia
semantical conventionality along with Spenglish color discourse!

These objections against the Eddington-Putnam claim that GC ha
bona fide analogues in every empirical domain are not intended to den
the existence of one or another genuine analogue but to deny only tha
GC may be deemed to be trivial on the strength of such relatively fev
bona fide analogues as may obtain. Putnam has given one example whicl
does seem to qualify as a bona fide analogue. This example differs from
his color case in that not merely the *name* given to a property but th
sameness of the property named is dependent on spatial position as fol
lows: when two bodies are at essentially the same place, their samenes
with respect to a certain property is a matter of fact, but when they ar
(sufficiently) apart spatially, no objective relation of sameness or differ
ence with respect to the given property obtains between them. And, ir
the latter case, therefore, it becomes a matter of convention whethe
sameness or difference is ascribed to them in this respect. Specifically
suppose that we do not aim at a definition of mass adequate to classica
physical theory and thus ignore Mach's definition of mass, which we dis
cussed earlier (cf. Section 2, part (iii)). Then we can consider Putnam'
hypothetical definition of "mass equality," according to which two bodie
balancing one another on a suitable scale at what is essentially the same
place in space have equal masses. Whereas on the Machian definition
mass equality obtains between two bodies *as a matter of fact* independent
ly of the extent of their spatial separation, on Putnam's definition such
separation leaves the relation of mass equality at a distance *indeterminate*. Hence, on Putnam's definition, it would be a matter of convention
whether (a) we would say that two masses which balance at a given place
remain equal to one another in respect to mass after being spatially separated, or (b) we would make the masses of two bodies space dependent
such that two masses that balance at once place would have different
masses when separated, as specified by a certain function of the coordinates. The conventionality arising in Putnam's mass example is *not* a consequence of GC but is logically independent of it. For it is *not* spatial
congruence of noncoinciding intervals but spatial *position* that is the
source of conventionality here.

In conclusion, we must persist therefore with Poincaré, Einstein, Reichenbach, and Carnap in attaching a very different significance to alterna-

tive metric geometries or chronometries as equivalent descriptions of the same facts than to alternate types of visual color discourse as equivalent descriptions of the same phenomenal data. By the same token, we must attach much greater significance to being able to render *factually different* geochronometric states of affairs by the *same* geometry or chronometry, coupled with appropriately *different* congruence definitions, than to formulating both Boyle's law and van der Waals' law, which *differ* in factual content, by the *same* law statement coupled with appropriately different semantical rules. In short, there is an important respect in which physical geochronometry is less of an empirical science than all or almost all of the nongeochronometric portions (ingredients) of other sciences.

6. The Bearing of Alternative Metrizability on the Interdependence of Geochronometry and Physics

(i) The Fundamental Difference between the LINGUISTIC *Interdependence of Geometry and Physics Affirmed by the Conventionalism of H. Poincaré and Their* EPISTEMOLOGICAL *(Inductive) Interdependence in the Sense of P. Duhem.*

The central theme of Poincaré's so-called conventionalism is essentially an elaboration of the thesis of alternative metrizability whose fundamental justification we owe to Riemann, and *not* [32, Section 5] the *radical* conventionalism attributed to him by Reichenbach [72, p. 36].

Poincaré's much-cited and often misunderstood statement concerning the possibility of always giving a Euclidean description of any results of stellar parallax measurements [63, pp. 81–86] is a less lucid statement of exactly the same point made by him with magisterial clarity in the following passage: "In space we know rectilinear triangles the sum of whose angles is equal to two right angles; but equally we know curvilinear triangles the sum of whose angles is less than two right angles. . . . To give the name of straights to the sides of the first is to adopt Euclidean geometry; to give the name of straights to the sides of the latter is to adopt the non-Euclidean geometry. So that to ask what geometry it is proper to adopt is to ask, to what line is it proper to give the name straight? It is evident that experiment can not settle such a question" [63, p. 235].

Now, the equivalence of this contention to Riemann's view of congruence becomes evident the moment we note that the legitimacy of identifying lines which are curvilinear in the usual geometrical parlance as "straights" is vouchsafed by the warrant for our choosing a new definition of congruence such that the previously curvilinear lines become geodesic of the new congruence. And we note that whereas the *original* geodesic in space exemplified the formal relations obtaining between Euclidean "straight lines," the *different* geodesics associated with the new metrization embody the relations prescribed for straight lines by the formal postulates of *hyperbolic* geometry. Awareness of the fact that Poincaré begins the quoted passage with the words "In [physical] space" enables us to see that he is making the following assertion here: the same *physical* surface or region of three-dimensional *physical* space admits of *alternative* metrizations so as to constitute a physical realization of either the formal postulates of Euclidean geometry or of one of the non-Euclidean abstract calculi. To be sure, syntactically, this alternative metrizability involves a formal intertranslatability of the *relevant portions* of these incompatible geometrical calculi, the "intertranslatability" being guaranteed by a "dictionary" which pairs off with one another the *alternative names* (or descriptions) of each physical path or configuration. But the essential point made here by Poincaré is *not* that a purely formal translatability obtains; instead, Poincaré is emphasizing here that a *given* physical surface or region of physical 3-space can indeed be a model of one of the non-Euclidean geometrical calculi no less than of the Euclidean one. In *this* sense one can say, therefore, that Poincaré affirmed the conventional or definitional status of *applied* geometry.

Hence, we must reject the following wholly syntactical interpretation of the above citation from Poincaré, which is offered by Ernest Nagel, who writes: "The thesis he [Poincaré] establishes by this argument is simply the thesis that choice of notation in formulating a system of pure geometry is a convention" [55, p. 261]. Having thus misinterpreted Poincaré's conventionalist thesis as pertaining only to formal intertranslatability, Nagel fails to see that Poincaré's avowal of the conventionality of physical or *applied* geometry is none other than the assertion of the *alternative metrizability of physical space* (or of a portion thereof). And, in this way, Nagel is driven to give the following unfounded interpretation of Poincaré's conception of the status of applied (physical) geometry: "Poincaré also argued for the definitional status of *applied* as well as of pure geometry. He maintained that, even when an interpretation is given to the primitive terms of a pure geometry so that the system is then converted into statements about certain physical configurations (for example, interpreting 'straight line' to signify the path of a light ray), no experi-

494

ment on physical geometry can ever decide against one of the alternative systems of physical geometry and in favor of another" [55, p. 261]. But far from having claimed that the geometry is still conventional even after the provision of a particular physical interpretation of a pure geometry, Poincaré merely reiterated the following thesis of alternative metrizability in the passages which Nagel [55, pp. 261–262] then goes on to quote from him: suitable alternative semantical interpretations of the term "congruent" (for line segments and for angles), and correlatively of "straight line," etc., can readily demonstrate that, subject to the restrictions imposed by the existing topology, it is always a live option to give either a Euclidean or a non-Euclidean description of any given set of physico-geometric facts. And since alternative metrizations are just as legitimate epistemologically as alternative systems of units of length or temperature, one can always, in principle, reformulate any physical theory based on a given metrization of space—or, as we saw in Section 4 above, of time—so as to be based on an alternative metric.

There is therefore no warrant at all for the following caution expressed by Nagel in regard to the feasibility of what is merely a reformulation of physical theory on the basis of a new metrization: ". . . even if we admit universal forces in order to retain Euclid . . . we must incorporate the assumption of universal forces into the rest of our physical theory, rather than introduce such forces piecemeal subsequent to each observed 'deformation' in bodies. It is by no means self-evident, however, that physical theories can in fact always be devised that have built-in provisions for such universal forces" [55, pp. 264–265]. Yet, precisely that fact is self-evident, and its self-evidence is obscured from view by the logical havoc created by the statement of a remetrization issuing in Euclidean geometry in terms of "universal forces." For that metaphor seems to have misled Nagel into imputing the status of an empirical hypothesis to the use of a nonstandard spatial metric merely because the latter metric is described by saying that we "assume" appropriate universal forces. In fact, our discussion in Section 4, part (i), has shown mathematically for the one-dimensional case of time how Newtonian mechanics is to be recast via suitable transformation equations, such as equation (4) there, so as to implement a remetrization given by $T = f(t)$, which can be described metaphorically by saying that all clocks are "accelerated" by "universal forces."

Corresponding remarks apply to Poincaré's contention that we can always preserve Euclidean geometry in the face of any data obtained from stellar parallax measurements: if the paths of light rays are geodesics on the customary definition of congruence, as indeed they are in the Schwarzschild procedure cited by Robertson [77], and if the paths of light ray are found parallactically to sustain non-Euclidean relations on that metrization, then we need only choose a different definition of congruence such that these same paths will no longer be geodesics and that the geodesic of the newly chosen congruence are Euclideanly related. From the standpoint of synthetic geometry, the latter choice effects a *renaming* of optical and other paths and thus is merely a *recasting of the same factual content in Euclidean language rather than a revision of the extra-linguistic content of optical and other laws. The remetrizational retainability of Euclideanism affirmed by Poincaré therefore involves a merely linguistic interdependence of the geometric theory of rigid solids and the optical theory of light rays.* And since Poincaré's claim here is a straightforward elaboration of the metric amorphousness of the continuous manifold of space, it is not clear how H. P. Robertson [77, pp. 324–325] can reject it as a "pontifical pronouncement" and even regard it as being in contrast with what he calls Schwarzschild's "sound operational approach to the problem of physical geometry." For Schwarzschild had rendered the question concerning the prevailing geometry *factual* only by the adoption of a particular spatial metrization based on the travel times of light, which does indeed turn the direct light paths of his astronomical triangle into geodesics.

Poincaré's interpretation of the parallactic determination of the geometry of a stellar triangle has also been obscured by Ernest Nagel's statement of it. Apart from encumbering that statement with the metaphorical use of "universal forces," Nagel fails to point out that the crux of the preservability (retainability) of Euclidean geometry lies in (i) the denial of the geodesicity (straightness) of optical paths which are found parallactically to sustain non-Euclidean relations on the customary metrization of line segments and angles, or at least in the rejection of the customary congruence for angles (cf. Section 3, part (iii)),[40] and (ii) the ability to

[40] Under the assumed conditions as to the parallactic findings, the optical paths might still be interpretable as Euclideanly-related geodesics on a new metrization, but only if the customary angle congruences were abandoned and changed suitably as part of the remetrization (cf. Sec. 3, part (iii)). In that case, the paths of light rays would be straight lines even in the Euclidean description obtained by the new metrization, but the optical laws involving angles would have to be suitably restated.

uarantee the existence of a suitable new metrization whose associated ;eodesics are paths which do exhibit the formal relations of Euclidean traights. For Nagel characterizes the retainability of Euclidean geometry ome what may by asserting that the latter's retention is effected "only »y maintaining that the sides of the stellar triangles are not really Eulidean [sic] straight lines, and he [the Euclidean geometer] will thereore adopt the hypothesis that the optical paths are deformed by some ields of force" [55, p. 263]. But apart from the obscurity of the notion of he deformation of the optical *paths*, the unfortunate inclusion of the vord "Euclidean" in this sentence of Nagel's obscures the very point vhich the advocate of Euclid is concerned to make in this context in the nterests of his thesis. And this point is *not*, as Nagel would have it, that he optical paths are not really *Euclidean* straight lines, a fact whose admission (assuming the customary congruence for angles) provided the tarting point of the discussion. Instead, what the proponent of Euclid s concerned to point out here is that the legitimacy of alternative metrizations enables him to offer a metric such that the optical paths do *not*]ualify as *geodesics* (straights) from the outset. For it is by denying alogether the geodesicity of the optical paths that the advocate of Euclid an uphold his thesis successfully in the face of the admitted prima-facie 10n-Euclidean parallactic findings.

The invocation of the conventionality of congruence to carry out remetrizations is not at all peculiar to Poincaré. For F. Klein's relative consistency proof of hyperbolic geometry via a model furnished by the inerior of a circle on the Euclidean plane [3, pp. 164–175], for example, is »ased on one particular kind of possible remetrization of the circular porion of that plane, projective geometry having played the heuristic role)f furnishing Klein with a suitable definition of congruence. Thus what rom the point of view of synthetic geometry appears as intertranslatbility via a dictionary appears as alternative metrizability from the point)f view of differential geometry.

There are two respects, however, in which Poincaré is open to criticism n this connection: (i) He maintained [63, p. 81] that it would always be egarded as most convenient to preserve Euclidean geometry, even at the)rice of remetrization, on the grounds that this geometry is the simplest inalytically [63, p. 65]. Precisely the opposite development materialized n the general theory of relativity: Einstein forsook the simplicity of the geometry itself in the interests of being able to maximize the simplicity

497

of the definition of congruence. He makes clear in his fundamental paper of 1916 that had he insisted on the retention of Euclidean geometry in a gravitational field, then he could *not* have taken "one and the same rod independently of its place and orientation, as a realization of the same interval" [21, p. 161]. (ii) Even if the simplicity of the geometry itself were the sole determinant of its adoption, that simplicity might be judged by criteria other than Poincaré's analytical simplicity. Thus, Menger has urged that from the point of view of a criterion grounded on the simplicity of the undefined concepts used, hyperbolic and not Euclidean geometry is the simplest.

On the other hand, if Poincaré were alive today, he could point to an interesting recent illustration of the sacrifice of the simplicity and accessibility of the congruence standard on the altar of maximum simplicity of the resulting theory. Astronomers have recently proposed to remetrize the time continuum for the following reason: as indicated in Section 4, part (i), when the mean solar second, which is a very precisely known fraction of the period of the earth's rotation on its axis, is used as a standard of temporal congruence, then there are three kinds of discrepancies between the actual observational findings and those predicted by the usual theory of celestial mechanics. The empirical facts thus present astronomers with the following choice: either they retain the rather natural standard of temporal congruence at the cost of having to bring the principles of celestial mechanics into conformity with observed fact by revising them appropriately, or they remetrize the time continuum, employing a less simple definition of congruence so as to preserve these principles intact. Decisions taken by astronomers in the last few years were exactly the reverse of Einstein's choice of 1916 as between the simplicity of the standard of congruence and that of the resulting theory. The mean solar second is to be supplanted by a unit to which it is nonlinearly related: the sidereal year, which is the period of the earth's revolution around the sun, due account being taken of the irregularities produced by the gravitational influence of the other planets [13].

We see that the implementation of the requirement of descriptive simplicity in theory construction can take alternative forms, because agreement of astronomical theory with the evidence now available is achievable by revising *either* the definition of temporal congruence or the postulates of celestial mechanics. The existence of this alternative likewise illustrates that for an axiomatized physical theory containing a

498

geochronometry, it is *gratuitous* to single out the postulates of the theory as having been prompted by *empirical* findings in contradistinction to deeming the *definitions of congruence* to be wholly a priori, or vice versa. This conclusion bears out geochronometrically Braithwaite's contention [4] that there is an important sense in which axiomatized physical theory does not lend itself to compliance with Heinrich Hertz's injunction to "distinguish thoroughly and sharply between the elements . . . which arise from the necessities of thought, from experience, and from arbitrary choice" [38, p. 8]. The same point is illustrated by the possibility of characterizing the factual innovation wrought by Einstein's abandonment of Euclidean geometry in favor of Riemannian geometry in the GTR in several ways as follows: (i) *Upon using the customary definition of spatial congruence*, the geometry near the sun is *not* Euclidean, contrary to the claims of pre-GTR physics. (ii) The geometry near the sun is *not* Euclidean on the basis of the *customary* congruence, but it *is* Euclidean on a suitably modified congruence definition which makes the length of a rod a specified function of its position and orientation.[41] (iii) *Within the confines* of the requirement of giving a *Euclidean* description of the nonclassical facts postulated by the GTR, Einstein recognized the *factually dictated* need to *abandon* the *customary* definition of congruence, which had yielded a Euclidean description of the classically assumed facts. Thus, the revision of the Newtonian theory made necessary by the discovery of relativity can be formulated as *either* a change in the postulates of geometric theory or a change in the correspondence rule for congruence.

We saw that Poincaré's remetrizational retainability of Euclidean geometry or of some other particular geometry involves a merely *linguistic* interdependence of the geometric theory of rigid solids and the optical theory of light rays. Preparatory to clarifying the important difference between Duhem's and Poincaré's conceptions of the interdependence of geometry and physics, we first give a statement of Duhem's view of the falsifiability and confirmability of an isolated explanatory hypothesis in science.

We must distinguish the following two forms of Duhem's thesis: (i) The logic of every disconfirmation, no less than of every confirmation, of a presumably empirical hypothesis H is such as to *involve at some stage*

---

[41] The function in question is given in [8, p. 58].

Adolf Grünbaum

or *other* an entire network of interwoven hypotheses in which H is in gredient rather than the separate testing of the component H. (ii) Th falsifiability of (part of) an *explanans* is unavoidably inconclusive in ever case: *no one constituent hypothesis H can ever be extricated* from th ever-present web of collateral assumptions so as to be *open to decisiv refutation* by the evidence as part of an *explanans* of that evidence, jus as no such isolation is achievable for purposes of verification.

Duhem seems to think that the latter contention is justified by th following twofold argument or schema of unavoidably inconclusive falsi fiability: (a) It is an elementary fact of *deductive* logic that if certai observational consequences O are entailed by the *conjunction* of H an a set A of auxiliary assumptions, then the *failure* of O to materialize en tails *not* the falsity of H by itself but only the weaker conclusion that H and A cannot *both* be true; the falsifiability of H is therefore *inconclusiv* in the sense that the *falsity* of H is *not deductively inferable* from th premise $[(H \cdot A) \rightarrow O] \cdot \sim O$. (b) The actual observational findings O' which are incompatible with O, *allow* that H be true while A is false, be cause they permit the theorist to preserve H with impunity as part of a *explanans* of O' by so modifying A that the *conjunction* of H and th revised version A' of A does explain (entail) O'. This preservability of H is to be understood as a retainability in *principle* and does not depend o the ability of scientists to propound the required set A' of collateral as sumptions at any given time.

Thus, according to Duhem, there is an inductive (epistemological) in terdependence and inseparability between H and the auxiliary assump tions. And there is claimed to be an ingression of a kind of a priori choic into physical theory: at the price of suitable compensatory modification in the remainder of the theory, any one of its component hypotheses H may be retained in the face of seemingly contrary empirical findings a part of an *explanans* of these very findings. According to Duhem, thi quasi a priori preservability of H is sanctioned by the far-reaching theo retical ambiguity and flexibility of the logical constraints imposed by th observational evidence [19].[42]

---

[42] Duhem's explicit disavowal of both decisive falsifiability and crucial verifiability of an *explanans* will *not* bear K. R. Popper's reading of him [66, p. 78]: Popper, who is an exponent of decisive falsifiability [66], misinterprets Duhem as allowing that test of a hypothesis may be decisively falsifying and as denying only that they may be cru cially verifying. Notwithstanding Popper's *exegetical* error, we shall see in Sec. 7 that

Duhem would point to the fact that in a sense to be specified *in detail* in Section 7, the physical laws used to correct a measuring rod for substance-specific distortions presuppose a geometry and comprise the laws of optics. And hence he would *deny*, for example, that either of the following kinds of *independent* tests of geometry and optics are feasible:

1. Prior to and independently of knowing or presupposing the geometry, we find it to be a law of optics that the paths of light coincide with the geodesics of the congruence defined by rigid bodies.

Knowing this, we then use triangles consisting of a geodesic base line in the solar system and the stellar light rays connecting its extremities to various stars to determine the geometry of the system of rigid-body geodesics: stellar parallax measurements will tell us whether the angle sums of the triangles are 180° (Euclidean geometry), less than 180° (hyperbolic geometry), or in excess of 180° (spherical geometry).

If we thus find that the angle sum is different from 180°, then we shall know that the geometry of the rigid-body geodesics is *not* Euclidean. For in view of our prior independent ascertainment of the paths of light rays, such a non-Euclidean result could *not* be interpreted as due to the failure of optical paths to coincide with the rigid-body geodesics.

2. Prior to and independently of knowing or presupposing the laws of optics, we ascertain what the geometry is relatively to the rigid-body congruence.

Knowing this we then find out whether the paths of light rays coincide with the geodesics of the rigid-body congruence by making a parallactic or some other determination of the angle sum of a light-ray triangle.

Since we know the geometry of the rigid-body geodesics independently of the optics, we know what the corresponding angle sum of a triangle whose sides are geodesics ought to be. And hence the determination of the angle sum of a light-ray triangle is then decisive in regard to whether the paths of light rays coincide with the geodesics of the rigid-body congruence.

In place of such independent confirmability and falsifiability of the geometry and the optics, Duhem affirms their *inductive* (epistemological) inseparability and interdependence. Let us consider the interpretation of

his thesis of the feasibility of decisively falsifying tests can be buttressed by a telling counterexample to Duhem's categorical denial of that thesis.

Feigl [28] has outlined a defense of the claim that isolated parts of physical theory can be confirmed.

observational parallactic data to articulate, in turn, the differences between Duhem's and Poincaré's conceptions of (a) the feasibility of alternative geometric interpretations of observational findings, and (b) the retainability of a particular geometry as an account of such findings.

(a) *The feasibility of alternative geometric interpretations of parallactic or other observational findings.* The Duhemian conception envisions scope for alternative geometric accounts of a given body of evidence only to the extent that these geometries are associated with alternative non-equivalent sets of physical laws that are used to compute corrections for substance-specific distortions. On the other hand, the range of alternative geometric descriptions of given evidence affirmed by Poincaré is far wider and rests on very different grounds: instead of invoking the Duhemian *inductive* latitude, Poincaré bases the possibility of giving *either* a Euclidean or a non-Euclidean description of the same spatio-physical facts on alternative metrizability. For Poincaré tells us [63, pp. 66–80] *that quite apart from any considerations of substance-specific distorting influences and even after correcting for these in some way or other, we are at liberty to define congruence—and thereby to fix the geometry appropriate to the given facts—either by calling the solid rod equal to itself everywhere or by making its length vary in a specified way with its position and orientation.* The particular case of the interpretation of certain parallactic data will give concrete meanings to these assertions.

The attempt to explain parallactic data yielding an angle sum different from 180° for a stellar light-ray triangle by different geometries which constitute live options *in the inductive sense of Duhem* would presumably issue in the following alternative between two theoretical systems. Each of these theoretical systems comprises a geometry G and an optics O which are epistemologically inseparable and which are inductively interdependent in the sense that the combination of G and O must yield the observed results:

(a) G_E: the geometry of the rigid body geodesics is Euclidean, and
O_1: the paths of light rays do *not* coincide with these geodesics but form a non-Euclidean system,

or

(b) G_{non-E}: the geodesics of the rigid-body congruence are *not* a Euclidean system, and
O_2: the paths of light rays *do* coincide with these geodesics, and thus they form a non-Euclidean system.

502

To contrast this Duhemian conception of the feasibility of alternative geometric interpretations of the assumed parallactic data with that of Poincaré, we recall that the physically interpreted alternative geometries associated with two (or more) different metrizations *in the sense of Poincaré* have precisely the same total factual content, as do the corresponding two sets of optical laws. For an alternative metrization in the sense of Poincaré affects only the *language* in which the facts of optics and the coincidence behavior of a transported rod are described: the two geometric descriptions respectively associated with two alternative metrizations are *alternative representations of the same factual content*, and so are the two sets of optical laws corresponding to these geometries. Accordingly, Poincaré is affirming a *linguistic* interdependence of the geometric theory of rigid solids and the optical theory of light rays. By contrast, in the Duhemian account, G_E and G_{non-E} not only *differ* in factual content but are logically incompatible, and so are O_1 and O_2. And on the latter conception, there is sameness of factual content *in regard to the assumed parallactic data* only between the *combined* systems formed by the two conjunctions (G_E and O_1) and (G_{non-E} and O_2).[43] Thus, the need for the combined system of G and O to yield the empirical facts, coupled with the avowed epistemological (inductive) *inseparability* of G and O lead the Duhemian to conceive of the *interdependence* of geometry and optics as *inductive* (epistemological).

Hence whereas Duhem construes the interdependence of G and O inductively such that the geometry by itself is *not* accessible to empirical test, Poincaré's conception of their interdependence allows for an empirical determination of G *by itself, if we have renounced* recourse to an alternative metrization in which the length of the rod is held to vary with its position or orientation. This is not, of course, to say that Duhem regarded alternative metrizations as such to be illegitimate.

It would seem that it was Poincaré's discussion of the interdependence of optics and geometry by reference to stellar parallax measurements which has led many writers such as Einstein [26], Eddington [20, p. 9], and Nagel [55, p. 262] to regard him as a proponent of the Duhemian thesis. But this interpretation appears untenable not only in the light of the immediate context of Poincaré's discussion of his astronomical example, but also, as we shall see briefly in Section 6, part (ii), upon taking account of the remainder of his writings. An illustration of the widespread

[43] These combined systems do *not*, however, have the same *over-all* factual content.

503

conflation of the linguistic and inductive kinds of interdependence of geometry and physics (optics) is given by D. M. Y. Sommerville's discussion of what he calls "the inextricable entanglement of space and matter." He says [84, pp. 209–210]:

A . . . "vicious circle" . . . arises in connection with the astronomical attempts to determine the nature of space. These experiments are based upon the received laws of astronomy and optics, which are themselves based upon the euclidean assumption. It might well happen, then, that a discrepancy observed in the sum of the angles of a triangle could admit of an explanation by some modification of these laws, or that even the absence of any such discrepancy might still be compatible with the assumptions of non-euclidean geometry.

Sommerville then quotes the following assertion by C. D. Broad:

All measurement involves both physical and geometrical assumptions, and the two things, space and matter, are not given separately, but analysed out of a common experience. Subject to the general condition that space is to be changeless and matter to move about in space, we can explain the same observed results in many different ways by making compensatory changes in the qualities that we assign to space and the qualities we assign to matter. Hence it seems theoretically impossible to decide by any experiment what are the qualities of one of them in distinction from the other.

And Sommerville's immediate comment on Broad's statement is the following:

It was on such grounds that Poincaré maintained the essential impropriety of the question, "Which is the true geometry?" In his view it is merely a matter of convenience. Facts are and always will be most simply described on the euclidean hypothesis, but they can still be described on the non-euclidean hypothesis, with suitable modifications of the physical laws. To ask which is the true geometry is then just as unmeaning as to ask whether the old or the metric system is the true one.

(b) *The retainability of a particular geometry as an account of observational findings.* The key to the difference between the geometric conventionalism of H. Poincaré and the geometrical form of the conventionalism of P. Duhem is furnished by the distinction between preserving a particular geometry (e.g., the Euclidean one) by a remetrizational change in the congruence definition, on the one hand, and intending to retain a particular geometry *without* change in that definition (or in other semantical rules) by an alteration of the factual content of the auxiliary

assumptions, on the other. More specifically, whereas Duhem's affirmation of the retainability of Euclidean geometry in the face of any observational evidence and the associated interdependence of geometry and the
remainder of physics are *inductive* (epistemological), the preservability
of that geometry asserted by Poincaré is *remetrizational*: Poincaré's conventionalist claim regarding geometry is that if the customary definition
of congruence on the basis of the coincidence behavior common to all
kinds of solid rods does not assure a particular geometric description of
the facts, then such a description can be guaranteed remetrizationally,
i.e., by merely choosing an appropriately different noncustomary congruence definition which makes the length of every kind of solid rod a specified *nonconstant* function of the *independent* variables of position and
orientation.

(ii) Exegetical Excursus: Poincaré's Philosophy of Geometry.

Einstein [22, 26] and Reichenbach [71, 73] have interpreted Poincaré
to have been a proponent of the Duhemian conception of the interdependence of geometry and physics. As evidence *against* this interpretation, I now cite the following two passages from Poincaré, the first being
taken from his rejoinder to Bertrand Russell and the second from *Science
and Hypothesis*:

The term "to preserve one's form" has no meaning by itself. But I confer a meaning on it by *stipulating* that certain bodies will be said to preserve their form. These bodies, thus chosen, can henceforth serve as instruments of measurement. But if I say that these bodies preserve their
form, it is because *I choose to do so* and not because experience obliges
me to do so.

In the present context I choose to do so, because by a series of observations ("constatations") . . . *experience has proven to me* that their
movements form a Euclidean group. I have been able to make these observations in the manner just indicated *without having any preconceived
idea concerning metric geometry.* And, having made them, I judge that
the convention will be convenient and I adopt it [65; the italics in the
second of these paragraphs are mine].[44]

No doubt, in our world, natural solids . . . undergo variations of form
and volume due to warming or cooling. But we neglect these variations
in laying the foundations of geometry, because, besides their being very
slight, they are irregular and consequently seem to us accidental [63, p. 76].

[44] For a documentation of the fact that Poincaré espoused a similarly empiricist
conception of the three-dimensionality of space, see [32, Sec. 5].

But one might either contest my interpretation here or conclude that Poincaré was inconsistent by pointing to the following passage by him:

Should we . . . conclude that the axioms of geometry are experimental verities? . . . If geometry were an experimental science, it would not be an exact science, it would be subject to a continual revision. Nay, it would from this very day be convicted of error, since we know that there is no rigorously rigid solid.

The axioms of geometry therefore are . . . conventions . . . Thus it is that the postulates can remain rigorously true even though the experimental laws which have determined their adoption are only approximative [63, pp. 64–65].[45]

The only way in which I can construe the latter passage and others like it in the face of our earlier citations from him is by assuming that Poincaré maintained the following: there are *practical* rather than *logical* obstacles which frustrate the complete elimination of perturbational distortions, and the resulting vagueness (spread) as well as the finitude of the empirical data provide scope for the exercise of a certain measure of convention in the adoption of a metric tensor.

This non-Duhemian reading of Poincaré accords with the interpretation of him in L. Rougier's *La Philosophie Géométrique de Henri Poincaré*. Rougier writes: "The conventions fix the language of science which can be indefinitely varied: once these conventions are accepted, the facts expressed by science necessarily are either true or false. . . . Other conventions remain possible, leading to other modes of expressing oneself; but the truth, thus diversely translated, remains the same. One can pass from one system of conventions to another, from one language to another, by means of an appropriate dictionary. The very possibility of a translation shows here the existence of an invariant. . . . Conventions relate to the variable language of science, not to the invariant reality which they express" [78, pp. 200–201].

7. The Empirical Status of Physical Geometry

(i) Einstein's Duhemian Espousal of the Interdependence of Geometry and Physics.

Our statement in Section 6 of the difference between Duhem's and Poincaré's conceptions of the interdependence of geometry and the remainder of physics calls for a critical appraisal of Duhem's thesis that the

[45] Similar statements are found in [63, pp. 79, 240].

falsification of a hypothesis as to the geometry of physical space is unavoidably inconclusive in isolation from the remainder of physics.

Since Duhem's argument was articulated and endorsed by Einstein a decade ago, I shall state and then examine Einstein's version of it.

We have noted throughout that physical geometry is usually conceived as the system of metric relations exhibited by transported solid bodies independently of their particular chemical composition. On this conception, the criterion of congruence can be furnished by a transported solid body for the purpose of determining the geometry by measurement only if the computational application of suitable "corrections" (or, ideally, appropriate shielding) has assured rigidity in the sense of essentially eliminating inhomogeneous thermal, elastic, electromagnetic, and other perturbational influences. For these influences are "deforming" in the sense of producing changes of varying degree in different kinds of materials. Since the existence of perturbational influences thus issues in a dependence of the coincidence behavior of transported solid rods on the latter's chemical composition, and since physical geometry is concerned with the behavior common to all solids apart from their substance-specific idiosyncrasies, the discounting of idiosyncratic distortions is an essential aspect of the logic of physical geometry. The demand for the computational elimination of such distortions as a prerequisite to the experimental determination of the geometry has a thermodynamic counterpart: the requirement of a means for measuring temperature which does not yield the discordant results produced by expansion thermometers at other than fixed points when different thermometric substances are employed. This thermometric need is fulfilled successfully by Kelvin's thermodynamic scale of temperature. But attention to the implementation of the corresponding prerequisite of physical geometry has led Einstein to impugn the empirical status of that geometry. He considers the case in which congruence has been defined by the diverse kinds of transported solid measuring rods as corrected for their respective idiosyncratic distortions with a view to then making an empirical determination of the prevailing geometry. And Einstein's thesis is that the very logic of computing these corrections precludes that the geometry itself be accessible to experimental ascertainment in isolation from other physical regularities.[46] Specifically, he states his case in the form of a dialogue [26, pp. 676–678] in which he

[46] For a very detailed treatment of the relevant computations, see [94, 60, 85, 45].

Adolf Grünbaum

attributes his own Duhemian view to Poincaré and offers that view in opposition to Hans Reichenbach's conception, which we discussed in Section 3. But, as we saw in Section 6, part (ii), Poincaré's text will not bear Einstein's interpretation. For we noted that in speaking of the variations which solids exhibit under distorting influences, Poincaré says, "we neglect these variations in laying the foundations of geometry, because, besides their being very slight, they are irregular and consequently seem to us accidental" [63, p. 76]. I am therefore taking the liberty of replacing the name "Poincaré" in Einstein's dialogue by the term "Duhem and Einstein." With this modification, the dialogue reads as follows:

Duhem and Einstein: The empirically given bodies are not rigid, and consequently cannot be used for the embodiment of geometric intervals. Therefore, the theorems of geometry are not verifiable.

Reichenbach: I admit that there are no bodies which can be *immediately* adduced for the "real definition" [i.e., physical definition] of the interval. Nevertheless, this real definition can be achieved by taking the thermal volume-dependence, elasticity, electro- and magneto-striction, etc., into consideration. That this is really and without contradiction possible, classical physics has surely demonstrated.

Duhem and Einstein: In gaining the real definition improved by yourself you have made use of physical laws, the formulation of which presupposes (in this case) Euclidean geometry. The verification, of which you have spoken, refers, therefore, not merely to geometry but to the entire system of physical laws which constitute its foundation. An examination of geometry by itself is consequently not thinkable. Why should it consequently not be entirely up to me to choose geometry according to my own convenience (i.e. Euclidean) and to fit the remaining (in the usual sense "physical") laws to this choice in such manner that there can arise no contradiction of the whole with experience? [26, pp. 676–678.]

By speaking here of the "real definition" (i.e., the coordinative definition) of "congruent intervals" by the corrected transported rod, Einstein is ignoring that the actual and potential physical meaning of congruence in physics cannot be given exhaustively by any one physical criterion or test condition. But here as elsewhere throughout this essay (cf. Section 5, part (ii)), we can safely ignore this open cluster character of the concept of congruence. For our concern as well as Einstein's is merely to single out one particular congruence class from among an infinitude of such alternative classes. And as long as our specification of that one chosen class is unambiguous, it is wholly immaterial that there are also other physical criteria (or test conditions) by which it could be specified.

508

Einstein is making two major points here: (1) In obtaining a physical geometry by giving a physical interpretation of the postulates of a formal geometric axiom system, the specification of the physical meaning of such theoretical terms as "congruent," "length," or "distance" is *not* at all simply a matter of giving an operational definition in the strict sense. Instead, what has been variously called a "rule of correspondence" (Margenau and Carnap), a "coordinative definition" (Reichenbach), an "epistemic correlation" (Northrop), or a "dictionary" (N. R. Campbell) is provided here *through the mediation of hypotheses and laws* which are collateral to the geometric theory whose physical meaning is being specified. Einstein's point that the physical meaning of congruence is given by the transported rod *as corrected theoretically* for idiosyncratic distortions is an illuminating one and has an abundance of analogues throughout physical theory, thus showing, incidentally, that strictly operational definitions are a rather simplified and limiting species of rules of correspondence (cf. Section 2, part (ii)). (2) Einstein's second claim, which is the cardinal one for our purposes, is that the role of collateral theory in the physical definition of congruence is such as to issue in the following *circularity*, from which there is no escape, he maintains, short of acknowledging the existence of an a priori element *in the sense of the Duhemian ambiguity*: the rigid body is not even defined without first decreeing the validity of Euclidean geometry (or of some other particular geometry). For *before* the *corrected* rod can be used to make an empirical determination of the *de facto* geometry, the required corrections must be computed via laws, such as those of elasticity, which involve Euclideanly calculated areas and volumes [83, 89]. But clearly the warrant for thus introducing Euclidean geometry *at this stage* cannot be empirical.

In the same vein, H. Weyl endorses Duhem's position as follows: "Geometry, mechanics, and physics form an inseparable theoretical whole [96, p. 67]. . . . Philosophers have put forward the thesis that the validity or non-validity of Euclidean geometry cannot be proved by empirical observations. It must in fact be granted that in all such observations essentially physically assumptions, such as the statement that the path of a ray of light is a straight line and other similar statements, play a prominent part. This merely bears out the remark already made above that it is only the whole composed of geometry and physics that may be tested empirically" [96, p. 93].

If Einstein's Duhemian thesis were to prove correct, then it would have to be acknowledged that there is a sense in which physical geometry *itself* does not provide a geometric characterization of physical reality. For by this characterization we understand the articulation of the system of relations obtaining between bodies and transported solid rods quite apart from their substance-specific distortions. And to the extent to which physical geometry is a priori in the sense of the Duhemian ambiguity, there is an ingression of a priori elements into physical theory to take the place of distinctively geometric gaps in our knowledge of the physical world.

(ii) Critique of Einstein's Duhemian Thesis.

I shall set forth my doubts regarding the soundness of Einstein's contention in three parts consisting of the following: (1) A critique of the general Duhemian schema of the logic of falsifiability, as presented in Section 6, part (i), in the form of assertions (a) and (b). (2) An analysis of the status of the Einstein-Duhem argument in the special case in which effectively no deforming influences are present in a certain region whose geometry is to be ascertained. (3) An evaluation of Einstein's version of Duhem as applied to the empirical determination of the geometry of a region which *is* subject to deforming influences.

1. Referring to Section 6, part (i), let us now consider the two parts (a) and (b) of the schema which the *stronger* form (ii) of the Duhemian thesis claims to be the *universal paradigm* of the logic of falsifiability in empirical science. Clearly, part (a) is valid, being a *modus tollens* argument in which the antecedent of the conditional premise is a conjunction. But part (a) utilizes the *de facto* findings O' only to the extent that they are *incompatible* with the observational expectations O derived from the conjunction of H and A. And part (a) is *not at all sufficient to show that the falsifiability of H as part of an explanans of the actual empirical facts O' is unavoidably inconclusive.* For neither part (a) nor other general logical considerations can guarantee the deducibility of O' from an explanans constituted by the conjunction of H and some *nontrivial* revised set A' of the auxiliary assumptions which is logically incompatible with A under the hypothesis H.[47]

[47] The requirement of nontriviality of A' requires clarification. If one were to allow O' itself, for example, to qualify as a set A', then, of course, O' could be deduced trivially, and H would not even be needed in the *explanans*. Hence a necessary condition for the *nontriviality* of A' is that H be required in addition to A' for the deduction of the *explanandum*. But, as N. Rescher has pointed out to me, this necessary

How then does Duhem propose to assure that there exists such a non-trivial set A′ for any one component hypothesis H *independently* of the domain of empirical science to which H pertains? It would seem that such assurance *cannot* be given on general logical grounds at all but that the existence of the required set A′ needs *separate* and *concrete* demonstration for each particular case. In short, even in contexts to which part (a) of the Duhemian schema is applicable—which is *not* true for *all* contexts, as we shall see—neither the premise

$$[(H \cdot A) \rightarrow O] \cdot \sim O$$

nor other general logical considerations entail that

$$(\exists A')[(H \cdot A') \rightarrow O'],$$

condition is not also sufficient. For it fails to rule out an A′ of the trivial form $\sim H \vee O'$ (or $H \supset O'$) from which O′ could *not* be deduced without H.

The unavailability of a formal sufficient condition for nontriviality is not, however, damaging to the critique of Duhem presented in this paper. For surely Duhem's illustrations from the history of physics as well as the whole tenor of his writing indicate that *he intends his thesis to stand or fall on the existence of the kind of A′ which we would all recognize as nontrivial in any given case.* Any endeavor to save Duhem's thesis from refutation by invoking the kind of A′ which no scientist would accept as admissible would turn Duhem's thesis into a most *unenlightening* triviality that no one would wish to contest. Thus, I have no intention whatever of denying the following compound formal claim: "if H and A entail O, the falsity of O does not entail the falsity of H, and there will always be *some kind of* A′ which, in conjunction with H, will entail O′."

Interestingly, G. Rayna has pointed out to me that there is *equivalence* between the following two claims:

(i) Duhem's thesis that for any H and O′,

$(\exists A')[(H \cdot A') \rightarrow O']$, where A′ is nontrivial,

and

(ii) the seemingly stronger assertion that there exist infinitely many significantly different $A_1', A_2', A_3', \ldots$ such that

$(H \cdot A_i') \rightarrow O', \qquad i = 1, 2, 3, \ldots$

To prove this equivalence, we consider any infinite set $B_1, B_2, B_3, \ldots$ of pairwise incompatible hypotheses of the kind that would qualify as a constituent of A′. Upon treating the conjunction $(H \cdot B_i)$ as one hypothesis to which Duhem's thesis can now be applied, we obtain

$(\exists B_i')[\{(H \cdot B_i) \cdot B_i'\} \rightarrow O'], \qquad i = 1, 2, 3, \ldots$

But this can be written as

$(\exists B_i')[\{H \cdot (B_i \cdot B_i')\} \rightarrow O'].$

Let $A_i' = B_i \cdot B_i'$ by definition. In view of the previously granted existence of the B_i, we then have

$(\exists A_i')[(H \cdot A_i') \rightarrow O'], \qquad i = 1, 2, 3, \ldots$

where the A_i' are significantly distinct, since the B_i, B_j were chosen incompatible for $i \neq j$. Q.E.D.

Adolf Grünbaum

where A' is nontrivial in the sense discussed in footnote 47. And hence Duhem's thesis that the falsifiability of an *explanans* H is unavoidably inconclusive is a *non sequitur*.

That the Duhemian thesis is not only a *non sequitur* but actually false is borne out, as we shall now see, by the case of testing the hypothesis that a certain *physical geometry* holds, a case of conclusive falsifiability which yields an important counterexample to Duhem's stronger thesis concerning falsifiability but which does justify the *weaker* form (i) of his thesis stated in Section 6, part (i).

This counterexample will serve to show that (1) by denying the feasibility of conclusive falsification, the Duhemian schema is a serious *misrepresentation* of the actual logical situation characterizing an important class of cases of falsifiability of a purported explanans, and that (2) the plausibility of Duhem's thesis derives from the false supposition that part (a) of the schema *is* always applicable *and* that its formal validity guarantees the applicability of part (b) of the schema.

2. If we are confronted with the problem of the falsifiability of the geometry ascribed to a region which is effectively free from deforming influences, then the *correctional* physical laws play no role as auxiliary assumptions, and the latter reduce to the claim that the region in question is, in fact, effectively *free* from deforming influences. And *if* such freedom can be affirmed *without* presupposing collateral theory, then the geometry alone rather than only a wider theory in which it is ingredient will be falsifiable. On the other hand, if collateral theory *were* presupposed here, then Duhem and Einstein might be able to adduce its modifiability to support their claim that the geometry *itself* is *not* conclusively falsifiable. The question is therefore whether freedom from deforming influences can be asserted and ascertained independently of (sophisticated) collateral theory. My answer to this question is Yes. For quite independently of the conceptual elaboration of such physical magnitudes as temperature whose constancy would characterize a region free from deforming influences, the absence of perturbations is certifiable for the region as follows: two solid rods of very *different* chemical constitution which coincide at one place in the region will also coincide everywhere else in it independently of their paths of transport. It would *not* do for the Duhemian to object here that the certification of two solids as quite *different* chemically is theory-laden to an extent permitting him to uphold his thesis of the inconclusive falsifiability of the geometry. For suppose that

512

observations were so ambiguous as to permit us to assume that two solids which appear strongly to be chemically different are, in fact, chemically identical in all relevant respects. If so rudimentary an observation were thus ambiguous, then no observation could ever possess the required univocity to be incompatible with an observational consequence of a *total theoretical system*. And if that were the case, Duhem could hardly avoid the following conclusion: "observational findings are always so unrestrictedly ambiguous as *not* to permit even the refutation of any given *total theoretical system*." But such a result would be tantamount to the absurdity that *any* total theoretical system can be espoused as true a priori. Thus, it would seem that if Duhem is to maintain, as he does, that a *total theoretical system* is refutable by confrontation with observational results, then he must allow that the coincidence of diverse kinds of rods independently of their paths of transport is certifiable observationally. Accordingly, the absence of deforming influences is ascertainable *independently* of any assumptions as to the geometry and of other (sophisticated) collateral theory.

Let us now employ our earlier notation and denote the geometry by "H" and the assertion concerning the freedom from perturbations by "A." Then, once we have laid down the congruence definition and the remaining semantical rules, the physical geometry H becomes conclusively falsifiable as an *explanans* of the posited empirical findings O'. For the actual logical situation is characterized *not* by part (a) of the Duhemian schema but instead by the schema

$$[\{(H \cdot A) \to O\} \cdot \sim O \cdot A] \to \sim H.$$

It will be noted that we identified the H of the Duhemian schema with the geometry. But since a geometric theory, at least in its synthetic form, can be axiomatized as a conjunction of logically independent postulates, a particular axiomatization of H could be decomposed logically into various sets of component subhypotheses. Thus, for example, the hypothesis of Euclidean geometry could be stated, if we wished, as the conjunction of two parts consisting respectively of the Euclidean parallel postulate and the postulates of absolute geometry. And the hypothesis of hyperbolic geometry could be stated in the form of a conjunction of absolute geometry and the hyperbolic parallel postulate.

In view of the logically compounded character of a geometric hypothesis, Professor Grover Maxwell has suggested that the Duhemian thesis may be tenable in this context if we construe it as pertaining *not* to the falsifiability of a geometry as a whole but to the falsifiability of its component subhypotheses in any given axiomatization. There are two ways in which this proposed interpretation might be understood: (1) as an assertion that *any one component subhypothesis* eludes conclusive refutation on the grounds that the empirical findings can falsify the set of axioms only as a whole, or (2) in any given axiomatization of a physical geometry there exists *at least one component subhypothesis* which eludes conclusive refutation.

The first version of the proposed interpretation will not bear examination. For suppose that H is the hypothesis of Euclidean geometry and that we consider absolute geometry as one of its subhypotheses and the Euclidean parallel postulate as the other. If now the empirical findings were to show that the geometry is hyperbolic, then indeed absolute geometry would have eluded refutation. But if, one the other hand, the prevailing geometry were to turn out to be spherical, then the mere replacement of the Euclidean parallel postulate by the spherical one could not possibly save absolute geometry from refutation. For absolute geometry alone is logically incompatible with spherical geometry and hence with the posited empirical findings.

If one were to read Duhem as per the very cautious *second* version of Maxwell's proposed interpretation, then our analysis of the logic of testing the geometry of a *perturbation-free* region could not be adduced as having furnished a counterexample to so mild a form of Duhemism. And the question of the validity of this highly attenuated version is thus left open by our analysis without any detriment to that analysis.

We now turn to the critique of Einstein's Duhemian argument as applied to the empirical determination of the geometry of a region which is subject to deforming influences.

3. There can be no question that when deforming influences are present, the laws used to make the corrections for deformations involve areas and volumes in a fundamental way (e.g., in the definitions of the elastic stresses and strains) and that this involvement presupposes a geometry, as is evident from the area and volume formulae of differential geometry, which contain the square root of the determinant of the components g_{ik} of the metric tensor [27, p. 177]. Thus, the empirical determination of the geometry involves the joint assumption of a geometry and of certain collateral hypotheses. But we see already that this assumption *cannot*

be adequately represented by the conjunction H · A of the Duhemian schema where H represents the geometry.

Now suppose that we begin with a set of Euclideanly formulated physical laws P_0 in correcting for the distortions induced by perturbations and then use the thus Euclideanly corrected congruence standard for *empirically* exploring the geometry of space by determining the metric tensor. *The initial stipulational affirmation of the Euclidean geometry G_0 in the physical laws P_0 used to compute the corrections in no way assures that the geometry obtained by the corrected rods will be Euclidean!* If it is non-Euclidean, then the question is, What will be required by Einstein's fitting of the physical laws to preserve Euclideanism and avoid a contradiction of the theoretical system with experience? Will the adjustments in P_0 necessitated by the retention of Euclidean geometry entail merely a change in the dependence of the length assigned to the transported rod on such *nonpositional* parameters as temperature, pressure, and magnetic field? Or could the putative empirical findings compel that the length of the transported rod be likewise made a nonconstant function of its *position* and *orientation* as *independent* variables in order to square the coincidence findings with the requirement of Euclideanism? The possibility of obtaining *non*-Euclidean results by measurements carried out in a spatial region uniformly characterized by standard conditions of temperature, pressure, electric and magnetic field strength, etc., shows it to be *extremely doubtful*, as we shall now show, that the preservation of Euclideanism could *always* be accomplished short of introducing *the dependence of the rod's length on the independent variables of position or orientation.*

Suppose that, relatively to the customary congruence standard, the geometry prevailing in a given region when *free* from perturbational influences is that of a strongly non-Euclidean space of spatially and temporally constant curvature. Then what would be the character of the alterations in the *customary* correctional laws which Einstein's thesis would require to assure the *Euclideanism* of that region relatively to the customary congruence standard under *perturbational* conditions? The required alterations would be *independently falsifiable*, as will now be demonstrated, because they would involve affirming that such coefficients as those of linear thermal expansion *depend on the independent variables of spatial position.* That such a space dependence of the correctional co-

efficients might well be necessitated by the exigencies of Einstein's Duhemian thesis can be seen as follows by reference to the law of linear
thermal expansion. In the usual version of physical theory, the first approximation of that law is given by

$$L = L_0(1 + a \cdot \triangle T).$$

If Einstein is to guarantee the Euclideanism of the region under discussion by means of logical devices that are consonant with his thesis, and
if our region is subject only to *thermal* perturbations for some time, then,
unlike the customary law of linear thermal expansion, the revised form
of that law needed by Einstein will have to bear the twin burden of effecting *both* of the following two kinds of superposed corrections: (1) the
changes in the lengths ascribed to the transported rod in different positions or orientations which would be required even if our region *were*
everywhere at the standard temperature, merely for the sake of rendering
Euclidean its otherwise *non*-Euclidean geometry, and (2) corrections
compensating for the effects of the *de facto* deviations from the standard
temperature, these corrections being the sole onus of the *usual* version
of the law of linear thermal expansion. What will be the consequences of
requiring the *revised* version of the law of thermal elongation to implement the *first* of these two kinds of corrections in a context in which the
deviation $\triangle T$ from the standard temperature is *the same* at different
points of the region, that temperature deviation having been measured
in the manner chosen by the Duhemian? Specifically, what will be the
character of the coefficients a of the *revised* law of thermal elongation
under the posited circumstances, if Einstein's thesis is to be implemented
by effecting the *first* set of corrections? Since the new version of the law
of thermal expansion will then have to guarantee that the lengths L assigned to the rod at the various points of *equal* temperature T *differ* appropriately, it would seem clear that logically possible empirical findings
could compel Einstein to make the coefficients a of solids *depend* on the
space coordinates.

But such a spatial dependence is *independently falsifiable*: comparison
of the thermal elongations of an aluminum rod, for example, with an
invar rod of essentially zero a by, say, the Fizeau method *might* well show
that the a of the aluminum rod is a characteristic of aluminum which is
not dependent on the space coordinates. And even if it were the case that
the a's are found to be space dependent, how could Duhem and Einstein

ssure that this space dependence would have the particular functional orm required for the success of their thesis?

We see that the required resort to the introduction of a spatial dependence of the thermal coefficients might well not be open to Einstein. Hence, in order to retain Euclideanism, it would then be necessary to remetrize the space in the sense of abandoning the customary definition of congruence, entirely apart from any consideration of idiosyncratic distortions and even after correcting for these in some way or other. But this kind of remetrization, though entirely admissible in other contexts, does not provide the requisite support for Einstein's Duhemian thesis! For Einstein offered it as a criticism of Reichenbach's conception. And hence it is the avowed onus of that thesis to show that the geometry by itself cannot be held to be empirical, i.e., falsifiable, even when, with Reichenbach, we have sought to assure its empirical character by choosing and then adhering to the usual (standard) definition of spatial congruence, which excludes resorting to such remetrization.

Thus, there may well obtain observational findings O', expressed in terms of a particular definition of congruence (e.g., the customary one), which are such that there does not exist any nontrivial set A' of auxiliary assumptions capable of preserving the Euclidean H in the face of O'. And this result alone suffices to invalidate the Einsteinian version of Duhem's thesis to the effect that any geometry, such as Euclid's, can be preserved in the face of any experimental findings which are expressed in terms of the customary definition of congruence.

It might appear that my geometric counterexample to the Duhemian thesis of unavoidably inconclusive falsifiability of an explanans is vulnerable to the following criticism: "To be sure, Einstein's geometric articulation of that thesis does not leave room for saving it by resorting to a remetrization in the sense of making the length of the rod vary with position or orientation even after it has been corrected for idiosyncratic distortions. But why saddle the Duhemian thesis as such with a restriction peculiar to Einstein's particular version of it? And thus why not allow Duhem to save his thesis by countenancing those alterations in the congruence definition which are remetrizations?"

My reply is that to deny the Duhemian the invocation of such an alteration of the congruence definition in this context is not a matter of gratuitously requiring him to justify his thesis within the confines of Einstein's particular version of that thesis; instead, the imposition of

517

this restriction is entirely legitimate here, and the Duhemian could hardly wish to reject it as unwarranted. For it is of the essence of Duhem's contention that H (in this case, Euclidean geometry) can always be preserved *not* by tampering with the semantical rules (interpretive sentences linking H to the observational base but rather by availing oneself of the alleged *inductive latitude* afforded by the ambiguity of the experimental evidence to do the following: (a) leave the factual commitments of H *unaltered* by retaining both the statement of H and the semantical rule linking its terms to the observational base, and (b) replace the set A by a set A' of auxiliary assumptions *differing in factual content* from A such that A and A' are logically incompatible under the hypothesis H. Now the factual content of a geometrical hypothesis can be changed either by preserving the original statement of the hypothesis while changing one or more of the semantical rules or by keeping all of the semantical rules intact and suitably changing the statement of the hypothesis (cf. Section 6, part (i)). We can see, therefore, that the retention of a Euclidean H by the device of changing through remetrization the semantical rule governing the meaning of "congruent" (for line segments) effects a retention *not* of the *factual commitments* of the original Euclidean H but only of its *linguistic trappings*. That the thus "preserved" Euclidean H actually repudiates the factual commitments of the *original* one is clear from the following: the *original* Euclidean H had asserted that the coincidence behavior common to all kinds of solid rods is Euclidean, *if* such transported rods are taken as the physical realization of congruent intervals; but the Euclidean H which survived the confrontation with the posited empirical findings only by dint of a *remetrization* is predicated on a denial of the very assertion that was made by the original Euclidean H, which it was to "preserve." It is as if a physician were to endeavor to "preserve" an a priori diagnosis that a patient has acute appendicitis in the face of a negative finding (yielded by an exploratory operation) as follows: he would redefine "acute appendicitis" to denote the healthy state of the appendix!

Hence, the confines within which the Duhemian must make good his claim of the preservability of a Euclidean H do *not* admit of the kind of change in the congruence definition which alone would render his claim tenable under the assumed empirical conditions. Accordingly, the geometrical critique of Duhem's thesis given in this paper does *not* depend for its validity on restrictions peculiar to Einstein's version of it.

Even apart from the fact that Duhem's thesis precludes resorting to an alternative metrization to save it from refutation in our geometrical context, the very feasibility of alternative metrizations is vouchsafed not by any general Duhemian considerations pertaining to the logic of falsifiability but by a property peculiar to the subject matter of geometry (and chronometry): the latitude for *convention* in the ascription of the spatial (or temporal) *equality* relation to intervals in the continuous manifolds of physical space (or time).

But what of the possibility of actually *extricating* the unique underlying geometry (to within experimental accuracy) from the network of hypotheses which enter into the testing procedure?

That contrary to Duhem and Einstein, the geometry itself may well be empirical, once we have renounced the kinds of alternative congruence definitions employed by Poincaré, is seen from the following possibilities of its successful empirical determination. After assumedly obtaining a non-Euclidean geometry G_1 from measurements with a rod corrected on the basis of Euclideanly formulated physical laws P_0, we can revise P_0 so as to conform to the non-Euclidean geometry G_1 just obtained by measurement. This retroactive revision of P_0 would be effected by recalculating such quantities as areas and volumes on the basis of G_1 and changing the functional dependencies relating them to temperatures and other physical parameters. Let us denote by "P_1'" the set of physical laws P resulting from thus revising P_0 to incorporate the geometry G_1. Since various physical magnitudes ingredient in P_1' involve lengths and durations, we now use the set P_1' to correct the rods (and clocks) with a view to seeing whether rods and clocks thus corrected will reconfirm the set P_1'. If not, *modifications* need to be made in this set of laws so that the functional dependencies between the magnitudes ingredient in them reflect the new standards of spatial and temporal congruence defined by P_1'-corrected rods and clocks. We may thus obtain a new set of physical laws P_1.

Now we employ this set P_1 of laws to correct the rods for perturbational influences and then determine the geometry with the thus corrected rods. Suppose the result is a geometry G_2 different from G_1. Then if, upon repeating this two-step process several more times, there is convergence to a geometry of constant curvature we must continue to repeat the two-step process an additional finite number of times until we find the following: the geometry G_n ingredient in the laws P_n providing the basis for per-

turbation corrections is indeed the same (to within experimental accuracy) as the geometry obtained by measurements with rods that have
been corrected via the set P_n. If there is such convergence at all, it will
be to the same geometry G_n even if the physical laws used in making the
initial corrections are not the set P_o, which presupposes Euclidean geometry, but a different set P based on some non-Euclidean geometry or other.
That there can exist only one such geometry of constant curvature G_n
would seem to be guaranteed by the identity of G_n with the unique underlying geometry G_t characterized by the following properties: (i) G_t would
be exhibited by the coincidence behavior of a transported rod if the *whole*
of the space were actually free of deforming influences, (ii) G_t would be
obtained by measurements with rods corrected for distortions on the
basis of physical laws P_t presupposing G_t, and (iii) G_t would be found to
prevail in a given relatively small, perturbation-free region of the space
quite independently of the assumed geometry ingredient in the correctional physical laws. Hence, if our method of successive approximation
does converge to a geometry G_n of *constant* curvature, then G_n would be
this unique underlying geometry G_t. And, in that event, we can claim to
have found empirically that G_t is indeed the geometry prevailing in the
entire space which we have explored.

But what if there is no convergence? It might happen that whereas convergence would obtain by starting out with corrections based on the set
P_o of physical laws, it would *not* obtain by beginning instead with corrections presupposing some particular *non*-Euclidean set P or *vice versa*:
just as in the case of Newton's method of successive approximation [15,
p. 286], there are conditions, as A. Suna has pointed out to me, under
which there would be no convergence. We might then nonetheless succeed as follows in finding the geometry G_t empirically, *if* our space is one
of constant curvature.

The geometry G_r resulting from measurements by means of a corrected
rod is a single-valued function of the geometry G_a assumed in the correctional physical laws, and a Laplacian demon having sufficient knowledge
of the facts of the world would know this function $G_r = f(G_a)$. Accordingly, we can formulate the problem of determining the geometry empirically as the problem of finding the point of intersection between the curve
representing this function and the straight line $G_r = G_a$. That there exists
one and only one such point of intersection follows from the existence
of the geometry G_t defined above, provided that our space is one of con-

tant curvature. Thus, what is now needed is to make determinations of he G_r corresponding to a number of geometrically different sets of corectional physical laws P_a, to draw the most reasonable curve $G_r = f(G_a)$ hrough this finite number of points (G_a, G_r), and then to find the point f intersection of this curve and the straight line $G_r = G_a$.

Whether this point of intersection turns out to be the one representing Euclidean geometry or not is beyond the reach of our conventions, baring a remetrization. And thus the least that we can conclude is that since empirical findings can greatly narrow down the range of uncertainty as to he prevailing geometry, there is no assurance of the *latitude* for the choice of a geometry which Einstein takes for granted. Einstein's Duhemian position would appear to be inescapable *only* if our proposed method of determining the geometry by itself empirically *cannot* be generalized in ome way to cover the general relativity case of a space of variable curvature and if the latter kind of theory turns out to be true.[48]

It would seem therefore that, contrary to Einstein, the logic of eliminating distorting influences prior to stipulating the rigidity of a solid body is not such as to provide scope for the ingression of conventions over and above those acknowledged in Riemann's analysis of congruence, and trivial ones such as the system of units used. *Mutatis mutandis*, an analogous conclusion can be established in regard to the application of corrections to provide a standard of isochronism for clocks.

8. Summary

The present essay has endeavored to answer the following multi-faceted question: In what sense and to what extent can the ascription of a particular metric geometry to physical space and the chronometry ingredient in physical theory be held to have an *empirical* warrant?

Our analysis of the logical status of the concept of a rigid body and of an isochronous clock leads to the conclusion that once the physical meaning of congruence has been stipulated by reference to a solid body and to a clock respectively for whose distortions allowance has been made com-

[48] The extension of our method to the case of a geometry of variable curvature is not simple. For in that case, the geometry G is no longer represented by a single scalar given by a Gaussian curvature, and our graphical method breaks down. Whatever the answer to this open question of the extensibility of our method, I have argued in another publication [31] that it is wholly misconceived to suppose with J. Maritain [49] that there are supra-scientific philosophical means for ascertaining the underlying geometry, if *scientific* procedures do not succeed in unraveling it.

putationally as outlined, then the geometry and the ascriptions of durations to time intervals is determined uniquely by the totality of relevan
empirical facts. It is true, of course, that even apart from experimenta
errors, not to speak of quantum limitations on the accuracy with whic
the metric tensor of *space-time* can be meaningfully ascertained by mea
urement [103, 82], no finite number of data can uniquely determine th
functions constituting the representations g_{ik} of the metric tensor in an
given coordinate system. But the criterion of *inductive* simplicity which
governs the free creativity of the geometer's imagination in his choice o
a particular metric tensor here is the same as the one employed in theor
formation in any of the nongeometrical portions of empirical science. An
choices made on the basis of such inductive simplicity are in principl
true or false, unlike those springing from considerations of descriptiv
simplicity, which merely reflect conventions.

BIBLIOGRAPHY

1. Barankin, E. W. "Heat Flow and Non-Euclidean Geometry," *American Mathe
matical Monthly*, 49:4–14 (1942).
2. Baruch, J. J. "Horological Accuracy: Its Limits and Implications," *America
Scientist*, 46:188A–196A (1958).
3. Bonola, R. *Non-Euclidean Geometry*. New York: Dover Publications, 1955.
4. Braithwaite, R. B. "Axiomatizing a Scientific System by Axioms in the Form o
Identification," in *The Axiomatic Method*, L. Henkin, P. Suppes, and A. Tarsk
eds. Amsterdam: North Holland Publishing Company, 1959. Pp. 429–442.
5. Brouwer, D. "The Accurate Measurement of Time," *Physics Today*, 4:7–1
(1951).
6. Brown, F. A., Jr. "Biological Clocks and the Fiddler Crab," *Scientific American
190:34–37 (1954).
7. Brown, F. A., Jr. "The Rhythmic Nature of Animals and Plants," *America
Scientist*, 47:147–168 (1959); "Living Clocks," *Science*, 130:1535–1544 (1959)
and "Response to Pervasive Geophysical Factors and the Biological Clock Prob
lem," *Cold Spring Harbor Symposia on Quantitative Biology*, 25:57–71 (1960)
8. Carnap, R. *Der Raum*. Berlin: Reuther and Reichard, 1922.
9. Carnap, R. Preface to H. Reichenbach's *The Philosophy of Space and Time
New York: Dover Publications, 1958.
10. Clemence, G. M. "Astronomical Time," *Reviews of Modern Physics*, 29:2–
(1957).
11. Clemence, G. M. "Dynamics of the Solar System," in *Handbook of Physics
E. Condon and H. Odishaw, eds. New York: McGraw-Hill, 1958. P. 65.
12. Clemence, G. M. "Ephemeris Time," *Astronomical Journal*, 64:113–115 (1959)
and *Transactions of the International Astronomical Union*, Vol. 10, 1958.
13. Clemence, G. M. "Time and Its Measurement," *American Scientist*, 40:26
(1952).
14. Clifford, W. K. *The Common Sense of the Exact Sciences*. New York: Dove
Publications, 1955.
15. Courant, R. *Vorlesungen über Differential und Integralrechnung*, Vol. 1. Berlin
Springer, 1927.

16. D'Abro, A. *The Evolution of Scientific Thought from Newton to Einstein.* New York: Dover Publications, 1950.
17. Denbigh, K. G. "Irreversibility and the Time Sense," forthcoming.
18. Dingler, H. "Die Rolle der Konvention in der Physik," *Physikalische Zeitschrift,* 23:47–52 (1922).
19. Duhem, P. *The Aim and Structure of Physical Theory.* Princeton: Princeton University Press, 1954.
20. Eddington, A. S. *Space, Time and Gravitation.* Cambridge: Cambridge University Press, 1953.
21. Einstein, A. "The Foundations of the General Theory of Relativity," in *The Principle of Relativity, a Collection of Original Memoirs.* New York: Dover Publications, 1952. Pp. 111–164.
22. Einstein, A. *Geometrie und Erfahrung.* Berlin: Springer, 1921.
23. Einstein, A. "On the Electrodynamics of Moving Bodies," in *The Principle of Relativity, a Collection of Original Memoirs.* New York: Dover Publications, 1952. Pp. 37–65.
24. Einstein, A. "On the Influence of Gravitation on the Propagation of Light," in *The Principle of Relativity, a Collection of Original Memoirs,* New York: Dover Publications, 1952. Pp. 97–108.
25. Einstein, A. "Prinzipielles zur allgemeinen Relativitätstheorie," *Annalen der Physik,* 55:241 (1918).
26. Einstein, A. "Reply to Criticisms," in *Albert Einstein: Philosopher-Scientist,* P. A. Schilpp, ed. Evanston: The Library of Living Philosophers, 1949. Pp. 665–688.
27. Eisenhart, L. P. *Riemannian Geometry.* Princeton: Princeton University Press, 1949.
28. Feigl, H. "Confirmability and Confirmation," *Revue Internationale de Philosophie,* 5:268–279 (1951). Reprinted in *Readings in Philosophy of Science,* P. P. Wiener, ed. New York: Scribner, 1953.
29. Fraenkel, A. A., and Y. Bar-Hillel. *Foundations of Set Theory.* Amsterdam: North Holland Publishing Company, 1958.
30. Goodhard, C. B. "Biological Time," *Discovery,* 18:519–521 (December 1957).
31. Grünbaum, A. "The A Priori in Physical Theory," in *The Nature of Physical Knowledge,* L. W. Friedrich, ed. Bloomington: Indiana University Press, 1960. Pp. 109–128.
32. Grünbaum, A. "Carnap's Views on the Foundations of Geometry," in *The Philosophy of Rudolf Carnap,* P. A. Schilpp, ed. La Salle, Ill.: Open Court, 1962.
33. Grünbaum, A. "A Consistent Conception of the Extended Linear Continuum as an Aggregate of Unextended Elements," *Philosophy of Science,* 19:288–306 (1952).
34. Grünbaum, A. "Logical and Philosophical Foundations of the Special Theory of Relativity," in *Philosophy of Science: Readings,* A. Danto and S. Morgenbesser, eds. New York: Meridian, 1960. Pp. 399–434.
35. Grünbaum, A. "Operationism and Relativity," *Scientific Monthly,* 79:228–231 (1954). Reprinted in *The Validation of Scientific Theories,* P. Frank, ed. Boston: Beacon Press, 1957. Pp. 84–94.
36. Grünbaum, A. "The Philosophical Retention of Absolute Space in Einstein's General Theory of Relativity," *Philosophical Review,* 66:525–534 (1957).
37. Grünbaum, A. "Relativity and the Atomicity of Becoming," *Review of Metaphysics,* 4:143–186 (1950).
38. Hertz, H. *The Principles of Mechanics.* New York: Dover Publications, 1956.
39. Hoagland, H. "Chemical Pacemakers and Physiological Rhythms," in *Colloid Chemistry,* Vol. V, J. Alexander, ed. New York: Reinhold, 1944. Pp. 762–785.

40. Hoagland, H. "The Physiological Control of Judgments of Duration: Evidence for a Chemical Clock," Journal of General Psychology, 9:267–287 (1933).
41. Hobson, E. W. The Theory of Functions of a Real Variable, Vol. 1. New York Dover Publications, 1957.
42. Hood, P. How Time Is Measured. London: Oxford University Press, 1955.
43. Jeffreys, H. The Earth. 3rd ed.; Cambridge: Cambridge University Press, 1952.
44. Klein, F. Vorlesungen über Nicht-Euklidische Geometrie. Berlin: Springer 1928.
45. Leclercq, R. Guide Théorique et Pratique de la Recherche Expérimentale. Paris Gauthier-Villars, 1958.
46. Lyons, H. "Atomic Clocks," Scientific American, 196:71–82 (February 1957)
47. McVittie, G. C. "Distance and Relativity," Science, 127:501–505 (1958).
48. Margenau, H., and G. M. Murphy. The Mathematics of Physics and Chemistry New York: Van Nostrand, 1943.
49. Maritain, J. The Degrees of Knowledge. New York: Scribner, 1959.
50. Milham, W. I. Time and Timekeepers. New York: Macmillan, 1929. A more recent edition appeared in 1941.
51. Milne, E. A. Kinematic Relativity. London: Oxford University Press, 1948.
52. Milne, E. A. Modern Cosmology and the Christian Idea of God. London Oxford University Press, 1952.
53. Møller, C. The Theory of Relativity. London: Oxford University Press, 1952
54. Newton, I. Principia, F. Cajori, ed. Berkeley: University of California Press 1947.
55. Nagel, E. The Structure of Science. New York: Harcourt, Brace and World 1961.
56. Northrop, F. S. C. The Meeting of East and West. New York: Macmillan 1946.
57. Northrop, F. S. C. "Whitehead's Philosophy of Science," in The Philosophy of Alfred North Whitehead, P. A. Schilpp, ed. New York: Tudor, 1941. Pp 165–207.
58. Page, L. Introduction to Theoretical Physics. New York: Van Nostrand, 1935
59. Page, L., and N. I. Adams. Electrodynamics. New York: Van Nostrand, 1940
60. Pérard, A. Les Mesures Physiques. Paris: Presses Universitaires de France, 1955
61. Poincaré, H. Dernières Pensées. Paris: Flammarion, 1913. Chs. 2 and 3.
62. Poincaré, H. "Des Fondements de la Géométrie, à propos d'un Livre de M Russell," Revue de Métaphysique et de Morale, 7:251–279 (1899).
63. Poincaré, H. The Foundations of Science. Lancaster, Pa.: Science Press, 1946
64. Poincaré, H. "La Mesure du Temps," Revue de Métaphysique et de Morale 6:1–13 (1898).
65. Poincaré, H. "Sur les Principes de la Géométrie, Réponse à M. Russell," Revue de Métaphysique et de Morale, 8:73–86 (1900).
66. Popper, K. R. The Logic of Scientific Discovery. London: Hutchinson, 1959
67. Price, D. J. "The Prehistory of the Clock," Discovery, 17:153–157 (April 1956)
68. Reichenbach, H. "Discussion of Dingler's Paper," Physikalische Zeitschrift 23:52–53 (1922).
69. Reichenbach, H. Experience and Prediction. Chicago: University of Chicago Press, 1938.
70. Reichenbach, H. Philosophic Foundations of Quantum Mechanics. Berkeley University of California Press, 1948.
71. Reichenbach, H. "The Philosophical Significance of the Theory of Relativity," in Albert Einstein: Philosopher-Scientist, P. A. Schilpp, ed. Evanston: Library of Living Philosophers, 1949.
72. Reichenbach, H. The Philosophy of Space and Time. New York: Dover Publications, 1958.

73. Reichenbach, H. *The Rise of Scientific Philosophy*. Berkeley: University of California Press, 1951.
74. Reichenbach, H. "Über die physikalischen Konsequenzen der relativistischen Axiomatik," *Zeitschrift für Physik*, 34:35 (1925).
75. Riemann, B. "Über die Hypothesen, welche der Geometrie zu Grunde liegen," in *Gesammelte Mathematische Werke*, H. Weber, ed. New York: Dover Publications, 1953. Pp. 272–287.
76. Robertson, H. P. "The Geometries of the Thermal and Gravitational Fields," *American Mathematical Monthly*, 57:232–245 (1950).
77. Robertson, H. P. "Geometry as a Branch of Physics," *Albert Einstein: Philosopher-Scientist*, P. A. Schilpp, ed. Evanston: Library of Living Philosophers, 1949. Pp. 313–332.
78. Rougier, L. *La Philosophie Géométrique de Henri Poincaré*. Paris: Alcan, 1920.
79. Russell, B. *The Foundations of Geometry*. New York: Dover Publications, 1956.
80. Russell, B. *Our Knowledge of the External World*. London: Allen and Unwin, 1926.
81. Russell, B. "Sur les Axiomes de la Géométrie," *Revue de Métaphysique et de Morale*, 7:684–707 (1899).
82. Salecker, H., and E. P. Wigner. "Quantum Limitations of the Measurement of Space-Time Distances," *Physical Review*, 109:571–577 (1958).
83. Sokolnikoff, I. S. *Mathematical Theory of Elasticity*. New York: McGraw-Hill, 1946.
84. Sommerville, D. M. Y. *The Elements of Non-Euclidean Geometry*. New York: Dover Publications, 1958.
85. Stille, U. *Messen und Rechnen in der Physik*. Brunswick: Vieweg, 1955.
86. Struik, D. J. *Classical Differential Geometry*. Cambridge, Mass.: Addison-Wesley, 1950.
87. Tarski, A. "What Is Elementary Geometry?" in *The Axiomatic Method*, L. Henkin, P. Suppes, and A. Tarski, eds. Amsterdam: North Holland Publishing Company, 1959. Pp. 16–29.
88. Taylor, G. I. "Tidal Friction in the Irish Sea," *Philosophical Transactions, Royal Society of London*, Series A, 220:1–33 (1920).
89. Timoshenko, S., and J. N. Goodier. *Theory of Elasticity*. New York: McGraw-Hill, 1951.
90. Veblen, O., and J. H. C. Whitehead. *The Foundations of Differential Geometry*, No. 29 of Cambridge Tracts in Mathematics and Mathematical Physics. Cambridge: Cambridge University Press, 1932.
91. von Helmholtz, H. *Schriften zur Erkenntnistheorie*, P. Hertz and M. Schlick, eds. Berlin: Springer, 1921.
92. Walker, A. G. "Axioms for Cosmology," in *The Axiomatic Method*, L. Henkin, P. Suppes, and A. Tarski, eds. Amsterdam: North Holland Publishing Company, 1959. Pp. 308–321.
93. Ward, F. A. B. *Time Measurement*. Part I, Historical Review. 4th edition; London: Royal Stationery Office, 1958.
94. Weinstein, B. *Handbuch der Physikalischen Massbestimmungen*. Berlin: Springer, Vol. 1, 1886, and Vol. 2, 1888.
95. Weyl, H. *Philosophy of Mathematics and Natural Science*. Princeton: Princeton University Press, 1949.
96. Weyl, H. *Space-Time-Matter*. New York: Dover Publications, 1950.
97. Whitehead, A. N. *The Concept of Nature*. Cambridge: Cambridge University Press, 1926. Ch. VI.
98. Whitehead, A. N. *Essays in Science and Philosophy*. New York: Philosophical Library, 1947.

99. Whitehead, A. N. *The Principle of Relativity*. Cambridge: Cambridge University Press, 1922. Ch. III.
100. Whitehead, A. N. *The Principles of Natural Knowledge*. Cambridge: Cambridge University Press, 1955.
101. Whitehead, A. N. *Process and Reality*. New York: Macmillan, 1929.
102. Whyte, L. L. "Light Signal Kinematics," *British Journal for the Philosophy of Science*, 4:160–161 (1953).
103. Wigner, E. P. "Relativistic Invariance and Quantum Phenomena," *Reviews of Modern Physics*, 29:255–268 (1957).

Time and the World Order

The aim of this essay is to develop a framework in terms of which some perennial puzzles about time and the temporal aspects might be resolved. The treatment is dialectical, consists, that is, in an attempt to fit standard positions' on the topics discussed into a sustained argument having the logical structure, if not the literary form, of a dialogue in which the participants develop and modify their views under the impact of the discussion. That the present essay falls far short of the ideal suggested by this description will be clear to anyone who ventures to begin it. It fails abjectly if construed as a comprehensive dialogue which begins in unreflective common sense and ends with all relevant puzzles resolved. It has, I hope, greater merit if viewed as an abstract of a series of excerpts from such a dialogue, a series which breaks into the discussion after it has long been under way and breaks off where it does because the dialogue is still going on.

The argument begins with some familiar puzzles about truth and time. The reader may well see through these puzzles at a glance. I hope he does, for they serve the purpose of introducing as directly and as simply as I know how the major themes which it is the purpose of this essay to explore, some of which are as baffling as any philosophy has to offer. I have "taken time seriously" since I cut my philosophical teeth on McTaggart's well-known paper on the unreality of time and the attempts of Broad and others to refute him. I soon discovered that the 'problem of time' is rivaled only by the 'mind-body problem' in the extent to which it inexorably brings into play all the major concerns of philosophy. Here, if anywhere, analysis without synopsis must be blind.

Among the topics I propose to discuss are the connections between truth, confirmability, and determinism; the philosophical and scientific significance of the three-valued logics; the relative priority of things and events; the status of time in the common-sense world, and of space-time

in macromechanics; the meaning of existence statements about episodes
and things and the sense in which even existence statements about ab
stract entities have a tense; the objectivity of becoming (with some re
marks on the significance, in this connection, of the relativity of simul
taneity). As is implied by the dialectical character of the treatment, these
topics make multiple appearances, and the 'conclusions' of one section
are often radically recast in another.

I. FACTS, EPISODES, AND THINGS
1. Truths about Other Times

I

Suppose that the following statements, made today (1958), are all true

(1) S was Φ_1 (in 1957).
(2) S is Φ_2 today (1958).
(3) S will be Φ_3 (in 1959).

According to one version of the correspondence theory of truth—a version
which it is my purpose to criticize at a later stage of my argument—the
above statements are true because each of them corresponds to a fact
Thus, (1) corresponds to the fact that S *was* Φ_1 in 1957; (2) to the fact
that S *is* Φ_2 today; and (3) to the fact that S *will be* Φ_3 in 1959. And, indeed, if these statements are true, it *is* a fact that S was Φ_1 in 1957; it *is* a
fact that S is Φ_2 today; it *is* a fact that S will be Φ_3 in 1959.

Notice that in mentioning each of these facts, whether the fact about
the past, the fact about the present, or the fact about the future, I wrote
in each case, "It *is* a fact that . . ." We say of an episode that it *took
place*, *is taking place*, or *will take place*. But if something is a fact, it *is* a
fact, even if the verb in the that-clause is in the past or future tense. Having written this, I must at once qualify it, for it would be a mistake to
suppose that we never use locutions of the form "It *was* a fact that . . ."
or "It *will be* a fact that . . ." Consider, for example,

(4) It *was* a fact (in 1957) that S *would be* Φ_2 in 1958.

But it is important to appreciate the kind of context in which statements
of this sort are appropriate. They can be typified by the following:

(5) (In 1957) Jones *thought* (*said, wrote*, etc.) that S would be
Φ_2 in 1958; and it was a fact that S would be Φ_2 in 1958.

In other words, the kind of context in which we use such locutions as

(In 1957) it was a fact that . . ." and "(In 1959) it will be a fact that
. ." are those in which we are viewing someone (it may be ourselves) as
thinking or asserting something at a time other than the present and
evaluating this thought or assertion. In the absence of concern for what
someone thought or said, or might have thought or said, *at a prior time,*
we should say, not

(6) (In 1957) it *was* a fact that S was Φ_1 in 1957

or

(7) (In 1959) it *will be* a fact that S will be Φ_3 in 1959

but, supposing the context to call for a fact statement at all,

(8) It *is* a fact that S was Φ_1 in 1957

and

(9) It *is* a fact that S will be Φ_3 in 1959.

II

When a fact statement is appropriate, and when the only temporal
"point of view" in question is that of the person who makes the statement
at the time of making it, the statement is always of the form "It *is* a fact
that . . ." regardless of the tense of the verb which appears in the that-
clause. It is this characteristic of what might be called "one-perspective"
fact statements which has tempted philosophers to hold that the 'is' of
'It is a fact that . . .' has to do with a timeless mode of being. Thus, it
has seemed proper to connect the 'was' of 'It *was* a fact that . . .' not
with the *factuality* of the fact, but with the temporal location of the person or persons whose point of view is being considered in a *two*-perspec-
tive fact statement. It is argued that just as one does not say

(10) Two plus two *was* equal to four

or

(11) Two plus two *will be* equal to four,

the 'is' of

(12) Two plus two *is* equal to four

being an 'is' which has been "detensed" (turned into a 'tenseless present')
by depriving it of its normal contrast with 'will be' and 'was,' so the 'is'
of a one-perspective statement of empirical fact is equally detensed, and
gives expression to a "timeless mode of being" like that attributed to num-

529

bers. The 'was' of 'It *was* a fact that . . .' where the latter is appropriate is then interpreted as an indication of the pastness of the second point of view of a two-perspective fact statement, rather than as a temporal qualification of the *factuality* of the state of affairs mentioned by the that clause.

This line of thought is not without its insights. But that the matter is not quite so simple becomes manifest if one turns one's attention to two perspective mathematical fact statements. For the above reasoning would lead one to expect that in cases where a *past* mathematical thinking or saying is being evaluated, it would be proper to make such statements as

(13) It *was* a fact that $2 + 2 = 4$.[1]

Thus, we would expect to find such statements as

(14) (In 1957) Jones thought that $2 + 2 = 4$; and, indeed, it *was* a fact that $2 + 2 = 4$.

But, of course, we immediately sense that something has gone wrong. For just as it is odd to say

(10) Two plus two *was* equal to four,

so it is odd to say, even in such a context as (14), ". . . it was a fact that $2 + 2 = 4$," to say it, that is, even where *two* perspectives are involved. We must surely say

(15) (In 1957) Jones thought that $2 + 2 = 4$ and, indeed, it *is* a fact that $2 + 2 = 4$.

The proponents of facts as 'timeless entities' may be expected to reply that this criticism actually supplies grist to their mill. They must grant, to be sure, that the two-perspective character of a two-perspective fact statement is, though *necessary to,* not a *sufficient* condition of the appropriateness of 'It *was* (*will be*) a fact that . . .' They will point out that the additional requirement seems to be that the that-clause be a *tensed* that-clause. And they can be expected to argue that the inappropriateness of "It *was* (*will be*) a fact that $2 + 2 = 4$," even in two perspective contexts, springs from the tenseless character of mathematical statements themselves, e.g., the tenseless character of 'Two plus two is equal to four.' It is this, they conclude, which accounts for the correctness of

(16) (In 1957) Jones thought that two plus two *is* equal to four and, indeed, it *is* a fact that two plus two *is* equal to four.

[1] I deliberately switch to arithmetical notation to take attention temporarily away from the problem of the tense—or tenselessness—of the mathematical that-clause.

But while these considerations may give aid and comfort to the idea that *mathematical* facts are timeless entities, they constitute an *ignoratio elenchi* as far as the central point at issue is concerned. For by conceding that the pastness of the second perspective in a two-perspective fact statement is not a sufficient condition of the appropriateness of "It *was* a fact that . . ." they have abandoned the ground on which they rested their claim that the 'was' pertains *not* to the *factuality* of the fact, but simply to the pastness of the second perspective. The question thus arises, can the proponents of the "time transcendence" of facts about temporal episodes turn to their advantage the second condition of the appropriateness of 'It *was* (*will be*) a fact that . . .'—namely, that the that-clause be a tensed that-clause? And to ask this question is to recognize that these philosophers have, in effect, argued that the failure to appreciate that facts about temporal episodes are as 'timeless' as mathematical facts springs from the assumption, according to them mistaken, that discourse about temporal episodes must be *tensed* discourse. We can almost hear them expostulate, "Surely to assume that time talk must be *tensed* talk is like assuming that space talk must be '*here-there*' talk!" And indeed one must grant that there is a sufficient parallel between tensed talk and 'here-there' talk to make this remark a telling one, if we were to admit that there is (in a relevant sense) space talk *about the world* which is not *at bottom* 'here-there' talk (or equivalent to it). Postponing the issues posed by the concluding clause of the preceding sentence to a much later stage in the argument, we notice that having made the above expostulation, the proponents of 'timeless facts' argue that the forms "It *was* a fact that . . ." and "It *will be* a fact that . . ." (which appear to imply that facts about episodes are temporal entities) can be dispensed with by the simple expedient of abandoning the language of tenses and replacing it by the use of detensed verbs (the 'tenseless present') together with dates. In other words, they propose to replace the three statements with which we began this paper by the following detensed statements,

(1') S is Φ_1 in 1957,
(2') S is Φ_2 in 1958,
(3') S is Φ_3 in 1959,

or, introducing the archaic 'be' to play the role of a detensed 'is,'

(1″) S be Φ_1 in 1957,
(2″) S be Φ_2 in 1958,
(3″) S be Φ_3 in 1959.

Since they obviously cannot be proposing that the latter are equivalent in meaning to the original statements, the claim must be that $(1')$, $(2')$, and $(3')$ are somehow "more basic" then the original statements, as having a core meaning which is somehow *prior to* the perspectival idiosyncrasies of speakers and thinkers, and constitutes the 'neutral' foundation on which the latter are somehow built. In this detensed language, the argument concludes, it makes no more sense to say

> (17) It *was* a fact (in 1957) that S be Φ_2 in 1958

than to say

> (13) It *was* a fact (in 1957) that $2 + 2 = 4$ (i.e., that 2 plus 2 *be* equal to 4).

Rather, just as we say

> $(15')$ (In 1957) Jones thought that $2 + 2 = 4$; and, indeed, it *(tenselessly)* is—that is, *be*—a fact that $2 + 2 = 4$

so, given this use of 'be,' we would be in a position to say

> (18) (In 1957) Jones thought that S be Φ_2 in 1958; and, indeed, it *be* a fact that S be Φ_2 in 1958.

In other words, the introduction of a postulated "stripped down" tenseless discourse about episodes, and, hence, the introduction of detensed that-clauses, would carry with it the consequence that the fact locutions appropriate to these detensed that-clauses would not be tensed locutions like 'It is (was, will be) a fact that . . .' but rather the tenseless present appropriate to fact statements in the domain of mathematics, i.e., by the above convention, "It *be* a fact that . . ." This "stripped down" locution would express the timeless mode of being shared, at bottom, according to this point of view, by historical and mathematical facts. The argument adds, as an afterthought, that (18) in its turn might be "stripped down" to read

> $(18')$ Jones *think* in 1957 that S be Φ_2 in 1958; and, indeed, it *be* a fact that S be Φ_2 in 1958

where 'think' like 'be' is in the 'tenseless present.'

III

That the above line of thought is profoundly mistaken is scarcely news. On the other hand, the task of exposing the numerous confusions on which it rests has not, in my opinion, been successfully completed; for though most of the relevant distinctions have been drawn, they have not yet been mobilized into a coordinated attack on the perennial nexus of puzzles pertaining to the existence of temporal facts. It is the aim of the present paper to make an attempt in this direction. But if the full story on the mistakes involved in the thesis of the 'timeless being of temporal facts' is a long one which is scarcely under way, a provisional measure of clarification can be gained by noting that one can find a place for a 'tenseless present' in the formulation of temporal statements without assimilating this tenseless present to the tenseless present of mathematical statements. For it would obviously be perfectly legitimate to introduce a use of 'is' in accordance with the schema

(19) x is Φ at t $\cdot \equiv \cdot$ *Either* x *was* Φ at t or x *is* Φ at t or x will be Φ at t.

Thus, using the archaic 'be' for this use of 'is,' we could introduce the statement 'Eisenhower be president in 1956' in terms of the equivalence

(20) Eisenhower be president in 1956 $\cdot \equiv \cdot$ Either Eisenhower *was* president in 1956 or Eisenhower *is* president in 1956 or Eisenhower *will be* president in 1956.

And if this convention were a recognized feature of English usage, Tom, in 1955, Dick, in 1956, and Harry, in 1957, could all agree in saying, "Eisenhower be president in 1956." Would they all be "making the same statement"? Only, of course, in the sense in which "Eisenhower will be president in 1956" (said in 1955) makes the same statement as "Eisenhower was president in 1956" (said in 1957). Thus, Tom, if pressed, would say

(21) Eisenhower *be* president in 1956 *because* Eisenhower *will be* president in 1956.

Whereas Dick and Harry, respectively, would say

(22) Eisenhower *be* president in 1956 *because* Eisenhower *is* president in 1956,
(23) Eisenhower *be* president in 1956 *because* Eisenhower *was* president in 1956.

These considerations make it clear that the 'neutral' tenseless present, thus introduced, would be quite other than the tenseless present of mathematical statements. This suggests that instead of construing tensed verbs as the *enrichment* of a neutral "stripped down" 'perspective-free' mode of making temporal assertions, the device of using 'perspective-neutral' sentences to make temporal statements may rest on and presuppose the tensed verbs of everyday temporal discourse.

IV

The argument to date suggests that whatever we are to do with the 'is' of

(21) It is a fact that $2 + 2 = 4$

we must take seriously the prima-facie tensed character of the 'is' in such statements as

(8) It *is* a fact that S was Φ_1 in 1957

and

(9) It *is* a fact that S will be Φ_3 in 1959.

If we then turn our attention to the pair

(22) It *is* a fact that S is Φ_2 today (1958)

and

(4) It *was* a fact (in 1957) that S would be Φ_2 in 1958,

we may easily be tempted to say that such facts are *temporal* entities which exist *at* times, and, indeed, to say that the 1957 fact that S *would be* Φ_2 was replaced by the 1958 fact that S *is* Φ_2. We may even be led to speculate whether it could be a fact *today* that S is Φ_2, without its having been a fact (in 1957) that S would be Φ_2.

Now there is a sound kernel of truth in the idea that facts about episodes are temporal entities. And if we focus our attention on 'one-perspective' fact statements, that is, fact statements which express the speaker's point of view at the time he makes them, it is tempting to put this by saying that a fact *quoad* ourselves now is a *present* entity, even though it is a fact *about* the past or *about* the future. And having said this, it may well occur to us that this 'insight' enables a resolution of a classic puzzle about truth. For suppose that we were committed to the idea (confused but endemic) that "the past and the future do not exist." We might well say with a sense of relief, "Thank goodness that among the things which exist are *present* facts *about* the past and *present* facts *about* the future. For these account for the truth or falsity of our thoughts and statements about the future and the past. For if there were no *present facts* about the past and the future, there would be nothing for these thoughts to correspond to, and they would be neither true nor false!"

It is not my purpose to dwell on error for its own sake. But it will, I believe, be worthwhile to reflect on the above misinterpretation of the correspondence theory of truth because of its connection with a familiar gambit, which develops as follows. Facts about the present are in a privileged position. For the fact that S is now Φ_2 has as its companion the episode[2] of S's being Φ_2. This episode exists *now*. On the other hand, neither the episode of S's being Φ_1 nor the episode of S's being Φ_3 exists now. The former episode *existed* in 1957, the latter *will exist* in 1959.

The next step in this line of thought is to argue (not implausibly) that the existence of the *episode* of S's being Φ_2 is more basic than the existence of the *fact* that S is Φ_2. Surely, it is said, episodes are the very stuff of the world; and even if they are not, they are surely presupposed by facts about them. What, then, of facts about the future and the past? There are such facts, and they are responsible for the truth of such of our statements about the future and the past as are true. But while the *facts* that S was Φ_1 and that S will be Φ_3 exist *now*, the corresponding episodes do *not* exist (though one *did* exist and the other *will* exist).

At this point, the argument, concentrating its attention on statements and facts about the future, takes a familiar turn. Since the episode of S's being Φ_3 does not now exist, it cannot account for the present existence of the fact that S will be Φ_3. Unless, therefore, we are going to abandon the idea that statements about the future are ever true, we must find some explanation of how there can be the fact that S will be Φ_3 in 1959 although the episode of S's being Φ_3 does not exist. Is there anything which *does* exist and can account for the fact that S will be Φ_3 in 1959? Yes, continues the argument, there is such a thing, namely the set of facts about the present which physically imply that S will be Φ_3 in 1959. For if we knew what these facts were, we could properly say

> (24) It *is* a fact that $S_1, S_2, S_3, \ldots, S_n$ are thus and so; *therefore* it *is* a fact that S will be Φ_3 in 1959

and

> (25) It *is* a fact that S will be Φ_3 in 1959, *because* it *is* a fact that $S_1, S_2, S_3, \ldots, S_n$ are thus and so.

This is summed up by saying that facts about the future exist as (*physically*) *implied by facts about the present*. The conclusion is then drawn

[2] The term 'episode' will be used, for the time being, in a broad sense in which no distinction is drawn among episodes, events, states, etc. These distinctions will be subsequently drawn to a degree of precision which suffices for the purposes of this paper.

that statements about the future are true or false only to the extent that the future is (physically) determined by the present, or, to put it negatively, to the extent that there are 'gaps' in the set of facts about the future which are implied by facts about the present, there are formulable statements and corresponding thinkable thoughts about the future which are neither true nor false.

We have reached the point at which the major confusions which make the above gambit possible must be cleared away before we can locate the element of truth it contains. But before we undertake this task, it is worth noting that if the argument is sound, it applies to the past as well as to the future. Thus, facts about the past would exist as (physically) implied by facts about the present state of the universe, and the truth of statements and thoughts about the past would rest on these implications. Now there is surely no greater a priori (as opposed to empirical-scientific) reason for supposing that the present uniquely determines the past ('retro-determinism') than for supposing that it uniquely determines the future ('antedeterminism'). Once, therefore, the 'practical' sense in which the past is 'determined'—there can be no action which is the bringing about of a past state of affairs as my lighting a match was the bringing about of a future state of affairs (surely an analytic statement as these words are ordinarily used)—is distinguished from the 'theoretical' sense in which it is not self-contradictory to say that there are facts about the past which are not 'determined by' ('in principle inferable from') the present, we see that the above analysis confronts us with the challenging idea that there may well be formulable statements and thinkable thoughts about the past which are neither true nor false. Indeed, by no means the least startling prima-facie implication of the analysis is that while the statement 'S is Φ_1' made in 1957 may well have been true, the corresponding statement, made today, 'S was Φ_1 in 1957' may be neither true nor false as neither (physically) implied by nor (physically) incompatible with the contemporary (1958) state of the universe. One might even begin to wonder whether, to schematize a medieval example, 'S was Φ_1 in 1957,' said in 1958, might be false, although 'S is Φ_1,' said in 1957, was true. Clearly something has gone wrong, and we must find out where.

V

In introducing the above line of thought, I pointed out that the idea that facts quoad ourselves now, that is, facts referred to by 'one-perspective' fact statements, are *present* entities has the virtue of taking seriously the present tense of "It *is* a fact that . . ." For that in the case of facts about episodes, at least, the 'is' of 'It is a fact that . . .' is indeed in the present tense is clear once one abandons the attempt to detense temporal statements. Thus, the 'will be' in the that-clause of

(9) It is a fact that S will be Φ_3 in 1959

is 'will be' by contrast to the present tense of the 'is.'

But what can it possibly mean to say that facts *quoad* ourselves now are *present* entities? And in what sense, if any, is the fact referred to by the 'two-perspective' fact statement

(4) It was a fact (*quoad* 1957) that S would be Φ_2 in 1958

a *past* entity? The answer involves a recognition of the intimate connection between

(8) It *is* a fact that S was Φ_1 in 1957,
(22) It *is* a fact that S is Φ_2 today (1958), and
(9) It *is* a fact that S will be Φ_3 in 1959

on the one hand, and

(26) The statement 'S was Φ_1 in 1957' *is* a true statement,
(27) The statement 'S is Φ_2 today' *is* a true statement, and
(28) The statement 'S will be Φ_3 in 1959' *is* a true statement

on the other. Each of the former three is a very close cousin, I might almost say a brother, of its counterpart in the latter trio.

Notice that I did not write, instead of (26), 'The *sentence* "S was Φ_1 in 1957" is a true *sentence*,' for in discussing problems pertaining to tense, it is essential that we avail ourselves of Strawson's distinction between *statements* and *sentences*. Thus, to refer to the statement 'S will be Φ_3 in 1959' is to refer to the sentence 'S will be Φ_3 in 1959' as used (indeed, as what *would be*—properly—*used*) at a certain time (or during a certain period of time), and when I say

(28) The statement 'S will be Φ_3 in 1959' *is* a true statement

the time in question is *now* (in a relevant sense of 'now').

Notice, next, that I can also say

(29) 'S will be Φ_3 in 1959' *was* a true statement,

in which case the reference is to the same sentence as used at a time before now; while if I say

(30) 'S will be Φ_3 in 1959' *will be* a true statement,

Wilfrid Sellars

the reference is to a future (but still pre-1959) use of this sentence. Thus, the 'is,' 'was,' and 'will be' of 'is a true statement,' 'was a true statement,' and 'will be a true statement' indicate the time at which the sentence in question would be properly used to make the statement characterized as true.

Note, next, that we say,

(4) It was a fact (quoad 1957) that S_2 would be Φ in 1958

and not

(31) It was a fact (quoad 1957) that S will be Φ_2 in 1958.

The explanation is to be found by reflecting on the parallel between

(32) That S would be Φ_2 in 1958 was a fact

and

(33) 'S will be Φ_2 in 1958' (said in 1957) was a true statement.

For whereas the that-clauses of

(34) It is a fact {that S was Φ_1 in 1957, that S is Φ_2 today, that S will be Φ_3 in 1959}

refer (in a manner to be discussed) to present uses of the sentences "S was Φ_1 in 1957," "S is Φ_2 today," and "S will be Φ_3 in 1959" (or their translations in any language), the that-clause, for example, of (4) refers to a prior use of the sentence "S will be Φ_2 in 1958" (or any of its translations), a reference which is manifest in (33). In short, the 'was' of 'It was a fact that' like the 'is' of 'It is a fact that' locates the time with respect to which the use of one or another of a specified set of mutually translatable sentences (including 'mental sentences') is being considered.

These considerations, incidentally, make it clear why the initial 'is' of

(21) It is a fact that $2 + 2 = 4$

is as tenseless as the '$=$' of the mathematical statement itself. For mathematical sentences, not being tensed, are appropriately used at any time and make "the same statement" on each occasion, whereas in the case of tensed statements, different sentences ("differently tensed counterparts") must be used to "make the same statement" at relevantly different times. Thus, "S will be Φ in 1958" (said in 1957) and "S is Φ_2 today (1958)" (said in 1958) "make the same statement." The 'is' of (21) does not serve the purpose of indicating that a present rather than a past or future use of the sentence "$2 + 2 = 4$" is under consideration.

538

VI

I wrote above that each of a trio of fact statements, for example,

(9) It is a fact that S will be Φ_3 in 1959,

is a cousin, perhaps a brother, of a certain truth statement, thus

(28) The statement 'S will be Φ_3 in 1959' is a true statement.

Let me now burn my bridges and say that they are identical twins, and, in general, that statements of the form

(35) It *is* a fact that p,

where 'p' represents a sentence, do not differ in sense from

(36) 'p' *is* a true statement *in our language.*

The reference to our language (now) is, of course, essential if we are to have any chance of circumventing Church's "translation argument" against linguistic interpretations of 'abstract entities.' And even with its inclusion, the thesis is a brutal oversimplification, as any isolated philosophical claim must be. I have discussed Church's argument on another occasion[3] and believe myself to have shown that and how it can be circumvented. But for the purposes of the present discussion I shall simply postulate that the above equation stands. For part of the case which can be made for it consists in the light it throws on the puzzles in which we are involved.

The actual crux of the matter is that if this "nominalistic" thesis be granted, then the equivalence

(37) The statement in (our language) L, "S will be Φ_3 in 1959," is true $\cdot \equiv \cdot$ it is a fact that S will be Φ_3 in 1959

is of a piece with

(38) We're here $\cdot \equiv \cdot$ we're here.

The classical correspondence theory of truth combines a fundamental insight with a fundamental error. It confuses between (and I oversimplify to make the point stand out)

(39) 'S will be Φ_3 in 1959' is true $\cdot \equiv \cdot$ S will be Φ_3 in 1959,

which is both nontrivial and true, and

(40) 'S will be Φ_3 in 1959' is true $\cdot \equiv \cdot$ it is a fact that S will be Φ_3 in 1959,

[3] "Grammar and Existence: A Preface to Ontology," *Mind*, 69:499–533 (1960); also "Truth and 'Correspondence,'" *Journal of Philosophy*, forthcoming (1962).

which has, in essence, the form

(41) $p \equiv p$.

The second point to be noted is that while the equivalence

(42) 'S will be Φ_2 in 1958' (said in 1957) was true $\cdot \equiv \cdot$ it was a fact (quoad 1957) that S would be Φ_2 in 1958

is as sound as two dollars, it does not illuminate what it is to characterize a past statement as true, for it shares the triviality of (40). It is therefore important to note that whereas the nontrivial semantical equivalence (39), which concerns a statement properly made now, involves a use on the right-hand side of the sentence[4] used to make the statement mentioned on the left-hand side and characterized as true, a nontrivial semantical equivalence concerning past statements cannot involve the use on the right-hand side of the sentence[5] used to make the statement mentioned on the left-hand side, and characterized as true. We must use on the right-hand side the appropriate differently tensed counterpart of this sentence.[6] A parallel point can be made concerning the truth of certain spatial statements; thus if Jones, yonder, says "The box is over here," the appropriate nontrivial semantical equivalence is

(43) 'The box is over here' (said by Jones yonder) is true $\cdot \equiv \cdot$ the box is over there.

Thus, to apply the semantical explication of truth to past statements we must place on the right-hand side the sentence which is the appropriate differently tensed counterpart of the sentence[7] used to make the original statement; for example,

(44) The statement 'S will be Φ_2 in 1958' (made in 1957) was true $\cdot \equiv \cdot$ S is Φ_2 today (1958).

I shall have more to say about differently tensed counterparts at a later stage in my argument. For the moment I shall limit myself to pointing out that the right-hand side of a semantical truth equivalence is always a statement in our language, here and now, so that even when the statement which is being characterized as true is past or future, the characterizing of it as true expresses our point of view here and now.

[4] Or, if it is in another language, the translation into our language of the sentence . . .

[5] See fn. 4.

[6] See fn. 4.

[7] See fn. 4.

2. Do Past and Future Episodes Exist?

VII

It will be remembered that the views we were examining in Section IV above made certain assumptions pertaining to *facts* and *episodes* in order to argue that statements about the future are true or false only if the states of affairs they express are physically implied by or physically incompatible with facts about the present state of the universe. In the meantime, we have shown that one of the ideas on which the argument rests, viz., that truth is a 'correspondence' between statements and facts, is a mistake. It is equally important to see that the other idea to which it appeals, viz., that facts about the present are privileged in that 'there are' present episodes, but no past or future ones, is also a mistake. It can readily be shown to rest on a confusion between

> (45) The (future) episode, E, does not 'exist'—i.e., is not taking place

and

> (46) The (future) episode, E, does not 'exist'—i.e., there is no such thing as this episode.

Episode E would, of course, simply not be future if it were taking place. But that there is no such thing as a future episode is surely false, and escapes being obviously so only because it is confused with the idea that no future episodes are taking place. To dispel this confusion, it is necessary to see how the language of 'episodes' or 'events' is related to simple tensed statements of the kind with which this paper began.

But first a terminological remark is in order. It will undoubtedly have been noticed that in the preceding sections the term 'episode' has, with a minimum of warning, been stretched to cover items which would not ordinarily be so designated. Thus, we would not ordinarily say that the statement 'The soup is salty' reports an episode, even though it does report something that "comes to pass." Thus, we distinguish, for example, between 'episodes' and 'states.' It is no easy task to botanize the various kinds of temporal statement, or to find a plausible term for the broader category to which both episodes ('the salting of the soup') and states ('the being salty of the soup') belong. Perhaps they might be lumped together under 'outcome.' For the time being, however, I shall avoid any discussion of states, and limit myself to episodes proper. I shall, therefore, modify the original statements, with which I began, to read

(1') S became Φ_1 in 1957,
(2') S is becoming Φ_2 today (1958),
(3') S will become Φ_3 in 1959,

and ask how such episode expressions as "S's becoming Φ_1," "S's becoming Φ_2," and "S's becoming Φ_3" are related to statements of these forms.

Actually, the relation between episode expressions and tensed statements which are about *things* rather than *episodes* is quite simple, and has been formulated with reasonable clarity by more than one philosopher.[8] Thus, the episode expression 'S's becoming Φ_1' is derivative from tensed statements to the effect that S is (or was or will be) Φ_1 in accordance with the following equivalence schema:

(47) S's becoming Φ_1 {is taking place, took place, will take place}
$\cdot \equiv \cdot$ S {is becoming, became, will become} Φ_1.

Thus we note that there are two kinds of *singular term* which can be derived from tensed statements of the kind represented on the right-hand side of (47): (a) *that-clauses*, thus,

(48) That S will become Φ_i,

and (b) *episode-expressions*, thus,

(49) S's becoming Φ_i.

We have already argued that singular terms of the former kind are a special kind of statement-mentioning device and are metalinguistic in character. This being so, we can appreciate the truth contained in the idea that *episodes* are more basic than *facts*; for *episode-expressions*, unlike *that-clauses, are in the object language.*

On the other hand, it is important not to be misled by this insight into supposing that episodes are the entities of which the world is 'made up,' for although it is correct to say that episode-expressions 'refer to extra-linguistic entities'—indeed, to episodes—the above account tells us that episodes are *derivative* entities and rest on the referring expressions which occur in tensed statements about things (or 'substances').

It is worth noting, in this connection, that though there is a necessary equivalence between the corresponding statements in the following two columns,

(48) It is the case *that S is*   S's becoming Φ_i is taking place
becoming Φ_i

[8] I have particularly in mind Hans Reichenbach's discussion of events and things in his *Introduction to Symbolic Logic*.

| It is the case *that S be-came* Φ_i | *S's becoming* Φ_i took place |
| It is the case *that S will become* Φ_i | *S's becoming* Φ_i will take place |

each statement on the left is differently related than its right-hand counterpart to the 'basic' statement from which it is derived. Again, corresponding to the right-hand statements we have the set of (respectively) equivalent statements,

(49) It is the case *that S's becoming* Φ_i *is taking place*,
It is the case *that S's becoming* Φ_i *took place*,
It is the case *that S's becoming* Φ_i *will take place*.

That it can be illuminating to play with the compounding of singular terms of these two varieties will become clear in the sections yet to come.

It is also worth noting—this time as an aside—that both types of singular term (that-clauses and episode-expressions) function in "predicative" implication statements.[9] We can say not only

(50) *That the litmus paper was put in acid* (physically) implies *that it turned red*

but also

(51) *The litmus paper's being put in acid* (physically) implied *its turning red*.

(Note the subtle difference in tense structure of these equivalent statements.) The fact that episode *expressions* occur as singular terms both in the fundamental, patently object language, contexts explicated in (47) and in such predicative implication statements as (51) gives prima-facie support to the idea that (physical) implication is a relation *in re* between events, an idea which finds no support in statements like (50) once the metalinguistic character of that-clauses is understood. This makes it doubly important to see that episode-expressions are grounded in tensed statements about things, where these statements, since they are not singular terms, must be *that*-ed (in effect, quoted) to serve as the subject of statements to the effect that something physically implies something else.

[9] By a 'predicative implication statement' I mean an implication statement in which the function '―――― implies ――――' plays the role of a predicate, taking singular terms as its substituends. These statements are to be contrasted with the contrived (but illuminating) form '―――― ⊃ ――――' in which statements rather than singular terms fill in the blanks to make 'material implication' statements.

VIII

I have pointed out that episode-expressions are introduced in terms of the equivalences represented by schema (47). The next step is to explore the relationships represented by the schema

(52) S's becoming Φ_1 is {present, past, future} $\cdot \equiv \cdot$ S's becoming Φ_1 {is taking place, has taken place, will take place}.

The corresponding statements on the left- and right-hand sides are clearly equivalent. But before we ask whether this equivalence is an identity of sense, let us note that the introduction of the adjectival expressions 'past,' 'present,' and 'future' makes possible, when combined with various tenses of the copulative 'is,' the introduction of the forms

(52) E *was* present (past, future)

and

(53) E *will be* present (past, future)

which, by the use of one overt tensed verb, make statements that would require the use of complex tenses if reformulated as statements about *things* rather than *episodes*. Thus, while

(54) S's being Φ_1 *is* future

is the counterpart of

(55) S will become Φ_1,

to get the counterpart of

(56) S's becoming Φ_1 *was* future (in 1900)

we must say something like

(57) (In 1900) S was (yet) to become Φ_1,

and to get the nonepisodic counterpart of

(58) S's becoming Φ_1 *will be* past (in 1960)

we must say something like

(59) (In 1960) S will (already) have been Φ_1.

These considerations call attention to the fact that there are several ways in which we can make 'two-perspective' temporal statements. Thus, compare

(60) It *was* a fact (quoad t) that S *would* now (1958) become Φ_1,
(61) (At t) S *was to become* Φ_1 today (1958),
(62) S's becoming Φ today (1958) *was future* (quoad t).

Of these, the first has, we have seen, roughly the form

(60′) 'S *will* become Φ_1 in 1958' (said at t) *was* a true statement.

But what of (62)? Surely it also has a metalinguistic force. That it is *equivalent* to

(63) 'S's becoming Φ_2 will take place in 1958' (said at t) was a true statement

is reasonably clear. I propose to argue that this equivalence is an identity of sense, and that, more generally,

(64) E {is, was, will be} past (at t) $\cdot \equiv \cdot$ 'E took place' (said at t) {is, was, will be} true

(65) E {is, was, will be} present (at t) $\cdot \equiv \cdot$ 'E is taking place' (said at t) {is, was, will be} true

(66) E {is, was, will be} future (at t) $\cdot \equiv \cdot$ 'E will take place' (said at t) {is, was, will be} true

If we combine these schemata with considerations relating to the moves from either

(67) **S** was true

or

(68) **S** *will be* true

to

(69) **S′** *is* true

where **S′** is the appropriate differently tensed counterpart of **S** which is used *now* to make the statement which **S** was used to make (in case (67)) or will be used to make (in case (68)), we have a direct route from statements of the forms represented by

(70) E {is, was, will be} {present, past, future}

to statements of the form

(71) E is taking place (has taken place, will take place)

and hence (in simple cases) to statements of the form

(72) S is becoming (has become, will become) Φ_1.

But more of this in a moment. For it remains to suggest that even the second of the three ways of making a two-perspective statement listed above, namely,

(61) (At t) S was to become Φ_1 today (1958)

has a metalinguistic component to its sense, and involves a tacit quoting of "S will become Φ in 1958" as appropriately tokened at the prior time t. For a temporal perspective is always a *cognitive* perspective, *the perspective of a user of temporal language.* But to bring out the metalinguistic component of (61) requires a deeper analysis of temporal expressions, and, in particular, at least a rudimentary account of how they are tied up with references to dates, moments, and periods of time.

IX

Let us consolidate some of the ground which has (tentatively) been won, by noting that the classical notion that 'there is' a series of events related by *earlier than* (or *precedes*), where the 'is' of

(73) E_1 is earlier than E_2

is in the 'tenseless present,' is a mistake *if it is supposed that this 'tenseless present' is logically independent of the use of tensed verbs.* Actually, of course, if 'earlier than' is to be a *temporal* predicate at all, the 'is' of 'is earlier than' can be 'tenseless' only as the 'be' of the sentence "Eisenhower be president in 1956," so contrived as to be suitable for making a statement and, in an appropriate sense, the same statement, whenever it is used, was 'tenseless' only by virtue of its stipulated equivalence to a disjunction of three sentences involving, respectively, the past, the present, and the future tenses of 'to be,' as in (20). In the case at hand, we have the equivalence

(74) E_1 precedes $E_2 \cdot \equiv \cdot$ *either* E_1 *is* present and E_2 future, *or* E_1 *was* present and E_2 future, *or* E_1 *will be* present and E_2 future.

Notice, of course, that in (74) the component "E_1 was present and E_2 future" must have the sense of "E_1 was present and E_2 was *at that time* future." In other words, this component must be construed as having the force of

(75) 'E_1 is taking place and E_2 will take place' was true

rather than

(76) 'E_1 is taking place' was true and 'E_2 will take place' was true.

Obviously the latter entails the former only if the statements "E is taking place" and "E will take place" are construed as made or to be made at the same time. In the case of (75) this proviso is unnecessary, since we have to do with one *conjunctive statement* rather than two statements.

X

Are there past and future episodes? That the answer is 'yes' is surely a foregone conclusion, but the logical niceties remain to be determined. To begin with, something must be said about the status of the very term 'episode.' That it is a common noun, and that "There are episodes" has the same general form as "There are lions," is clear. But more than this we can say that 'episode,' like 'property' and 'relation,' is a 'category word'; and to say this is to say that like the latter pair it is the counterpart in the material mode of a logical pigeonhole for a certain class of expressions in our language. Thus,

(77) E is an episode

tells us no more about E than is exhibited by

(78) E is taking place or has taken place or will take place

and serves to indicate that the singular term represented by 'E' is the sort of term which belongs in this type of context. Thus, to say that there are episodes is, in effect, to say that something[10] either is taking place, has taken place, or will take place. And as saying this it is equivalent to (though it does not have the same sense as) a statement to the effect that something is either *present, past,* or *future.* These statements can be put *logistice* as follows:

(79) (Ex) x is taking place **v** x has taken place **v** x will take place;

(80) (Ex) x is present **v** x is past **v** x is future.

If, now, we introduce into the latter the categorizing function, 'x is an episode,' to obtain

(81) (Ex) x is an episode · x is present · **v** · x is an episode · x is past · **v** · x is an episode · x is future,

we are in a position to note the different roles played by the first and second occurrences of 'is' in each of the disjuncts. For while the second 'is' in each case is in the present tense in a full-blooded sense (i.e., the context admits 'was' and 'will be' as well as 'is'), the first is not. The first 'is' is in the 'tenseless present,' as is the 'is' in all categorizing statements, thus

(82) Triangularity is a quality.

---

[10] 'Something' here does not, of course, mean *some thing,* i.e., *some continuant* or *substance.* It is the ordinary language equivalent of so-called existential quantification and as such moves from category to category depending on context, just as the reversed 'E' of the 'existential' operator is appropriately combined with variables of all types, thus, '(Ex) . . . x . . .'; '(Ef) . . . f . . .'

We may, indeed, say

> (83) E *was* an episode,

but this has the force of

> (84) E *is* an episode · E has taken place.

In other words, the past tense is connected not with the *categorizing*, but with the temporal location of the categorized entity.[11]

On the other hand, the second 'is' in each disjunct of (81) is full-bloodedly in the present tense, for the functions

> (82) x *is* {present, past, future}; ['x {is taking, has taken, will take} place' *is* true]

are, as we have seen, to be contrasted with

> (83) x *was* {present, past, future} (at t);
> ['x {is taking, has taken, will take} place' (said at t) *was* true]

and

> (84) x *will be* {present, past, future} (at t);
> ['x {is taking, has taken, will take} place' (said at t) *will be* true].

Thus, to say that some episodes are past or that some episodes are future (or, for that matter, that some episodes are present) is simply to affirm one of the disjuncts in the statement which is the explicitly categorized form of the assertion that there are such things as episodes. For

> (85) There are episodes

has, as we have seen, the force of

> (86) (Ex) x is a present episode ∨ x is a past episode ∨ x is a future episode,

which we may read

> (87) *Something* is either a present episode or a past episode or a future episode

and is equivalent to

> (88) *Something* is a present episode ∨ *something* is a past episode ∨ *something* is a future episode.

[11] I shall return to the topic of the role of the present tense in categorizing statements at a later stage in the argument.

Our discussion of the difference between the 'is' of 'is an episode' and the 'is' of 'is present (past, future)' makes it clear that the 'are' in

(89) Some episodes are {present, past, future}; [(Ex) x is a {present, past, future} episode]

is *full-bloodedly in the present tense*, for while this 'are' contains the 'tenseless present' of the categorizing function, 'x is an episode,' it also contains the full-blooded present tense of the functions 'x is present,' 'x is past,' and 'x is future.' Thus the statements

(90) There are {present, past, future} episodes

contrast with

(91) There were {present, past, future} episodes;
[There are episodes which were {present, past, future}]

and

(92) There *will be* {present, past, future} episodes;
[There are episodes which *will be* {present, past, future}]

which are based on the functions 'x *was* present (past, future)' and 'x *will be* present (past, future).'

It is important to see that

(93) There are past episodes

and

(94) There were past episodes

make quite different statements, as do

(95) There are future episodes

and

(96) There *will be* future episodes.

For this puts us in a position to see that to suppose that the correct way of talking about the existence of past and future episodes is by saying

(94) There were past episodes

and

(96) There *will be* future episodes

is to make a simple mistake. It is, indeed, incorrect to say either

(97) Past episodes are taking place

or

(98) Future episodes are taking place.

549

Here the tense of 'to be' must agree with the temporal adjective applied to the episodes. On the other hand, the 'are' of (93) and (95), as traceable to the functions 'x is past' and 'x is future,' and hence as giving expression to our temporal point of view in making these statements, contrasts radically with the 'were' of (94) and the 'will be' of (96) which are traceable to functions ('x was past,' 'x will be future') each of which locates (from our point of view) a second and different point of view from which the episodes in question might have been, or might yet be, viewed

Thus, to make 'single-perspective' existence statements about episodes we must use the present tense,

(90) There are {present, past, future} episodes

and hence make statements which, by virtue of their relationship to simply tensed statements about changeable things, are, in elementary cases, equivalent, respectively, to

(91) (ES) (Ef) S {is becoming, became, will become} f.

3. Time and Temporal Relations in a World of Things

XI

The above analysis throws light, I believe, on a number of venerable and well-worn puzzles. Thus, philosophers have been prone to ask, "How can two successive events be temporally related if when the earlier event exists, the later does not yet exist, and when the later event exists, the earlier event no longer exists?" Surely, it has been argued, the terms of any relation must 'coexist.' I will not take the time to apply the above considerations to this elementary confusion. What is of somewhat greater interest, however, is that our analysis throws light on the sense in which 'there are' temporal relations at all. For while there clearly are temporal relations between events, the latter (we have argued) have a derivative status in the sense that statements about events are, in principle, translatable into statements about changeable things. If we put this somewhat misleadingly by saying that 'ultimately' or 'in the last analysis' there are no such things as events, we must also say that *'ultimately' or 'in the last analysis' there are no such things as temporal relations.*

It is impossible, however, to make sweeping statements about temporal relations without coming to grips with topics of the most central importance which have simply been bypassed in the argument to date. Thus it will undoubtedly have been noticed by readers who are sensitive to classical issues in the philosophy of time that I have permitted myself to pass back and forth from tensed statements about things which make no explicit reference to *location in time*, to tensed statements about things which contain a reference to a *moment* or *period* of time, thus, "S was Φ_i at t." Now concepts pertaining to time, and moments or periods of time, are *metrical* concepts, and involve *logical individuals*, whether derivative (as we have construed episodes to be) or *primitive* for which metrical relationships have been defined. It is time, therefore, that we faced the fact that if we are going to take *things* as our only primitive logical individuals, we must find a *nonrelational* way of talking about *changing things* by the use of tensed verbs which provides a logical basis for statements about topological and metrical *relations* between *events* when it is translated into the derived framework of episodes and events which we have been concerned to analyze. For once the transition from tensed talk about things to a topologically ordered framework of events has been made, we will have established contact with the many excellent explorations of the constructibility of concepts pertaining to time and its periods and moments on a basis consisting of topologically characterized relationships among events, which exist in the literature of the subject.

There are, roughly, two ways in which the step from a relationally ordered system of events to time, its periods and moments, has been conceived. (1) There is the idea that concepts pertaining to time are explicitly definable in terms of such a relation between events as *overlapping*, thus Whitehead's account in terms of the 'Method of Extensive Abstraction'; (2) there is the idea that time has the status of a quasi-theoretical entity the ultimate particulars of which are *moments*. According to the latter interpretation, metrical relationships between periods and moments of time would be 'idealized' counterparts of empirically ascertainable metrical relationships between episodes pertaining to everyday (and scientific) things. It is the latter approach which I would defend. It is therefore incumbent on me to explain what I mean by characterizing time as a *quasi*-theoretical entity.

Actually, it is misleading to use the term 'theoretical' in this context at all. For all that *time* has in common with *population of molecules* is the existence of rules for coordinating statements concerning empirically ascertainable metrical relations between episodes pertaining to the things of everyday life and science, with statements locating these episodes, relatively to other episodes, in time, that is, with statements having the characteristic syntax of statements "about time." There remains the essential difference that time is introduced as a *metrical framework* rather than, as in the case of molecules, as part of the content of the world. Needless to say, the fact that we can say that 'time' refers to time, i.e., talk semantically about expressions such as 'time,' 'the year 1900,' 't₀,' etc. throws no light whatever on the status of time, since it simply gives expression to the fact that temporal expressions have a use.

"But," it will be said, "even granting that something like the position you have been sketching can stand the gaff, you have not yet shown how metrical relations between empirically ascertainable episodes can be derivative from *nonrelational* temporal facts concerning things. For, as you yourself have insisted, if things are the only basic individuals, then all relational temporal facts pertaining to episodes must rest on nonrelational temporal facts pertaining to things." The answer to this challenge consists in calling attention to such locutions as

(92) Nero fiddled *while* Rome burned

and noting that '*while*' *is a connective which connects statements* and remembering that *statements are not singular terms*. In other words, the answer is simply that we must not *equate* statements involving temporal connectives such as 'while' with statements formulating temporal relations between episodes, thus

(93) Nero's fiddling coincided with Rome's burning.

Nor are statements of this kind to be *equated* with statements explicitly mentioning periods of time, thus

(94) Nero fiddled during the period of time in which Rome burned.

On the other hand, it must be granted that these temporal connectives are free from involvement with the framework of time only in a hypothetically primitive use. For tensed discourse with these connectives, but without the framework of time, would constitute a most primitive picture of the world.

XII

Assuming that some such account of the dating of episodes is true, the next point to note is that the rules coordinating time and its moments with discourse about changing things permit us to speak not only of epi-odes,[12] but also of the periods and moments of time, as past, present, or future, as having been past, present, or future, and as going to be past, present, or future. And the application of 'is (was, will be) present (past, future)' to moments clearly rests on their application to events in a way which is roughly indicated by the formula

(97) t_i {is, was, will be} present $\cdot \equiv \cdot$ (E) E occupies $t_i \supset$ E {is, was, will be} present

which, for simplicity's sake, has been put in terms of moments and momentary events.

Again, the concept of one moment as preceding another moment in the continuum of time is exhibited by the formula

(98) t_i precedes $t_j \cdot \equiv \cdot t_i$ is present and t_j future or, t_i was present and t_j future or, t_i will be present and t_j future.

We can also introduce 'now' as an expression referring to the present moment, thus,

(99) Now $= (\iota t)$ t is present.

It is essential to note the tensed character of the 'is' on the right-hand side, for this formula highlights the fundamental role played by tensed verbs in temporal discourse. We can put this point roughly by saying that 'now' is to be understood in terms of 'is,' not 'is' in terms of 'now' construed as a basic demonstrative. Or, more accurately, this is the account of 'now' we must give if we are to construe our language as one in which the basic logical individuals are changeable things. For we shall subsequently be exploring the logic of a framework which, while not that of ordinary discourse and, indeed, an invention of the philosophers, is, if consistently developed, a legitimate alternative to the framework of things. The basic logical individuals of this new framework are 'events' in a sense

[12] It is, perhaps, worth noting that 'S remained Φ' generates the dull or 'null' episode of S's remaining Φ, 'null' in the sense that remaining Φ is a limiting case of a going's on which is a change. Notice, also, that whereas the idea of a momentary episode (change) is, strictly speaking, nonsense, the idea of a momentary state is not. The latter is introduced by the equivalence between

(95) x remains Φ throughout p

and

(96) x is Φ at every moment of p.

These statement forms, of course, presuppose the coordination of episode talk with time talk.) The idea of episodes as consisting of a continuum of momentary states would be a reconstruction, in terms of the framework of time, of the idea of an episode on which this framework rests.

of this term which is radically different from 'episode' or 'event' as the
occur in the framework of things. In this 'event' framework, 'now' wil
play a radically different role. That the two frameworks are, in an impo
tant sense, equivalent and easily confused will be found to account fo
many of the recurring ontological puzzles concerning time and change.

Another important schema is the following, which relates 'was true
and 'will be true' to 'is true':

> (100) 'E_1 is taking place' (said at) t_1 {is, was, will be} true $\cdot \equiv$
> 'E_1 {is taking place, took place, will take place} at t_1' (said
> now) is true.

With these equivalences behind us, we are in a position to introduc
the metaevent expression 'E_1's being present,' which we can abbreviate a
'$E_{pr}[E_1]$' in terms of the schema

> (101) $E_{pr}[E_1]$ {is taking, has taken, will take} place at $t_1 \cdot \equiv \cdot$ E
> {is, was, will be} present at t_1.

(Note that 'E_1 is present at t_1' has the force of 'E_1 is now present.' We hav
not yet introduced the detensed form 'E_1 "is" present at t_1.' This can, o
course, be done as follows, using as before 'be' for the detensed 'is,'

> (102) E_1 be present at $t_1 \cdot \equiv \cdot E_1$ is present at t_1 or, E_1 was presen
> at t_1 or, E_1 will be present at t_1,

where it is a necessary truth that

> (103) E_1 be present at $t_1 \supset : (t) \, t \neq t_1 \supset -(E_1$ be present at t

i.e., there is only one time, the time which it occupies, at which a (mo
mentary) event "is" (be) present, as there is only one time (now) a
which a (momentary) event is present.)

Continuing the line of thought initiated by schema (101) we add

> (104) $E_{pr}[E_1]$ is (now) {present, past, future} $\cdot \equiv \cdot$ '$E_{pr}[E_1]$ {i
> taking, has taken, will take} place' (said now) is true.

From these equivalences, together with certain considerations which hav
not been spelled out, but are reasonably straightforward, we can derive

> (105) $E_{pr}[E_1]$ {is taking, has taken, will take} place at $t_1 \cdot \equiv \cdot$ E
> {is taking, has taken, will take} place at t_1

and

> (106) E_1 is {present, past, future} (now) $\cdot \equiv \cdot E_{pr}[E_1]$ is {pres
> ent, past, future} (now).

To spell out the relationship in a particular case, note that

(107) $E_{pr}[E_i]$ is future $\cdot \equiv \cdot$ '$E_{pr}[E_i]$ will take place' is true

$\cdot \equiv \cdot E_{pr}[E_i]$ will take place

$\cdot \equiv \cdot (Et)$ E_i will be present at t

$\cdot \equiv \cdot (Et)$ 'E_i is taking place' (said at t) *will be* true

$\cdot \equiv \cdot (Et)$ 'E_i will take place at t' (said *now*) *is* true

$\cdot \equiv \cdot E_i$ is future.

The steps which have not yet been adequately clarified (though intuitively sound) are those which involve quantification over moments. For these steps tacitly involve such principles as

(108) 1 E_i is now future $\supset$: (t) t precedes now $\supset$ E_i was future at t

2 E_i is now past $\supset$: (t) t follows now $\supset$ E_i will be past at t

3 E_i is now present $\supset$: (t) t precedes now $\supset$ E_i was future at t

$\supset$: (t) t follows now $\supset$ E_i will be past at t

4 E_i is now future $\supset$: (Et) t follows now $\supset$ E_i will be present at t

5 E_i is now past $\supset$: (Et) t precedes now $\supset$ E_i was present at t.

It is therefore time to note that the temporal relation between the meta-events $E_{pr}[E_i]$ and $E_{pr}[E_j]$ like the relation between E_i and E_j is 'timeless' only as involving a disjunction of tenses, thus,

(109) $E_{pr}[E_i]$ precedes $E_{pr}[E_j]$ $\cdot \equiv \cdot E_{pr}[E_i]$ *is* present $\cdot$ $E_{pr}[E_j]$ *is* future, or $E_{pr}[E_i]$ *was* present $\cdot E_{pr}[E_j]$ *was* future, or $E_{pr}[E_i]$ *will be* present $\cdot E_{pr}[E_j]$ *will be* future.

And the cumulative force of the principles and equivalences set down above is to make intelligible the 'analytic' character of the principle that "earlier events become present before later events," i.e., that, to use the traditional (and dangerous) metaphor, the "bulls-eye of the present moves along the series of events from earlier to later," for it is a consequence of these principles and equivalences that

(110) E_i precedes E_j $\supset$ $E_{pr}[E_i]$ precedes $E_{pr}[E_j]$.

It is also important to note, for future reference, that if it is granted tha

> (111) E is (*now*) present ⊃ : (t) t precedes *now* ⊃ E was f
> ture at t

and that

> (112) E *was* future at t ⊃ : 'E *is* future' (said at t) *was* true
> (113) 'E *is* future' (said at t) *was* true · ≡ · 'E *will* take plac
> (said at t) *was* true
> (114) 'E *will* take place' (said at t) *was* true · ≡ · 'S *will* be
> (said at t) *was* true

where E = S's being Φ); and, finally that

> (115) 'S *will* be Φ' (said at t) *was* true · ≡ · It *was* a fact (*quoa*
> t) that S would be Φ at a time subsequent to t

then the very structure of time talk, as we have laid it out, involves that

> (116) S *is* Φ · ≡ · It *was always* a fact that S would be Φ
> · ≡ · 'S will be Φ' (said at any time before now) wa
> *always* true.

4. Do Past and Future Things Exist?

XIII

The explorations of the preceding sections make possible a clarificatio
of certain concepts pertaining to *becoming*. The first thing to note is tha
if we have in mind by 'becoming,' becoming Φ, then both things an
events can be the subjects of becoming, thus

> (116) S {is becoming, became, will become} red
> E {is becoming, became, will become} more and mor
> past.[13]

The next thing to note is that whereas both *things* and *events* can b
come Φ, only *things* become in the sense of *come into being*. Many of th
puzzles pertaining to becoming rest on a failure to see that in no ordinar
sense do *events* come into being. To equate

> (117) E is becoming present

---

[13] It is not my purpose to analyze statements of these forms where the subject is a
event and the predicate other than a temporal predicate. That there are such stat
ments is clear, thus 'E is becoming less and less approved.' There is, indeed, a sen
in which events, as McTaggart points out, can change only with respect to tempor
characteristics, but this means only that other changes are to be analyzed in terms
change with respect to temporal characteristics, so that the claim is true only in th
sense in which "only primitives exist."

with

      (118) E is coming to be (or coming into existence)

is either to make a mistake, or to stipulate a *new* use for 'come to be' ('come into existence'). For, as we have seen, there is an *elementary* sense in which *there are* future events (and past events) as well as present events. Thus, if

      (119) E shall come to pass

is construed (as, I suspect, it is often construed) as

      (120) E shall come to be (come to exist)

rather than

      (121) E shall come to take place (i.e., will take place)

one is likely to think that the fact that E "shall come to pass" implies that *there is no such event as E until it takes place*. (Of course, E will not be *present* until it takes place.) It is easy to fall into the absurdity of supposing that the fact that future events "shall come to pass" implies that there are no future events.

We shall be discussing in a moment the coming into being and the passing away of *things*. Before we take up this crucial topic, the topic of 'absolute becoming' (as contrasted with 'becoming Φ'), let us remind ourselves that the most puzzling of traditional puzzles about becoming have to do with its status as 'objective' or 'subjective.' Thus, it is asked, would there be becoming if there were no knowing minds immersed in the temporal order? To this latter question, the answer implied by our analysis is, in a certain sense, No. But to say this at this stage is not to say that becoming is in any usual sense 'subjective,' but merely to remind ourselves that however 'objective' temporal statements may be in the sense of belonging to intersubjective, rational discourse, they are irreducibly 'token-reflexive.' After all, the world would contain no *heres* and *theres* if it included no users of 'here' and 'there'—or their equivalents. And to think of a possible world as containing *heres* and *theres* is, in effect, to imagine oneself in it, using 'here' and 'there.' [14]

It is tempting to suppose that the expressions 'here' and 'there' are the spatial counterparts of 'now' and 'then.' And, of course, in a sense this is

[14] This generates the question, 'Can we describe in nontensed terms what a world must be like for tensed talk to be appropriately used in it?' The present essay is, in a certain sense, a preface to an exploration of this fundamental issue in the philosophy of time. See also Sec. 8 below.

true. But to be impressed by the similarity between 'x is *here*' and 'x *now*' is to run the danger of overlooking an important difference, namel the association, in the latter case, of the *temporal* predicate 'now' wit the *temporal* present tense of 'to be,' as contrasted with the associatio in the former case, of the *spatial* predicate 'here' with the *temporal* pre ent tense of 'to be.' To put the point bluntly, our language does not con tain spatial tenses.

But before we assess the significance of this difference, let us press th similarities. Thus, just as the temporal dimension of discourse presen us with what we have called 'differently tensed counterparts,' the spati dimension presents us with what can be called 'differently located counte parts.' If S_1 is *here* (where *we are*) and S_2 is *there*, the statements, whic we now make,

> (122) S_1 is here
> (123) S_2 is there

have, as counterparts, the statements, made by someone over there,

> (124) S_1 is there
> (125) S_2 is here

so that we can say not only,

> (126) (From here) it is a fact that S_1 is here
> (127) (From here) it is a fact that S_2 is there

but also (albeit somewhat forcedly)

> (128) (From there) it is a fact that S_1 is 'there'
> (129) (From there) it is a fact that S_2 is 'here'

that is, both

> (130) 'S_1 is here' (said here) is a true statement
> (131) 'S_2 is there' (said here) is a true statement

and

> (132) 'S_1 is there' (said there) is a true statement
> (133) 'S_2 is here' (said there) is a true statement.

Suppose, now, someone were to argue as follows: "The distinction be tween *here* and *there* is, in an important sense, 'subjective,' for while dis course about *here* and *there* is rational, intersubjective discourse, ther would be no such thing as *here* or *there* if there were no language user using expressions having the force of 'here' and 'there.' On the othe hand," he continues, "spatial relationships, like 'between,' 'colinear with, perpendicular to,' etc., are objective, and would be in the world even if there had been no language users. These *objective* relations are presupposed by the distinction between *here* and *there*, for it is because language users stand in the objective relations they do to other people and things that they can make proper use of the expressions 'here' and 'there.'" Having thus prepared the way, he goes on to expostulate, "Surely the same is true in the case of time. Must there not be relations independent of the distinction between *now* and *then* which are presupposed by this distinction, and are as objective as the spatial relations mentioned above?"

The above argument is as old as the hills, and it points up a familiar dilemma. On the one hand there is the fact that if our argument to date is sound, all temporal concepts contain an irreducibly 'subjective' element (though the term 'subjective' must be used, as we have seen, with caution); on the other hand, the analogy of spatial discourse suggests that the 'token-reflexive' aspects of temporal discourse rest on 'purely objective' temporal relations which would 'be there' even if there were no language users. Is there any way out of this dilemma?

It might seem that not only is there a way out, but we have already taken it. For have we not undercut it by arguing that events have a derivative status? In the framework of things, it is *things* which stand in spatial relations (though they do so at a time), whereas it is *events* which stand in temporal relations. Have we not shown that there is a 'level of being more basic than the level of events' and have we not therefore shown that even if all temporal relations between events contain an irreducibly 'subjective' element, nevertheless the existence of *changing things* is objective? Can we not say with Bergson that the framework of temporal relationships rests upon a nonrelational mode of becoming (his *durée*, more adequately categorized)? The answer is, of course, that even if we have shown that something remotely analogous to what Bergson had in mind is true, to establish the derivative status of the relational framework of events is by no means to find a 'purely objective' foundation for temporal facts. For even if statements of the form 'E_1 overlaps (overlapped, will overlap) E_2' rest on statements of the form represented by 'Nero fiddled while Rome burned,' we are still confronted by the irreducibly tensed, and hence, in the sense in which we are using the term, 'subjective' character of such statements.

Nevertheless, as I hope to show, the step of locating the 'radical sub-

559

jectivity' of 'becoming' in tensed statements about things rather than in relational statements about events is a significant advance toward dissolving the puzzles. In addition to its role in facilitating the next steps in the argument, it serves the immediate purpose of making us aware of the limitations of analogies drawn from space. For it brings us back to the difference pointed out above (p. 558) between 'x is here' and 'x is now', by highlighting the irreducible role of tensed verbs in the expression of temporal facts.

XIV

Further progress requires a scrutiny of the logical connections between the concept of *existence* and concepts pertaining to temporal location. It can best be introduced by taking a closer look at exactly *how* the analogy of space generates puzzles with respect to the status of *becoming*. We have all heard arguments of the following sort: "The basic individuals of the framework of time are events. They are the domain of a purely objective relation of *earlier than*. 'Now' is a token-reflexive expression tokens of which, occurring in the system of events, give expression to, and have as an essential part of their sense, their location in the system. The distinction between *now* and *then* exists only with respect to (from the point of view of) linguistic events in the system. The objective status of the events as a temporal order is independent of the distinction between *now* and *then*, and, indeed, the temporal order is prior to and embraces the perspectival facts which are constituted by the occurrence of token-reflexive linguistic events. The temporal order exists in a sense which is independent of, and prior to, any use of token-reflexive expressions. Statements to the effect that such and such events *exist* are *tenseless* statements. To say that a certain event *will exist* is to say that it *exists* (in a tenseless sense) *and is later than now*. After all," the argument concludes, "it is well known that the concept of existence is nothing more than the *existential* operator, which is no more tensed than 'or' or 'not.'"

But while this argument has proved persuasive, it leaves us puzzles. The idea of a *tenseless* existence of events *tenselessly* related by *earlier than* has a flavor of absurdity, if not of self-contradiction. To view the status of *now* and *then* as a matter of the presence in a tenselessly existing relational order of tokens of 'now' and 'then' is to run counter to the idea, at least as persuasive as the above argument, that to say of two events that one is earlier than the other is to *use*, and not merely to *mention* a tem

oral token-reflexive expression. And if this latter idea is sound, how can
ᴛere be such a thing as a token-reflexive-free statement to the effect that
ᴛere exist events which are related as earlier to later?

But the above remarks only heighten the tension, and provide no relief.
ᴀnd a review of the situation makes it clear that the tension is focused in
ᴛe clash between "Statements to the effect that such and such events
xist are tenseless statements" and "The idea of a tenseless existence of
ᴠents . . . has a flavor of absurdity . . ." Clearly this conflict can only
ᴇ resolved by a careful analysis of the concept of existence as it relates
ᴏ entities in time.

Let us first discuss the concept of existence as it appears in the frame-
ᴠork of things. In this framework, it will be remembered, the basic logi-
al individuals are things (or substances or continuants), and the names
ᴏf the language refer to these individuals. While things are referred to by
ᴀmes, the fundamental form of event expressions *in the thing framework*
ᴵ indicated by the following: 'S's being Φ,' 'S's becoming Φ,' 'S's V-ing
ᴏr being V-ed)' (where 'V' represents an appropriate verb). Both 'S' and
ᴔ's being Φ' are *singular terms*, but their statuses within this category are
ᴀdically different. We have already had quite a bit to say about the 'exist-
ᴎce' of events and, indeed, of past, present, and future events within the
ᴋamework of things. It is time we said something about the 'existence' of
ᴛhings themselves.

Let me put my finger on the essential point at the very beginning.
ᴇxistence statements about things are as irreducibly tensed as statements
ᴀbout the qualitative and relational vicissitudes of things. Thus paralleling

$$(134) \quad S \{\text{is, was, will be}\} \Phi$$

ᴠe have

$$(135) \quad S \{\text{exists, existed, will exist}\}.$$

ᴴow are these latter statements to be understood?

Here we run head on into the fact that it is widely thought (indeed,
ᴀken for granted) that there are no such statements, or, more accurately,
ᴛhat there are no such statements *if S is construed as a name*. It is, in short,
ᴀlmost dogma that existence statements are statements having the form
(Ex) . . . x . . .' and that the difference between general existence
ᴛtatements and singular existence statements lies in the presence or ab-
ᴕence of a uniqueness condition. I wish to contend, for reasons which I

561

have developed in another context,[15] that the truth is the very opposit
of this dogma, and that existence statements invariably are of the form

(136) N exist(s)

where 'N' is either a *proper* name, in which case the statement is a singu
lar existence statement, or a *common* name, in which case the statemen
is a general existence statement.

We have all been brought up to recognize that the argument

(137) Lions exist
Leo is a lion
Therefore Leo exists

exhibits a misunderstanding of the logical form of 'Lions exist.' The latte
does not stand to 'Leo exists' as 'All men are mortal' stands to 'Socrate
is mortal.' On the other hand, the fallacy does not consist in forming
nonsense sentence 'Leo exists' under the impression that such a paralle
exists. The sentence 'Leo exists' makes perfectly good sense, and is no
to be construed as the mistake of putting a *name* (instead of a variable
after the 'existential operator' or reversed 'E.' Nor is the fallacy adequatel
exposed by rendering it, *logistice*, as

(138) (Ex) x is a lion
Leo is a lion
Therefore, (Ex) x = Leo.

For while 'Lions exist' would not be true unless '(Ex) x is a lion' wer
true, and while 'Leo exists' would not be true unless '(Ex) x = Leo' wer
true, these logistical expressions do not represent the sense of the origina
existence statements.

How are these existence statements to be understood? Let me begi
with the rough suggestion that ('S' being a proper name)

(139) S exists (did exist, will exist)

has the sense of

(140) Something satisfies (satisfied, will satisfy) the criteria fo
being called S,

where the criteria include a uniqueness condition. Notice that an every
day rendering of (140) runs

(141) There is such a (unique) thing as S.

[15] See "Grammar and Existence: A Preface to Ontology."

f this suggestion is sound, it springs to the eyes that there is a general re-
emblance (such as we would naturally expect) between 'Leo exists' and
Lions exist,' for the latter would have, roughly, the sense of

(142) Something satisfies the criteria for being called a lion

r

(143) There are such things as lions.

Notice that both (142) and (143) are in the present tense, as is 'Lions
xist.' It is important to realize that we are dealing with the schema

(144) Ks {exist, existed, will exist} $\cdot \equiv \cdot$ Something {satisfies,
satisfied, will satisfy} the criteria for being called a K $\cdot \equiv \cdot$ There
{are, were, will be} such things as Ks.

The crucial point is that statements of the form

(145) (Ex) . . . x . . .

re not as such in any ordinary sense existence statements. They corre-
pond to existence statements, where they have the force of

(146) (Ex) x is properly called (an) N

where 'N' is a proper or common name. Most of the puzzles about existence, and, in particular, puzzles about the existence of abstract entities,
re rooted in a crude equation of *existence* statement with '*existentially*'
quantified statement.

If one thinks that 'S exists' has the force of '(Ex) x = S,' one will conclude that things have a 'tenseless' existence, for the function '(Ex) . . .
. . .' is not a tensed function. If, for example, one construes

(147) Eisenhower exists

i.e., 'There is such a person as Eisenhower') as

(148) (Ex) x = Eisenhower,

one will conclude that the existence of Eisenhower is a tenseless existence,
and go on, perhaps, to contrast his tenseless existence with the tensed
truths which record his history. The truth of the matter, however, is that
Eisenhower exists' is, as it seems to be, a tensed statement, and has the
orce of

(149) (Ex) x satisfies the criteria for being called Eisenhower.

On the other hand, 'Napoleon existed' has the force of

(150) (Ex) x satisfied the criteria for being called Napoleon.

563

Similarly, 'Men will exist' has the force of

(151) (Ex) x *will satisfy* the criteria for being called a man.

It is, therefore, a radical mistake to suppose that in the framework of things the basic individuals of the framework have a tenseless existence, with tenses playing a role only at the level of predication about them.

The above line of thought is reinforced by the following consideration. Once we realize that 'existence' is not to be confused with 'existential' quantification, we are in a position to note that whereas such radically different existence statements as

(147) Eisenhower exists

and

(152) Triangularity exists,

not to mention

(153) Lions exist

and

(154) Numbers exist,

have in common the general form

(155) (Ex) x satisfies the criteria for being called (an) N,

there is a radical difference between the first and second member of each pair, a difference which concerns the nature of the *criteria*. And once we reflect on these differences we note that whatever may ultimately be true of (152) and (154), the existence statements concerning Eisenhower and lions essentially involve a relation to the person making the statement. For to say that Eisenhower exists is to imply that he belongs to a system (world) which includes us as knowers (i.e., language users). In other words, such statements as that Eisenhower exists have an intimate logical connection with statements which give expression to their own location in the framework to which belongs the referent of the statement (in this case Eisenhower), i.e., token-reflexive statements. And the token-reflexive statements in question are those which formulate the nexus of observation and inference in terms of which the claim that there is something which satisfies the criteria for being called Dwight D. Eisenhower would be justified.

Again, even though proper names are not shorthand for definite descriptions, they have a *sense* which is properly formulated as a definite description. And the use of the proper name presupposes the truth of the Russell sentence which is the foundation of the description. Thus, the sense of 'S' is given by expressions of the form '$(\iota x)fx$.' It would, however, be a mistake to conclude that the sense of 'S exists' is given by 'E! (ιx) fx.' For while 'S exists' is in some sense equivalent to 'E!(ιx) fx,' the former makes *explicit* something which is only implicit in the latter, and *what* it makes explicit is the claim that the framework (the language) to which both 'S' and 'E!(ιx) fx' belong is our language in its straightforward or primary use, and that the things and states of affairs of which it speaks are *our companions*, so to speak. For if we reflect on the difference between *fictional* names (e.g., 'Oliver Twist') and the criteria which constitute their sense (say, '(ιx) Fx') and '*real*' names (e.g., 'Dwight D. Eisenhower') and the criteria which constitute *their* sense (say, '(ιx) Gx') we see that it is not enough to say that the difference between them consists in the fact that we are entitled to say 'E!(ιx) Gx' but not 'E! (ιx) Fx.' For in *fictional* contexts we are as entitled to say 'E!(ιx) Fx' as to use the name 'Oliver Twist.' Obviously, then, the crux of the concept of (actual) existence is to be found in the distinguishing traits of the *real life* as contrasted with the *make-believe* or fictional use of language. Thus, to explicate 'S exists' it is not sufficient to call attention to its equivalence to 'E!(ιx) fx' or to emphasize that a tensed verb is lurking in the function 'fx' or to emphasize that being *the thing which is f* is the criterion for being called S. One must also make explicit the *real-life* character of the latter statement. And this can be done only by making explicit its connection with our activity as *knowers*, rather than as *storytellers*. And it is clearly a reasonable step in this direction to suggest the equivalence

(156) S exists $\cdot \equiv \cdot$ S belongs to a system of things which includes *this*.

It is not, however, my purpose on this occasion to explore in further detail the relation of token-reflexive expressions to the concept of observation, or to analyze the concept of observation. For my purposes it is sufficient to note that if the above equivalence is, as I suggest, a necessary one, then it follows that existence statements have a different sense on each occasion of their use in the sense in which 'this' has a different sense on each (relevantly) different occasion of its use. And we have been led

once again to recognize the essential token reflexivity of existence state
ments.

xv

In the thing framework, then, statements asserting the existence o
named individuals are fully tensed statements. The nontensed form '(Ex
x = S' is not the logistical formulation of an existence statement, thoug
of course if S exists, then it is necessarily true, *logistice*, that (Ex) x = S
We can, if we wish, introduce a tenseless form

(157) S 'exists'

in terms of a disjunction of tenses, thus,

(158) S 'exists' · ≡ · S exists or S existed or S will exist.

(Notice, however, that though (157) is, in a sense, tenseless, it expresse
the temporal point of view of the speaker, so that two people who avai
themselves at relevantly different times of this *sentence* will be makin
the same statement only in the sense in which differently tensed counter
parts make the same statement. The fundamentally tensed character c
(157) would also become manifest if this contrived verb were to be use
in two-perspective statements.)

Again,

(159) There are future things

is to be understood as a derived statement which rests on

(160) S is future · ≡ · 'S will exist' is true

and, hence, on

(161) S will exist.

Here we find a crucial difference between *things* and *events* (in the thin
framework), for, as we saw,

(95) There are future episodes

does not rest on

(162) E will *exist*

but rather on

(163) E will *take place*

which is equivalent to a statement of the form

(164) S will V.

5. Relativity and the Objectivity of Becoming

XVI

Philosophers who have taken relativity seriously—and for our purposes we can limit ourselves to the Special Theory—have often wondered how it can be reconciled with the idea that there is such a thing as *becoming*, whether 'qualitative' (i.e., coming to be Φ) or 'absolute' (i.e., coming to be (exist)). "For surely," they are prone to argue, "relativity theory has an event ontology, and pictures the world as a continuum of events for which the distinction between 'past,' 'present,' and 'future' is relative not only to a *now* (which is obvious), but to a set of coordinates which is only one among many sets of coordinates, each of which is an equally authentic structuring of the world into one temporal and three spatial dimensions. And," they continue, "if we call 'objective' that which is a matter of inter-subjective reasoned agreement, then according to the picture of the world painted by relativity theory, neither spatial distances nor temporal intervals are objective, but only the space-time separations of events (and their character as 'space-like' or 'time-like'), for only the space-time separations of events are invariant with respect to the measurements of all 'galilean' observers. How," they conclude, "can *becoming* be *objective* if time itself is not objective, but dissolves into a multitude of times each of which is a 'shadow,' to use Minkowski's metaphor, of a more basic reality (i.e., space-time)?"

The (imaginary, but representative) philosopher from whom the above is quoted is, of course, seriously confused. His confusions, however, are aided and abetted by most of the existing philosophical 'clarifications' of relativity—and our philosopher has at least the merit of taking his role as a philosopher seriously. Much of the groundwork has already been laid in the previous sections for a dispelling of these confusions. But certain considerations have been left to one side and must now be discussed before this groundwork can be used.

The reader who is familiar with the philosophy of measurement will undoubtedly have noted that our previous discussion of events and episodes in the framework of things has been tacitly built on the assumption that the events in which the things of this world participate constitute a four-dimensional continuum which, in its turn, is a temporal continuum of spatial, three-dimensional continua. And, indeed, we have been taking for granted, rather than exploring, the *metrical* character of this con-

tinuum of events.[16] When, therefore, it is said that the structuring of
the continuum of events into three spatial and one temporal dimension
is relative to a given system of world lines (galilean frame), we must ask
ourselves in what sense the continuum of events is 'prior' to its metriciz
ing by an actual (or hypothetical) observer who belongs to this frame—
at least during the process of establishing the congruences which consti
tute this metricizing. And to ask this is also to ask in what sense space
time intervals are 'prior' to the variety of their separations into spatial dis
tances and temporal intervals with respect to different galilean frames
For most of the confusions about the Special Theory concern the relative
'reality' of (a) the continuum of events; (b) space-time; (c) our space
and our time.

Given that the 'four-dimensional continuum of events' has been 'metri
cized' by an observer in one galilean frame, and representing this metriciz
ing as a cutting up of the continuum into a temporal series of three
dimensional spatial cross sections, the Lorentz transformations provide
us with a way of calculating the metricizing of this continuum into a tem
poral series of three-dimensional spatial cross sections with respect to any
other galilean frame. And, of course, these cross sections will consist o
different sets of events in the two metricizings where the observers are
in uniform relative motion with respect to each other. In a typical space
time diagram, the metricizing of the continuum with respect to the firs
frame is represented as in Diagram I. The system of particles $(P_1, \ldots$
$P_n)$ which constitutes the frame is represented (more accurately, of course
their histories are represented) in the diagram by straight lines parallel to
the T axis. And the second galilean frame, moving with respect to the
first, is represented (its history is represented) by a system of parallel
straight lines at an angle to the T axis, $P_1', P_2', \ldots, P_n'$.

A number of points are to be noted at once. (1) The metricizing is to
be understood as a system of direct and indirect measurements, i.e., con
gruence relations between certain sets of events belonging to the frame
(the 'measurings') and other sets of events (the 'measured' events). The
metrical character of the system of particles constituting the frame, in
cluding the observer, is, of course, as much a function of these measure
ments as the metrical character of any event or string of events belonging
to the continuum. (2) The point O represents the origin of a particular

---

[16] In the following remarks I shall make the usual simplification of things and
events into particles and motions.

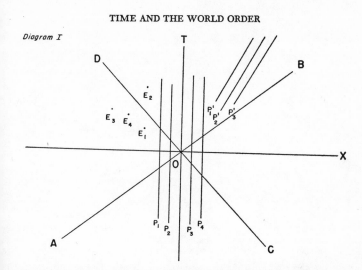

Diagram I

set of coordinates which is selected from the infinite set of alternative sets of coordinates which are subject only to the requirement that any T axis (for this metricizing) be parallel to T, O being an arbitrarily selected event. (3) The X axis represents the spatial dimension of the metricizing, and any two points at equal distance above or below the X axis represent events which are simultaneous with respect to this metricizing. (4) The lines AB and CD represent the histories of photons 'passing through' O. All pairs of events which can belong to the history of one particle must be connectable by a line which is such that a parallel to it through O falls within the angle DOB. Thus, E_1 and E_2 can belong to the history of one particle, whereas E_3 and E_4 cannot.

Consider, next, the situation represented by Diagram II. Suppose that E_5 and E_6 belong to the history of an observer S' in uniform motion relative to the original observer S. Suppose, furthermore, that O is an event in the history of S, and that the T axis coincides with the history of S (at least in so far as the measurements defining the metrics represented by this diagram are concerned). We can now represent a second metricizing of the continuum of events, a metricizing with respect to measurements belonging to the history of S', by choosing the same origin, but drawing the new T axis (T'), and the new X axis (X') at angles to the original axes. T' will be parallel to E_5E_6 and X' will be symmetrically located on the other side of the line AOB. We could, of course, have drawn our original diagram to represent the metricization of the con-

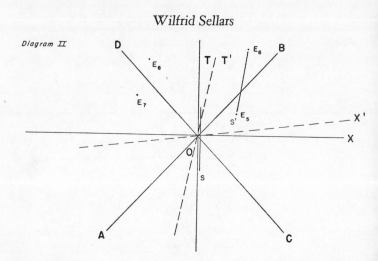

Diagram II

tinuum by S′, in which case the situation with respect to the axes would be reversed, T′ and X′ being at right angles, and T and X forming an acute angle in the upper right-hand quadrant, as do T′ and X′ above.

It is often said that in such diagrams as these, the intersecting lines AOB and COD (the 'cone' constituted by actual and possible 'paths' of photons passing near S at O) divide the continuum of events into three regions, a 'past,' a 'future,' and a 'present' (or 'elsewhere'); see Diagram III. This, however, is a most dangerous way of putting a sound point. The correct way of putting the point is to say that the region mislabeled

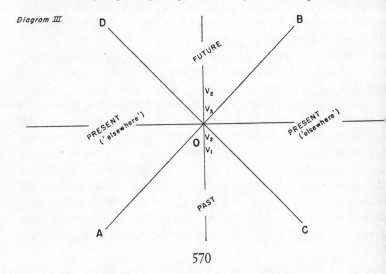

Diagram III

570

'future' is the region containing the events which are properly labeled 'future' not only by S (speaking at O) but by any observer belonging to a galilean frame in relative uniform motion with respect to S, and who, at the moment of speaking properly, calls O 'now.' Or, abstracting for the moment from the distinction between 'past,' 'present,' and 'future,' we can say that any event in the region labeled 'future' in the diagram will be classified as later than O by all galilean observers who cut the pie of events into temporal series of spatial cross sections. Similarly, to label the regions contained within AOD and COB 'present' is properly speaking to say that for every event in these regions there is a possible observer passing near S at O with a permissible relative velocity who would properly label it as 'present' if, at the moment of speaking, he properly labels O 'now,' and would properly say that the event in question is neither earlier nor later than O.

Now what these considerations amount to is simply that the metricizing of a set of events into a three-dimensional spatial array and the metricizing of spatially related events into a one-dimensional temporal array are *not independent operations*. It brings out clearly the fact that a framework of events structured into a past, a present, and a future is a metrical framework, as is a framework of events structured into a series of three-dimensional cross sections related by *earlier than*.

In what sense are space-time intervals "more real" than lapses of time and spatial distances? Only in the sense that the space-time interval between two events is an invariant quantity with respect to the Lorentz transformations, that is, with respect to all metricizings into a temporal order of spatially related events. To suppose that it is in any other sense "more real" is, as we shall see, analogous to supposing that events as *standing in the earlier-later relations* (with respect to a given metricization) are "more real" than events as *past, present, or future* (in a given metricization) because *earlier than* is invariant with respect to "the changing location of the 'now.'"

Confusion is twice confounded when it is supposed that the 'cone' represented by the angle DOB constitutes the "edge of becoming." For just as it is a mistake to think of the area within the angle as "the future with respect to S at O" (except in a very Pickwickian (derived) sense of 'future'), so it is a mistake to think of the series of light ray 'cones' whose vertices lie along T (at $V_1, V_2, \ldots, V_n$) as stratifying events into layers

of those which have just become with respect to V_1, that is, as representing "the moving surface of becoming."

We must now remind ourselves that although we have permitted ourselves to speak above without qualification of a framework of events, these
events have a derivative status in the sense that singular terms referring
to events are contextually introduced in terms of sentences involving
singular terms referring to things. And we must remind ourselves that in
the framework of things it is *things* which come to be and cease to be,
and that the event which is the coming to be or the ceasing to be of a
thing itself neither comes to be nor ceases to be but (like all events)
simply takes place. On the other hand, all metricizings in the framework
of things is a matter of the locating of *events*, including the events which
are the coming to be and ceasing to be of things. Thus, while neither

> (165) S was Φ at t

nor

> (166) S came into being at t

says that S *is* a series of events which includes, at time t, earlier than now,
an event which, in the case of (165) is of a certain kind, and, in the case
of (166) is the first event in the series, nevertheless these statements
specify *when* S was Φ and *when* S came into being, with reference to a
metric system of events. Thus (165) and (166) are necessarily equivalent
to certain statements of the form

> (167) E took place at t

but do not have the same sense as these statements. For (167) represents
statements which, made explicit, have the forms

> (168) S's being Φ took place at t
> (169) S's coming to be took place at t.

These considerations remind us that a sound account of time must combine a recognition that statements of the form

> (170) S is (was, will be) Φ *at t*

involve a framework of events, i.e., of 'individuals,' which *take place*, with
the recognition that the framework of events, made explicit, shows its
dependence on the framework of tensed statements about things.

I have already[17] sketched an account of time which indicates how *relational* statements pertaining to *events* are grounded in nonrelational state-

[17] Above, Sec. 2.

ments pertaining to things. All that is needed to provide the sought-for clarification of the status of the "continuum of events" differently "cut up" by S and S′ is to recognize that metrical relations between events are simply a special case of relations between events, and are similarly grounded in the use of temporal connectives which combine tensed statements about things, as contrasted with relational predicates which have event expressions as their terms.

Once we fully appreciate the fact that while a thing has as its counterpart in a framework of events a series or 'string' of events, it is not to be identified with this series or string, and once we appreciate the metrical character of time and temporal relations, we see that while it is in a sense correct to say, in the context of relativity theory, that two observers in relative uniform motion "measure the same events" but "order them differently with respect to spatial and temporal relations," it can also be very misleading. For it implies that events are logically prior to the changing things which are metricized by the use of clocks and rods. Rather we should say that just as certain metrical quantities are invariant with respect to all galilean metricizings of changing things as spatial and temporal, so certain topological features are invariant with respect to all metricizings of these same things without qualification. And it is therefore essential to note that topologically characterized events, instead of being the concrete reality of the world process, are simply abstract features common to all metrical pictures of the world. The temptation to think of the continuum of events topologically conceived apart from specific metrics as the basic reality which includes these metrics as specific patterns of topological relationship is a mislocation of the fact that metrical discourse about events is rooted in premetrical tensed discourse in which we talk about doing this or that while (before, after) other things do this or that in our immediate practical environment.

Now the burden of my remarks is, to put it simply, that it is a radical mistake to suppose that what really is, is that which is common to metrical frames. In other words, it is not correct to identify real with invariant features of metrical pictures of the world. For the status of these invariant features is surely an ontologically secondary one; they are abstract facts about the changing things which are the primary realities. On the other hand, these metrical pictures do not falsify, they simply are what they are in accordance with their own logic. It is therefore by no means inconsistent with the above to say that a consideration of abstract topological facts about the histories of things may throw light upon the puzzles of becoming. For while the simultaneity or successiveness of two events which belong to different things is a metrical property of these events in relation to a frame of reference, so that the "same" events may be simultaneous relative to one frame and successive with respect to another, two events belonging to the same thing have a topological order which is invariant with respect to all metrical frames. Thus, while the total set of copresent events (relative to S) is not the same set of events which are copresent for S', the temporal order of the events which belong to one and the same thing is (with qualifications which pertain to the idealization of things into punctiform particles) frame invariant, as is the order in which these events become less and less future, then present, then more and more past.

It is often said that we must avoid 'spatializing' time. Statements to this effect are invariably confused, for in so far as they imply that we should not think of time in metrical terms they are actually a contradiction. But they do contain insights which account for their vitality. These are the insights that changing things are not to be identified with their histories, that time as a measure of events is also a measure of things, and that the foundation of temporal discourse is the use of tensed verbs and nonrelational temporal connectives.

XVII

Attention has been called to three modes of invariance between different metricizings of 'the' events happening to the things of the world: (1) the topological invariance discussed above; (2) the metrical invariance specified by the Lorentz transformations; (3) the invariance, within a metrical frame, of the earlier-later relation with respect to different nows. It remains only to spell out in somewhat more detail the fact that these invariances constitute three levels of abstraction from the primary mode of existence of a framework of events, which is their existence as divided (with respect to a particular frame) into a particular now with its correlative future and past. The earlier-later relation has its primary mode of being as earlier-later in the context of a specific past-present-future.

It is but another way of making this point to call attention to the fact that temporal statements exist primarily as statement episodes, which have, as such, an irreducibly 'token-reflexive' character. A temporal picture of the world does not have the form of Diagram IV, nor even that of Diagram V, for while the latter recognizes that 'reality' is that which

574

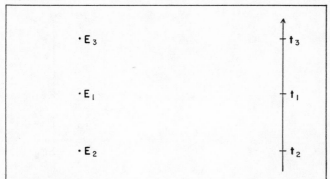

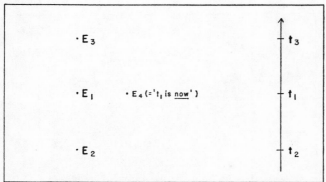

includes, in addition to nonlinguistic events, correct *utterances of* 'this,' 'here,' 'now,' etc., it may blind us to the fact that reality is that which includes *this* (not merely 'this'), *here* (not merely 'here'), and *now* (not merely 'now'). And reality is the same for two observers only as one person's *this* is another person's *that*, and one person's *now* is another's (or, subsequently, the same person's) *then*.

Thus, a person who *uses* that which is shown in Diagram VI is giving a temporal picture of the world.[18] And while all true (ideal) pictures of the world will agree in the linguistic episodes they *mention*, they will *use* relevantly different token reflexives to do so.

[18] And to picture a 'possible' world (as opposed to merely setting up the language for picturing it) is to tell a consistent story by the *make-believe* or *storytelling* use of token reflexives, and, therefore, to *pretend* that one is in it.

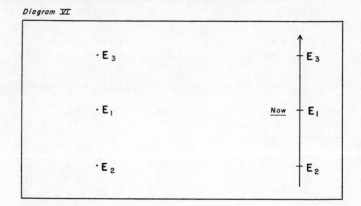

Diagram VI

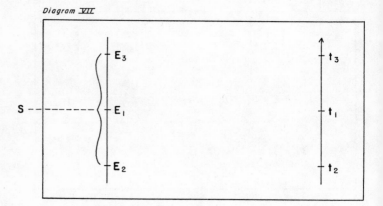

Diagram VII

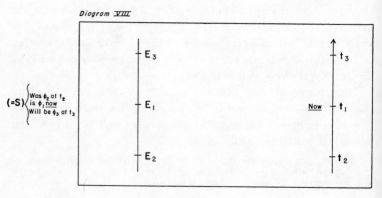

Diagram VIII

How do *things* fit into a temporal picture of the world? Not being events or strings of events (represented, respectively, by points and lines) they must, in a sense, lie outside the above picture (as, we shall see, must time). The general sort of thing which must be done is clear. We must represent things by symbols outside the rectangle, connecting them by, say, broken lines, representing a relation of *belonging to*, to their histories. (See Diagram VII.) More than this, we must include in the picture statements of the form 'S was Φ at t,' which can be done as shown in Diagram VIII.

Finally, we must distinguish between time and events pertaining to time; for only the latter belong in the event framework represented by the rectangle. Thus we must distinguish between a moment, t, and the event of the moment's being present with respect to a given perspective and, above all, between the event of the moment's being present with respect to a given perspective and the event of the moment's being *present*. The latter, of course, is the essential feature of a temporal picture of the world. This gives us Diagram IX as a primary picture of the world. In this picture, while the metrical character of temporal discourse is emphasized, the lower left-hand corner contains the irreducible element of *tensed discourse about things* which is the heart of the picture.

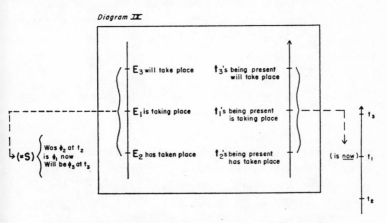

Diagram IX

6. The Problem Recast: an Ontology of 'Events'?

XVIII

Now it might be argued that although the basic individuals of ordinary discourse and, indeed, of physical science, are continuants or changing

577

things, and although in this framework events have the derivative status
exhibited by the forms 'S's becoming Φ,' 'S's V-ing' (or 'being V-ed'),
'S's being ψ,' we can perfectly well conceive of a framework in which the
basic individuals are the counterparts of the point-instant events of a
sophisticated thing framework, and the counterparts of things are 'genidentical' series of basic individuals, their property of belonging together
in one 'world-line' being defined in terms of spatio-temporal order and
empirical laws as these latter, in their turn, appear when recategorized to
fit this new framework.[19] If we call this new framework the framework
of 'events,' we see at once that the differences between it and the framework of substances or things will be systematic and pervasive. For the
basic sentences of the new framework will be quite unlike the basic sentences of the framework we have been exploring to date. And this difference at the level of basic sentences will infect the entire framework.[20]
It is obvious, to begin with, that the names of the 'event' framework could
not be translated into the names of the thing framework; that the same
is true of the primitive predicates of the two frameworks is equally clear
after a moment's reflection. What is more interesting, and to the point,
is that in this new type of framework temporal facts would be more like
spatial facts in that the role of tensed verbs, if introduced at all, would
be derivative from that of token-reflexive predicate expressions combined
with a tenseless copula. That is, just as spatial discourse in the thing
framework contains such sentences as

(171) S is here

where the copula, though temporal, has no spatial sense, and the spatial-
reflexive role is played by 'here,' so, in this new framework,

(172) x is now

would combine a nontensed 'is' with the temporal-reflexive 'now.' In other
words, representing this nontensed role of 'is' by 'be,' the new framework
would involve a more thoroughgoing parallelism between

[19] It is essential to note that the concept of a world line in a four-dimensional
continuum of events finds a proper, though highly derived, place in the framework
of things. It has too often been supposed by philosophers (and physicists) who discuss the Special Theory of Relativity that the Minkowski mathematical apparatus
automatically carries with it a commitment to an ontology of 'events.' This, as is
shown by the argument of the preceding sections, is simply a mistake.

[20] For an evaluation, all things considered, of this systematic difference, see Sec. 9
below.

(173) x be *here*

and

(174) x be *now*

than is characteristic of the framework of things.[21]

Let us permit this suggestion to grow before we evaluate it. Notice, in the first place, that whereas in the thing framework event expressions are derived expressions, the entities which in the new framework are the counterparts of events—let us call them 'events'—are *nameable*. Thus, 'x_1' is a name, and not shorthand for a noun phrase. Note, next, that in the 'event' framework the basic form of discourse pertaining to the becomingness of things will be *relational*, as contrasted with the nonrelational character of basic tensed discourse about things. Only at the derived level of events did we find terms for temporal relations in the frame-

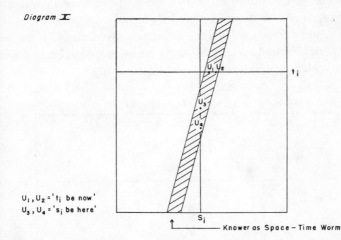

Diagram **X**

$U_1, U_2 = $ 't_i be now'
$U_3, U_4 = $ 's_i be here'

Knower as Space–Time Worm

[21] Although this parallelism is complete for the purposes of the present, introductory remarks on the framework we are analyzing, there remains a radical difference between 'here' and 'now' which touches the heart of the difference between the spatial and the temporal dimensions of discourse. This difference concerns the fact that no two utterances of 'now' *by one speaker* are strictly simultaneous, whereas two utterances of 'here' can be at the same place. Thus, two utterances of the form 't_i be now'—by one speaker—must refer to different times, whereas two utterances of the form 's_i be here' can well refer to the same place. (See Diagram X.) It might be thought that the fact that a spatially large speaker might say 't_i be now' out of each corner of his big mouth restores the symmetry. But, of course, to make a Kantian point, these two statements would not be statements *by the same knower*, unless that knower makes the *conjunctive* statement, in which case one of the utterances of 't_i be now' would be redundant.

work of things. It would not be surprising, therefore, to find those who elaborate the ontology of 'events' to believe that a simple distinction between the 'subjective' or 'perspectival' relational properties of 'events,' such as *after now* and *before now*, and an underlying 'objective' system of temporal relations—'events' as the domain of *earlier than*—will do for time what the distinction between the perspectival spatial attributes of things (at a time) and the 'objective' system of spatial relations they presuppose does for space.

In order to worm our way into this framework and to see what form our puzzles about becoming and passage take in the ontology of 'events,' let us, postponing relativity considerations, conceive of the world as a four-dimensional continuum of 'events' construed as basic particulars which are the terms of spatial and temporal relations. Representing a system of spatially related events by a line in the horizontal dimension, Diagram XI gives us an initial picture of the world as a system of 'events' (a, b, c, and d are 'events' at different places and/or times; O is the 'history' of an observer belonging to the system, o_1, o_2, and o_3 being, respectively, contemporaneous with a, b, and c).

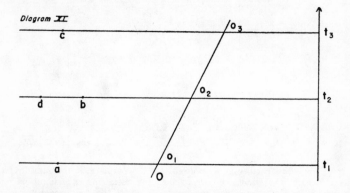

Diagram XI

It is assumed, of course, that the spatial cross sections (represented by horizontal lines) are the domain of a relation R, roughly *earlier than*, which has the properties necessary to define an open, continuous *order of between-ness*. It is also assumed that the set of cross sections constitutes a *series* as well as an *order of between-ness*. (Cf. the distinction between the *series* of numbers and the mere *between-ness* of points on a line apart from perspectival considerations.) It is also assumed that the occurrence of linguistic and conceptual 'events' in the 'history' of O constitute a *series* as well as an *order of between-ness*. And, finally, it is assumed that the temporal expression 'now' gives expression to the place of 'statement' events containing it in the series which makes up the 'history' of O.

If, now, we begin by focusing our attention on the system of 'events' as the domain of 'objective' temporal relations it seems reasonable to suppose that these 'objective' temporal relations are formulated by statements which are completely lacking in token reflexivity, so that token-reflexive statements can be laid aside for subsequent reflection. It is tempting, in other words, to suppose that there is a set of non-token-reflexive statements which simply formulate the 'objective' temporal relations which are the presupposition of the use of token-reflexive statements such as 'a be before *now*' 'b be *now*,' and 'c be after *now*.' We therefore focus our attention on such statements as

(175) a be Φ,
(176) a be R to b,
(177) b be (spatially) between c and d,
(178) a be at t_1,
(179) b be at t_2.

We would, if we took the above tack, construe the fact that O can make a statement of any of these forms at any time, and that if it is true as made at one time, it is true as made at any other time, as indicating that these statements have the same sense at all times, exactly as '$2 + 2 = 4$' has the same sense at all times. We would construe the token-reflexive statements

(180) a be at $t_1 \cdot t_1$ be before *now*,
(181) b be at $t_2 \cdot t_2$ be *now*,
(182) c be at $t_3 \cdot t_3$ be after *now*,

which are true[22] only if they are made in the appropriate temporal relation to a, b, and c, respectively, as built on the non-token-reflexive foundation provided by the 'objective' statements.

Thus, if someone were to suggest that in addition to the 'objective' form of predication,

(183) x be Φ

[22] Note that to characterize these statements as true (as made at a certain time) is to project ourselves imaginatively into the framework and regard O as one of us.

we also need the form

(184) x be Φ *now*

to be abbreviated as

(185) x *is* Φ

where 'is,' unlike 'be,' is a tensed verb, we might well reply that such a form might well be *introduced*, provided we are not misled by it. The parallel introduction of

(186) x *was* Φ

as equivalent to

(187) x be Φ before *now*

would bring this danger to a head, for there would be the temptation to suppose that

(188) a *was* Φ

stands to

(189) a *is* Φ

as, *in the thing language,*

(190) S *was* Φ

stands to

(191) S *is* Φ.

But, whereas things exist *throughout* periods of time and can change, 'events' exist *at* moments of time and, in a sense, cannot change. Thus, whereas (191) is compatible with

(192) Not-(S *was* Φ)

in the sense that S may well have come to be Φ from a state in which it was not Φ, the only way in which (189) can be compatible, in an event framework, with

(193) Not-(a *was* Φ)

is by virtue of the latter's containing the assertion that a be not a *past* event. Thus,

(194) a *is* Φ · = · a be Φ · a be at *now*
(195) Not-(a *was* Φ) · = · not-(a be Φ · a be before *now*).

Consequently, the truth of (193) follows from the truth of (189).

This can be put more simply by saying that there is a sense in which 'events' do not change except with respect to perspectival temporal characteristics. We have already seen that with certain necessary qualifications something like this is true of events and episodes in the framework of things. This makes it doubly necessary to bear in mind the essential difference between the 'events' of the 'event' frame, and the *events* of a thing frame. Thus, while both the 'a' of the 'event' framework and the 'E₁' of the thing framework are singular terms, their statuses within this category are radically different.

Following up the above line of thought we would conclude that perspectival temporal statements are built on 'objective' temporal statements not only in the sense that they have the form

(196) x be at t · t be {before now, now, after now}

but in the more radical sense that the conjunct 'x be at t' is independent of any reference to a now. We would also conclude that the attribution of an empirical characteristic to an 'event' in the past, present, or future is to be construed as the conjunction of a perspectival temporal statement locating the time occupied by the event with respect to the speaker speaking, and a tenseless statement of attribution, thus:

(197) c *will be* Φ_3 · = · c be Φ_3 · the time of c be after *now*
(198) a *was* Φ_1 · = · a be Φ_1 · the time of a be before *now*.

7. More on 'To Exist'

XIX

The question to which all the preceding is but an introduction concerns the sense in which, in an 'event' framework, past and future 'events' can be said to *exist*. We saw that in the thing framework it is perfectly legitimate to say

(90) There *are* {past, present, future} episodes

and we distinguished carefully between

(95) There *are* future episodes

and

(96) There *will be* future episodes

and, correspondingly, between

(93) There *are* past episodes

and

(94) There *were* past episodes

by pointing out that the second member of each of the latter pairs is a two-perspective statement, resting, respectively, on the forms

(199) E *will be* future (*quoad then*)

and

(200) E *was* past (*quoad then*)

as contrasted with

(201) E *is* future (*quoad now*)

and

(202) E *is* past (*quoad now*)

We also saw that if we were to reformulate (93) and (95) to read 'Past episodes exist' and 'Future episodes exist,' and if we were to suppose that in the case of episodes the various tensed forms of 'to exist' are equivalent to the corresponding tensed forms of 'to take place' we could easily be led to deny that either past or future events exist. For the obviously true statement 'Future events are not (yet) taking place' would be taken to entail 'Future events do not (yet) exist.' The solution of this perplexity lies, as we saw, in the fact that the maxim "The existence of events is their taking place" is sound only if, instead of proceeding as above, we take 'There are episodes' to be equivalent to 'Something is taking place, or has taken place or will take place,' and, in accordance with the schema

(203) Φ Ks exist $\cdot = \cdot$ Ks exist and some of them are Φ

construe

(95) There are future episodes

as

(204) There are episodes and some of them are future

where the form

(205) E *is* future

is equivalent (via "'E will take place' is true") to

(206) E *will* take place.

We also saw that to say that there are future episodes is equivalent to saying something like

(91) (ES)(Ef) S {is becoming, became, will become} f.

Now in an 'event' framework, the statement

(207) There are future 'events'

584

annot be reduced to tensed statements about things. 'Events' being the
basic particulars of this frame, we come to rock bottom with the very first
tep, that is, the step to

(208) There are 'events' and some of them are future,

which, in the language proposed for the frame, would *seem* to amount to

(209) (Ex) x *be* an 'event' · x *be* later than *now*

where the statement

(210) (Ex) x *be* an 'event'

s a *tenseless* affirmation of existence. If so, the fundamental form of an
existence statement in the language of 'events' would be

(210) There *be* 'events'

and this statement would affirm, in a purely 'objective' manner, the existence of entities which, in relation to different points of view, would have
he perspectival ('subjective') characteristics of pastness, presentness, and
uturity.

On the other hand, we are strongly tempted to say that in an 'event'
framework, future 'events' are those which *don't exist* but *will exist*, while
past 'events' are those which *don't exist* but *have existed*. And the ground
of this temptation lies in the fact that in the thing framework the existence of things finds its expression in *tensed* statements, for if *in one way*
events' in the 'event' framework correspond to *events* in the thing framework, in another way they correspond to *things*, being the basic individuals of their frame. Thus we are tempted on the one hand to say that
statements in the framework of 'events' which affirm the existence of
events' are purely 'objective,' and, on the other hand, to say that they
must have a perspectival character or 'subjectivity' corresponding to the
existed,' 'exist,' and 'will exist' of the thing framework.

To resolve this puzzle, we must take a closer look at those statements
n the thing framework which 'affirm the existence of things.' Part of the
trouble lies in the ambiguity of the phrase 'statements which affirm the
existence of things.' Consider, once again, the statement 'Napoleon ex-
sted' or 'There *was* such a person as Napoleon.' If our general account of
the force of existence statements was correct, these statements concern
the being satisfied of name criteria, so that

(211) Something {satisfied, satisfies, will satisfy} the 'N'-criteria

is the general form of an existence statement. But to satisfy criteria is to
have the criterion characteristics (or an appropriate selection of them,
they do not constitute a set of separately necessary and jointly sufficient
conditions). Thus we must break the formula down into

(212) Something $\{$*was, is, will be*$\}$ Φ_1, . . ., Φ_n · to be Φ_1, . .
Φ_n *is* to be appropriately called N

and in this expanded formula, the tenses of the first conjunct are straight
forward and pose no special problem in this connection. *But* the tense o
the 'is' in the second conjunct is more interesting, for, as a statemen
about the criteria for a *name*, it makes a reference to a language; and i
we reflect on which language it refers to, the answer can only be our lan
guage *now*. For whereas

(213) Something *was* appropriately called Napoleon

does not entail that we call anything Napoleon,

(214) Napoleon *existed*

does imply that we call something (i.e., Napoleon) Napoleon. Thu
(214) has the force of

(215) (Ex) x was Φ_1, . . ., Φ_n · 'Φ_1,' . . ., 'Φ_n' in our languag
are the criteria for the singular term 'Napoleon'

and, in general,

(216) N $\{$existed, exists, will exist$\}$ · $\equiv$ · (Ex) x $\{$was, is, will be
Φ_1, . . ., Φ_n · 'Φ_1,' . . ., 'Φ_n' in our language *are* the criteria fo
the proper name 'N'

and

(217) Ns $\{$existed, exist, will exist$\}$ · $\equiv$ · (Ex) x $\{$was, is, will be
Φ_1, . . ., Φ_n · 'Φ_1,' . . ., 'Φ_n' in our language *are* the criteria fo
the common name 'N.'

One more remark on existence statements in the thing framework be
fore we apply our results to 'events.' It concerns the peculiar status o
what can be called 'categorial existence statements,' thus, 'Things exist,
'qualities exist,' 'relations exist,' etc. It is important to catch the difference
between category words and ordinary common names; for example, be
tween 'thing' and 'lion.' One might expect that 'things exist' has the
sense of

(218) (Ex) x satisfies the criteria for being called a thing.

But 'thing' is not a common name of things, nor 'quality' a common name of qualities. Rather, as Carnap has correctly emphasized,

(219) . . . is a thing

has the force of

(220) '. . .' is a thing word in L

where, I would emphasize, L is understood to be our language as used in (219), so that (219) presupposes the existence in our language (now) of a domain of thing words and statements of the form

(221) W is a thing word in our language (now).

We must, therefore, be careful to avoid inferring that the move from

(222) Leo is a thing

to

(223) (Ex) x is a thing

is a simple parallel to that from

(223) Leo is a lion

to

(224) (Ex) x is a lion

or that (223) is simply a very abstract example of the form

(225) (Ex) x is a K

or that because '(Ex) x is a lion' is correlated with 'Lions exist' and, like the latter, is a tensed statement, contrasting with '(Ex) x was a lion' and '(Ex) x will be a lion,' '(Ex) x is a thing' is correlated with 'Things exist,' or that if it is, it can meaningfully have a past or future tense.

Perhaps the best thing to say at this point is that the traditional view that categories are *summa genera* is a mistake, and that it is this mistake which underlies the idea that statements of the form

(226) Something is a thing

or, *logistice*,

(227) (Ex) x is a thing

are well formed. It is, therefore, worth noting that even if we reject these forms[23] and, correspondingly, restrict existence statements to statements

[23] Which is not to say that such expressions as 'white thing,' in general, expressions of the form 'f-thing,' may not be introduced contextually in terms of the equivalence,
(228) x ε White thing · ≡ · x ε (all the x's such that x is white).

involving proper or common names (excluding category words) there still remains a role—often confused with the above—for the sentences

(229) Some things {have existed, exist, will exist},

for these forms of words may be used to express the idea that there are true statements of the forms

(230) S {has existed, exists, will exist}
(231) Ks {have existed, exist, will exist}

where 'S' is the proper name of a thing, and 'K' is a thing-kind expression e.g., 'lion.'[24] Thus to say that "future things are things which *will* exist' is to express the equivalence schemata,

(232) S is a future thing ≡ (S is a thing) · S will exist
(233) Ks are future things ≡ (Ks are things) · Ks will exist

where 'Ks are things' has the force of 'K is a thing-kind expression' and tells us what kind of expression belongs in the blank of

(234) . . . is a K.

8. Existence and Tense

xx

Now if this analytic framework is correct, it follows that in the language of 'events' the expression

(235) Events exist

construed as an existence statement containing the category word 'event' is, if permitted at all, of the form

(236) Individuals exist

and, as resting on singular statements such as

(237) a is an individual,

is in the material mode of speech, and has no past or future tenses, for, as we saw, it gives expression to the *present* use of 'a' in our language. On

[24] Notice that if it is granted that abstract singular terms are metalinguistic references to expressions in our language, e.g., 'Triangularity' to 'triangular,' 'dissolution' to 'to dissolve,' 'mankind' to 'man,' then the idea that abstract entities have a timeless existence has its source in the fact that to refer to triangularity is to refer to our *present* use of 'triangular' and that no *past* or *future* use of this word is relevant to "the existence of triangularity." Thus, both the criteria and that to which the criteria apply are *present*, and necessarily so, in the case of abstract singular terms and abstract common nouns, whereas in the case of singular terms and common nouns which name individuals, though the *criteria* are present, the *nominata* need not be.

he other hand, (235) can also be construed as giving expression to the
lea that there are true statements of the form

(238) a exists

nd

(239) Ks exist

vhere 'a' and 'K' are singular and common names of 'events.'

In these terms our problem is to determine whether, in the 'event'
ramework, existence statements of the latter type (i.e., the type repre-
ented by (238) and (239), rather than (235) construed as a categorial
xistence statement) have the force of tensed statements. In other words,
o use the language we have constructed for the 'event' framework, is the
asic form of an existence statement,

(240) a be existent

r

(240) a be existent {before now, now, after now}?

At first sight it might appear that the former is the appropriate form,
or, we might argue, the existence statement in question is to be equiva-
ent to

(241) (Ex) x be $\Phi_1, \ldots, \Phi_n \cdot '\Phi_1,' \ldots, '\Phi_n'$ be our criteria now
for 'a'

n which case the only way in which 'now' enters in is with respect to
he criteria, and not with respect to the time of the item which satisfies
hem. Thus, we might argue, whether the 'event' be past or future, the
:riteria will be now-criteria, so that (241) consists of a 'purely objective'
:omponent, '(Ex) x be $\Phi_1, \ldots, \Phi_n$,' which specifies the satisfying of the
:riteria, and a perspectival component specifying the criteria. The con-
:lusion we would draw is that the basic form of existence statements con-
:erning individual events is that illustrated by (240), which is equally
ppropriate whether a be before now, now, or after now.[25]

The above reasoning, based as it is on the idea of a purely objective
:omponent, may conceal a dangerous error. It all depends on how the
:riteria for 'a,' represented by 'Φ_1,' $\ldots$, 'Φ_n' are conceived. If it is tacitly

[25] Though, of course, taking (240) as basic, we could introduce the form repre-
ented by

(242) a {existed, exists, will exist}

s equivalent to

(243) a be existent $\cdot$ a be {before now, now, after now}.

589

presupposed that these criteria include relational properties involvin
temporal priority or succession, and hence, leaving aside speculation
about time-dimensionally closed worlds, a relation capable of defining
temporal serial order, then the fat is in the fire. For, as we have alread
seen, 'earlier than,' far from being logically independent of the distinctio
between *past, present,* and *future,* is inextricably bound up with them
The only way in which this error can be avoided is by substituting a *non
temporal counterpart* of 'earlier than' in specifying the criteria for 'a,
*where by speaking of this counterpart as nontemporal I mean only tha
it does not belong to the circle of concepts (earlier, later, past, present
future, now, then, etc.) which together make up the framework of ord
nary temporal discourse.* It is not my purpose to deny that such a counter
part can be found.[26] My aim is simply to emphasize that 'earlier than
can define the 'objective' order taken for granted by the above reasonin
only if it is given a new use in which (a) it is freed from its dependenc
on egocentric discourse, and (b) as a necessary condition of this, it is made
to stand for a theoretical construct in thermodynamics by means of whicl
an *intrinsic*[27] serial order can be defined which would be the *explanatio
of temporal order in the ordinary sense* much as theoretical chemistr
gives an explanation of the solubility of table salt in water. If these con
siderations are sound, then the idea that, in an 'event' framework, event
have a timeless *existence* in which they stand in objective *temporal* rela
tions and constitute a system which includes the perspectival distinction
of pastness, presentness, and futurity as properties relative to points o
view located within the system is a mistake. It is the mistake of assumin
that a primary *temporal* picture of the world can be one which does no
use but only *mentions* the term 'now.'

Let me unpack this point by offering an analysis of the status of 'inter
subjectivity' in the 'event' framework. The basic intersubjective entitie
of the frame are, of course, the individual 'events' themselves. Thus, jus
as *thing-names* in the framework of things can have the same referent
for all observers regardless of their frames of reference or the location o
their *nows*—even though the *criteria* employed by these observers are th

[26] Recent attempts by Reichenbach and others to find a physical counterpart of th
serial character of 'earlier than' are explored by Adolf Grünbaum (to whom I am in
debted for many helpful comments and criticisms) in his essay for the present volume
These attempts are to be contrasted with McTaggart's attempt to construct a meta
physical counterpart.

[27] Intrinsic in the sense in which the series of integers has an intrinsic order.

same' only in the sense that they transform into each other and/or are differently tensed counterparts—so 'event' names can have the same refrents for all observers communicating with one another in terms of a given framework of 'events.'

Again, just as a primary picture of the world in the framework of things is a *tensed* picture, and detensed pictures have their roots in tensed pictures, so, in the case of an 'event' framework, a primary temporal picture is a picture with a *now*. And even if one observer's *now* is another observer's *then*, or one observer's simultaneous cross sections of the world are another observer's sets of differently dated 'events' (see Diagram XII), each of their *now*-pictures is a primary picture, and the purely topological pic-

Diagram XII

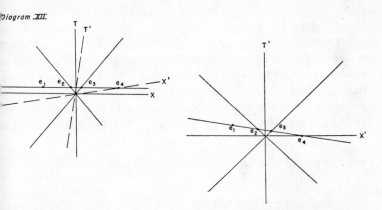

ture (which includes the measurements performed by S and S' as topological facts) which is common to them is not *the primary picture* of the world construed as a system of 'events,' but merely a topological abstraction common to the various primary pictures; and the topologically formulated location of individual events in the topological picture is merely the topologically invariant features of the criteria which identify these 'events' in a primary picture.

It follows that an 'objective' *temporal* picture exists only as an abstraction common to a class of 'subjective' (perspectival) *temporal* pictures, and that

(244) a be before b

is *not* logically prior to, and *does not make sense without*,

(245) Either a be present now · b be future now or a be present

before *now* · b be future *then* or a be present after *now* · b be future *then*.[28]

If we apply these considerations to existence statements, we see that *both* conjuncts of (241) are, *as belonging to a primary temporal picture of the world*, irreducibly 'tensed.' And to make this 'tensed' (token-reflexive) character explicit is to formulate our analysis of existence statements pertaining to individual 'events' as the equivalence

(246) a be existent {before now, now, after now} · ≡ · (Ex) x be {before now, now, after now} and x be Φ_1, . . ., Φ_n and 'Φ_1,' . . ., 'Φ_n' be our criteria now for 'a.'

"But why," it may be asked, "need existence statements belong to a primary picture? Why may they not belong to a level of abstraction which, though rooted in a primary picture, does not make its rootedness in our *present primary picture* manifest?" The answer, surely, is that, as was pointed out in the case of things, only a primary picture with its explicit *now* makes clear the *nonfictional* character of the statement, its rootedness in the real-life activities of *observation* and *inference*. It is failure to appreciate this dimension of existence statements which has made plausible the idea that singular existence statements have the form illustrated by

(247) (Ex) x = a

and that general existence statements have the form illustrated by

(248) (Ex) x ε K.

And it is a failure to think through the implications of the idea that existence statements concern, at bottom, the *having application of specific proper and common names* which has made plausible the idea that things (in the thing framework) and 'events' (in the 'event' framework) have a tenseless mode of existence which is prior to the 'perspectival' characteristics of pastness, presentness, and futurity. In short, if one confuses between existence statements and 'existentially' quantified statements, one will think that '(Ex) x = a' (in the 'event' framework) and '(Ex) x = S' (in the thing framework) make existence statements, and interpret the fact that any observer at any time can make 'existentially' quantified state-

[28] Note that to grasp the *directedness* of time is to use token-reflexives in *primary pictures of the world*. A scientific account of this consciousness must therefore be a scientific account of the mechanisms involved in *token-reflexive* verbal and conceptual activity. Needless to say, this scientific account of temporal discourse *as a fact in the world* must not be confused with the grasping of the directedness of time which is the *use*, rather than the *mention*, of token-reflexive temporal expressions.

nents of these forms, the domain of which is the *totality* of 'events' or things *irrespective of their perspectival characters as past, present, or future (with respect to the speaker speaking)* to mean that the existence of 'events' or things is a purely nonperspectival fact.

And if one makes these mistakes, one will suppose, if one works with framework of 'events,' that the world has, as its primary mode of being, existence as a topologically ordered system of 'events' (topological space-time) which contains (a) a multitude of (relative) subsystems of metrical relationships characterized by quantitative invariances (which constitute metrical space-time), and (b) a multitude of perspectival facts concerning pastness, presentness, or futurity (with respect to particular points of view defined with respect to given metrical systems) as special (and complicated) topological facts about these 'events.' And to suppose that an -temporal existence of such a topological matrix of metrical temporal relations is the primary mode of being of the world is (a) to reify once again (cf. XVI above) an abstraction common to *temporal* pictures; (b) to suppose that the non-perspectival relationships which are thus conceived to underlie all temporal perspectives are in a full—rather than analogical—ense *temporal* relationships.

The *existence* of the world as well as of the 'events' which make it up s irreducibly perspectival. The *structure* of the world as a *temporal* structure is irreducibly perspectival—though not, as we have seen, 'subjective' in any pejorative sense. The non-perspectival structure which, as realists, we conceive to underlie and support perspectival temporal discourse is, as yet, a partially covered promissory note the cash for which is to be provided not by metaphysics (McTaggart's C-series), but by the advance of science (physical theory of time).

9. Things or 'Events'?

XXI

I shall conclude Part I of this essay on time and the world order with a brief discussion of this question: Which does the world *really* consist of, *things* or 'events'? It will come, I hope, as no surprise that my answer is that in one sense these are not genuine alternatives, while in another he answer is obviously *things*.

Questions as to the reality of time arise, in part, from puzzles concerning the place of temporal expressions in a *primary picture of the world*, and, in part, from puzzles concerning what it is that *primarily* exists. Now in the thing framework it is *things* which primarily exist, and in the 'event' framework it is 'events' which primarily exist. The contrast, in each case is between the items which are *named* (by both proper and common names) and the items which are either contextually introduced (e.g. events in the thing framework, and 'things' in the 'event' framework) or are at bottom *linguistic entities* (thus *qualities, relations, facts*).

Only in a framework of 'events' are there *primitive* temporal relations for only in an 'event' framework is the sort of singular term required by temporal relations *basic*. Thus it is essential to note that in an 'event' framework temporal relations hold between items which *have existed and no longer exist* on the one hand, and items which *are yet to exist on* the other. In a thing framework, the corresponding temporal facts concern episodes which *have taken place*, on the one hand, and episodes which *are yet to take place on the other*. If one confuses existence statements with 'existential' quantification, and if one confuses the 'events' of an 'event' framework and the *events* of a thing framework, one will immediately generate the puzzle 'How can temporal relations obtain between an item which exists and one which doesn't exist? Doesn't "aRb" entail "(Ex)(Ey) xRy"?'

The more one appreciates the systematic character of the difference between the framework of things and the framework of 'events,' the more one comes to realize that the latter framework is *in the first instance* simply a reaxiomatization of the former, and differs from it only as a Euclidean geometry axiomatized with one set of primitives differs from one which has been axiomatized with another set of primitives. Which is the 'correct' axiomatization? If this is interpreted as a question concerning the structure of 'ordinary' temporal discourse, it seems to me to be perfectly clear that the basic individuals of this universe of discourse are things and persons—in short the 'substances' of classical philosophy. It is, I believe, only when the deeper levels of physical theory are approximated that the framework of 'events'-in-Time ceases to be a mere reaxiomatization of the framework of changing-things-in-Time and becomes, instead, the anticipation of a scientific (rather than metaphysical) framework in which the temporal dimension of the macro-world finds its adequate theoretical counterpart in a dimension of physical content.

However this may be, our analysis has made clear that a primary picture of the world is always a perspectival picture, and would be a perspectival

icture even if the theoretical framework envisaged above were to become
ur primary conceptual structure (i.e., we were to perceive the world in
erms of it). We have already seen, however, that the irreducibly perspec-
ival character of primary pictures and of the *existence* of the pictured
vorld is compatible with the non-perspectival (though perspective em-
racing) *character* of the latter.

II. DETERMINISM AND TRUTH

10. Decidability and Truth: Toward a Three-Valued Logic?

XXII

I have argued (Part I, pp. 554ff) that

(249) 'E is taking place' (said *now*) *is* true

ntails

(250) 'E will take place' (said *then*) *was* true

nd hence, where E = S's becoming Φ_2,

(251) S is becoming Φ_2

ntails

(252) 'S will become Φ_2' (said *then*) *was* true

nd, in particular, that

(253) S is becoming $\Phi_2 \supset$: (t) t is before now $\supset$ 'S will be Φ_2'
$\qquad$ (said at t) *was* true

r, to use the language of facts,

$\qquad\qquad \supset$: (t) t is before now $\supset$ it was (quoad
$\qquad\qquad$ t) a fact that S would become Φ_2.

Briefly, I called attention to the fact that the very language of time carries
with it the idea that if we are in a position to say

(254) **S** (said now) *is* true

hen, *ipso facto*, we are in a position to say

(255) **S'** (said before now) *was* true

nd

(256) **S"** (said after now) *will be* true

where **S**, **S'**, and **S"** are what we have called differently tensed counterparts.
To make the same point in the language of facts, if it *is* a fact that p, then
t was (with respect to all times before now) a fact that p' and *will be*

(with respect to all times after *now*) a fact that p'', where the sentence represented by 'p,' 'p',' 'p'',' respectively, are also differently tensed counterparts.

It follows that if we are in a position to say

(257) 'S will be Φ_3' *is* true

which is, of course, to be in a position to say

(258) S *will be* Φ_3

we are, *ipso facto*, in a position to say

(259) 'S is Φ_3' will be true
(260) S's being Φ_3 is future
(261) It will be a fact that S is Φ_3.[29]

On the other hand, it has been argued, as is well known, that in certain cases, at least, properly formed statements of the form

(258) S will be Φ

are neither true nor false. It would follow from the above considerations that in these cases not only

(257) 'S will be Φ' *is* true

but also

(259) 'S is Φ' *will be* true

and

(264) 'S will be Φ' *was* true

are also neither true nor false. But before we explore the significance of the fact that to say

(265) 'S will be Φ' *is* neither true nor false

is to be committed to saying

(266) 'S is Φ' *will be* neither true nor false

let us take a closer look at the reasons which have been offered in support of the claim that (certain) statements about the future are neither true

[29] As far as my ear for, and knowledge of, the English language is concerned, there is no clear convention for indicating that the occasion of S's being Φ_3 in question contemporaneous with the statement which *will* (from our present point of view) affirm it, rather than with our statement (261). Perhaps we may distinguish between

(262) It will be a fact that S will be Φ_3

and

(263) It will be a fact that S will be yet to be Φ_3

and use the former for the sense desired.

nor false. They are, as we have already noted, reasons which purport to connect truth or falsity with physical implication. In particular, they attempt to show that

(267) '. . .' is {true, false}

entails

(268) That . . . is {physically implied by, physically incompatible with} the present state of the universe.

Our explorations have made it clear that one line of thought which has been offered in support of this contention is unsound. That line of thought was based on the idea that to say of a statement that it is true is to say that it corresponds to a fact.[30] It proceeded to argue that (a) episodes are more basic than facts; (b) future episodes do not yet exist; hence (c) facts about the future exist only as physically implied by facts about the present. This line of thought was undercut by showing that while there is a perfectly legitimate sense in which episodes are more basic than facts, the idea that 'future episodes do not yet exist' is radically confused. Future episodes are simply those which have yet to take place.

If, then, there are future episodes, i.e., certain episodes will take place, the corresponding statements about the future are true and their contradictories false. On the other hand, to say that there are future episodes is to say something like

(271) (ES)(Ef)(Et) S will be f at t.

Thus the claim that there are facts about the future, the claim that certain statements about the future are true, and the claim that there are future episodes all have their ground in common or garden variety statements in the future tense.

But if the line of thought we have been criticizing won't do, there are other considerations which point to a connection between empirical truth and causal implication. For surely, it will be said, to make a statement is

---

[30] I pointed out that while it is perfectly correct to say that true statements correspond to facts, this amounts to no more than saying that true statements translate into true statements. In short, the 'correspondence theory of truth' in its traditional form confuses between

(269) S (in L) is true $\cdot \equiv \cdot$ (Ep) S (in L) means p $\cdot$ p

and

(270) S (in L) is true $\cdot \equiv \cdot$ (E that p) $\cdot$ S corresponds to that p $\cdot$ that p is a fact.

For a more adequate formulation of this point see "Truth and 'Correspondence,'" referred to in fn. 3 above.

Wilfrid Sellars

to imply ('contextually' or 'pragmatically') that one has good reasons fo
what one asserts, and that the statement is to be withdrawn if these rea
sons fail to withstand criticism. It needs only a hasty move or two (usuall
aided and abetted by the confusions already exposed) to conclude that
statement isn't a statement unless there are (in principle) good reason
for or against the statement.[31] The philosopher who takes this line con
strues the implication between making a statement and having reasons
as resting on a more basic connection between the statement-making rol
and there being reasons, reasons which, if known, could be used to sup
port (or attack) the statement. Thus the form of words "S will be Φ'
would express a statement if and only if there were reasons, R_1, which,
one but knew them, would authorize either

(272) $R_1 \cdot R_2, \ldots, R_n$. Therefore, S will be Φ.

or

(273) $R_1 \cdot R_2, \ldots, R_n$. Therefore, S will not be Φ.

Finally, he argues that these reasons from which, if we but knew them
'S will (not) be Φ' could be inferred must, in the last analysis, concern
the present state of the universe, and contends that "S will be Φ" make
a statement if and only if the present state of the universe 'determines
either that S will be Φ or that S will not be Φ.[32] In short, his contention
is that 'statements' which are, in this sense 'undecidable' are not state
ments at all, though they are closer to being statements than forms o
words which fail to make statements for such familiar reasons as that the
contain a name or demonstrative which fails to denote, or contain non
sense syllables, or inappropriately combine otherwise meaningful expres
sions. In effect, then, our philosopher is proposing to distinguish withi
the class of forms of words which if they fail to make statements do s
only because they are—in his sense—undecidable (let us call them 'puta
tive statements') two subclasses: (a) statements proper; (b) a comple

---

[31] It would clearly be absurd to say that a person hasn't made a statement unles
he has good reasons for what he says. The statement-making character of a particula
utterance is 'intersubjective' and independent of the soundness of the reasons whic
the speaker would adduce to support it.

[32] If one takes into account probability arguments and demands only that the trut
or falsity of a statement, S, which asserts that p, presupposes that the present state o
the universe imply that p (or that not p) either 'deterministically' or 'probabilistically
then the position becomes highly paradoxical, to say the least, for then a certain de
gree of probability would divide statements from non-statements (or, if one prefers
true or false statements from statements which are neither true nor false), namely
the minimum degree which permits the conclusion '. . . therefore, probably p.'

mentary class of putative statements which are not statements proper (let us call them *quasi-statements*). He then tells us that only statements proper are either true or false, quasi-statements being neither true nor false.

He now argues that if we always knew whether or not a given putative statement was 'decidable,' and, hence, a statement proper, we could appropriately use a two-valued logic. Furthermore, he argues that if we had reason to accept the thesis of determinism (retro-determinism as well as ante-determinism) we would be in a position to accept all putative statements as statements, so that the above distinctions would be 'academic.' He then points out that the thesis of determinism is challenged by most if not all contemporary philosophers of science, and concludes that we should *build our logic around putative statements and make it three-valued*. Putative statements would be either true, false, or middle ('undecidable'). Statements, on the other hand, would be either true or false.

XXIII

Let us begin our examination of the above line of thought by pointing out that few if any philosophers have held that not all statements about the past are either true or false,[33] as contrasted with the multitude who have held the corresponding thesis about the future. What would they say if confronted with the above argument? Would they reject the claim that *true or false* implies 'decidability' (in principle) and argue that statements about the past are either true or false whether or not they are (in principle) inferable from the present state of the universe? Or would they argue that *since* every statement about the past is either true or false, retro-determinism (by a kind of 'transcendental deduction') must be true? Needless to say, whichever line they took would be equally applicable to statements about the future.

We shall pick up this thread at a later stage in the argument. For the moment it has served the purpose of introducing an argument against the suggestion under consideration which highlights its paradoxical character, and in doing so brings out its broader implications. According to that suggestion, it runs, we can no longer assume that if the form of words

(2) S is now Φ_2 (in 1958)

[33] We have reverted, for the moment, to more traditional terminology in which 'statement' has, roughly, the sense of 'putative statement' as introduced by our fictitious philosopher.

(said today) makes a statement, then the corresponding form of words

(274) S will be Φ_2 in 1958

(said in the same language in 1950) made a statement at all, let alone a true one. For if the state of the world in 1957 did not (physically) imply that S would be Φ_2 in 1958, then, on the view in question, the second form of words failed to make a statement. Surely, however, the argument concludes, it is a desideratum of any explication of the nexus of epistemic terms pertaining to statements that not only the same form of words may be used by two people to make the same statement irrespective of the difference in value of the reasons by which they would support it, but also that what I have been calling *differently tensed counterparts* (e.g., 'S will be Φ_2 in 1958' said in 1950, and 'S is now Φ_2 (in 1958)' said in 1958) *both make a statement if either makes a statement.* Of course the statements must indeed be differently tensed counterparts. This means, roughly, that the only relevant difference between them is one of tense. There must be no difference in meaning between the 'S' and 'Φ_2' of the first and the 'S' and 'Φ_2' of the second. 'S' must refer to the same individual on each occasion; and, obviously, if 'Φ_2' contained a covert reference to circumstances contemporary to the time of utterance—if, for example, it were definable as 'approved by our graduating class'—the statements would not be counterparts. Let us call predicates which satisfy this condition "tense-indifferent predicates.'

Now it is a framework trait of ordinary discourse, in so far as it pertains to the physical properties of physical objects, that tense-indifferent predicates are the rule rather than the exception, just as it is a framework trait of ordinary ordinary discourse (though not of ordinary scientific discourse) that shape, size, and weight are independent of relative motion. Yet even if it should be in some sense a regulative ideal of scientific investigation that the basic concepts of an ultimately satisfactory explanation of physical phenomena be tense-indifferent, it is by no means necessary that the same should be true of physics-on-the-way. It is arguable that the development of quantum theory, by implicitly introducing a reference to the occurrence of a certain kind of event (a 'measurement') into its fundamental concepts, has, in effect, constructed relation-ized[34] cousins of pre-quantum-theory concepts, just as relativity kinematics has replaced

[34] I use this barbarism because the term 'relativized' has acquired a specific meaning in contemporary physical theory.

pre-relativity magnitudes by relationized counterparts. And it is also arguable that the 'paradoxes' of quantum theory spring from a failure to reflect this implicit relationizing in the explicit syntax of the theory. (It was
a stroke of good fortune that the modern revolution in kinematics was
guided by a philosopher-scientist who from the beginning reshaped the
language of kinematics to give explicit recognition to the conceptual
changes it involved.) If so, then the required revision of the language of
QM would bring with it a radical modification of the role of differently
tensed counterparts.

The notion of a regulative ideal of scientific investigation is a difficult
one to pin down. Clearly it implies a distinction between *regulative ideals*
and such highly general *inductions* of whatever 'order' as may emerge at
any given stage of scientific development. When it is said, for example,
that a deterministic world picture is a regulative ideal of scientific investigation, it is implied that determinism, thus conceived, is not an inductive
conclusion which could be confirmed by evidence (itself consisting of
inductive conclusions) available at one time, and subsequently disconfirmed by new evidence. I have discussed this topic in another place,[35]
and shall simply indicate my conviction that something like the Kantian
conception of ideals of reason can be defended. My present concern is to
point out that the hope of classical physics that its particulate concepts,
which share the tense-indifference of the molar concepts of classical mechanics in which they were rooted (by means of highly qualified analogies), would permit the formulation of deterministic laws has proved
incapable of realization.

Needless to say, if at some future date QM should be given a deterministic substructure, this substructure might well be constructed from tense-
indifferent predicates. In any event, it is essential to note that though the
revision of the language of QM suggested in the previous paragraph but
one would, in a sense, diminish the role of differently tensed counterparts by introducing into its basic state descriptions a reference to individual measurement episodes, it would, of course, not *eliminate* differently tensed counterparts altogether, any more than the language of
relativity kinematics finds no place for statements which do not characterize kinematic situations in relation to a specific frame of reference. It
simply means that differently tensed counterparts would occur at a de-

[35] See the concluding sections of my paper on "Counterfactuals" in the second
volume of *Minnesota Studies in the Philosophy of Science*.

rived level of the language and would have a more complicated logical structure.

Now there is all the difference in the world between the *scientific* (methodological) thesis—that the fact that the failure of the fundamental magnitudes of classical physics to yield deterministic laws, together with the fact that these concepts when placed in the context of QM generate paradoxes and anomalies if their classical heritage is taken seriously *with or without the assumption of strict (nonstatistical) determinism,* is to be met by relationizing these concepts in a way which destroys their tense-indifference—and the *metaphysical* thesis—that *terms which are granted to be tense-indifferent* can make a statement when put together at one time in one tense, and yet fail to make a statement when put together at a different time in a different (but appropriate) tense. The scientific thesis is a legitimate move in the strategy of man's intellectual warfare against nature; one which involves no commitment on issues of ontology. The metaphysical thesis is, on the other hand, a simple mistake. For if anything has emerged from the explorations which make up the first part of this essay, it is that it is 'analytic of' the framework of tensed discourse and tense-indifferent terms that if a tensed sentence constructed from them can be used at a certain time to make a statement, its differently tensed counterparts, used at appropriate times, also make statements, and, indeed, in an appropriate sense, the *same* statement.

It would seem, then, that if we had to choose between the view which says that of two differently tensed counterparts one may make a statement and be either true or false while the other is neither true nor false *as not making a statement,* and the view that if one makes a statement they both make statements, but that one of these statements might be neither true nor false, we should choose the latter. I wrote, "If we had to choose . . ." because it is by no means clear that such a choice is forced upon us. Certainly the notion of a statement which is neither true nor false is a startling one. Perhaps, however, it can be fitted into a larger framework which would put us at our ease. Perhaps it cannot.

XXIV

Before we take leave of the view that a 'putative' statement which is not 'in principle' 'decidable' is not a genuine statement, it is important to draw some distinctions. It will not have escaped the reader's attention that the term 'decidable' is highly ambiguous. There is all the difference in the world between 'decidable by us now on the basis of directly obtainable evidence which we are conceptually prepared to recognize as such,' 'decidable by us now if we had adequate knowledge of the present state of the universe,' 'decidable by those of us who have lived or will live at some time or other on the basis of directly available evidence which they are conceptually prepared to recognize as such,' etc., etc. The changes can be rung on the qualifications of the 'decider' and his status as actual or hypothetical. It is well known that over the past few decades the idea has gained ground that a putative statement of fact which is not 'in principle' decidable in *some* sense of 'decidable' is neither true nor false *as not making a statement.* But it seems to be clear that philosophers who take this line have, by and large, not used the term 'decidable' in the sense in which it implies the *existence of evidence* which is 'in principle' 'obtainable' by us, whatever latitude be taken in defining this 'in principle obtainability.' Rather they have tended to explicate this decidability in terms of 'observation predicate' and 'syntactically well-formed expressions constructed from observation predicates.' It is one thing to say that "S was Φ at t" is meaningful if and only if a primary picture (a tensed story which refers to ourselves *here* and *now*) which includes this form of words is equivalent to a primary picture which contains (in addition to purely logical devices) only observation predicates, predicates definable in terms of observation predicates, properly introduced theoretical predicates, and proper names the presuppositions of which are reasonably held to obtain. It is quite another thing to say that this form of words is meaningful if and only if *there is in principle obtainable evidence*—evidence obtainable by *us*—which would tell us whether "S was Φ at t" is true (or false); that is, it is quite another thing to say that the form of words in question is meaningful if and only if the present state of the universe is such that there is a realizable program of investigation starting from here and now which would provide the premises for an inference concerning its truth or falsity. The difference between these two conceptions has not always been clearly noted, however obvious it may be when thus brutally expressed.[36]

[36] Thus Ayer's account of the meaningfulness of statements about the past in the first edition of *LTL* would seem to be rooted in a tacit commitment to decidability in the latter sense. For he demands that certain future observations be 'possible' which imply either the statement about the past or its contradictory. It must be ad-

11. Decidability and Truth: A Radical View

xxv

Let us turn our attention, therefore, to the philosopher who argues (whether on well- or ill-conceived grounds) that statements about the past or the future are neither true nor false, but *statements* none the less, if they are undecidable in the radical sense that the state of affairs which they formulate is neither physically implied by nor physically incompatible with the *present* state of the universe. He argues, in brief, that 'true or false' entails 'decidable *now*'—'decidable' *in principle*, whatever exactly this may mean. Statements which are not decidable he calls 'middle.' Thus, 'middle' entails 'neither true nor false,' and 'true or false' entails 'not middle.'

Now, unless our philosopher is careful, he is likely to fall into a trap. The trap concerns the slogan "Once true (middle, false), always true (middle, false)." For this slogan turns out, on examination, to contain a hidden ambiguity, and to be in one sense admissible, but in another sense not. The ambiguity arises from the existence of differently tensed counterparts. Suppose, for example, that Jones (in 1900) says

> (275) S *will be* Φ_2 in 1958

and Smith (in 1958) says

> (276) S *is* Φ_2 (in 1958).

There is, we have seen, an important sense in which Jones and Smith are making the same statement. Suppose, now, that this statement was undecidable *quoad* 1900, but *is* decidable *quoad* 1958. We (1958) are tempted to say

> (277) 'S is Φ_2 in 1958' (said today) *is* true or false

but that

> (278) 'S will be Φ_2 in 1958' (said in 1900) *was* neither true nor false, but middle.

To do so, however, would be to make a simple mistake. The statement

> (279) 'S is Φ_2 in 1958' (said today) is true or false *if and only if* 'S is Φ_2 in 1958' (said today) is decidable *now*

mitted, however, that if his further requirement (however odd) that statements about the past 'have the same sense' as these verifying or falsifying statements about the future is taken literally, then Ayer's account turns out to be a variety of the 'well-formed story' version of the criterion, a version modified to require that *all basic primary pictures of the world be stories about the future alone!*

falls under the equivalence schema

> (280) '. . .' (said today) is T or F $\cdot \equiv \cdot$ '. . .' (said today) is decidable *now*.

But what equivalence schema shall we lay down for the differently tensed counterparts of '. . .' which it would have been appropriate to utter in the past? In terms of our example, which of the following equivalences is correct:

> (281) 'S will be Φ_2' (said earlier) *was* T or F $\cdot \equiv \cdot$ 'S will be Φ_2' (said earlier) *was* decidable *then*

or

> (282) 'S will be Φ_2' (said earlier) *was* T or F $\cdot \equiv \cdot$ 'S will be Φ_2' (said earlier) *is* decidable *now*

or is, perhaps, neither of these equivalences correct? To resolve this puzzle, it must first be noted that if Jones told me yesterday that it would rain today, and it does, and I therefore characterize his statement as true, I do so *not* by saying "Jones's statement *is* true," but rather "Jones's statement *was* true." The 'was' serves to indicate that the statement being characterized as true occurred in the past; *but the 'point of view' expressed by the characterizing of it as true is that of myself now.*

It appears, then, that where the point of view of the statement

> (283) '. . .' (said earlier) *was* true

is that of the later speaker who uses this form of words, as it will be unless the context is an oblique one, the 'was' refers to the time at which the statement '. . .' would appropriately have been made, and not to the point of view with respect to which it is characterized as true. If so, then the right-hand side of the equivalence which is to express the desired connection between 'true or false' and decidability will concern decidability with respect to the speaker who makes the truth statement at the time he makes it. Shall we, then, formulate the equivalence as

> (284) '. . .' (said earlier) was T or F $\cdot \equiv \cdot$ '. . .' (said earlier) *is* decidable *now*?

This won't *quite* do, though we are getting warm, for strictly speaking it is only a statement appropriately made *now* which can properly be said to be decidable *now*. This means that 'decidable now' unlike 'true or false' is properly predicated not of the earlier statement but of its re-tensed counterpart appropriate to the present. Thus, in terms of our example, the desired equivalence would read:

> (285) 'S will be Φ_2' (said earlier) was true or false $\cdot \equiv \cdot$ 'S is Φ_2' (said now) is decidable now

or, in general

> (286) S' (said earlier) was true or false $\cdot \equiv \cdot$ S (said now) is decidable now,

where S and S' are differently tensed counterparts.

What, then, it may be asked, is the equivalence which relates 'true or false' to 'decidable then'? It clearly can't be

> (287) S' (said earlier) was true or false $\cdot \equiv \cdot$ S' (said earlier) was decidable then,

for this together with the above equivalence (286) would entail

> (288) S (said now) is decidable now $\cdot \equiv \cdot$ S' (said then) was decidable then,

an equivalence which, according to the assumptions of the present discussion, does not obtain. But what, then? The answer must surely be that *the question itself is a mistake*. There can be no equivalence of the form

> (289) '. . .' (said then) was T or F $\cdot \equiv \cdot$ '. . .' (said then) was decidable then

to do the job done by (280) for the reason, already emphasized, that the predicate 'true' is an *endorsing* expression (and 'false' the opposite of an endorsing expression) which expresses the point of view of the speaker. Thus, if *decidability quoad X* is relevant at all to 'true or false,' it must be decidability *quoad* the speaker speaking. In short, the decidability relevant to 'true or false' would have to be 'decidability now.'

There is, however, an equivalence statement which is easily confused with the above and which seems to formulate as close a connection between 'true or false' and 'was decidable then' as can be asserted without nonsense, namely,

> (290) '. . .' was *rightly said to be* T or F $\cdot \equiv \cdot$ '. . .' was decidable then.

Notice, however, that this time the expression 'true or false' occurs in an oblique context, and does not express the point of view of the person making the equivalence statement.

To sum up, suppose that S, S', and S" are three differently tensed counterparts of which **S** is the one appropriately used today, **S'** the one appropriately used in the past, and **S''** appropriately used in the future, then

(291) **S** is true $\cdot \equiv \cdot$ **S'** was true $\cdot \equiv \cdot$ **S''** will be true

and, always on the assumption, which, for the moment, we are not questioning, that 'true or false' presupposes 'decidable *now*'

(292) **S** is T or F $\cdot \equiv \cdot$ **S'** was T or F $\cdot \equiv \cdot$ **S''** will be T or F $\cdot \equiv \cdot$ **S** is decidable *now*

whereas

(293) **S'** was decidable *then* $\cdot \equiv \cdot$ **S'** was *rightly said to be* T or F
　　S'' will be decidable then $\cdot \equiv \cdot$ **S''** will be rightly said to be T or F

but neither

(294) **S'** was T or F $\cdot \equiv \cdot$ **S'** was decidable *then*
　　S'' will be T or F $\cdot \equiv \cdot$ **S''** *will be* decidable *then*.

We can now separate out the sense in which even on the assumption of an equivalence between 'true or false' and 'decidable *now*' we could say "Once true (false, middle), always true (false, middle)" from the sense in which we could not. For given that 'S is Φ' is decidable *now*,

(295) 'S *is* Φ' is true $\cdot \equiv \cdot$ 'S *will be* Φ' was true $\cdot \equiv \cdot$ 'S *was* Φ' *will be* true,

which gives us the sense in which "Once true, always true." On the other hand, given that 'S will be Φ' *was* not decidable *then* we can say

(296) 'S will be Φ' *would have wrongly been said to be* true or false

and, specifically,

(297) "'S will be Φ' is true" *was* wrongly said

whereas we can say,

(298) 'S is Φ' *is* rightly said to be true or false

and, specifically,

(299) "'S is Φ' is true" *is* rightly said,

which gives us the sense in which "Once true, always true" would, on the assumptions we have been considering, be false.

607

Now it might be thought that

(300) "'S *will be* Φ_2' is true" was wrong[37]

must be false, given that

(301) 'S *is* Φ_2' is true

on the ground that if (301) is the case, then,

(302) 'S *will be* Φ_2' was true

so that

(303) "'S *will be* Φ_2' is true" was true

and if *true*, "'S *will be* Φ_2' is true" (said at the earlier time at which the question of S' being Φ_2 in 1958 was not decidable) could not have been 'wrong' or 'incorrect.' But, of course, the move from (302) to (303) is open to challenge on the ground that it begs the question by assuming that "'S will be Φ_2' is true" was 'right' or 'correct.' The move in question must not be confused with that from

(302) 'S will be Φ_2' was true

to

(304) "'S will be Φ_2' was true" *is* true

which does not, at least directly, involve this assumption. The difference between the two moves—that from (302) to (303) on the one hand and that from (302) to (304) stands forth clearly if we make proper use of the semantic 'definition' of truth. Thus (304) takes us directly to (302), and thence to (301), and finally to

(305) S is Φ_2

simpliciter. It is thus clear that at every stage in this series the *endorsing* expressed by the word 'true' in every occurrence is our endorsing *now*. On the other hand, the endorsing expressed by the first occurrence of 'true' in (303) is that of the pastlings, and to assert (303) is not only to concur in their endorsement, but to grant that their endorsement was 'right' or 'correct' in the sense under question. If this presupposition does not obtain, (303)—the challenge continues—must be withdrawn, and is

---

[37] I have been using 'wrong' rather than 'incorrect' because the literature of normative expressions is full of variously drawn distinctions between 'objectively wrong,' 'subjectively wrong,' 'putatively wrong,' etc., and I wish to mobilize an awareness of the puzzles which generate these distinctions. It will not have escaped the reader that a 'strong' sense of 'objectively wrong' is required by the view we are examining.

no longer available as a *premise* from which the 'rightness' of the pastlings' characterization of 'S will be Φ_2' as true can be inferred.

XXVI

Before we consider a more telling objection to the idea that statements which are undecidable at the time at which they are made are incorrectly said (at *that* time) to be either true or false, let us develop this idea in terms of certain formulas which seem to give it proper expression:

A-1. '**S** *is* decidable *now*' does not imply '**S**' *was* decidable *then* (in the past)' nor does it imply '**S**'' *will* be decidable *then* (in the future)'

where **S**, **S**', and **S**'' are appropriately tensed counterparts.

B-1. **S** *is* decidable now $\cdot \equiv \cdot$ **S** *is* true $\cdot$ **v** $\cdot$ **S** *is* not true
B-2. **S** *is* decidable now $\cdot \equiv \cdot$ **S**' *was* true $\cdot$ **v** $\cdot$ **S**' *was* not true
B-3. **S** *is* decidable now $\cdot \equiv \cdot$ **S**'' *will be* true $\cdot$ **v** $\cdot$ **S**'' *will not be* true
C-1. 'S *is* Φ' *is* true $\cdot \equiv \cdot$ S is Φ
C-2. 'S *was* Φ' *will be* true $\cdot \equiv \cdot$ S is Φ
C-3. 'S *will be* Φ' *was* true $\cdot \equiv \cdot$ S is Φ
C-4. 'S *is* Φ' *is* false $\cdot \equiv \cdot$ S is not Φ
C-5. 'S *was* Φ' *will be* false $\cdot \equiv \cdot$ S is not Φ
C-6. 'S *will be* Φ' *was* false $\cdot \equiv \cdot$ S is not Φ

If, now, we introduce the predicate 'middle' in terms of the equivalence

D-1. 'S *is* Φ' *is* middle $\cdot \equiv \cdot$ 'S is Φ' *is* not decidable *now*

we have

D-2. 'S *will be* Φ' *was* middle $\cdot \equiv \cdot$ 'S is Φ' *is* not decidable *now*
D-3. 'S *was* Φ' *will be* middle $\cdot \equiv \cdot$ 'S is Φ' *is* not decidable *now*

and, abbreviating 'middle' by 'M,'

E-1. **S** *is* M $\cdot \equiv \cdot \sim$ (**S** is true **v** **S** is false)
E-2. **S**' *was* M $\cdot \equiv \cdot \sim$ (**S**' was true **v** **S**' was false)
E-3. **S**'' *will be* M $\cdot \equiv \cdot \sim$ (**S**'' will be true **v** **S**'' will be false)

Hence,

F-1. **S** *is* M $\cdot \equiv \cdot$ **S**'' *was* M $\cdot \equiv \cdot$ **S**'' *will be* M

and

G-1. 'S *is* Φ' *is* M $\cdot \equiv \cdot$ 'S *will be* Φ' *was* M $\cdot \equiv \cdot$ 'S *was* Φ' *will be* M $\cdot \equiv \cdot \sim$ (S *is* Φ **v** S *is not* Φ)

With these formulas in front of us, let us consider a more telling objection to the idea that statements which are undecidable at the time at which

which they are made are 'incorrectly' said at that time to be either true or false. The objection has as its premise the principle that where **S** is properly used, '**S** is true' is also properly used. It proceeds to point out that if "'S will be Φ_2' is true" was incorrect, it follows, given this principle, that 'S will be Φ_2' itself (said then) must also have been incorrect. But this, surely, implies that our pastlings had no way of 'correctly' formulating a prediction about the state of S in 1958. For they could 'correctly' say neither 'S will be Φ_2' nor 'S will not be Φ_2.' Yet the very thesis itself insists that statements which are undecidable at the time at which they are made are *statements*, nonetheless, and hence, presumably, 'correctly' made, even though they cannot be 'correctly' characterized as true or as false. The thesis seems, therefore, to be confronted by the following dilemma. Either it grants that 'S will be Φ_2' was correct, in which case it must surely grant that "'S will be Φ_2' is true" was equally correct; or it denies that 'S will be Φ_2' was correct, in which case it must surely grant that that 'S will be Φ_2' was not a statement. In either case the thesis would collapse.

Before commenting on this dilemma, let us consider a possible line of reply to the initial formulation of the objection. This formulation was that the pastlings would have no 'right' way in which to make a prediction about the state of S in 1958. The reply to be considered is that while they can say neither 'S will be Φ' nor 'S will not be Φ,' they can perfectly well avail themselves of a third form of statement, namely,

$$(305) \text{ S will ? be } \Phi_2$$

where '?' is a monary connective introduced by the equivalence

$$\text{H-1. S will ? be } \Phi \cdot \equiv \cdot \sim [\text{S will be } \Phi \cdot \mathbf{v} \cdot \sim (\text{S will be } \Phi)]$$

or, in general

$$\text{H-2. ? } \mathbf{S} \cdot \equiv \cdot \sim (\mathbf{S} \, \mathbf{v} \, \mathbf{S}).$$

And, indeed, it can be seen that unless the object language contains tautologies of the form '$\mathbf{S} \, \mathbf{v} \sim \mathbf{S} \, ? \, \mathbf{S}$,' rather than '$\mathbf{S} \, \mathbf{v} \sim \mathbf{S}$' the semantical equivalences C-1 and C-4 would lead to the logical truth of

$$(306) \text{ 'S is } \Phi \text{' is true} \cdot \mathbf{v} \cdot \text{'S is } \Phi \text{' is false}$$

and hence exclude the 'middle' which it is desired to include.

If, however, we introduce this new connective, we note that we are committed to the equivalence

$$\text{G-1. S is ? } \Phi \cdot \equiv \cdot \text{'S is } \Phi \text{' is M.}$$

And it is at once clear that this equivalence differs from

C-1. S is Φ · $\equiv$ · 'S is Φ' is true

and

C-4. S is not Φ · $\equiv$ · 'S is Φ' is false

in that the right-hand side of G-1 contains the explicitly pragmatic notion of undecidability, whereas the notions of truth and falsity, though they may well *in some sense* imply decidability *in some sense*, do not seem to have as straightforward a relation to decidability as the thesis demands. Thus, the relation between

(307) **S** is true or **S** is false

and

(308) **S** is decidable now

cannot be identity of sense, for the sense of

(309) '. . .' is true

is given by the schema

(310) '. . .' is true · $\equiv$ · . . .

with the result that the sense of

(311) '. . .' is true or '. . .' is false

is given by the schema

(312) '. . .' is true or '. . .' is false · $\equiv$ · . . . or $\sim$. . .

Thus even the sense of

(313) 'S is ? Φ' is true or 'S is ? Φ' is false

would be given by its equivalence to

(314) S is ? Φ or S is not ? Φ

rather than by

(315) 'S is ? Φ' is true or 'S is ? Φ' is false · $\equiv$ · 'S is ? Φ' *is decidable now,*

and the sense of "'S is ? Φ' is true" by the equivalence

(316) 'S is ? Φ' is true $\equiv$ S is ? Φ.

Continuing, for the moment, to beat around the bushes, we notice that whatever one is to make of the thesis under examination, one would have to be careful about the interpretation of the equivalence

(317) 'S will be Φ' *was* M · $\equiv$ · S *is* ? Φ

611

which emerges from G-1 together with H-1. For it is easy to forget that

(318) 'S will be Φ' was M

has a sense which is given by the equivalence D-3, which concerns decidability *now*, and to confuse (318) with

(319) 'S will be Φ' was undecidable *then*.

Thus the following two equivalences obtain:

H-1. 'S will be Φ' was M ≡ S is ? Φ
H-2. 'S will ? be Φ' was true ≡ S is ? Φ

but neither

(320) 'S will be Φ' was undecidable *then* ≡ 'S will be Φ' was M

nor

(321) 'S will be Φ' was undecidable *then* ≡ 'S will ? be Φ' was T.

Surely the thesis that a statement which is undecidable at the time at which it was made is incorrectly said at that time to be either true or false rests on a failure to note that

(322) A statement which *was* undecidable *was* neither true nor false

doesn't follow from

(323) A statement which *is* undecidable is neither true nor false

and it was by exposing the fallacy of this inference that we were led to reformulate the claim intended by (322) to read

(324) A statement which *was* undecidable was 'incorrectly' said to be either true or false.

12. Decidability and Truth: Conclusion

XXVII

Let us return, now, to the dilemma which was posed as an objection to the thesis under examination. What does it amount to? and is there any way out? The answer is to be found by noting that it is simply not true that the form

(305) S will ? be Φ

provides the pastlings with a way of making a statement about the future of S. For if this form is introduced in terms of the equivalence H-1, it can make a statement only if the equivalent on the right-hand side of H-1 makes a statement, and hence only if both 'S will be Φ' and 'S will not be Φ' made statements. Thus, granting the assumption of the objection, namely that where S is correct, 'S is true' is also correct, it follows that 'S will ? be Φ_2,' used by the pastlings, made a statement if and only if 'S will be Φ_2' made a statement, and, hence, only if "'S will be Φ_2' is true" made a statement. Thus the introduction of the statement form 'S is ? Φ' does not provide a means of reconciling the claims of the thesis with the idea that the pastlings can make a prediction about the state of S in 1958 with respect to the property of being Φ_2.

Shall we conclude that the thesis that a statement which is undecidable when it is made is incorrectly said *at that time* to be either true or false collapses into the thesis that such a form of words does not make a statement at all, but is rather a "pseudo-statement," a thesis which we have already—but perhaps prematurely—dismissed? I think we must, for I see no escape from the principle that where S makes a statement 'S is true' also makes a statement. And it seems to be clear that to deny that 'S will be Φ' was 'right' or 'correct' differs only verbally from denying that it made a statement.

But if it is admitted that 'S will be Φ' made a statement, and, consequently, that "'S will be Φ' is true" made a statement and was itself 'correct,' we are surely forced to deny that the truth or falsity of statements is, in the sense envisaged, dependent on their decidability or non-decidability at the time at which they are made. For if

(325) 'S will be Φ_2' *is* true

made a statement as used by the pastlings, surely it made the same statement as

(302) 'S will be Φ_2' *was* true

which we, today (1958) might say. But if (302) has any connection with decidability of the kind envisaged by the thesis, it is, we have seen, with decidability *now*, rather than with decidability *then*. And if (325) made the same statement as (302) makes today, then it made the same statement as

(301) 'S is Φ_2' *is* true

(made today), so that unless the decidability of a statement made at one time entails the decidability of all its differently tensed counterparts, when they are appropriately used, (301) *can't* be equivalent to

(325) 'S is Φ_2' is decidable now and not false.

XXVIII

At this point the proponents of the thesis may retreat to a weaker form of the decidability thesis by insisting that instead of connecting 'true or false' with 'decidable *at a certain time* (i.e., the time of utterance),' all they have meant to do is connect it with 'decidable *ever*'—i.e., with the property of not being physically implied by *any* evidence *in principle* available to *anybody* at *any time*. That this would get them out of the above difficulty is clear. Has it any other virtues?

The thesis then reduces, roughly, to the idea that a statement is neither true nor false if it is at no time decidable. This, however, does not make sense as the idea that no state of the universe, past, present, or future, physically implies that the episode in question occurs or does not occur. For one state of the universe even logically implies that the episode in question occurs or does not occur. It must therefore be the idea that *no evidence, obtainable at any time in a certain way* (by the use of perception and 'in principle' definable instruments) *physically implies the occurrence or non-occurrence of the episode in question.*

Clearly such a modified thesis is in the neighborhood of the familiar conception of empirical meaningfulness (and in *this* sense of truth or falsity) as syntactical well-formedness from observational primitives which was characterized in an earlier section. Yet it does not coincide with it, and amounts to the idea that a form of words can be in this sense meaningful, and belong to a primary picture (state description) of the world, *and yet be an empirically undecidable element in the picture.*

Now there do seem to be such 'statements,' specifically in the field of QM. Should they be called 'pseudo-statements'? Or should we, indeed, call them statements which are neither true nor false, but middle? Surely the answer lies in the fact that we cannot separate a form of words for making statements from the conceptual framework to which it belongs, and we must take seriously the evolution of conceptual frameworks. The above 'in principle definability' is a framework conception, and, in this sense *quoad* the speaker speaking. Thus, once we take seriously the difference between the *semantical* predicates 'true' and 'false' and the *pragmatic* predicate 'undecidable' and ponder on the consequent difference between 'S is ? Φ' (construed as the equivalent of "'S is Φ' is undecidable") on the one hand, and 'S is Φ' and 'S is not Φ' on the other, we see that the alternatives mentioned above are false alternatives. What we must rather say is that while to have reason to suppose that a statement in our frame is radically undecidable is to have a radical reason for refusing to say that it is true or that it is false, and, consequently, a reason for saying that it is 'middle' if 'middle' is construed as a pragmatic rather than a semantic predicate, the heart of the matter is that such statements are symptoms of the breakdown of a conceptual frame, a breakdown which requires its reconstruction along lines which would eliminate the possibility of such 'in principle' undecidable statements. This is the task which confronts contemporary quantum mechanics, and the program of this paper as a whole has been to show that this task belongs to neither logic nor semantics, but to empirical science.

XXIX

A concluding postscript: If 'M' is to be on a par with 'T' and 'F,' its sense must be given by some such schema as I-3 in this set:

I-1. 'S is Φ' is T $\equiv$ S is Φ
I-2. 'S is Φ' is F $\equiv$ (S is Φ)
I-3. 'S is Φ' is M $\equiv$? (S is Φ)

In other words, if 'M' is to be a genuinely *semantical* alternative to 'T' and 'F,' the sense of "'S is Φ' is M" must be given by its equivalence to 'S is ? Φ,' where the latter is to be understood in terms of the tautology

(326) S is Φ or S is not Φ or S is ? Φ

as a tautology in a language with a three-valued logic. For while its semantical character could be ensured by equating 'M' with 'neither T nor F,' this move would be pointless unless its object-language counterpart, '?,' was interpreted as an expression in an object language with a genuinely 'three-valued' structure, i.e., in terms of the 'three-valued' tautology

(327) $S \vee \sim S \vee ? S$

(where the symbols '$\vee$,' '$\sim$,' and '?' play the new role appropriate to a three-valued logic), and not in terms of a two-valued contradiction, thus,

(328) $? S = \sim (S \vee \sim S)$.

In any event, 'M' cannot simply mean undecidable and yet be a *semantical* concept on a par with 'true' and 'false.' This makes it clear that there is no simple connection between *the existence of in principle undecidable statements*, and *the problem of three-valued logics*. For in a three-valued language, to *decide* would be to decide whether S is Φ or S is not Φ or

S is ? Φ, and *only because of this* to decide *whether* 'S is Φ' is *true* or 'S is Φ' is *false* or 'S is Φ' is *middle*. It is just as radical a mistake to suppose that 'true' and 'false' would have *exactly* their ordinary meanings (in our two-valued language), as to suppose that 'not' and 'or' would be anything but cousins of these terms as we use them.

INDEXES

Name Index

Abu'l-Barakat, 55n
Agassi, J., 70n
Alexander, H. Gavin, 111n, 256n
Anaximander, 52, 74
Aristarchus, 49
Aristotle, 28, 52, 53, 56, 85, 89, 378
Austin, J. L., 86
Ayer, A. J., 603n, 604n

Bacon, Roger, 73
Bar-Hillel, Y., 154n, 427n
Barker, Stephen F., 60n, 112n, 138n, 140, 156n
Bergmann, G., 7ff, 21, 241n, 262n, 285
Bergson, H., 271, 559
Berkeley, G., 67
Bernoulli, J., 138n
Black, Max, 199n, 399
Bohm, David, 32, 66
Bohr, N., 3, 43, 49, 88
Born, Max, 44–45
Boyle, R., 79, 108, 242
Braithwaite, R. B., 102n, 158n, 172, 173, 256, 499
Brecht, Bertolt, 71–72
Bridgman, P. W., 429
Broad, C. D., 504, 527
Brodbeck, May, 7n, 101n, 173, 232
Bromberger, S., 109, 113n
Brouwer, L. E. J., 427
Bruno, G., 87
Bunge, M., 56n
Burnet, J., 52

Campbell, N. R., 99n, 509
Cantor, G., 413, 466
Carnap, Rudolf, 22, 23n, 35, 36, 38n, 41–42, 51, 111n, 137, 138, 139n, 140n, 145, 152n, 153, 154n, 158n, 160, 230, 273n, 278–79, 304, 400–1, 404, 406, 426, 432ff, 492, 509, 587
Cavendish, H., 188

Charles, J. A. C., 79, 108
Chomsky, N., 394n
Church, A., 539
Clagett, M., 52n, 53, 55n, 56n
Clifford, W. K., 406, 428
Cohen, M. R., 85n, 98n
Conant, J. B., 37, 188, 190
Copernicus, N., 49
Craig, W., 15–19
Cramér, H., 129, 130

d'Abro, A., 414n, 445
de Broglie, L., 9, 45
De Finetti, B., 159n
Descartes, R., 31, 89
Dewey, John, 31n, 70n
Dingler, Hugo, 451–52
Donagan, Alan, 262–71 passim
Drabkin, I. E., 85n
Dray, W., 100n, 174n, 235n
Duhem, P., 101n, 161, 493ff

Eddington, A., 30, 406, 485ff, 503
Einstein, Albert, 66, 364, 368ff, 406, 411ff, 418, 421ff, 428n, 442, 453, 465ff, 492, 497, 503, 505, 506ff
Epicurus, 85n
Euclid, 594
Euler, L., 475

Feigl, Herbert, 3n, 4n, 5n, 7n, 11n, 16n, 23n, 32, 85n, 101n, 103, 173n, 212n, 214n, 232n, 273n, 398n, 402n, 403n, 463, 501n
Fermi, E., 77
Feyerabend, P. K., 3n, 13, 14, 22n, 214, 260n, 485ff
Ford, Henry, 190
Fraenkel, A. A., 427
Frank, P., 85
Franz, R., 108
Fuerth, R., 65n

619

Name Index

Galileo, 37, 46–47, 49, 54, 89, 92, 101, 108, 113, 114, 213n, 232, 240, 260, 264
Gandhi, M., 196–97
Gardiner, P. L., 172n, 232n, 235n, 240n, 242n, 246n, 247n, 252n, 255n, 262n, 265n, 267n, 269n, 271n
Goodman, Nelson, 63n, 102n, 121n, 124n, 156n
Grice, A. P., 360
Grünbaum, Adolf, 172, 400, 413n, 414n, 590n
Gulden, Samuel, 405n

Hanson, N. R., 3n, 39n, 142n, 143n, 214n, 258n, 259n, 260n, 403n
Hegel, G. W. F., 68n
Heisenberg, W., 3, 9, 49n, 89
Helmholtz, H. von, 471n, 472
Hempel, C. G., 16, 17, 19, 28, 34, 41n, 44, 47, 48n, 50n, 60n, 62n, 64, 78, 92, 98n, 101n, 112n, 172ff, 184n, 186, 190, 194, 198, 204, 209, 210, 213, 221, 226ff, 231n, 232n, 243, 246, 265–66, 267n, 268, 273n, 382
Hertz, H., 201, 499
Hilbert, D., 477
Hobson, E. W., 413n
Hooke, R., 221
Hume, David, 11, 12, 63n, 88, 232, 265, 358, 372ff
Huygens, C., 67, 221

Jammer, M., 85n
Jeffrey, R. C., 158n, 160

Kant, E., 29, 88, 579n, 601
Kepler, J., 45n, 49, 92–93, 101n, 108, 179n, 264, 267
Keynes, J. M., 137
Klein, F., 417n, 491, 497
Körner, S., 32, 48n, 124n
Kolmogoroff, A., 129n, 130n
Kraft, V., 47n, 48n
Kuhn, T. S., 32, 49n, 60n

Lagrange, J. L., 475
Landau, L. D., 45
Leibniz, G. W., 67, 371, 378
Leucippus, 52
Lewis, C. I., 136n, 138n
Lie, S., 473, 484, 491
Lifschitz, E. M., 45
Lobachevski, N., 372
Locke, John, 11

Lorentz, H. A., 568, 571, 574
Luce, R. D., 152n, 158n, 159n

Mach, E., 44, 62, 67
McLaurin, C., 73
McTaggart, J. E., 527, 556n, 590n, 593
Maier, A., 56n
Malcolm, Norman, 21
Mandelbaum, M., 236n
Margenau, H., 509
Maritain, J., 521n
Matson, W. I., 73–74
Maupertuis, P. L. M. de, 221
Maxwell, Grover, 4n, 11n, 16n, 22n, 23n, 32, 212n, 214n, 398n, 405n, 453, 513–14
Maxwell, J. C., 28, 370
Mayo, B., 26n
Meehl, P. E., 244n
Mendel, G., 141
Menger, K., 498
Mill, J. S., 232, 358, 374n, 384
Milne, E. A., 418–19, 465ff
Minkowski, H., 567, 578n
Mises, R. von, 166
Moore, G. E., 262
Morgenstern, O., 152n
Morris, E., 36n

Nagel, Ernest, 3, 5n, 16, 20, 28, 32–33, 34n, 46–47, 55, 58, 76, 77n, 78ff, 83n, 98n, 111n, 112n, 214, 256n, 440ff, 494ff, 503
Nernst, W., 223
Neurath, O., 35
Newman, E., 405n
Newton, I., 28, 45n, 46–47, 54–55, 57, 67, 68n, 74, 85–86, 88–89, 92–93, 101, 107, 143n, 179, 198, 213, 225, 240, 242, 247, 371–72, 408ff, 414, 419, 422, 431, 443, 449, 453, 472, 475ff, 499
Neyman, J., 159n
Northrop, F. S. C., 410–11, 479n, 509

Ohm, G. S., 108
Oppenheim, P., 28, 34, 47, 60n, 78, 92, 98, 137n, 173, 174–75, 176, 190, 194, 198, 204, 209–10, 213, 221, 232n

Page, L., 470–71
Panofsky, E., 49
Pap, A., 401, 403n
Parmenides, 52
Perrin, J., 66
Plato, 31, 49, 68n, 72, 90

620

Name Index

621

Subject Index

Subject Index

Subject Index

Subject Index